S0-BAC-663

STUDENT SOLUTIONS MANUAL

PHYSICS

FOR SCIENTISTS AND ENGINEERS

SECOND EDITION

IRVIN A. MILLER
Drexel University

STUDENT SOLUTIONS MANUAL

PHYSICS

FOR SCIENTISTS AND ENGINEERS
SECOND EDITION

Paul M. Fishbane
University of Virginia

Stephen Gasiorowicz
University of Minnesota

Stephen T. Thornton
University of Virginia

PRENTICE HALL Upper Saddle River, NJ 07458

Acquisition Editor: *Alison Reeves*
Production Editor: *Kimberly Knox*
Production Supervisor: *Joan Eurell*
Assistant Editor: *Wendy Rivers*
Production Coordinator: *Ben Smith*

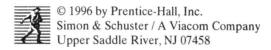 © 1996 by Prentice-Hall, Inc.
Simon & Schuster / A Viacom Company
Upper Saddle River, NJ 07458

All rights reserved. No part of this book may be
reproduced, in any form or by any means,
without permission in writing from the publisher.

Printed in the United States of America

10 9 8 7 6 5 4 3 2 1

ISBN 0-13-231697-8

Prentice-Hall International (UK) Limited, *London*
Prentice-Hall of Australia Pty. Limited, *Sydney*
Prentice-Hall Canada, Inc., *Toronto*
Prentice-Hall Hispanoamericana, S.A., *Mexico*
Prentice-Hall of India Private Limited, *New Delhi*
Prentice-Hall of Japan, Inc., *Tokyo*
Simon & Schuster Asia Pte. Ltd., *Singapore*
Editora Prentice-Hall do Brasil, Ltda., *Rio de Janeiro*

Contents

Contents

Part II
Discussions of Odd-numbered Questions

CHAPTER 1

5. We use 10^{10} m for an atomic diameter.
   For a person 6 ft tall, we have $h = (6\text{ ft})(0.305\text{ m/ft})/(10^{-10}\text{ m/atom}) = \boxed{1.8 \times 10^{10}\text{ atoms.}}$

7. We have 1400 Lire = \$1, or \$1/1400 Lire = 1. We let the units tell us whether the conversion should be in the numerator or the denominator:
   Price = $(1000\text{ Lire/kg})(\$1/1400\text{ Lire}) = \boxed{\$0.71\text{/kg.}}$

11. Density $= \text{mass/volume} = \text{mass}/\frac{4}{3}\pi R^3$
    $= [(7.35 \times 10^{22}\text{ kg})/\frac{4}{3}\pi(1.74 \times 10^6\text{ m})^3](10^3\text{ g/kg})(1\text{ m}/10^2\text{ cm})^3 = \boxed{3.33\text{ g/cm}^3.}$

17. Because the volume is the product of the three dimensions, the % uncertainties add.
    For each dimension we can find the % uncertainty:
    $w = 1.25 \pm 0.03\text{ m} = 1.25\text{ m} \pm (0.03/1.25)(100) = 1.25\text{ m} \pm 2\%;$
    $\ell = 0.5 \pm 0.1\text{ m} = 0.5\text{ m} \pm (0.1/0.5)(100) = 0.5\text{ m} \pm 2 \times 10\%;$
    $h = 0.137 \pm 0.028\text{ m} = 0.137\text{ m} \pm (0.028/0.137)(100) = 0.137\text{ m} \pm 2.0 \times 10\%.$
    Thus the volume is
    $V = w\ell h = (1.25\text{ m})(0.5\text{ m})(0.137\text{ m}) \pm (0.2 + 2 + 2.0) \times 10\%.$
    If significant figures are ignored, we get $V = 0.0856\text{ m}^3 \pm 4.2 \times 10\% = 0.0856 \pm 0.036\text{ m}^3;$
    however, there is only one significant figure in the % uncertainty, so the numerical uncertainty
    is $\pm 0.04\text{ m}^3$, and we finally get $\boxed{V = 0.09 \pm 0.04\text{ m}^3.}$
    You might compare this to the results using the extreme values of the dimensions.

19. In terms of the diameter, the area is $A = \pi D^2 = \pi DD$. When multiplying, % uncertainties add,
    so $\pm\%A = \%D + \%D = 2(\%D)$ or $5\% = 2(\%D)$ and thus $\%D = \boxed{2.5\%.}$

23. For $L = h/m_e c$ we can write the dimensions as $[L] = [h][m_e]^{-1}[c]^{-1}$, or $[L] = [h][M]^{-1}[LT^{-1}]^{-1}$.
    Thus $[h] = [L][M][LT^{-1}] = \boxed{[ML^2T^{-1}].}$

25. (a) For $F = Ame^{-\alpha r}/r^3$, the argument of the exponential function can have no dimension,
        thus $[\alpha r] = [\alpha][r]$, or $1 = [\alpha][L]$, and $\boxed{[\alpha] = [L^{-1}]\quad (\text{m}^{-1}).}$
    (b) Then $[F] = [A][m][r]^{-3}$ or $[MLT^{-2}] = [A][M][L]^{-3}$ and $\boxed{[A] = [L^4T^{-2}]\quad (\text{m}^4/\text{s}^2).}$

33. The volume of a droplet is $\frac{4}{3}\pi(0.5 \times 10^{-4}\text{ m})^3 = 0.5 \times 10^{-12}\text{ m}^3$. The volume of water needed will depend on
    the size of the city. If we assume an area of $(30\text{ km})^2$, the volume of water required is
    $(1\text{ cm})(30\text{ km})^2(10^3\text{ m/km})^2(1\text{ m}/100\text{ cm}) = 9 \times 10^6\text{ m}^3.$
    The number of droplets is
    $N = (9 \times 10^6\text{ m}^3)/(0.5 \times 10^{-12}\text{ m}^3/\text{droplet}) \approx \boxed{10^{19}\text{ droplets.}}$

35. The average speed of an automobile over a year is $(20{,}000\text{ mi/yr})(1\text{ yr}/365\text{ days})(1\text{ day}/24\text{ h}) = 2\text{ mi/h}.$
    If we estimate the average highway speed as 40 mi/h, then the number of automobiles on the road at any
    one moment is
    $N = (100 \times 10^6\text{ automobiles})(2\text{ mi/h})/(40\text{ mi/h}) \approx \boxed{5 \times 10^6\text{ automobiles.}}$

41. Because she is always a constant distance from the center of the lake, her position can be described by the clockwise angle ϕ from the north-south line drawn from the center to the starting point. Her direction of travel will be tangent to the circle with a magnitude of 2 m/s at an angle ϕ above the west direction.

43. If we take **i** as east and **j** as north, the four displacements in paces are
    $$4j, \ 6\cos 45°\, i + 6\sin 45°\, j, \ 2i \ \text{and} \ -5i.$$
    The resultant displacement is their sum:
    $$\mathbf{R} = 4j + 6(0.707)i + 6(0.707)j + 2i - 5i = \boxed{1.2i + 8.2j \ \text{paces} \quad (8.3 \ \text{paces}, 82° \ \text{north of east}).}$$

51. (a) $\mathbf{A} = -4i + 2j$, $\mathbf{B} = -i + 4j$, $\mathbf{C} = 2i + 2j$, $\mathbf{D} = 5i - 3j$.
    (b) $2\mathbf{A} + \mathbf{C} - \mathbf{D} = 2(-4i + 2j) + (2i + 2j) - (5i - 3j)$ $\mathbf{B} + \mathbf{C}/2 = (-i + 4j) + (2i + 2j)/2$ $\mathbf{D} - \mathbf{B} = (5i - 3j) - (-i + 4j)$
    $\qquad\qquad = \boxed{-11i + 9j.}$ $\qquad\qquad\qquad\qquad = \boxed{5j.}$ $\qquad\qquad\qquad = 6i - 7j.$

    $$|\mathbf{D} - \mathbf{B}| = \sqrt{6^2 + 7^2} = \boxed{9.2.}$$

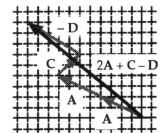
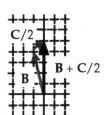
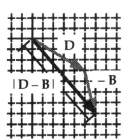

53. For the dimensions of the proposed relation, we have $[\tau] = [\ell]^r[g]^s$, or
    $$[T] = [L]^r[LT^{-2}]^s = [L]^{+r+s}[T]^{-2s}.$$
    By equating the exponents on the two sides for each dimension, we obtain $r + s = 0$, and $-2s = 1$. These equations are satisfied by $r = 1/2$, and $s = -1/2$, so $\boxed{\tau = (\text{dimensionless constant})(\ell/g)^{1/2}.}$

59. The time is distance traveled/speed:
    $$t = d/c = (2.4 \times 10^5 \ \text{mi})(1.60 \times 10^3 \ \text{m/mi})/(2.998 \times 10^8 \ \text{m/s}) = \boxed{1.3 \ \text{s.}}$$

63. A density of 1 g/cm^3 means 18 g of water occupies 18 cm^3, so the volume of one molecule is
    $18 \ \text{cm}^3/(6.02 \times 10^{23} \ \text{molecules}) = 3.0 \times 10^{-23} \ \text{cm}^3$.
    With one molecule in a cube, the volume must be L^3, so $L = (3.0 \times 10^{-23} \ \text{cm}^3)^{1/3} \approx \boxed{3 \times 10^{-8} \ \text{cm.}}$

69. We will estimate 1×10^6 persons with each person using 10 gal/day and each truck carrying 1×10^4 gal. Then the number of trucks needed in a day is
    $$N = (10^6 \ \text{persons})(10 \ \text{gal/day/person})/(10^4 \ \text{gal/truck}) \approx \boxed{10^3 \ \text{truck/day.}}$$
    If a bath requires 30 gal, this would require an average of an additional 10 gal/day, so the number of trucks would double to $\boxed{2 \times 10^3 \ \text{trucks/day.}}$

73. The number of molecules in cylinder = (total mass of air)/(mass of a molecule). Because the mass of air is the mass of water (density of 1 g/cm^3) in a cylinder 30 ft high with cross-section of 1 cm^2, we have

$$N = [(30 \text{ ft})(12 \text{ in/ft})(2.54 \text{ cm/in})(1 \text{ cm}^2)(1 \text{ g/cm}^3)]/[(1.6)(18 \text{ g})/6.0 \times 10^{23} \text{ molecules}]$$
$$= 1.9 \times 10^{25} \text{ air molecules above 1 cm}^2.$$

The total number is the number above 1 cm^2 times the surface area of the earth in cm^2:

$$N_{\text{total}} = N(4\pi R^2) = (1.9 \times 10^{25} \text{ molecules/cm}^2)(4\pi[(6.4 \times 10^6 \text{ m})(10^2 \text{ cm/m})]^2) \approx 10^{44} \text{ molecules.}$$

75. (a) The two vectors can be obtained from the diagram:

$$\mathbf{u} = u \cos\theta \, \mathbf{i} + u \sin\theta \, \mathbf{j}$$
$$\mathbf{v} = -v \sin\theta \, \mathbf{i} + v \cos\theta \, \mathbf{j}, \text{ or}$$
$$\mathbf{v} = +v \sin\theta \, \mathbf{i} - v \cos\theta \, \mathbf{j}$$

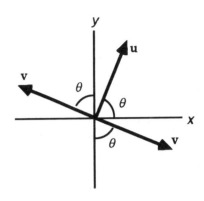

(b) $v_x u_y + v_y u_x = (\mp v \sin\theta)(u \cos\theta) + (\pm v \cos\theta)(u \sin\theta)$
$$= uv(\mp \sin\theta \cos\theta \pm \sin\theta \cos\theta) = 0.$$

79. Because the mass $\propto$ volume $\propto L^3$, we can form a ratio for the two birds:

$$m_{\text{flamingo}}/m_{\text{stilt}} = (L_{\text{flamingo}}/L_{\text{stilt}})^3 \quad \text{or} \quad (2.5 \times 10^3 \text{ g})/(150 \text{ g}) = (L_{\text{flamingo}}/20 \text{ cm})^3,$$

from which we get $L_{\text{flamingo}} \approx$ ⟶ 50 cm.

81. With the three physical attributes we can form the following relation:

$t_0 = (\text{constant}) \lambda^\alpha \ell^\beta \tau^\gamma$, which leads to $[t_0] = [\lambda]^\alpha [\ell]^\beta [\tau]^\gamma$, or
$$[T] = [ML^{-1}]^\alpha [L]^\beta [MLT^{-2}]^\gamma = [M]^{\alpha+\gamma} [L]^{-\alpha+\beta+\gamma} [T]^{-2\gamma}.$$

By equating the exponents on the two sides for each dimension, we obtain $\alpha + \gamma = 0$, $-\alpha + \beta + \gamma = 0$, and $-2\gamma = 1$.

These equations are satisfied by $\alpha = 1/2$, $\beta = 1$, and $\gamma = -1/2$, so $t_0 = (\text{a constant}) \ell (\lambda/\tau)^{1/2}$.

CHAPTER 2

For problems where it is necessary to specify the characteristics of the motion of an object, a coordinate system must be selected, with emphasis on the origin and positive direction.

3. The total distance will be the sum of the distances for each leg:
    distance = (3)(2)(42 m) = 252 m.
    Because the student returns to the starting point, we have
    displacement = 0,
    since final position = initial position.

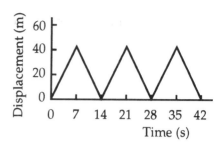

5. (a) For the first automobile the times for each leg and the elapsed times are
    $\Delta t_1 = \Delta x_1 / v_1 = 15 \text{ km}/(75 \text{ km/h}) = 0.20 \text{ h} = 12 \text{ min}; \; t_1 = 12 \text{ min}.$
    $\Delta t_2 = 25 \text{ min}; \; t_2 = 37 \text{ min}.$
    $\Delta t_3 = \Delta x_3 / v_3 = 40 \text{ km}/(100 \text{ km/h}) = 0.4 \text{ h} = 24 \text{ min}; \; t_3 = 61 \text{ min}.$
    $\Delta t_4 = 5 \text{ min}; \; t_4 = 66 \text{ min}.$
    $\Delta t_5 = \Delta x_5 / v_5 = 55 \text{ km}/(60 \text{ km/h}) = 0.92 \text{ h} = 55 \text{ min}; \; t_5 = 121 \text{ min}.$
    (b) From the plot, we see that the two automobiles meet twice.
    The position of the second auto as a function of time is
    $x_2 = [(74 \text{ km/h})/(60 \text{ min/h})](t - 25 \text{ min}).$
    At $t_2 = 37 \text{ min}$, $x_2 = 15 \text{ km} = x_1$; the two autos meet at
     $t_2 = 37 \text{ min and } x = 15 \text{ km.}$
    Because they meet during the last leg, we need the position of the first auto during this leg:
    $x_1 = [(- 60 \text{ km/h})/(60 \text{ min/h})](t - 66 \text{ min}) + 55 \text{ km}.$
    By setting $x_1 = x_2$, we have
    $[(- 60 \text{ km/h})/(60 \text{ min/h})](t - 66 \text{ min}) + 55 \text{ km} = [(74 \text{ km/h})/(60 \text{ min/h})](t - 25 \text{ min}).$
    We solve this equation for t to get $t = 68 \text{ min}$. We find the meeting location from
    $x_1 = [(- 60 \text{ km/h})/(60 \text{ min/h})](68 \text{ min} - 66 \text{ min}) + 55 \text{ km} = 53 \text{ km}.$
    Thus they meet at $t = 68 \text{ min and } x = 53 \text{ km.}$

7.

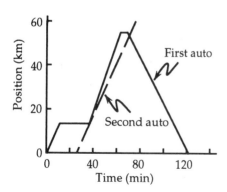

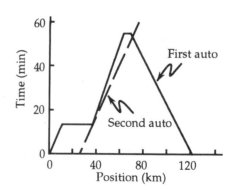

From the second plot we see that the first car travels infinitely fast at $t = 15$ min and at $t = 55$ min and then travels backward in time, obviously not possible. The second car travels at constant speed.

The correct first plot shows the position as a function of time and is the one used most often.

13. If we add the distances estimated from the plot for the different trips, we get a total distance of
    48 km + 24 km + 5 km + 2.5 km + ... = $\boxed{79.5+ \text{ km.}}$
    To calculate the distance, we can say that the insect will fly at a constant speed of 240 km/h until the trains collide. We find the time until the trains collide from their relative velocity:
    $$\Delta t = \Delta x / v_{\text{relative}}$$
    $$= 80 \text{ km} / [(80 \text{ km/h}) + (160 \text{ km/h})] = \tfrac{1}{3} \text{ h} = 20 \text{ min.}$$
    The distance the insect travels in this time is
    $$v_{\text{insect}} \Delta t = (240 \text{ km/h})(\tfrac{1}{3} \text{ h}) = \boxed{80 \text{ km.}}$$

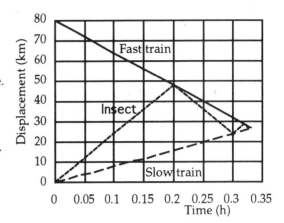

15. From $x = 0.02t^3 - 0.1t^2 + 2t$ cm, we can get the speed: $v = dx/dt = 0.06t^2 - 0.2t + 2$ cm/s.
    Because this is one-dimensional motion along the x-axis, we have
    $$v_1 = 0.06(1)^2 - 0.2(1) + 2 = + 1.9 \text{ cm/s}; \quad \boxed{\mathbf{v}_1 = + 1.9\mathbf{i} \text{ cm/s};}$$
    $$v_5 = 0.06(5)^2 - 0.2(5) + 2 = + 2.5 \text{ cm/s}; \quad \boxed{\mathbf{v}_5 = + 2.5\mathbf{i} \text{ cm/s};}$$
    $$v_{10} = 0.06(10)^2 - 0.2(10) + 2 = + 6.0 \text{ cm/s}. \quad \boxed{\mathbf{v}_{10} = + 6.0\mathbf{i} \text{ cm/s}.}$$
    The magnitude of the average velocity for the 10 s is
    $$v_{\text{av}} = \Delta x / \Delta t$$
    $$= \{[0.02(10)^3 - 0.1(10)^2 + 2(10)] \text{ cm} - \{[0.02(0)^3 - 0.1(0)^2 + 2(0)] \text{ cm}\} / (10 - 0) \text{ s} = 3.0 \text{ cm/s};$$
    $$\boxed{\mathbf{v}_{\text{av}} = + 3.0\mathbf{i} \text{ cm/s.}}$$
    The formula is unrealistic for large times because the t^2 term in v will eventually produce a velocity that cannot be achieved by an ant.

21. We will take the origin as the location at $t = 0$ s. Because the automobile starts from rest, we have
    $$x = x_0 + v_0 t + \tfrac{1}{2}at^2 = \tfrac{1}{2}at^2.$$
    The separation of the marks is the difference in the displacements, which we write:
    at $t = 8$ s: $x_8 = \tfrac{1}{2}a(8 \text{ s})^2$; at $t = 12$ s: $x_{12} = \tfrac{1}{2}a(12 \text{ s})^2$.
    Then $x_{12} - x_8 = 64 \text{ m} = \tfrac{1}{2}a[(12 \text{ s})^2 - (8 \text{ s})^2]$, which gives $a = \boxed{1.6 \text{ m/s}^2.}$

23. We get the velocity and acceleration by differentiation of the displacement:
    $$x = A \sin(\pi t / 12);$$
    $$v = dx/dt = A [\cos(\pi t / 12)] (\pi / 12) = \boxed{(A\pi/12) \cos (\pi t / 12);}$$
    $$a = dv/dt = (A\pi / 12) [- \sin(\pi t / 12)](\pi / 12) = \boxed{- (A\pi^2 / 144) \sin(\pi t / 12).}$$

27. We see from $a = A - (v/t_0)$ that the acceleration is a function of velocity and thus time. At $t = 0$, $a = A - (v_0 / t_0)$. If this is positive, the velocity will increase, which causes a decrease in the acceleration. The rate of change of the velocity (the slope of the v-t curve) will decrease. Eventually the acceleration becomes zero, at which point the velocity becomes constant (called a terminal velocity). After a long time $a = 0$, so $A - (v_{\text{terminal}} / t_0) = 0$, or $v_{\text{terminal}} = At_0$.
    The graphs assume that $A > (v_0 / t_0)$.

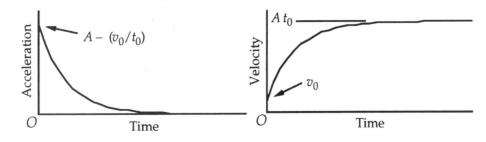

35.
(a)
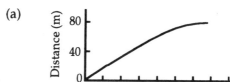
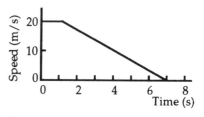

(b) Before the brakes are applied, the car moves with constant velocity, so we find the distance traveled before the brakes are applied from

$$x_1 = v_0 t_1 = (20 \text{ m/s})(1.2 \text{ s}) = 24 \text{ m}.$$

We use the remaining distance to find the constant acceleration while the brakes are applied:

$$v^2 = v_0{}^2 + 2a(x_2 - x_1);$$
$$0 = (20 \text{ m/s})^2 + 2a(80 \text{ m} - 24 \text{ m}), \text{ which gives}$$
$$\boxed{a = -3.6 \text{ m/s}^2.}$$

41. We choose the origin at the initial passing point and change units of the speeds:

$$75 \text{ mi/h} = (75 \text{ mi/h})(5280 \text{ ft/mi})(1 \text{ h/3600 s}) = 110 \text{ ft/s};$$
$$85 \text{ mi/h} = 125 \text{ ft/s}.$$

Because the speeder is traveling at constant velocity, its position is

$$x_s = x_0 + v_s t = 0 + (110 \text{ ft/s})t.$$

During the period of constant acceleration of the police car, its average velocity is $\frac{1}{2}(v_0 + v_f)$. Thus the distance traveled by the police car while accelerating until t_1 is

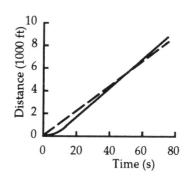

$$x_{p1} = v_{av} t_1 = [\tfrac{1}{2}(0 + 125 \text{ ft/s})](13 \text{ s}) = 813 \text{ ft}.$$

Thereafter it travels at constant velocity so its position is

$$x_p = x_{p1} + v_p(t - t_1) = 813 \text{ ft} + (125 \text{ ft/s})(t - 13 \text{ s}).$$

The police car will overtake the speeder when $x_s = x_p$;

$$(110 \text{ ft/s})t = 813 \text{ ft} + (125 \text{ ft/s})(t - 13 \text{ s}).$$

The solution gives (c) $t = \boxed{54 \text{ s}}$ and (b) $x_p = x_s = x_p = (110 \text{ ft/s})(54 \text{ s}) = \boxed{5940 \text{ ft} \ (1.13 \text{ mi}).}$

47. We use a coordinate system with the origin at the ground and up positive. Thus the elevator starts with $y_{0elev} = 0$ and $v_{0elev} = 0$.

At the instant the object is dropped (not thrown!), it has the position and velocity of the elevator:

$$y_{elev} = y_{0elev} + v_{0elev}t_1 + \tfrac{1}{2}at_1{}^2 = 0 + 0 + \tfrac{1}{2}at_1{}^2 = \tfrac{1}{2}a(3 \text{ s})^2 = (4.5 \text{ s}^2)a;$$
$$v_{elev} = v_{0elev} + at_1 = 0 + at_1 = a(3 \text{ s});$$

Thus for the object's falling motion: $y_0 = (4.5 \text{ s}^2)a$, $v_0 = (3 \text{ s})a$; $a_o = -9.8 \text{ m/s}^2$.

When the object hits the ground, $y = 0$, so we have

$$y - y_0 = v_0 t_2 + \tfrac{1}{2}a_o t_2{}^2;$$
$$0 - (4.5 \text{ s}^2)a = (3 \text{ s})a(3.5 \text{ s}) + \tfrac{1}{2}(-9.8 \text{ m/s}^2)(3.5 \text{ s})^2, \text{ which gives } a = \boxed{4.0 \text{ m/s}^2.}$$

The position of release is

$$y_0 = (4.5 \text{ s}^2)a = (4.5 \text{ s}^2)(5.0 \text{ m/s}^2) = \boxed{18 \text{ m}.}$$

53. We estimate 10 ft for each story. After a free fall of 97 stories $\approx$ 970 ft, the speed is found from

$$v^2 = v_0{}^2 + 2a(x - x_0) = 0 + 2(32 \text{ ft/s})(970 \text{ ft}), \text{ which gives } v \approx 250 \text{ ft/s}.$$

If we take this to be the average speed for the rest of the fall, the time to fall the last 3 stories is

$$\Delta t \approx 30 \text{ ft}/(250 \text{ ft/s}) \approx 0.1 \text{ s}. \text{ This is not much time to say anything!}$$

57. We use a coordinate system with the origin at the ground and up positive and label the first rock A and the second rock B.
    Rock A starts 20 m above the ground with a positive velocity. We find the time for rock A to hit the ground from
    $$y_A = y_{0A} + v_{0A}t_A + \tfrac{1}{2}at_A^2 \; ;$$
    $$0 = 20 \text{ m} + (10 \text{ m/s})t_A + \tfrac{1}{2}(-9.8 \text{ m/s}^2)t_A^2.$$
    This quadratic equation has two roots, but only the positive one is physically possible: $t_A = 3.28$ s.
    Rock B starts 20 m above the ground with zero velocity. We find the time for rock B to hit the ground from
    $$y_B = y_{0B} + v_{0B}t_B + \tfrac{1}{2}at_B^2;$$
    $$0 = 20 \text{ m} + 0 + \tfrac{1}{2}(-9.8 \text{ m/s}^2)t_B^2, \text{ which gives a positive answer of } t_B = 2.02 \text{ s}.$$
    For the rocks to hit the ground at the same time, rock B must be released
    $$\Delta t = t_A - t_B = \boxed{1.26 \text{ s}} \text{ after rock } A.$$

59. We use a coordinate system with the origin at the ground and up positive.
    For the fall of the ball we have
    $$v_1^2 = v_0^2 + 2a(y_1 - y_0);$$
    $$v_1^2 = 0 + 2(-9.8 \text{ m/s}^2)(0 - 25 \text{ m}), \text{ which gives } v_1 = \boxed{-22 \text{ m/s.}}$$
    (We use the negative answer because we know that the ball is moving downward.)
    For the rebound, at the top of the motion the velocity is zero, so we have
    $$v_2^2 = v_1^2 + 2a(y_2 - y_0);$$
    $$0 = (22 \text{ m/s})^2 + 2(-9.8 \text{ m/s}^2)(y_2 - 0); \text{ which gives } y_2 = \boxed{25 \text{ m,}} \text{ as expected.}$$
    For a bounce of 20 m we have
    $$v_4^2 = v_3^2 + 2a(y_4 - y_0);$$
    $$0 = v_3^2 + 2(-9.8 \text{ m/s}^2)(20 \text{ m} - 0), \text{ which gives } v_3 = \boxed{+20 \text{ m/s.}}$$
    (We use the positive answer because we know that the ball is moving upward.)

65. Because the acceleration is a function of time, $a = \alpha t^2$, we rearrange the definition of acceleration, $dv = a\, dt$, and obtain the speed by integrating:
    $$\int_0^{v_f} dv = \int_0^t a\,dt = \int_0^t \left(\alpha t^2\right) dt, \text{ which gives } v_f = \frac{\alpha}{3}t^3. \text{ Thus } \boxed{t = \left(\frac{3v_f}{\alpha}\right)^{1/3}.}$$

67. If 20 transfers take 15 s, the time for one transfer is $\tau = 0.75$ s. To juggle two objects, one must be in the air while the transfer is made. Thus one will be in the air for τ, which means a time of $\tau/2$ to reach the highest point and $\tau/2$ to fall back to the hand. We use a coordinate system with the origin at the highest point and down positive.
    We call the distance an object must fall h_2 when two objects are juggled.
    From $y = y_0 + v_0 t + \tfrac{1}{2}at^2$; $h_2 = 0 + 0 + \tfrac{1}{2}g(\tau/2)^2$, we get $h_2 = \tfrac{1}{2}(9.8 \text{ m/s}^2)(0.75 \text{ s}/2)^2 = \boxed{0.7 \text{ m.}}$
    To juggle three objects, two must be in the air during a transfer so each one must be in the air during 2 transfers, for a time of 2τ, or fall for a time of τ. Thus $h_3 = \tfrac{1}{2}g(\tau)^2 = \tfrac{1}{2}g(9.8 \text{ m/s}^2)(0.75 \text{ s})^2 = \boxed{3 \text{ m.}}$

71. We use a coordinate system with the origin at the throw and up positive. Then $v_0 = 7$ m/s.
    For the constant acceleration we have
    $$v = v_0 + at = v_0 - gt;$$
    -4 m/s $= 7$ m/s $- g(7$ s), which gives $g = \boxed{1.6 \text{ m/s}^2.}$
    At the highest point the velocity will be zero, so we have
    $$v^2 = v_0^2 + 2a(y - y_0);$$
    $0 = (7 \text{ m/s})^2 + 2(-1.6 \text{ m/s}^2)h$, which gives $h = \boxed{15 \text{ m.}}$

73. We use a coordinate system with the origin at the top of the snowbank with down positive.
    For the constant acceleration, to find the distance to come to a stop, we have
    $$v^2 = v_0^2 + 2a(y - y_0);$$
    $0 = (40 \text{ m/s})^2 + 2(-50)(9.8 \text{ m/s}^2)(y - 0)$, which gives $y = \boxed{1.6 \text{ m.}}$
    To find the time we have
    $$v = v_0 + at;$$
    $0 = 40 \text{ m/s} + (-50)(9.8 \text{ m/s}^2)t$, which gives $t = \boxed{0.08 \text{ s.}}$

CHAPTER 3

For problems where it is necessary to specify the characteristics of the motion of an object, a coordinate system must be selected, with emphasis on the origin and positive directions.

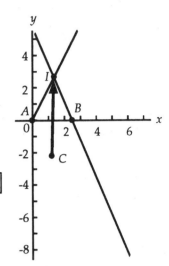

5. We find the algebraic equation for a straight line from the slope m and the y-intercept b: $y = mx + b$.
    For the first line from point A, we have
    $$y_1 = [(4.0 \text{ km} - 0)/(2.0 \text{ km} - 0)]x_1 + 0 = 2.0\, x_1 .$$
    For the second line from point B, we have
    $$y_2 = [(-8 \text{ km} - 0)/(6.0 \text{ km} - 2.5 \text{ km})]x_2$$
    $$+ [(-8 \text{ km} - 0)/(6.0 \text{ km} - 2.5 \text{ km})](0 - 2.5 \text{ km}) = -2.3\, x_2 + 5.7 \text{ km}.$$
    The intersection of the two lines occurs when $x_1 = x_2 = x_I$ and $y_1 = y_2 = y_I$:
    $$2.0\, x_I = -2.3 x_I + 5.7 \text{ km},$$
    which gives $x_I = 1.3$ km and $y_I = 2.7$ km or $\boxed{\mathbf{r}_I = (1.3 \text{ km})\mathbf{i} + (2.7 \text{ km})\mathbf{j} .}$
    Given that $\mathbf{r}_C = (1.2 \text{ km})\mathbf{i} - (2.2 \text{ km})\mathbf{j}$, then $\boxed{\mathbf{r}_I - \mathbf{r}_C = (0.1 \text{ km})\mathbf{i} + (4.9 \text{ km})\mathbf{j} .}$

9. The distance from the origin is found from $d^2 = x^2 + y^2$.
    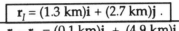
    Given $\mathbf{r} = (4 \text{ m}) \cos(\pi t/T)\mathbf{i} - (4 \text{ m}) \sin(\pi t/T)\mathbf{j}$, for the various times we have
    $$\mathbf{r}_{T/3} = (4 \text{ m}) \cos(\pi T/3T)\mathbf{i} - (4 \text{ m}) \sin(\pi T/3T)\mathbf{j} = \boxed{(2.0 \text{ m})\, \mathbf{i} - (3.5 \text{ m})\mathbf{j} ;}$$
    $$d = [(2.0 \text{ m})^2 + (3.5 \text{ m})^2]^{1/2} = \boxed{4.0 \text{ m}.}$$
    $$\mathbf{r}_{T/2} = (4 \text{ m}) \cos(\pi T/2T)\mathbf{i} - (4 \text{ m}) \sin(\pi T/2T)\mathbf{j} = \boxed{-(4.0 \text{ m})\mathbf{j} ; \; d = 4.0 \text{ m.}}$$
    $$\mathbf{r}_{2T} = (4 \text{ m}) \cos(\pi T/T)\mathbf{i} - (4 \text{ m}) \sin(\pi T/T)\mathbf{j} = \boxed{(4.0 \text{ m})\mathbf{i}; \; d = 4.0 \text{ m.}}$$
    The angle is found from $\tan \theta = y/x = -4 \sin(\pi t/T)/4 \cos(\pi t/T) = -\tan(\pi t/T)$, so $\boxed{\theta(t) = -\pi t/T.}$
    The particle is traveling CW around a circle of radius 4.0 m.

15. We choose a coordinate system with the origin at Malibu, the x-axis east, and the y-axis north.
    The position of the whale is
    $$\mathbf{r}_W = [-(5.0 \text{ km}) \cos 45° - (7.0 \text{ km/h})(\cos 45°)t]\mathbf{i} +$$
    $$[(5.0 \text{ km}) \sin 45° - (7.0 \text{ km/h})(\sin 45°)t]\mathbf{j}$$
    $$= [-(3.5 \text{ km}) - (4.9 \text{ km/h})t]\mathbf{i} + [(3.5 \text{ km}) - (4.9 \text{ km/h})t]\mathbf{j} ;$$
    and the position of the boat is
    $$\mathbf{r}_B = -[(30 \text{ km/h})(\cos \theta)t]\mathbf{i} + [(30 \text{ km/h})(\sin \theta)t]\mathbf{j},$$
    where θ is the angle north of west that the boat is heading.
    The interception occurs when $\mathbf{r}_W = \mathbf{r}_B$, or
    $$[-(3.5 \text{ km}) - (4.9 \text{ km/h})t]\mathbf{i} + [(3.5 \text{ km}) - (4.9 \text{ km/h})t]\mathbf{j} = -[(30 \text{ km/h})(\cos \theta)t]\mathbf{i} + [(30 \text{ km/h})(\sin \theta)t]\mathbf{j}.$$
    From this we get two equations relating θ and t:
    $$-(3.5 \text{ km}) - (4.9 \text{ km/h})t = -(30 \text{ km/h})(\cos \theta)t , \text{ and } (3.5 \text{ km}) - (4.9 \text{ km/h})t = (30 \text{ km/h})(\sin \theta)t.$$
    If we divide by t and add the two equations, we get an equation for θ:
    $$\sin \theta = (1 - \cos^2 \theta)^{1/2} = \cos \theta - 0.327, \text{ which gives } \theta = 31.6°. \text{ When we use this value in one of the}$$
    equations, we get $t = 0.17$ h = 10 min.
    The velocity of the boat is
    $$\mathbf{v}_B = [(30 \text{ km/h})][-(\cos \theta)\mathbf{i} + (\sin \theta)\mathbf{j}] = (-25.5\mathbf{i} + 15.7\mathbf{j}) \text{ km/h} = \boxed{30 \text{ km/h}, 31.6° \text{ north of west,}}$$
    and the interception position is $\boxed{\mathbf{r} = (-4.3\mathbf{i} + 2.7\mathbf{j}) \text{ km.}}$

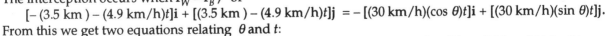

17. We choose the coordinate system shown in the diagram, with the
    origin directly below the joint midway between the tips. From
    the symmetry of the motion, we need to consider only the motion
    of the right-hand tip.

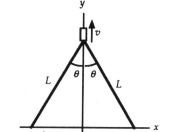

   (a) The position of the tip, which is moving at constant velocity, is
$$x = x_0 - v_0 t = L \sin \theta - v_0 t = 0.5L - v_0 t .$$
      Because the length of the arm is constant, the joint is at
$$y = [L^2 - x^2]^{1/2} = [L^2 - (0.5L - v_0 t)^2]^{1/2}$$
      For the joint $x = 0$, so its velocity is
$$\mathbf{v} = (dy/dt)\mathbf{j} = \tfrac{1}{2}(0.5L - v_0 t)2v_0 \mathbf{j} \ /[L^2 - (0.5L - v_0 t)^2]^{1/2}$$
$$= [0.5(0.015 \text{ m}) - (0.03 \text{ m/s})t](0.03 \text{ m/s})\mathbf{j} \ /\{(0.015 \text{ m})^2 - [0.5(0.015 \text{ m}) - (0.03 \text{ m/s})t]^2\}^{1/2}$$
$$\boxed{= \ [(0.00225 \text{ m}^2/\text{s}) - (0.0009 \text{ m}^2/\text{s}^2)t]\mathbf{j} \ /[(0.0169 \text{ m}^2) - (0.0045 \text{ m}^2/\text{s})t - (0.0009 \text{ m}^2/\text{s}^2)t^2]^{1/2}.}$$

   (b) When $t = 0$, $x = 0.5L$, so we have
$$\mathbf{v} = [(0.00225 \text{ m}^2/\text{s}) - 0]\mathbf{j} \ /[(0.0169 \text{ m}^2) - 0 - 0]^{1/2} = \boxed{(0.017 \text{ m/s})\mathbf{j} .}$$
      When $x = 0$, $0.5L - v_0 t = 0$, so we have
$$\mathbf{v} = \boxed{0.}$$
      Note that, if the tips could move through each other, this is where the motion of the joint
      would reverse.

19. By differentiating the position vector $\mathbf{r} = [4 \cos(\pi t/T)\mathbf{i} - 4 \sin(\pi t/T)\mathbf{j}]$ m with respect to time, we get
$$\boxed{\mathbf{v} = [-(4\pi/T) \sin(\pi t/T)\mathbf{i} - (4\pi/T) \cos(\pi t/T)\mathbf{j}] \text{ m/s.}}$$
   The angle of the velocity vector with respect to the x-axis is found from
$$\tan \phi = v_y/v_x = [-(4\pi/T) \cos(\pi t/T)]/[-(4\pi/T) \sin(\pi t/T)] = \cot(\pi t/T) = \tan[(\pi/2) - (\pi t/T)],$$
   so $\boxed{\phi = (\pi/2) - (\pi t/T).}$
   From Problem 9 the angle of the position vector is $\theta = -\pi t/T$, so $\phi - \theta = (\pi/2) - (\pi t/T) - (-\pi t/T) = \pi/2$;
   $\mathbf{r}$ and $\mathbf{v}$ are perpendicular.

23. We are given $h = H - ut - (u/B) e^{-Bt}$.
    Because the exponent $-Bt$ must be dimensionless, $[B] = [t]^{-1} = \ \boxed{[T^{-1}].}$
    We get the velocity from $v = dh/dt = -u - (u/B) e^{-Bt}(-B) = u[-1 + e^{-Bt}]$.
    At $t = 0$, $v = \boxed{v_0 = 0;}$ as $t \to \infty$, $\boxed{v \to -u.}$
    We get the acceleration from $a = dv/dt = u[e^{-Bt}(-B)] = -Bu \, e^{-Bt}$.
    At $t = 0$, $a = \boxed{a_0 = -Bu;}$ as $t \to \infty$, $\boxed{a \to 0.}$

27. We choose a coordinate system with $y = 0$ at the ground and up positive.
    For the first rock, $y_1 = y_{01} + v_{01}t_1 + \tfrac{1}{2}a_1 t_1^2 = 0 + (21 \text{ m/s})t_1 + \tfrac{1}{2}(-9.8 \text{ m/s}^2)t_1^2$.
    The second rock is thrown 3 s later, or $t_2 = t_1 - 3$ s.
    For the second rock, $y_2 = y_{02} + v_{02}t_2 + \tfrac{1}{2}a_2 t_2^2 = 0 + (21 \text{ m/s})(t_1 - 3 \text{ s}) + \tfrac{1}{2}(-9.8 \text{ m/s}^2)(t_1 - 3 \text{ s})^2.$
   (a) The rocks will meet when $y_1 = y_2$;
$$21t_1 - 4.9t_1^2 = 21t_1 - 63 - 4.9t_1^2 + 29.4t_1 - 44.2, \text{ which gives } \boxed{t_1 = 3.64 \text{ s.}}$$
   (b) The height at which they meet is $y_1 = (21 \text{ m/s})(3.64 \text{ s}) + \tfrac{1}{2}(-9.8 \text{ m/s}^2)(3.64 \text{ s})^2 = \boxed{11.5 \text{ m}} = y_2.$
   (c) The velocities when they meet are
$$v_1 = v_{01} + at_1 = 21 \text{ m/s} + (-9.8 \text{ m/s}^2)(3.64 \text{ s}) = \boxed{-14.7 \text{ m/s (down)}} \text{ and}$$
$$v_2 = v_{02} + a(t_1 - 3 \text{ s}) = 21 \text{ m/s} + (-9.8 \text{ m/s}^2)(3.64 \text{ s} - 3.00 \text{ s}) = \boxed{+14.7 \text{ m/s (up)}.}$$

29. We choose a coordinate system with the origin at the takeoff point, with x horizontal and y vertical. The horizontal motion will have constant velocity. We find the time required for the jump from the horizontal motion:

$x = x_0 + v_{0x}t$;

$9.5 \text{ m} = 0 + (9.0 \text{ m/s})t$, which gives $t = 1.06$ s.

Because this is the time to return to the ground, the vertical motion is

$y = y_0 + v_{0y}t + \frac{1}{2}a_yt^2$;

$0 = 0 + v_{0y}(1.06 \text{ s}) + \frac{1}{2}(-9.8 \text{ m/s}^2)(1.06 \text{ s})^2$, which gives $\boxed{v_{0y} = 5.2 \text{ m/s.}}$

33. The horizontal motion will have constant velocity, v_{0x}. When the projectile lands, $y = y_0$.
We find the vertical velocity from

$v_y^2 = v_{0y}^2 + 2a_y(y - y_0) = v_{0y}^2 + 0 = v_{0y}^2$, so $v_y = -v_{0y}$.

Because the angle is determined by the direction of the velocity, $\tan\theta = v_y/v_{0x}$, the projectile will make an angle $\boxed{40° \text{ below the horizontal}}$ when it lands.

37. The initial speed of the arrows can be found from the maximum range, which occurs when $\theta_0 = 45°$:

$R = v_0^2 \sin(2\theta_0)/g$; $R_{max} = v_0^2/g = 350 \text{ m}$, or $v_0^2 = (9.80 \text{ m/s}^2)(350 \text{ m})$, which gives $v_0 = 58.6$ m/s.

At the launch angle of 55°, the range is

$R = (v_0^2/g)\sin(2\theta_0) = R_{max}\sin[2(55°)] = (350 \text{ m})(0.939) = \boxed{329 \text{ m.}}$

39. (a) The ball passes the goal posts when it has traveled the horizontal distance of 15 m:

$x = v_{0x}t = v_0(\cos\theta_0)t$; $15 \text{ m} = (15 \text{ m/s})\cos 37° \, t$, which gives $\boxed{t = 1.25 \text{ s.}}$

    (b) To see if the kick is successful, we must find the height of the ball at this time:

$y = y_0 + v_{0y}t + \frac{1}{2}a_yt^2 = 0 + (15 \text{ m/s})\sin 37° (1.25 \text{ s}) + \frac{1}{2}(-9.8 \text{ m/s}^2)(1.25 \text{ s})^2 = 3.6 \text{ m.}$

Thus the kick is $\boxed{\text{unsuccessful}}$ and passes $\boxed{0.4 \text{ m below the bar.}}$

49. The speed is $v = (55 \text{ mi/h})(1.61 \times 10^3 \text{ m/mi})(1 \text{ h}/3600 \text{ s}) = 24.4$ m/s.
The restriction on the centripetal acceleration when moving in a curve of radius R is

$a = v^2/R \leq 0.1g$, so $R \geq v^2/0.1g = (24.4 \text{ m/s})^2/0.1(9.8 \text{ m/s}^2) = \boxed{610 \text{ m.}}$

57. If we use a coordinate system with x in the direction of travel and y up, the velocity of the rain with respect to the car is $\mathbf{v}_B = \mathbf{v}_A - \mathbf{u}$, where

$\mathbf{v}_A = -v_A\mathbf{j}$ is the velocity of the rain with respect to the ground and

$\mathbf{u} = 80\mathbf{i}$ km/h is the velocity of the car with respect to the ground.

If $\theta = 58°$ is the angle on the window with respect to the vertical, then

$-v_B\cos\theta\,\mathbf{i} - v_B\sin\theta\,\mathbf{j} = (-v_A\mathbf{j} - 80\mathbf{i})$ km/h.

Equating the two components gives us two equations for v_A and v_B: $v_A = v_B\cos\theta$, and 80 km/h $= v_B\sin\theta$.

Dividing the two equations, we get $v_A/(80 \text{ km/h}) = \cot\theta = \cot 58°$, or $\boxed{v_A = 50 \text{ km/h.}}$

59. During the first leg, in order to fly due south, the airplane must head southwest, as in the diagram.
   If $\mathbf{v}_p$ is the velocity of the airplane with respect to the air,
      $\mathbf{v}_1$ the velocity of the airplane with respect to the ground, and
      $\mathbf{v}_w$ the velocity of the wind with respect to the ground, then
      $\mathbf{v}_1 = \mathbf{v}_p + \mathbf{v}_w$.

   First leg

   Because $v_p = 900$ km/h and $v_w = 120$ km/h, from the diagram we get
      $\sin\theta_1 = v_w/v_p = (120\ \text{km/h})/(900\ \text{km/h})$, or $\theta_1 = 7.7°$.
   Then $v_1 = v_p \cos\theta_1 = (900\ \text{km/h})\cos 7.7° = 892$ km/h.
   The distance traveled in the first leg is $d_1 = (892\ \text{km/h})(2\ \text{h}) = 1784$ km.
   During the second leg, the airplane must turn more toward the west, as shown.   **Second leg**
   If $\mathbf{v}_2$ is the new velocity of the airplane with respect to the ground, then
      $\mathbf{v}_2 = \mathbf{v}_p + \mathbf{v}_w$.
   The two component equations can be obtained from the diagram:
      $v_p \sin\theta_2 - v_w = v_2 \sin 45°$; $(900\ \text{km/h})\sin\theta_2 - 120\ \text{km/h} = v_2 \sin 45°$, and
      $v_p \cos\theta_2 = v_2 \cos 45°$; $(900\ \text{km/h})\cos\theta_2 = v_2 \cos 45°$.
   Because $\sin 45° = \cos 45°$, we can write this as
      $(900\ \text{km/h})\sin\theta_2 - 120\ \text{km/h} = (900\ \text{km/h})\cos\theta_2$, which reduces to $\sin\theta_2 = \cos\theta_2 + 0.133$.
   By squaring both sides, we get $1 - \cos^2\theta_2 = \cos^2\theta_2 + 0.266\cos\theta_2 + (0.133)^2$, which has the solution
      $\cos\theta_2 = 0.637$ or $\theta_2 = 50.4°$.
   Then $v_2 = (900\ \text{km/h})(\cos 50.4°)/0.707 = 811$ km/h.
   The distance traveled during the second leg is $d_2 = (811\ \text{km/h})(3\ \text{h}) = 2433$ km.
   (a) The average speed is (total distance)/(total time) = $(1784\ \text{m} + 2433\ \text{m})/(5\ \text{h}) = \boxed{843\ \text{km/h.}}$
   (b) To find the average velocity, we must first find the displacement, expressed in the
      usual coordinate system:
         $\mathbf{r} = -(2433\ \text{m})(0.707)\mathbf{i} + [(-2433\ \text{m})(0.707) - 1784]\mathbf{j} = -(1720\ \text{km})\mathbf{i} - (3505\ \text{km})\mathbf{j}$.
      The average velocity is $\mathbf{v} = \mathbf{r}/(5\ \text{h}) = -(344\mathbf{i} + 701\mathbf{j})$ km/h or $\boxed{781\ \text{km/h, }26°\text{ W of S.}}$
   (c) From part (b) the final position is
         $\mathbf{r} = \boxed{-(1720\ \text{km})\mathbf{i} - (3505\ \text{km})\mathbf{j},\quad \text{or}\quad 3900\ \text{km, }26°\text{ W of S.}}$

63. We choose a coordinate system with the origin at the tee, x horizontal and y up.
   The horizontal motion is $x = v_{0x}t_1$; $105\ \text{m} = v_0 \cos 65° \, t_1$, or $v_0 t_1 = 248$ m.
   The vertical motion is $y = y_0 + v_{0y} + \frac{1}{2}a_y t_1^2$; $-8.0\ \text{m} = 0 + v_0 \sin 65° \, t_1 + \frac{1}{2}(-9.8\ \text{m/s}^2)t_1^2$.
   (a) When $v_0 t_1$ is substituted in the second equation, we get an equation for t_1 which gives $t_1 = 6.90$ s.
      From the first equation we then get $\boxed{v_0 = 36\ \text{m/s at } 65°.}$
   (b) Because $v_y = 0$ at the maximum height, we can find the time to reach this height from
         $v_y = v_{0y} + a_y t_2$;
         $0 = (36\ \text{m/s})\sin 65° + (-9.8\ \text{m/s}^2)t_2$, which gives $t_2 = 3.33$ s.
      (Note that this is less than $\frac{1}{2}t_1$ because of the slope of the ground.)
      The height above the tee is $y_{max} = 0 + (36\ \text{m/s})\sin 65° \,(3.33\ \text{s}) + \frac{1}{2}(-9.8\ \text{m/s}^2)(3.33\ \text{s})^2 = 54.3$ m.
      The height above the green is $y_{max} + 8.0\ \text{m} = \boxed{62.3\ \text{m.}}$

65. The coordinate system is shown on the diagram. If $v_B = 10$ km/h is the
speed of the boat with respect to the water, then its velocity with respect
to the shore is

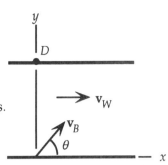

$$\mathbf{v}_s = (v_B \cos \theta + v_W)\mathbf{i} + v_B \sin \theta \mathbf{j}.$$

The position vector from its starting point is

$$\mathbf{r} = \mathbf{v}_s t = (v_B \cos \theta + v_W)t\mathbf{i} + (v_B \sin \theta)t\mathbf{j}, \quad \text{or}$$

$$\mathbf{r} = \boxed{(10 \cos \theta + 6)t\mathbf{i} + (10 \sin \theta)t\mathbf{j},} \quad \text{with } r \text{ in kilometers and } t \text{ in hours.}$$

To land at the point D, we have $x = 0$ or $v_B \cos \theta + v_W = 0$;

$$\cos \theta = -(6 \text{ km/h})/(10 \text{ km/h}), \text{ which gives } \theta = \boxed{127°.}$$

As expected, we must head upstream.

The trip time is found from $y = v_B \sin \theta \, t$;

$$150 \text{ m} = [(10 \times 10^3 \text{ m/h})/(3600 \text{ s/h})] \sin 127° \, t, \text{ which gives } t = \boxed{68 \text{ s.}}$$

69. Because there are four balls, each of which requires 0.5 s for the transfer, the time for a complete cycle is
2.0 s, 0.5 s of which the ball is in the hands. The time in the air for each ball is 2.0 s – 0.5 s = 1.5 s, which
means that it takes 0.75 s for a ball to reach the maximum height.

(a) Because the speed is zero at the maximum height, we can find the initial speed from

$$v_y = v_{0y} + a_y t; \quad 0 = v_0 + (-9.8 \text{ m/s}^2)(0.75 \text{ s}), \text{ which gives } v_0 = \boxed{7.35 \text{ m/s.}}$$

(b) The juggler must be releasing a ball when he catches one.

A ball is just being thrown: $\boxed{y = 0.}$

A ball is at $y = y_0 + v_{0y}t + \frac{1}{2}a_y t^2 = 0 + (7.35 \text{ m/s})(0.5 \text{ s}) + \frac{1}{2}(-9.8 \text{ m/s}^2)(0.5 \text{ s})^2 = \boxed{2.45 \text{ m going up.}}$

A ball is at $y = y_0 + v_{0y}t + \frac{1}{2}a_y t^2 = 0 + (7.35 \text{ m/s})(1.0 \text{ s}) + \frac{1}{2}(-9.8 \text{ m/s}^2)(1.0 \text{ s})^2 = \boxed{2.45 \text{ m coming down.}}$

A ball has just been caught: $\boxed{y = 0.}$

(c) For 5 balls the cycle time is 2.5 s and the time in the air for each ball is 2.0 s.

Using the analysis from part (a), the initial speed must be $v_0 = g(1.0 \text{ s}) = 9.8 \text{ m/s}$.

The maximum height is reached in half the time in the air:

$$y = y_0 + v_{0y}t + \frac{1}{2}a_y t^2 = 0 + (9.8 \text{ m/s})(1.0 \text{ s}) + \frac{1}{2}(-9.8 \text{ m/s}^2)(1.0 \text{ s})^2 = \boxed{4.9 \text{ m.}}$$

CHAPTER 4

The analysis of force diagrams is usually simplified if one of the axes of the coordinate system is chosen to be in the direction of the acceleration. It is important to specify all of the forces acting on the selected object. Most forces require contact between the object and its surroundings, so look at contact points for possible forces. Of course, the force of gravity does not require contact.

15. We change the speed units: $(100 \text{ km/h})(10^3 \text{ m/km})/(3600 \text{ s/h}) = 27.8 \text{ m/s}$.
   We simplify the system to two forces acting on the car: the forward force from the road (assumed constant) and the drag force (which depends on the speed, but is simplified to be a different constant for each of trials).
   Thus, $\Sigma F = F_{road} - F_{drag} = ma = m\,\Delta v/\Delta t$.
   Without streamlining we have
   $$F_{road} - F_{drag1} = ma_1 = m\,\Delta v/\Delta t_1 = (1150 \text{ kg})(27.8 \text{ m/s})/11 \text{ s} = 2.90 \times 10^3 \text{ N}.$$
   With streamlining we have
   $$F_{road} - F_{drag2} = ma_2 = m\,\Delta v/\Delta t_2 = (1150 \text{ kg})(27.8 \text{ m/s})/9 \text{ s} = 3.55 \times 10^3 \text{ N}.$$
   If we subtract the two equations, we get $F_{drag1} - F_{drag2} = \boxed{6.5 \times 10^2 \text{ N}.}$

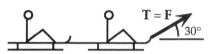

17. The tension in the pulled rope must be equal to the force the father exerts: $\boxed{T = F.}$
   We need to look at horizontal forces only.
   If we take both sleds as the object, we get the force diagram shown. Then for horizontal motion, we have
   $$\Sigma F = T\cos 30° = (m + m)a = 2ma, \text{ so } a = T(\cos 30°)/2m = 0.433T/m.$$
   If we take the second sled as the object we get the force diagram shown. Then
   $$\Sigma F = T_2 = ma, \text{ so } T_2 = m(0.433T/m) = \boxed{0.433F.}$$

25. (a) If we look at the five cars as a system, the only horizontal force is the force between the engine and the first car. We have
   $$\Sigma F_x = F = m_{cars}a_x\,;$$
   $$4.5 \times 10^4 \text{ N} = 5(3.0 \times 10^4 \text{ kg})a, \text{ which gives}$$
   $$a = \boxed{0.30 \text{ m/s}^2.}$$
   This is the acceleration of the engine and each of the cars.

   (c) If we look at the first car, the horizontal forces are the couplings shown in the diagram.
   We have $\Sigma F_x = ma_x$, or $F - F_{12} = m_1 a$:
   $$(4.5 \times 10^4 \text{ N}) - F_{12} = (3.0 \times 10^4 \text{ kg})(0.30 \text{ m/s}^2), \text{ which gives}$$
   $$F_{12} = \boxed{3.6 \times 10^4 \text{ N backward.}}$$

(b)

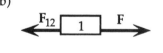

27. From Newton's third law, the force that charge 2 exerts on charge 1 is equal and opposite to the force that charge 1 exerts on charge 2: $\mathbf{F}_{12} = -\mathbf{F}_{21}$. Similarly, $\mathbf{F}_{13} = -\mathbf{F}_{31}$.
   Thus $\mathbf{F}_{net1} = \mathbf{F}_{12} + \mathbf{F}_{13} = -\mathbf{F}_{21} - \mathbf{F}_{31}$
   $$= (-2\mathbf{i} + 3\mathbf{j} - \mathbf{k}) \text{ N} + (+3\mathbf{i} - 2\mathbf{j} + 3\mathbf{k}) \text{ N} = \boxed{(\mathbf{i} + \mathbf{j} + 2\mathbf{k}) \text{ N}.}$$

29. Professor B will be able to tell that he is accelerating because he will feel the force exerted by the seat back. (At very small accelerations, this may not be evident.) If Professor B simply looks at the change in position of Professor A, he would think that Professor A is accelerating. This apparent acceleration would be 2 m/s^2 in the $-x$-direction. This is not a real acceleration, because there is no horizontal force on Professor A.

33. We simplify the normal forces from the ground as a single force. Note that there is no separate force from the engine, which is a force internal to the system chosen.

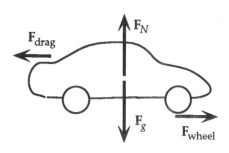

37. We displace the forces slightly to clarify the diagram. Note that the person cannot exert any force on the ball after it leaves the hand.

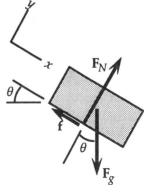

39.

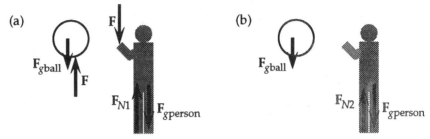

(d) It is the net force that determines the acceleration. In addition to the backward force from the wagon, there is a forward force from the earth on the horse's hooves.

41. As shown, we choose a coordinate system with the *x*-direction down the plane. The forces are due to gravity and contact with the plane:

$$\mathbf{F}_g = F_g \sin\theta\,\mathbf{i} - F_g \cos\theta\,\mathbf{j},$$
$$\mathbf{F}_N = F_N\,\mathbf{j},$$
$$\mathbf{f} = -f\,\mathbf{i}.$$

47. (a) The two forces are due to gravity and contact with the floor. Because there is no acceleration, the sum of the two forces must be zero:

$$\boxed{F_g = mg \text{ down} \quad \text{and} \quad F_N = F_g = mg \text{ up.}}$$

(b) We choose up as the positive direction. From $\Sigma \mathbf{F} = m\mathbf{a}$ we can write

$F_N - F_g = mg$, which gives $F_N = F_g + mg = 2mg$.

$$\boxed{\text{The forces are } mg \text{ down and } 2mg \text{ up.}}$$

(c) Since the acceleration is now g down, no contact force with the floor is needed;

$$\boxed{F_N = 0 \text{ and the only force is } mg \text{ down.}}$$

51. We can find the acceleration of the object by differentiating the displacement:

$v = dx/dt = d(At^{3/2})/dt = \tfrac{3}{2}At^{1/2};$

$a = dv/dt = d(\tfrac{3}{2}At^{1/2})/dt = \tfrac{3}{4}At^{-1/2}.$

This acceleration is produced by the net force:

$F_{\text{net}} = ma = \tfrac{3}{4}mAt^{-1/2} = \tfrac{3}{4}(2.0 \text{ kg})(6.0 \text{ m/s}^{3/2})t^{-1/2} = \boxed{(9.0 \text{ N} \cdot \text{s}^{1/2})t^{-1/2}.}$

57. The board exerts a normal force up on the washer. Since the speed is constant, from the force diagram for the washer we can write

$\Sigma F_y = ma_y = 0;\ T_1 + T_2 + F_N - m_Wg = 0.$

The washer exerts a normal force down on the board. Since the speed is constant, from the force diagram for the board we can write

$\Sigma F_y = ma_y = 0;\ T_1 + T_2 - F_N - m_Bg = 0.$

If we assume that $T_1 = T_2 = T$, when F_N is eliminated, we find that

$$\boxed{T = (m_Wg + m_Bg)/4.}$$

If there is an upward acceleration, the equations become

washer: $T_1 + T_2 + F_N - m_Wg = m_Wa;$

board: $T_1 + T_2 - F_N - m_Bg = m_Ba.$

Again, if we assume that $T_1 = T_2 = T$, we find that

$$\boxed{a = [4T - (m_Wg + m_Bg)]/(m_W + m_B).}$$

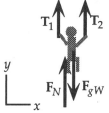

59. (a) To accelerate forward, the engine must produce a force at the wheels which pushes $\boxed{\text{backwards}}$ on the earth. The reaction to this force on the engine will be $\boxed{2 \times 10^4 \text{ N}}$ forward.

(b) If we look at the entire train, the only horizontal force is the force on the engine. We have $\Sigma F_x = ma_x$, or $F = m_{\text{train}}a$:

$2 \times 10^4 \text{ N} = (10^5 \text{ kg})a$, which gives $a = \boxed{0.20 \text{ m/s}^2.}$

This is the acceleration of the engine and each of the cars. If we look at the three cars, the only horizontal force is the force of the coupling with the engine as shown in the diagram.

We have $\Sigma F_x = ma_x$, or $F_1 = m_{\text{cars}}a$:

$F_1 = (6 \times 10^4 \text{ kg})(0.20 \text{ m/s}^2) = \boxed{1.2 \times 10^4 \text{ N forward.}}$

(c) If we look at the first car, the horizontal forces are the couplings shown in the diagram.

We have $\Sigma F_x = ma_x$, or $F_1 - F_2 = m_1a$:

$(1.2 \times 10^4 \text{ N}) - F_2 = (2 \times 10^4 \text{ kg})(0.20 \text{ m/s}^2)$, which gives $F_2 = 8 \times 10^3 \text{ N}$ backward.

If we look at the second car, the horizontal forces are the couplings shown in the diagram.

We have $\Sigma F_x = ma_x$, or $F_2 - F_3 = m_2a$:

$(8 \times 10^3 \text{ N}) - F_3 = (2 \times 10^4 \text{ kg})(0.20 \text{ m/s}^2)$, which gives $F_3 = 4 \times 10^3 \text{ N}$ forward.

The forces on the second car are

$$\boxed{F_2 = 8 \times 10^3 \text{ N forward and } F_3 = 4 \times 10^3 \text{ N backward; its acceleration is } 0.20 \text{ m/s}^2.}$$

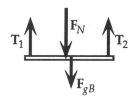

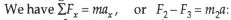

63. We take up as positive, so $\mathbf{F}_g = -mg\,\mathbf{j}$, $\mathbf{F}_d = Av^2\mathbf{j}$, and $\mathbf{v} = -v\,\mathbf{j}$.
    (a) We do a dimensional analysis of $F_d = Av^2$:
        $[F_d] = [A][v^2]$,
        $[MLT^{-2}] = [A][LT^{-1}]^2$, which gives $\boxed{[A] = [ML^{-1}]}$ with units of kg/m.
    (b) The net force produces the acceleration of the parachutist:
        $dv/dt = \Sigma F/m = (Av^2 - mg)/m = \boxed{(Av^2/m) - g.}$
    (c) As the speed increases, the drag force increases until the acceleration becomes zero. At constant

        velocity: $dv/dt = 0$; $(A/m)v_t^2 - g = 0$, which gives $\boxed{v_t = \sqrt{mg/A}\,.}$

65. (a) The forces are
        force of gravity: mg (down),
        normal force of ground: F_N (up),
        friction force of ground: f (forward),
        wind resistance: F_w (backward).
    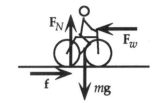
    (b) Because the speed is constant: $\Sigma\mathbf{F} = 0$.
    (c) In the bicycle's frame, the speed of the cyclist is $v_{cc} = 0$; the speed of the wind is $v_{wc} = v_w + v$.
        Because the speed is constant (zero): $\Sigma\mathbf{F} = 0$.
    (d) In the frame of the air, the speed of the cyclist is $v_{cw} = v + v_w$; the speed of the wind is $v_{ww} = 0$.
        Because the speed is constant: $\Sigma\mathbf{F} = 0$.

67. We choose a coordinate system with the x-axis in the direction of the acceleration.
    For the system of the frame and plumb bob, the
    tension in the rope is an internal force. From
    the force diagram for this system, we can write
        $\Sigma F_x = ma_x$:
        $(m_B + m)g\sin\theta = (m_B + m)a$; and
        $\Sigma F_y = ma_y$:
        $F_N - (m_B + m)g\cos\theta = 0$.
    From the x-equation we find
        $a = g\sin\theta$.
    From the force diagram for the system of the
    plumb bob, we can write
        $\Sigma F_x = ma_x$:
        $m_B g\sin\theta + T\sin(\phi - \theta) = m_B a$; and
        $\Sigma F_y = ma_y$:
        $T\cos(\phi - \theta) - m_B g = 0$.
    Using $a = g\sin\theta$ in the x-equation, we find $\sin(\phi - \theta) = 0$, or $\phi - \theta = 0$, so $\boxed{\phi = \theta.}$

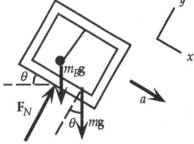
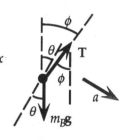

CHAPTER 5

The analysis of force diagrams is usually simplified if one of the axes of the coordinate system is chosen to be in the direction of the acceleration. It is possible and generally convenient to select a different coordinate system for each object to analyze the force diagram. It is important to specify all of the forces acting on the selected object. Most forces require contact between the object and its surroundings, so look at contact points for possible forces. Of course, the force of gravity does not require contact.

5. We find the acceleration from the equation of motion of the puck:
    $\Sigma F = ma$; 4.0i N = (0.10 kg)$\mathbf{a}$, which gives $\mathbf{a}$ = 40i m/s^2.
    Because the acceleration has no y-component, v_y does not change.
    We can find v_x for a speed of v = 6 m/s:
    $v^2 = v_x{}^2 + v_y{}^2$; (6.0 m/s)^2 = $v_x{}^2$ + (3.0 m/s)^2, which gives v_x = 5.2 m/s.
    With constant acceleration in the x-direction, we find the time required for the velocity change from
    $v_x = v_{0x} + a_x t$: 5.2 m/s = 1.4 m/s + (40 m/s^2)t, which gives t = $\boxed{0.095 \text{ s.}}$

7. The mass of the bottom section of the uniform rod, with d the distance from the top, is $m' = (M/L)(L - d)$. With the bottom section as our system, for the y-direction we can write: $\Sigma F = m'a$; $T_d - m'g = 0$, so
    $T_d = (M/L)(L - d)g$;
    thus T_d is maximum at the top and zero at the bottom,
    (a) When d = 2.0 m,
        T_2 = [(5.6 kg)/(3.5m)](3.5 m – 2.0 m)(9.8 m/s^2) = $\boxed{24 \text{ N.}}$
    (b) When d = 3.0 m,
        T_3 = [(5.6 kg)/(3.5 m)](3.5 m – 3.0 m)(9.8 m/s^2) = $\boxed{7.8 \text{ N.}}$

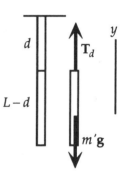

9. We choose a coordinate system with the x-axis up the plane.
    (a) In equilibrium, there is no acceleration of the block.
        We write $\Sigma F = ma$ from the force diagram for the block:
        y-component: $F_N - F \sin \theta - mg \cos \theta = 0$,
        x-component: $F \cos \theta - mg \sin \theta = 0$,
            $F \cos 15° - $ (25 kg)(9.8 m/s^2) $\sin 15° = 0$, which gives
        $\boxed{F = 66 \text{ N.}}$
    (b) With a larger force, the block will accelerate up the plane,
        so the x-equation becomes
            x-component: $F \cos \theta - mg \sin \theta = ma$,
                3(66 N) $\cos 15° - $ (25 kg)(9.8 m/s^2) $\sin 15° = $ (25 kg)a, which gives $\boxed{a = 5.1 \text{ m/s}^2 \text{ up.}}$

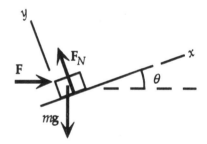

13. The direction of the static friction force will oppose the impending motion of the block M and with a magnitude that cannot exceed $\mu_s F_N$.

For the largest value of M, the block is on the verge of slipping down the plane, so the static friction force will be up the plane and maximum, $f_s = f_{s,\text{max}} = \mu_s F_N$.

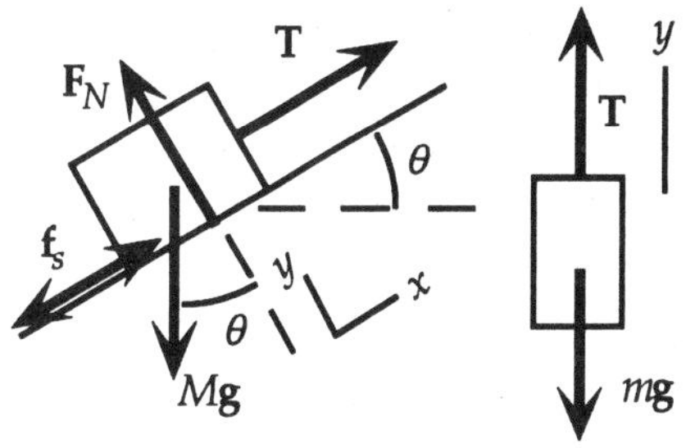

From the force diagram, with the block M as the system, we can write $\Sigma \mathbf{F} = M\mathbf{a}$:

x-component: $T - M_{\text{max}} g \sin\theta + f_{s,\text{max}} = 0$;
y-component: $F_N - M_{\text{max}} g \cos\theta = 0$; or
$$F_N = M_{\text{max}} g \cos\theta .$$

From the force diagram, with the block m as the system, we can write $\Sigma \mathbf{F} = m\mathbf{a}$:

y-component: $T - mg = 0$; or $T = mg$.

The x-equation becomes

$mg - M_{\text{max}} g \sin\theta + \mu_s M_{\text{max}} g \cos\theta = 0$, or
$M_{\text{max}}(\sin\theta - \mu_k \cos\theta) = m$;
$M_{\text{max}}(\sin 30° - 0.20 \cos 30°) = 3.0 \text{ kg}$, which gives $\boxed{M_{\text{max}} = 9.2 \text{ kg.}}$

For the smallest value of M, the block is on the verge of slipping up the plane, so the static friction force will be down the plane and maximum, $f_s = f_{s,\text{max}} = \mu_s F_N$. The only change will be in the x-equation. From the force diagram, with the block M as the system, we can write $\Sigma \mathbf{F} = M\mathbf{a}$:

x-component: $T - M_{\text{min}} g \sin\theta - f_{s,\text{max}} = 0$;
$mg - M_{\text{min}} g \sin\theta - \mu_s M_{\text{min}} g \cos\theta = 0$, or
$M_{\text{min}}(\sin\theta + \mu_k \cos\theta) = m$;
$M_{\text{min}}(\sin 30° + 0.20 \cos 30°) = 3.0 \text{ kg}$, which gives $\boxed{M_{\text{min}} = 4.5 \text{ kg.}}$

If $M = 6$ kg, which is between the two extreme values, the block will remain at rest. We assume the static friction force will be up the plane. The x-equation becomes

$mg - Mg \sin\theta + f_s = 0$;
$(3.0 \text{ kg})(9.8 \text{ m/s2}) - (6.0 \text{ kg})(9.8 \text{ m/s}^2) \sin 30° + f_s = 0$, which gives $\boxed{f_s = 0.}$

17. The force diagrams for each of the masses and the movable pulley are shown. Note that we take down as positive and the indicated accelerations are relative to the <u>fixed</u> pulley.
A downward acceleration of m_1 means an upward acceleration of the movable pulley. If we call a_r the acceleration of m_2 with respect to the movable pulley, we have
$$a_2 = a_r - a_1 \quad \text{and} \quad a_3 = -a_r - a_1,$$
because the acceleration of m_3 with respect to the movable pulley must be the negative of m_2's acceleration with respect to the pulley.
If the mass of the pulley is negligible, we write $\Sigma F_y = ma_y$:
$$2T_2 - T_1 = (0)(-a_1), \text{ so } 2T_2 = T_1.$$
For each of the masses, for $\Sigma F_y = ma_y$ we get
mass m_1: $m_1 g - T_1 = m_1 a_1,$
mass m_2: $m_2 g - T_2 = m_2 a_2 = m_2(a_r - a_1),$
mass m_3: $m_3 g - T_2 = m_3 a_3 = m_3(-a_r - a_1).$
We have four equations for the four unknowns:
$T_1, T_2, a_r,$ and a_1.
After some careful algebra, we get
$$a_1 = [(m_1 m_2 + m_1 m_3 - 4m_2 m_3)/(m_1 m_2 + m_1 m_3 + 4m_2 m_3)]g;$$
$$a_r = [2m_1(m_2 - m_3)/(m_1 m_2 + m_1 m_3 + 4m_2 m_3)]g;$$
$$T_1 = [8m_1 m_2 m_3/(m_1 m_2 + m_1 m_3 + 4m_2 m_3)]g; \text{ and}$$
$$T_2 = [4m_1 m_2 m_3/(m_1 m_2 + m_1 m_3 + 4m_2 m_3)]g.$$
We can now find the other accelerations:
$$a_2 = [(m_1 m_2 - 3m_1 m_3 + 4m_2 m_3)/(m_1 m_2 + m_1 m_3 + 4m_2 m_3)]g;$$
$$a_3 = [(-3m_1 m_2 + m_1 m_3 + 4m_2 m_3)/(m_1 m_2 + m_1 m_3 + 4m_2 m_3)]g.$$
If $m_2 = m_3 \ne m_1$:
$a_r = 0;$
$a_1 = [(m_1 - 2m_2)/(m_1 + 2m_2)]g$
$T_1 = [4m_1 m_2/(m_1 + 2m_2)]g;$
$T_2 = [2m_1 m_2/(m_1 + 2m_2)]g;$
$a_2 = a_3 = -a_1 = [(2m_2 - m_1)/(m_1 + 2m_2)]g.$
Thus m_2 and m_3 have the same acceleration as the pulley. Note that neither tension is equal to an mg! None of the masses are in equilibrium.

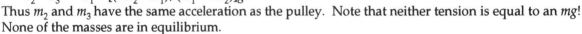

21. We assume that the static friction force is the maximum, without skidding. (Skidding would mean kinetic friction is present, which is usually smaller than the maximum static friction.)
The acceleration can be found from the car's one-dimensional motion:
$$v = v_0 + at; \quad 0 = 25 \text{ m/s} + a(4.2 \text{ s}), \text{ which gives}$$
$$a = -5.6 \text{ m/s}^2.$$
We write $\Sigma \mathbf{F} = m\mathbf{a}$ from the force diagram for the car:
x-component: $-f_s = -\mu_s F_N = ma$
y-component: $F_N - mg = 0.$
Thus $F_N = Mg$ and $\mu_s = -a/g = -(-5.6 \text{ m/s}^2)/(9.8 \text{ m/s}^2) = \boxed{0.61.}$

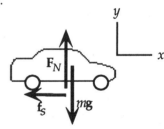

25. The kinetic friction will be up the plane to oppose the motion.
We write $\Sigma \mathbf{F} = m\mathbf{a}$ from the force diagram for the skier, with $\mathbf{a} = 0$:
x-component: $Mg \sin \theta - f_k = 0;$
y-component: $F_N - Mg \cos \theta = 0.$
Thus $Mg \sin \theta = f_k = \mu_k F_N = \mu_k Mg \cos \theta$, which gives
$$\mu_k = (\sin \theta)/(\cos \theta) = \tan \theta = \tan 22° = \boxed{0.40.}$$

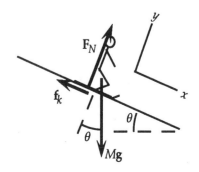

29. (a) The constant forces will produce a constant acceleration, so for the motion of the box we can write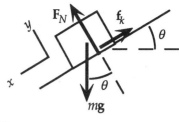

$x = x_0 + v_0 t + \frac{1}{2}at^2$;

$1.2 \text{ m} = 0 + 0 + \frac{1}{2}(0.3 \text{ m/s}^2)t^2$, which gives

$t = \boxed{2.8 \text{ s.}}$

(b) We write $\Sigma \mathbf{F} = m\mathbf{a}$ from the force diagram for the box:

x-component: $mg \sin \theta - f_k = ma$, which gives

$f_k = mg \sin \theta - ma$

$= (0.500 \text{ kg})(9.8 \text{ m/s}^2) \sin 30° - (0.500 \text{ kg})(0.3 \text{ m/s}^2)$

$= \boxed{2.3 \text{ N.}}$

(c) From the force diagram for the box, $\Sigma \mathbf{F} = m\mathbf{a}$:

y-component: $F_N - Mg \cos \theta = 0$, which gives

$F_N = Mg \cos \theta = (0.500 \text{ kg})(9.8 \text{ m/s}^2) \cos 30° = 4.24 \text{ N.}$

Thus $\mu_k = f_k/F_N = (2.3 \text{ N})/(4.24 \text{ N}) = \boxed{0.54.}$

31. Until the box moves, friction is static, opposing the impending motion, and the acceleration is zero.

(a) We write $\Sigma \mathbf{F} = m\mathbf{a}$ from the force diagram for the box:

x-component: $F \cos \theta - f_s = 0$;

y-component: $F \sin \theta + F_N - mg = 0$.

The box is on the verge of moving when the static friction force reaches its maximum value:

$f_s = f_{s,\text{max}} = \mu_s F_N$.

When this is put into the x-component equation, we can solve for F to get

$F = \mu_s mg/(\cos \theta + \mu_s \sin \theta)$

$= 0.75(50 \text{ kg})(9.8 \text{ m/s}^2)/(\cos \theta + 0.75 \sin \theta) = \boxed{370/(\cos \theta + 0.75 \sin \theta) \text{ N.}}$

(b) To find the angle at which F is a minimum, we set $dF/d\theta = 0$:

$dF/d\theta = \mu_s mg \, (-1)(- \sin \theta + \mu_s \cos \theta)/(\cos \theta + \mu_s \sin \theta)^2 = 0.$

This is true when $\sin \theta_{\text{min}} = \mu_s \cos \theta_{\text{min}},$ or

$\tan \theta_{\text{min}} = \mu_s = 0.75;$ which gives $\theta_{\text{min}} = \quad 37°.$

The force F is then

$F_{\text{min}} = \mu_s mg/[\cos(\tan^{-1} \mu_s) + \mu_s \sin(\tan^{-1} \mu_s)]$

$= (370 \text{ N})/[0.80 + 0.75(0.60)] = \boxed{2.9 \times 10^2 \text{ N.}}$

A minimum value occurs because at small angles the normal force, F_N, and thus the friction force, f_s, are large, which requires a large force F. At angles near 90°, F_N is small, creating a small friction force; however, the small horizontal component of F requires a large magnitude of F. There is an angle where the decrease in F_N is balanced by the decrease in horizontal component, and thus a minimum value of F.

35. (a) We write $\Sigma \mathbf{F} = m\mathbf{a}$ from the force diagram for the top mass:

x-component: $0 = m_1 a_1$;

y-component: $F_{N1} - m_1 g = 0$.

We write $\Sigma \mathbf{F} = m\mathbf{a}$ from the force diagram for the bottom mass:

x-component: $F = m_2 a_2$;

y-component: $F_{N2} - F_{N1} - m_2 g = 0$.

Thus we have $F_{N1} = m_1 g$; $F_{N2} = F_{N1} + m_2 g$; and

$$\boxed{a_1 = 0; \quad a_2 = F/m_2.}$$

As expected, the top mass does not accelerate because there is no horizontal friction force at the contact with the lower mass.

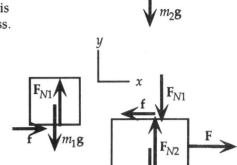

(b) Because the top mass must move with the bottom mass, we can treat the combination as a single system.

We write $\Sigma \mathbf{F} = m\mathbf{a}$ for the combined masses:

x-component: $F = (m_1 + m_2)a$;

Thus $a = \boxed{a_1 = a_2 = F/(m_1 + m_2).}$

(c) Assuming the top mass is sliding to the left, we have

$F_{N1} = m_1 g \mathbf{j}$ (up) and $\mathbf{f}_1 = \mu_k m_1 g \mathbf{i}$ (right), so

$$\boxed{\mathbf{F}\text{(lower on upper)} = \mu_k m_1 g \mathbf{i} + m_1 g \mathbf{j}.}$$

(d) The y-component equations are the same as in part (a).

We write $\Sigma \mathbf{F} = m\mathbf{a}$ from the force diagram for the top mass:

x-component: $\mu_k m_1 g = m_1 a_1$.

We write $\Sigma \mathbf{F} = m\mathbf{a}$ from the force diagram for the bottom mass:

x-component: $F - \mu_k m_1 g = m_2 a_2$.

Thus $\boxed{a_1 = \mu_k g}$ and $\boxed{a_2 = (F - \mu_k m_1 g)/m_2.}$

43. Because $F_D = \frac{1}{2}\rho A C_D v^2$, at terminal speed the drag force must balance the force of gravity, so we have

$mg = \frac{1}{2}\rho A C_D v^2$.

Because the mass of the skydiver does not change and the maximum area corresponds to the minimum speed, we can write

$mg = \frac{1}{2}\rho A_{max} C_{Dmin} v_{min}^2 = \frac{1}{2}\rho A_{min} C_{Dmax} v_{max}^2$.

Thus $C_{Dmax}/C_{Dmin} = (A_{max}/A_{min})(v_{min}/v_{max})^2 = (1.5)[(40 \text{ m/s})/(60 \text{ m/s})]^2 = \boxed{0.67.}$

49. If the automobile does not skid, the friction is static, with $f_s \le \mu_s F_N$.

At high speed, the automobile will tend to slide up the incline, so f_s will be down the incline, which will provide a greater centripetal force. Note that we use a coordinate system with the x-axis in the direction of the centripetal acceleration.

We write $\Sigma \mathbf{F} = m\mathbf{a}$ from the force diagram for the auto:

x-component: $F_N \sin\theta + f_s \cos\theta = ma = mv^2/R$;

y-component: $F_N \cos\theta - f_s \sin\theta - mg = 0$.

The speed is maximum when $f_s = f_{s,max} = \mu_s N$.

From the y-equation we get

$F_N \cos\theta - \mu_s F_N \sin\theta = mg$, or $F_N = mg/(\cos\theta - \mu_s \sin\theta)$.

From the x-equation we get

$v_{max}^2/R = g(\sin\theta + \mu_s \cos\theta)/(\cos\theta - \mu_s \sin\theta)$.

$v_{max}^2 = (150 \text{ m})(9.8 \text{ m/s}^2)(\sin 18° + 0.3 \cos 18°)/(\cos 18° - 0.3 \sin 18°)$, which gives $\boxed{v_{max} = 32 \text{ m/s.}}$

At low speed, the automobile will tend to slide down the incline, so f_s will be up the incline.

The speed is minimum when $f_s = f_{s,max} = \mu_s F_N$.

If we change the sign of f_s in the equations, we get

$F_N = mg/(\cos\theta + \mu_s \sin\theta)$ and $v_{min}^2/R = mg(\sin\theta - \mu_s \cos\theta)/(\cos\theta + \mu_s \sin\theta)$.

Thus $v_{min}^2 = (150 \text{ m})(9.8 \text{ m/s}^2)(\sin 18° - 0.3 \cos 18°)/(\cos 18° + 0.3 \sin 18°)$, which gives

$$\boxed{v_{min} = 5.8 \text{ m/s.}}$$

51. While the stone does not slide on the turntable, the static friction force provides the centripetal acceleration:
$$f_s = ma_r = mR\omega^2.$$
Thus a larger friction force is needed at greater ω and larger R. At the critical distance for the maximum speed of rotation the friction force is maximum:
$f_s = f_{s,max} = \mu_s mg$, so we have
$\mu_s mg = mR\omega^2$, or
$\mu_s = R\omega^2/g = (0.26\ \text{m})[(45\ \text{rev}/\text{min})(2\pi\ \text{rad}/\text{rev})/(60\ \text{s}/\text{min})]^2/(9.8\ \text{m}/\text{s}^2) = \boxed{0.59.}$

55. The tension in the rope supports the hanging mass and provides the centripetal acceleration of the puck.
We write $\Sigma\mathbf{F} = m\mathbf{a}$ from the force diagram for the hanging mass, with down positive:
$m_1 g - T = m_1 a_1 = 0$; which gives
$T = m_1 g = (1.00\ \text{kg})(9.8\ \text{m}/\text{s}^2) = 9.80\ \text{N}.$
For the rotating puck, the tension provides the centripetal acceleration, $\Sigma F_r = ma_r$:
$T = m_2 v^2/r$; $9.80\ \text{N} = (0.400\ \text{kg})v^2/(0.80\ \text{m})$, which gives
$v = \boxed{4.43\ \text{m}/\text{s}.}$

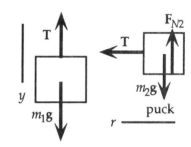

59. The centripetal acceleration is provided by a component of the normal force. We choose a coordinate system with one axis in the direction of the acceleration.
We write $\Sigma\mathbf{F} = m\mathbf{a}$ from the force diagram for the block:
x-component: $F_N \cos\theta = mr\omega^2$;
y-component: $F_N \sin\theta - mg = 0$.
Combining these we get $r = g/(\omega^2 \tan\theta)$.
Then　$h = r/\tan\theta = g/(\omega^2 \tan^2\theta)$
　　　　$= (9.8\ \text{m}/\text{s}^2)/[(6\ \text{rad}/\text{s})^2 \tan^2 50°]$
　　　　$= \boxed{0.19\ \text{m}.}$

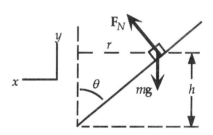

63. The tension in the string provides the centripetal acceleration and the friction force, which opposes the motion, provides a tangential acceleration.
We write $\Sigma\mathbf{F} = m\mathbf{a}$ from the force diagram for the puck:
x-component: $-T = -mv^2/R$;
y-component: $-f_k = m(dv/dt)$;
z-component: $F_N - mg = 0$.

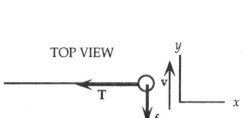

TOP VIEW

(a) At $t = 0$ s:
$T = mv_0^2/R = (0.2\ \text{kg})(10\ \text{m}/\text{s})^2/(0.5\ \text{m}) = \boxed{40\ \text{N}.}$
(b) From the z-equation, $F_N = mg$, so $f_k = \mu_k F_N = \mu_k mg$.
From the y-equation we can find the tangential acceleration:
$-0.2(0.2\ \text{kg})(9.8\ \text{m}/\text{s}^2) = (0.2\ \text{kg})(dv/dt)$, which gives $dv/dt = a_t = -1.96\ \text{m}/\text{s}^2$.
For the tangential motion of one revolution with circumference C we can write
$v^2 = v_0^2 + 2a_t(C - 0) = (10\ \text{m}/\text{s})^2 + 2(-1.96\ \text{m}/\text{s}^2)[2\pi(0.5\ \text{m}) - 0]$, which gives
the speed after one revolution: $v = 9.4\ \text{m}/\text{s}.$
The tension after one revolution is
$T = mv^2/R = (0.2\ \text{kg})(9.4\ \text{m}/\text{s})^2/(0.5\ \text{m}) = \boxed{35\ \text{N}.}$

69. Because there is no friction, we need to consider horizontal forces only.
    We write $\Sigma\mathbf{F} = m\mathbf{a}$ from the force diagram for the system of the three masses:
    x-component: $F = (m_1 + m_2 + m_3)a$;
    1.5 N = (0.3 kg + 0.4 kg + 0.2 kg)a, which gives

$$\boxed{a = 1.7 \text{ m/s}^2.}$$

(a) We write $\Sigma\mathbf{F} = m\mathbf{a}$ from the force diagram for m_1:
    x-component: $T_1 = m_1a$;
    $T_1 = (0.3 \text{ kg})(1.7 \text{ m/s}^2) = \boxed{0.5 \text{ N.}}$

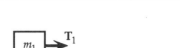

(b) We write $\Sigma\mathbf{F} = m\mathbf{a}$ from the force diagram for m_3:
    x-component: $F - T_2 = m_3a$;
    1.5 N $- T_2 = (0.2 \text{ kg})(1.7 \text{ m/s}^2)$, which gives $T_2 = \boxed{1.2 \text{ N.}}$

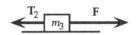

71. (a) Because the length of the rope is constant, when m_1 moves
        up Δx_1, the segment above m_1 decreases by Δx_1, and
        each segment above m_2 must increase by one-half that amount:
        $$\boxed{\Delta x_2 = -\tfrac{1}{2}\Delta x_1 \quad (- \text{ indicates the opposite direction}).}$$

    (b) If we differentiate with respect to time twice, we get
        $v_2 = -\tfrac{1}{2}v_1; \quad a_2 = -\tfrac{1}{2}a_1.$
        We write $\Sigma\mathbf{F} = m\mathbf{a}$ from the force diagram for m_1:
        x-component: $T - m_1g = m_1a_1.$
        We write $\Sigma\mathbf{F} = m\mathbf{a}$ from the force diagram for m_2:
        x-component: $2T - m_2g = m_2a_2 = -\tfrac{1}{2}m_2a_1.$
        By eliminating T between these equations, we get
        $$\begin{aligned} a_1 &= -2g[(2m_1 - m_2)/(4m_1 + m_2)] \\ &= -2(9.8 \text{ m/s}^2)[2(1.2 \text{ kg}) - 1.8 \text{ kg}]/[4(1.2 \text{ kg}) + 1.8 \text{ kg}] \\ &= \boxed{-1.78 \text{ m/s}^2 \text{ (down)};} \end{aligned}$$
        $a_2 = -\tfrac{1}{2}a_1 = \boxed{+0.89 \text{ m/s}^2 \text{ (up)}.}$

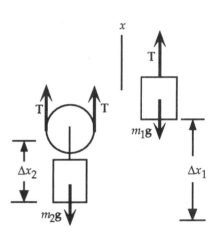

    (c) For the tension we have
        $T = m_1(g + a_1) = (1.2 \text{ kg})(9.8 \text{ m/s}^2 - 1.78 \text{ m/s}^2) = \boxed{9.6 \text{ N.}}$

73. The force that prevents slipping is an upward friction force.
    The normal force provides the centripetal acceleration.
    We write $\Sigma\mathbf{F} = m\mathbf{a}$ from the force diagram for the person:
    x-component: $F_N = mR\omega^2$;
    y-component: $f_s - mg = 0.$
    At the critical condition where slipping begins,
    $f_s = f_{s,\text{max}} = \mu_sF_N.$
    Thus we have $\mu_s(mR\omega^2) = mg$, which gives
    $$\boxed{\omega = (g/\mu_s R)^{1/2}.}$$

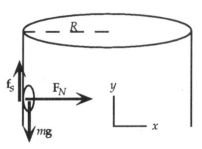

79. At the top of the loop, both the normal force and the weight are
    downward and they provide the centripetal acceleration.
    We write $\Sigma\mathbf{F} = m\mathbf{a}$ from the force diagram for the motorcycle:
    y-component: $F_N + mg = mv^2/R.$
    The speed v will be minimum when the normal
    force is minimum. The normal force can only push
    away from the ramp, that is, with our coordinate
    system it must be positive, so $F_{N\text{min}} = 0.$
    Thus we have $v_{\text{min}}^2 = gR,$ or
    $$\begin{aligned} v_{\text{min}} &= (gR)^{1/2} \\ &= [(9.8 \text{ m/s}^2)(12 \text{ m})]^{1/2} = \boxed{11 \text{ m/s}} \quad (\approx 24 \text{ mi/h}). \end{aligned}$$

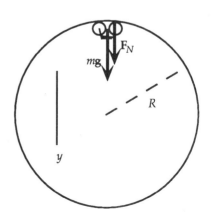

CHAPTER 6

To determine the work done on a system, it is necessary to know all of the forces acting on the system. The force diagram, as used in the preceding chapters, will help in determining which forces do work. Work is a scalar, so it is necessary to have the correct sign.

5. We find the work done on the crate from the increases in its kinetic energy:
$$W = \Delta K = \tfrac{1}{2}mv^2 - \tfrac{1}{2}mv_0^2 = \tfrac{1}{2}(66 \text{ kg})[(63 \text{ km/h})(10^3 \text{ m/km})/(3600 \text{ s/h})]^2 - 0 = \boxed{1.0 \times 10^4 \text{ J.}}$$

7. Because there is no acceleration, we have $F_N = mg$ and $F = f_k = \mu_k mg$.
    (a) The work done by the man is
    $$W_p = F \Delta x = 0.4(40 \text{ kg})(9.8 \text{ m/s}^2)(1.5 \text{ m}) = \boxed{2.4 \times 10^2 \text{ J.}}$$
    (b) The friction force also does work (negative).
    (c) Because there is no change in the kinetic energy of the refrigerator, the net work is zero:
    $$W_{\text{net}} = W_P + W_f = \Delta K = \boxed{0.}$$

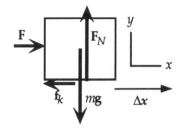

13. To find the work done we need to find the distance the load moves, which we can find by analyzing the forces on the worker and on the load, $\Sigma F_y = ma_y$:
    worker: $m_w g - T = (75 \text{ kg})(9.8 \text{ m/s}^2) - T = (75 \text{ kg})a$;
    load: $T - m_b g = T - (42 \text{ kg})(9.8 \text{ m/s}^2) = (42 \text{ kg})a$.
    By eliminating T, we find $a = 2.8 \text{ m/s}^2$.
    Then the distance traveled is $y = \tfrac{1}{2}at^2 = \tfrac{1}{2}(2.8 \text{ m/s}^2)(2.0 \text{ s})^2 = 5.6 \text{ m}$.
    Because the worker moves down while the load moves up, the total work done by gravity is
    $$\begin{aligned} W_g &= m_w g \, \Delta y - m_b g \, \Delta y \\ &= (75 \text{ kg} - 42 \text{ kg})(9.8 \text{ m/s}^2)(5.6 \text{ m}) = \boxed{1.8 \times 10^3 \text{ J.}} \end{aligned}$$

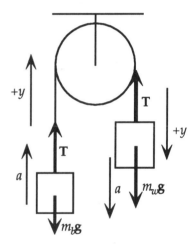

17. Because the net work must be zero, the work that the child must do will have the same magnitude as the work done by gravity. For each block this work is mg times the distance the center is raised (zero for the first block, one block-height for the second block, etc.).
    $$W_{\text{set1}} = (0.036 \text{ kg})(9.8 \text{ m/s}^2)(0 + 1 + 2)(0.12 \text{ m}) = \boxed{0.127 \text{ J.}}$$
    $$W_{\text{set2}} = (0.018 \text{ kg})(9.8 \text{ m/s}^2)(0 + 1 + 2 + 3 + 4 + 5)(0.06 \text{ m}) = \boxed{0.159 \text{ J.}}$$
    $$W_{\text{set3}} = (0.009 \text{ kg})(9.8 \text{ m/s}^2)(0 + 1 + 2 + 3 + 4 + 5 + 6 + 7 + 8 + 9 + 10 + 11)(0.03 \text{ m}) = \boxed{0.175 \text{ J.}}$$

23. Because the scalar product will be zero for two orthogonal vectors, we evaluate
    $$\mathbf{u} \cdot \mathbf{v} = (3\mathbf{i} - 4\mathbf{j} + 7\mathbf{k}) \cdot (-2\mathbf{i} + 3\mathbf{j} + z\mathbf{k}) = -6 - 12 + 7z = 0, \text{ which gives } \boxed{z = 2.57.}$$

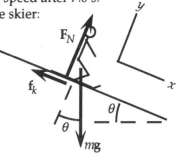

27. The net work equals the increase in kinetic energy, so we want to find the speed after 7.0 s. We can find the acceleration from $\Sigma F = ma$, using the force diagram for the skier:

    y-component: $F_N = mg \cos \theta$;

    x-component: $mg \sin \theta - \mu_k mg \cos \theta = ma$.

    From this we get

    $a = g(\sin \theta - \mu_k \cos \theta) = (9.8 \text{ m/s}^2)[\sin 18° - (0.12) \cos 18°] = 1.9 \text{ m/s}^2$.

    The final speed is $v = v_0 + at = 0 + (1.9 \text{ m/s}^2)(7.0 \text{ s}) = 13 \text{ m/s}$.

    We find the net work from the change in kinetic energy:

    $W_{net} = \Delta K = \tfrac{1}{2}m(v^2 - v_0^2) = \tfrac{1}{2}(72 \text{ kg})[(13 \text{ m/s})^2 - 0] = \boxed{6.1 \times 10^3 \text{ J.}}$

    This net work is from the positive work done by gravity and the negative work done by friction.

31. The scalar product of two vectors is the magnitude of one times the projection of the other. Because the magnitude of the unit vector is 1, we find the magnitude of the projection of $\mathbf{A} = 3\mathbf{i} - 2\mathbf{j}$ onto the unit vector $\mathbf{e} = -0.6\mathbf{i} + 0.8\mathbf{j}$ from

    $|\mathbf{A} \cdot \mathbf{e}| = |(3)(-0.6) + (-2)(0.8)| = \boxed{3.4.}$

33. The work is $W_F = \mathbf{F} \cdot \Delta \mathbf{r} = (2\mathbf{i} - 5\mathbf{j}) \text{ N} \cdot [(5\mathbf{i} - 4\mathbf{j} + 5\mathbf{k}) - (7\mathbf{i} - 8\mathbf{j} + 2\mathbf{k})] \text{ m}$, which gives

    $W_F = [(2)(5 - 7) + (-5)(-4 + 8) + (0)(5 - 2)] \text{ J} = \boxed{-24 \text{ J.}}$

37. Because the force is variable, we must integrate: $W = \int F \, dx$, or

    $$W = \int_0^{2.0 \text{ m}} (g_1 x - g_2 x^3) \, dx = g_1 \tfrac{1}{2} x^2 \Big|_0^{2.0 \text{ m}} - g_2 \tfrac{1}{4} x^4 \Big|_0^{2.0 \text{ m}} = \boxed{2g_1 - 4g_2.}$$

39. The work of the spring changes the kinetic energy of the pellet:

    $$W_{sp} = \int_L^0 (-kx) \, dx = -\tfrac{1}{2}k(0)^2 + \tfrac{1}{2}k(L)^2 = \tfrac{1}{2}mv^2 - 0; \quad \text{or} \quad \tfrac{1}{2}kL^2 = \tfrac{1}{2}mv^2; \quad .$$

    $\tfrac{1}{2}(60 \text{ N/m})(0.07 \text{ m})^2 = \tfrac{1}{2}(4 \times 10^{-3} \text{ kg})v^2$, which gives $\boxed{v = 8.6 \text{ m/s.}}$

41. The force required to stretch the spring must be opposite to the spring force and thus is

    $F = +k_1 x + k_2 x^3$.

    The work done by this variable force is

    $$W = \int_{0.10 \text{ m}}^{0.20 \text{ m}} F \, dx = \int_{0.10 \text{ m}}^{0.20 \text{ m}} (k_1 x + k_2 x^3) \, dx = k_1 \tfrac{1}{2} x^2 \Big|_{0.10 \text{ m}}^{0.20 \text{ m}} + k_2 \tfrac{1}{4} x^4 \Big|_{0.10 \text{ m}}^{0.20 \text{ m}}$$

    $$= (5.0 \text{ N/m})\tfrac{1}{2}\left[(0.20 \text{ m})^2 - (0.10 \text{ m})^2 \right] + (15 \text{ N/m}^3)\tfrac{1}{4}\left[(0.20 \text{ m})^4 - (0.10 \text{ m})^4 \right] = \boxed{8.1 \times 10^{-2} \text{ J.}}$$

43. There is no change in elevation, so the work done by gravity will be zero. Even though the path and the variation of the force applied by the child may be complicated, we find the net work from the change in kinetic energy:

    $W_{net} = \Delta K = \tfrac{1}{2}m(v_f^2 - v_i^2) = \tfrac{1}{2}m(r_f \omega_f)^2 - 0$

    $= \tfrac{1}{2}(0.085 \text{ kg})[(1.45 \text{ m})(2 \text{ rev/s})(2\pi \text{ rad/rev})]^2 - (0.01 \text{ m/s})^2] = \boxed{14 \text{ J.}}$

49. We are given $|F(x)| = Ax^2$ with the force always toward the origin.
   (a) For the motion in this part the force is toward positive x, so the work is

$$W_a = \int_{-5.0\,cm}^{0} \left(+Ax^2\right) dx = +\frac{1}{3} Ax^3 \Big|_{-5.0\,cm}^{0} = +\frac{1}{3}\left(1500\,\text{N/m}^2\right)\left[(0)^3 - \left(-0.05\,\text{m}\right)^3\right] = \boxed{6.3 \times 10^{-2}\ \text{J}.}$$

   (b) In addition to the work from part (a), there is additional work done while the force is toward negative x, which is

$$W_{b2} = \int_{0}^{5.0\,cm} \left(-Ax^2\right) dx = -\frac{1}{3} Ax^3 \Big|_{0}^{5.0\,cm} = -\frac{1}{3}\left(1500\,\text{N/m}^2\right)\left[(0.05\,\text{m})^3 - (0)^3\right] = -6.3 \times 10^{-2}\ \text{J}.$$

   Thus the total work now is $6.3 \times 10^{-2}\,\text{J} - 6.3 \times 10^{-2}\,\text{J} = \boxed{0.}$

   (c) Now the force is toward negative x, so the work is

$$W_c = \int_{5.0\,cm}^{2.0\,cm} \left(-Ax^2\right) dx = -\frac{1}{3} Ax^3 \Big|_{5.0\,cm}^{2.0\,cm} = -\frac{1}{3}\left(1500\,\text{N/m}^2\right)\left[(0.02\,\text{m})^3 - (0.05\,\text{m})^3\right] = \boxed{5.9 \times 10^{-2}\ \text{J}.}$$

   (d) Now the force is toward positive x, so the work is

$$W_d = \int_{-2.0\,cm}^{-5.0\,cm} \left(+Ax^2\right) dx = +\frac{1}{3} Ax^3 \Big|_{-2.0\,cm}^{-5.0\,cm} = +\frac{1}{3}\left(1500\,\text{N/m}^2\right)\left[\left(-0.05\,\text{m}\right)^3 - \left(-0.02\,\text{m}\right)^3\right] = \boxed{-5.9 \times 10^{-2}\ \text{J}.}$$

51. For a variable force, the work is found by integrating:

$$W_x = \int_{0}^{x} \left(F_0 + Cx'\right) dx' = \left(F_0 x' + \frac{1}{2} C x'^2\right)\Big|_{0}^{x} = F_0 x + \frac{1}{2} C x^2.$$

   (a) $W_1 = F_0 x + \frac{1}{2}Cx^2 = (5\,\text{N})(1\,\text{m}) + \frac{1}{2}(-2\,\text{N/m})(1\,\text{m})^2 = \boxed{4\,\text{J}.}$
       $W_2 = (5\,\text{N})(2\,\text{m}) + \frac{1}{2}(-2\,\text{N/m})(2\,\text{m})^2 = \boxed{6\,\text{J}.}$
       $W_3 = (5\,\text{N})(3\,\text{m}) + \frac{1}{2}(-2\,\text{N/m})(3\,\text{m})^2 = \boxed{6\,\text{J}.}$
       $W_4 = (5\,\text{N})(4\,\text{m}) + \frac{1}{2}(-2\,\text{N/m})(4\,\text{m})^2 = \boxed{4\,\text{J}.}$

   (b) To find where the total work will be zero, we have $W = 0 = F_0 x + \frac{1}{2}Cx^2 = x\left(F_0 + \frac{1}{2}Cx\right)$, which gives
       $x = 0$ and $x = -2F_0/C = -2(5\,\text{N})/(-2\,\text{N/m}) = \boxed{+5\,\text{m}.}$

   (c) In one dimension, a force that is a function of position only is $\boxed{\text{conservative.}}$

55. We are given $F(x) = C|x|$, which can be expressed as
$F(x) = C(-x)$ for negative x-values and
$F(x) = C(+x)$ for positive x-values.
For a variable force we integrate to find the work.
(a) This motion has positive and negative values for x, so we use two segments for the path:

$$W_a = \int_{-4.0\,cm}^{0} C(-x)\,dx + \int_{0}^{4.0\,cm} C x\,dx = -\frac{1}{2}Cx^2 \Big|_{-4.0\,cm}^{0} + \frac{1}{2}Cx^2 \Big|_{0}^{4.0\,cm}$$

$$= -\frac{1}{2}C(N/cm)(100\,cm/m)\left\{\left[(0)^2 - (-0.04\,m)^2\right] - \left[(0.04\,m)^2 - (0)^2\right]\right\} = \boxed{0.16C \text{ J.}}$$

If $F = Cx$, the direction of the force is given by the sign of x, so we get

$$W = \int_{-4.0\,cm}^{4.0\,cm} C x\,dx = +\frac{1}{2}Cx^2 \Big|_{-4.0\,cm}^{4.0\,cm}$$

$$= +\frac{1}{2}C(N/cm)(100\,cm/m)\left\{\left[(0.04\,m)^2 - (-0.04\,m)^2\right]\right\} = \boxed{0.}$$

(b) Because the motion has only positive values of x, both force laws give the same result:

$$W_b = \int_{0}^{8.0\,cm} C x\,dx = +\frac{1}{2}Cx^2 \Big|_{0}^{8.0\,cm}$$

$$= +\frac{1}{2}C(N/cm)(100\,cm/m)\left\{\left[(0.08\,m)^2 - (0)^2\right]\right\} = \boxed{0.32C \text{ J.}}$$

63. Because the velocity is uniform, for $\Sigma F = ma$ for the sled, we can write
x-component: $F - \mu_k F_N = 0$;
y-component: $F_N - mg = 0$, which gives
$F = \mu_k mg = (0.03)(5000\,N) = 150\,N$.
The maximum power will produce the maximum speed:
$P_{max} = Fv_{max}$;
$(1\,hp)(746\,W/hp) = (150\,N)v_{max}$, which gives $\boxed{v_{max} = 5.0\,m/s.}$
On an incline, we have:
x-component: $F - mg\sin\theta - \mu_k F_N = 0$;
y-component: $F_N - mg\cos\theta = 0$, which gives
$F = mg(\sin\theta + \mu_k\cos\theta) = (5000\,N)(\sin 5° + 0.03\cos 5°) = 585\,N$.
The maximum power will produce the maximum speed on the incline:
$P_{max} = Fv_{max}$;
$(1\,hp)(746\,W/hp) = (585\,N)v_{max}$, which gives $\boxed{v_{max} = 1.3\,m/s.}$
The big change is due to the (negative) work done by gravity.

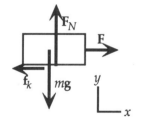

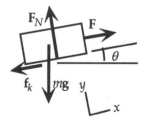

67. (a) At constant speed the force provided by the bicyclist must balance the wind resistance: $F = Av^2$.
We find the speed from the long-term power output:
$P = Fv = Av^3$; $(0.4\,hp)(746\,W/hp) = (0.08\,kg/m)v^3$, which gives $v = \boxed{16\,m/s.}$
(b) Because the time is short, we assume the maximum power output:
$t = W/P = mg\,\Delta y/P = (100\,kg)(9.8\,m/s^2)(2\,m)/(5\,hp)(746\,W/hp) = \boxed{0.5\,s.}$
(c) We assume a power of 1 hp:
$t = W/P = mg\,\Delta y/P = (75\,kg)(9.8\,m/s^2)(12\,m)/(1\,hp)(746\,W/hp) = \boxed{12\,s.}$

77. From the uniform acceleration, we find the speed attained by the rocket:
$v = v_0 + at = 0 + (2.0 \text{ m/s}^2)(33 \text{ s}) = 66 \text{ m/s}.$
 (a) The work done by the rocket engine changes the kinetic energy:
$W = \Delta K = \frac{1}{2}(1 \times 10^4 \text{ kg})[(66 \text{ m/s})^2 - 0] = \boxed{2.2 \times 10^7 \text{ J.}}$
 (b) Because the kinetic energy change is opposite to that in part (a), the work done will be
$W = \boxed{-2.2 \times 10^{-7} \text{ J.}}$
 The difference in times means that the power is different.

79. (a) The three forces acting on the block are
$mg = (2.6 \text{ kg})(9.8 \text{ m/s}^2) = 26 \text{ N down.}$
$F_N = mg \cos \theta = (26 \text{ N}) \cos 32° = 22 \text{ N perpendicular to plane (up).}$
$f_k = \mu_k F_N = (0.25)(22 \text{ N}) = 5.4 \text{ N parallel to plane (down).}$
 (b) We find the work done by each force from its component
 along the plane:
$W_g = -mg \sin \theta \, \Delta x = -(26 \text{ N}) \sin 32° (1.3 \text{ m}) = \boxed{-18 \text{ J.}}$
$W_N = \boxed{0.}$
$W_f = -f_k \Delta x = (5.4 \text{ N})(1.3 \text{ m}) = \boxed{-7.0 \text{ J.}}$
 (c) From the work-energy theorem, $W_{net} = \Delta K$, we get
$W_g + W_N + W_f = \Delta K;$
$-18 \text{ J} + 0 - 7.0 \text{ J} = 0 - \frac{1}{2}(2.6 \text{ kg})v_0^2$, which gives $\boxed{v_0 = 4.4 \text{ m/s.}}$

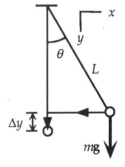

81. Because the work done is independent of the path, we can first go
 horizontally to the vertical line and then vertically down. For the
 horizontal motion, the work done by gravity is zero ($\mathbf{F}_g \perp \Delta \mathbf{x}$). For
 the vertical motion, we have
$W_g = mg \, \Delta y = mg(L - L \cos \theta) = mgL(1 - \cos \theta)$
$= (4.0 \text{ kg})(9.8 \text{ m/s}^2)(1.0 \text{ m})(1 - \cos 30°) = \boxed{5.3 \text{ J.}}$

83. (a) Because the tension in the rope is perpendicular to the motion, $\boxed{W_T = 0.}$
 (b) With $F_N = mg$, we have $f_k = \mu_k mg$. The friction force is opposite to the velocity and thus tangent to the
 path. The work done by the friction force in one revolution is
$W_f = -\mu_k mg(2\pi r) = -0.02(0.2 \text{ kg})(9.8 \text{ m/s}^2)(2\pi)(0.8 \text{ m}) = \boxed{-0.20 \text{ J.}}$
 (c) The net work on the puck changes its kinetic energy:
$W_{net} = \Delta K; \quad -0.20 \text{ J} = K_f - \frac{1}{2}(0.2 \text{ kg})(10 \text{ m/s})^2$, which gives $K_f = \boxed{9.8 \text{ J.}}$

85. (a) With $f_k = 0$ and no change in the speed (assumed very small) of
 the mass, we have
$W_{net} = \Delta K = 0;$
$W_T + W_g = W_T + (-mg)H = 0$, which gives
$W_T = mgH$, independent of the angle θ.
 (b) With friction present, we have $W_{net} = W_T + W_f + W_g = 0;$
$W_T - \mu_k mg \cos \theta (H/\sin \theta) - mgH = 0$, which gives
$\boxed{W_T = mgH(1 + \mu_k \cot \theta).}$

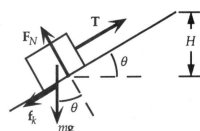

CHAPTER 7

It is necessary to choose the zero reference level for potential energy. The sign of each potential energy term will be determined relative to the chosen reference level.

1. (a) We choose the potential energy to be zero at the ground ($y = 0$), so $U = mgy$.
    Then $\Delta U = mg\,\Delta y = (10\text{ kg})(9.8\text{ m/s}^2)(2\text{ m}) = \boxed{196\text{ J.}}$
    (b) The net work changes the kinetic energy:
    $$W_{net} = \Delta K = \tfrac{1}{2}mv^2 - 0 = \tfrac{1}{2}(10\text{ kg})(2.4\text{ m/s})^2 = \boxed{29\text{ J.}}$$
    The work done by the unknown force changes the total energy:
    $$W_F = \Delta K + \Delta U = 29\text{ J} + 196\text{ J} = \boxed{225\text{ J.}}$$

7. (a) A constant force is a conservative force. We use the reference point of $x = 0$ and obtain
    $$U(x) = U(0) - \int_0^x F\,dx' = 0 - F\int_0^x dx' = -Fx = -(8\text{ N})x = \boxed{-8x\text{ J, with }x\text{ in m.}}$$
    (b) We find the constant mechanical energy by finding the value at $x = -1$ m:
    $$E = K + U = \tfrac{1}{2}mv^2 - (8\text{ N})x = \tfrac{1}{2}(5\text{ kg})(+2\text{ m/s})^2 - (8\text{ N})(-1\text{ m}) = \boxed{+18\text{ J.}}$$
    (c) Because energy is conserved, we have
    $$E = K_2 + U_2 = \tfrac{1}{2}mv_2{}^2 - (8\text{ N})x_2;$$
    $$18\text{ J} = \tfrac{1}{2}(5\text{ kg})v_2{}^2 - (8\text{ N})(3\text{ m}),\text{ which gives } v_2 = \boxed{4.1\text{ m/s.}}$$

11. With no friction the energy is conserved, so we have
    $$E = K_1 + U_1 = K_2 + U_2;\quad \tfrac{1}{2}mv_1{}^2 + \tfrac{1}{2}kx_1{}^2 = \tfrac{1}{2}mv_2{}^2 + \tfrac{1}{2}kx_2{}^2;$$
    $$\tfrac{1}{2}(0.528\text{ kg})(3.85\text{ m/s})^2 + 0 = 0 + \tfrac{1}{2}(26.7\text{ N/m})x^2,\text{ which gives } x = \boxed{0.541\text{ m.}}$$
    With a rough surface, there will be work done by the friction force, $f = -\mu_k F_N$, with $F_N = mg$. Thus
    $$W_f = \Delta K + \Delta U;\quad -\mu_k mg\,x = (\tfrac{1}{2}mv_2{}^2 - \tfrac{1}{2}mv_1{}^2) + (\tfrac{1}{2}kx_2{}^2 - \tfrac{1}{2}kx_1{}^2)$$
    $$-(0.411)(0.528\text{ kg})(9.8\text{ m/s}^2)x = [0 - \tfrac{1}{2}(0.528\text{ kg})(3.85\text{ m/s})^2] + [\tfrac{1}{2}(26.7\text{ N/m})x^2 - 0].$$
    The solutions of the quadratic equation are $x = 0.468$ m, -0.627 m. From our expression for the work done by friction, which must be negative, we select the positive result: $x = \boxed{0.468\text{ m.}}$

13. We choose $y = 0$ at the relaxed position of the spring, with up positive, and denote the height of release by h and the magnitude of the compression of the spring by Δy. The change in elevation for the compression is $-\Delta y$. Because the energy is conserved, we have
    $$E = K + U_g + U_{spring} = \text{constant};\quad 0 + mgh + 0 = 0 + mg(-\Delta y) + \tfrac{1}{2}k(-\Delta y)^2;$$
    $$(15\text{ kg})(9.8\text{ m/s}^2)(6.0\text{ m}) = (15\text{ kg})(9.8\text{ m/s}^2)(-\Delta y) + \tfrac{1}{2}(10^4\text{ N/m})(\Delta y)^2.$$
    This is a quadratic equation for Δy, from which we get $\Delta y = 0.43$ m, -0.41 m.
    From our choice of Δy as a magnitude, we select the positive value: $\boxed{\Delta y = 0.43\text{ m.}}$
    For the man, the numbers become
    $$(60\text{ kg})(9.8\text{ m/s}^2)(1.5\text{ m}) = (60\text{ kg})(9.8\text{ m/s}^2)(-\Delta y) + \tfrac{1}{2}(10^4\text{ N/m})(\Delta y)^2,\text{ which gives } \boxed{\Delta y = 0.48\text{ m.}}$$
    For the larger mass the greater decrease in gravitational potential energy requires a greater spring potential energy.

19. For the given force, $F(x) = + kx$, we find the potential energy:

$U = - \int kx \, dx = - \frac{1}{2}kx^2$, with $U = 0$ when $x = 0$.

From the release near the origin, $x = 0$, we find the energy:

$E = - \frac{1}{2}kx^2 + \frac{1}{2}mv^2 = - \frac{1}{2}k(0)^2 + \frac{1}{2}m(0)^2 = 0$.

(a) When $x > 0$, $F > 0$ and the mass will move $\boxed{\text{to the right;}}$ U decreases and K increases.

(b) From $E = - \frac{1}{2}kx^2 + \frac{1}{2}mv^2 = 0$, we get $v^2 = (k/m)x^2$, or $\boxed{v = (k/m)^{1/2}x.}$

(c) When $x < 0$, $F < 0$ and the mass will move $\boxed{\text{to the left;}}$ U decreases and K increases; $\boxed{v = - (k/m)^{1/2}x.}$

23. The potential energy is $U(x) = \alpha x^4 + \beta x^2$.

The equilibrium positions ($F = 0$) have $dU/dx = 0$:

$dU/dx = 4\alpha x^3 + 2\beta x = 0$, from which we get

$\qquad x = 0, \quad$ and

$\qquad x = \pm(-\beta/2\alpha)^{1/2} = \pm[-(-3.0\ \text{J/m}^2)/2(26\ \text{J/m}^4)]^{1/2} = \boxed{\pm 0.24\ \text{m.}}$

From the sketch of $U(x)$, we see that

$\qquad x = 0$ is unstable   and $x = \pm 0.24$ m is stable.

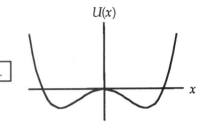

27. We choose $y = 0$ at the sea. If there is no drag, energy is conserved:

$E = \frac{1}{2}mv_i^2 + mgy_i = \frac{1}{2}mv_f^2 + mgy_f$;

$\frac{1}{2}m(125\ \text{m/s})^2 + m(9.8\ \text{m/s}^2)(68\ \text{m}) = \frac{1}{2}mv_f^2 + m(9.8\ \text{m/s}^2)(0)$, which gives $v_f = \boxed{130\ \text{m/s.}}$

Note that this is independent of the mass.

When fired at an angle, the gravitational energy change is the same; the final speed will be the $\boxed{\text{same.}}$

The direction of the velocity when the cannonball hits the water will be different.

29. We choose $y = 0$ at the release point. If we neglect air resistance, energy is conserved:

$E = \frac{1}{2}mv_1^2 + mgy_1 = \frac{1}{2}mv_2^2 + mgy_2$.

(a) The kinetic energy depends on the speed, not the direction:

$\qquad \frac{1}{2}mv_0^2 + 0 = \frac{1}{2}mv_h^2 + mgh$, or $v_h^2 = v_0^2 - 2gh$, which gives $\boxed{v_h = \sqrt{v_0^2 - 2gh} \ .}$

(b) For vertical motion, the speed at the highest point is zero:

$\qquad v_h = 0 = \sqrt{v_0^2 - 2gh}$, which gives $v_0 = \boxed{\sqrt{2gh} \ .}$

(c) When the ball is thrown at an angle θ, the speed at the highest point is horizontal and equal to the initial horizontal component: $v_h = v_{0x} = v_0 \cos \theta$; thus

$\qquad v_h = v_{0x} = \sqrt{v_0^2 - 2gh}$, or $v_{0x}^2 = v_0^2 - 2gh$.

Because $v_0^2 = v_{0x}^2 + v_{0y}^2$, we have $v_{0y}^2 = (v_0 \sin \theta)^2 = 2gh$.

Thus $v_0 = \sqrt{2gh} \ /\sin 45° = \boxed{2\sqrt{gh} \ .}$

35. We choose $y = 0$ at the bottom.

(a) For the motion from the initial point to point a there is no friction, so from energy conservation we have

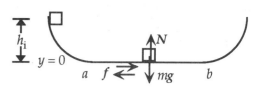

$$K_i + U_i = K_a + U_a; \quad 0 + mgh_i = \tfrac{1}{2}mv_a^2 + 0;$$
$$m(9.8 \text{ m/s}^2)(0.15 \text{ m}) = 0 + \tfrac{1}{2}mv_a^2,$$

which gives $v_a = \boxed{1.7 \text{ m/s.}}$

(b) For the motion from point a to point b, from the presence of friction we have

$$W_f = \Delta K + \Delta U; \quad -\mu_k mg\, \Delta x = \tfrac{1}{2}mv_b^2 - \tfrac{1}{2}mv_a^2 + 0;$$
$$-(0.21)m(9.8 \text{ m/s}^2)(0.30 \text{ m}) = \tfrac{1}{2}mv_b^2 - \tfrac{1}{2}m(1.7 \text{ m/s})^2, \text{ which gives } v_b = \boxed{1.3 \text{ m/s.}}$$

(c) For the motion from the initial release point to the stopping point, we let D represent the total horizontal distance over which the friction force acts. Because the friction force always opposes the motion, its work will always be negative, so we have

$$W_f = \Delta K + \Delta U;$$
$$-\mu_k mg\, D = (0 - 0) + (0 - mgh_i),$$

which gives $D = h_i/\mu_k = (0.15 \text{ m})/0.21 = 0.714 \text{ m} = 71.4 \text{ cm}.$

Because the distance from a to b is 30 cm, this corresponds to

$$71.4 \text{ cm} - 2(30 \text{ cm}) = \boxed{11.4 \text{ cm from point } a.}$$

39. (a) Because the potential energy $U(r) = -GMm/r$ depends only on the radius, the force will be radial and we find it from

$$F_r = -dU/dr = -[-GMm(-1/r^2)] = -GMm/r^2, \text{ so the force is } \boxed{GMm/r^2 \text{ toward } M.}$$

(b) The force provides the centripetal acceleration, which relates the speed to the radius:

$$F_r = ma_r; \quad GMm/r^2 = mv^2/r, \text{ which gives } v^2 = GM/r.$$

The kinetic energy is $K = \tfrac{1}{2}mv^2 = \boxed{GMm/2r.}$

(c) The total energy is

$$E = K + U = (GMm/2r) - (GMm/r) = \boxed{-GMm/2r.}$$

43. The work done by the nonconservative drag forces changes the energy of the parachutist:

$$W_{nc} = \Delta(K + U) = (K_f + U_f) - (K_i + U_i) = (\tfrac{1}{2}mv_f^2 + 0) - (0 + mgh_i)$$
$$= \tfrac{1}{2}(75 \text{ kg})(5.0 \text{ m/s})^2 - (75 \text{ kg})(9.8 \text{ m/s}^2)(85 \text{ m}) = \boxed{-6.2 \times 10^4 \text{ J.}}$$

47. The time between hits on a barrier is $\Delta t = (10^{-14} \text{ m})/(10^6 \text{ m/s}) = 10^{-20} \text{ s}.$

The average number of hits before tunneling through, as measured by S, is $N = 1/e^{-S} = e^{+72} = 1.9 \times 10^{31}.$

The average time to escape is $N\, \Delta t = (1.9 \times 10^{31})(10^{-20}) = \boxed{1.9 \times 10^{11} \text{ s}} \; (\approx 6 \times 10^3 \text{ yr}).$

49. The change in potential energy depends only on the vertical displacement:

$$mg\, \Delta y = (60 \text{ kg})(9.8 \text{ m/s}^2)(1.0 \text{ m}) \approx \boxed{5.9 \times 10^2 \text{ J.}}$$

53. (a) Because $F(x) = +ax + bx^3 + cx^4$ is a one-dimensional force that depends only on position, it is $\boxed{\text{conservative.}}$

   (b) To test $\mathbf{F} = Ax^2\mathbf{i} + Bxy\mathbf{j}$, we find the work for a displacement from $(0, 0)$ to $(1, 1)$ for the two paths indicated in the diagram:

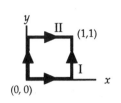

   $$W_\mathrm{I} = \int_{0,0}^{1,0} F_x\,dx + \int_{1,0}^{1,1} F_y\,dy = \int_{0,0}^{1,0} Ax^2\,dx + \int_{1,0}^{1,1} Bxy\,dy$$

   $$= \frac{1}{3}Ax^3\Big|_{0,0}^{1,0} + \frac{1}{2}Bxy^2\Big|_{1,0}^{1,1} = \left(\frac{A}{3} - 0\right) + \left(\frac{B}{2} - 0\right) = \frac{A}{3} + \frac{B}{2}.$$

   $$W_\mathrm{II} = \int_{0,0}^{0,1} F_y\,dy + \int_{0,1}^{1,1} F_x\,dx = \int_{0,0}^{0,1} Bxy\,dy + \int_{0,1}^{1,1} Ax^2\,dx$$

   $$= \frac{1}{2}Bxy^2\Big|_{0,0}^{0,1} + \frac{1}{3}Ax^3\Big|_{0,1}^{1,1} = (0 - 0) + \left(\frac{A}{3} - 0\right) = \frac{A}{3}.$$

   Because the work depends on the path, the force is $\boxed{\text{not conservative.}}$
   (Be careful! You may have to try a number of paths. See Problem 69 for a more general approach.)

57. We choose $y = 0$ at the bottom of the loop. With no friction, energy is conserved. The initial (and constant) energy is
   $$E = \tfrac{1}{2}mv_1^2 + mgh_1 = 0 + mgh_1 = mgh_1.$$
   We find the speed at a height h from
   $$\tfrac{1}{2}mv^2 + mgh = mgh_1; \quad v = \sqrt{2g(h_1 - h)}.$$

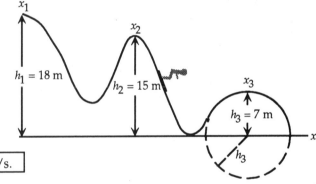

   (a) The skier starts from rest: $\boxed{v_1 = 0.}$

   $$v_2 = \sqrt{2g(h_1 - h_2)}$$
   $$= \sqrt{2(9.8\ \text{m/s}^2)(18\ \text{m} - 15\ \text{m})} = \boxed{7.7\ \text{m/s.}}$$
   $$v_3 = \sqrt{2g(h_1 - h_3)}$$
   $$= \sqrt{2(9.8\ \text{m/s}^2)(18\ \text{m} - 7\ \text{m})} = \boxed{14.7\ \text{m/s.}}$$

   (b) At x_3 the force of gravity is down and the normal force must be up (the track can only push on the skier). These two provide the centripetal acceleration: $mg - F_N = mv_3^2/R$, so the normal force is
   $$F_N = m[g - (v_3^2/R)]$$
   $$= m\{g - [2g(h_1 - h_3)/h_3]\}$$
   $$= mg[3 - (2h_1/h_3)] = mg\{3 - [2(18\ \text{m})/7\ \text{m}]\}, \text{ which is negative.}$$
   Because the normal force cannot be negative (which would correspond to a pull on the skier),
   $\boxed{\text{the skier leaves the surface.}}$
   We set F_N to its minimum value (zero) to find the maximum value of h_1 at which the skier leaves:
   $$F_N = 0; \quad (2h_1/h_3) - 3 = 0, \text{ which gives}$$
   $$h_1 = 3h_3/2 = 3(7\ \text{m})/2 = \boxed{10.5\ \text{m.}}$$
   (Note that h_2 will also have to be reduced.)

59. We choose $y = 0$ at the release position. Because the tension is always perpendicular to the displacement, it does no work, so we can apply conservation of energy.

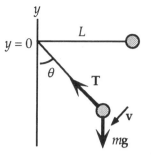

(a) From the release point to the lowest position, we have
   $K_i + U_i = K_f + U_f$;
   $0 + 0 = \frac{1}{2}mv_a^2 + mg(-L)$, which gives
   $v_a^2 = 2gL = 2(9.8 \text{ m/s}^2)(1.0 \text{ m}) = \boxed{4.4 \text{ m/s.}}$

(b) From the release point to an angle θ with the vertical, we have
   $K_i + U_i = K_f + U_f$;
   $0 + 0 = \frac{1}{2}mv_b^2 + mg(-L\cos\theta)$, which gives
   $v_a^2 = 2gL\cos\theta = 2(9.8 \text{ m/s}^2)(1.0 \text{ m})\cos 45° = \boxed{3.7 \text{ m/s.}}$

(c) The tension and the appropriate component of the weight must provide the centripetal acceleration. At an angle θ with the vertical, we have
   $T - mg\cos\theta = mv^2/L$.
   Using the result from part (b), we get
   $T - mg\cos\theta = 2mg\cos\theta$, or $T = 3mg\cos\theta$.
   At the bottom of the swing, $\theta = 0°$, so $T_a = 3(0.20 \text{ kg})(9.8 \text{ m/s}^2)\cos 0° = \boxed{5.9 \text{ N.}}$
   When $\theta = 45°$, we get $T_b = 3(0.20 \text{ kg})(9.8 \text{ m/s}^2)\cos 45° = \boxed{4.2 \text{ N.}}$

61. (a) Because the proton is much heavier than the electron, we take the proton to remain stationary. From the discussion in Section 7–3, this central force (a function of r only) is $\boxed{\text{a conservative force.}}$

(b) Because the force is attractive, its direction is opposite to that of **r**. We find the potential energy:

$$U(r) = U(s) - \int_s^r F(r')\ dr' = U(s) - \int_s^r -\frac{C}{r'^2}\ dr' = U(s) - \frac{C}{r} + \frac{C}{s}.$$

For such a force we take $s = \infty$ and $U(\infty) = 0$. The potential energy is then $\boxed{U(r) = -C/r.}$

(c) From energy conservation we have
   $K_i + U_i = K_f + U_f$;
   $0 + 0 = \frac{1}{2}(9.1 \times 10^{-31})v^2 + [-(2.3 \times 10^{-28} \text{ kg} \cdot \text{m}^3/\text{s}^2)/(1.2 \times 10^{-12} \text{ m})]$,
   which gives $v = \boxed{2.1 \times 10^7 \text{ m/s.}}$

65. Because the length of the rope is constant, when m_1 moves down Δy_1, the two segments above m_1 both increase by Δy_1, and the segment above m_2 must decrease by twice that amount:
   $\Delta y_2 = -2\Delta y_1$ ($-$ indicates the opposite direction).
   If we differentiate with respect to time, we have $v_2 = -2v_1$.
   With $y = 0$ at the ground, from energy conservation we have
   $K_i + U_i = K_f + U_f$; $0 + m_1gh_1 = \frac{1}{2}m_1v_1^2 + \frac{1}{2}m_2v_2^2 + m_2gh_2$;
   $m_1gh = \frac{1}{2}m_1v_1^2 + \frac{1}{2}m_2(-2v_1)^2 + m_2g(2h)$,   or
   $v_1^2 = 2gh(m_1 - 2m_2)/(m_1 + 4m_2)$
      $= 2(9.8 \text{ m/s}^2)(0.8 \text{ m})[(5.0 \text{ kg}) - 2(2.0 \text{ kg})]/[(5.0 \text{ kg}) + 4(2.0 \text{ kg})]$,
   which gives $\boxed{v_1 = 1.1 \text{ m/s.}}$

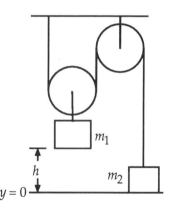

CHAPTER 8

As in the prior analysis of forces to determine the acceleration, it is important to select the system and understand the external forces that are acting to know whether linear momentum is conserved.

5. (a) During the firing we use momentum conservation for the system of rifle and bullet:

$p_{ri} + p_{bi} = p_{rf} + p_{bf}$;

$0 + 0 = (7 \text{ kg})v_{rf} + (10 \times 10^{-3} \text{ kg})(700 \text{ m/s})$, from which we get $v_{rf} = \boxed{-1.0 \text{ m/s}.}$

(b) The energy transmitted to the shoulder comes from the decrease in kinetic energy of the rifle:

$\Delta E = -\Delta K = -(0 - \tfrac{1}{2}mv^2) = \tfrac{1}{2}(7 \text{ kg})(-1.0 \text{ m/s})^2 = \boxed{3.5 \text{ J}.}$

7. (a) $\mathbf{p} = m_1\mathbf{v}_1 + m_2\mathbf{v}_2 = (2.4 \text{ kg})[(-2.0\mathbf{i} - 3.5\mathbf{j}) \text{ m/s}] + (1.6 \text{ kg})[(1.8\mathbf{i} - 1.5\mathbf{j}) \text{ m/s}] = \boxed{(-1.9\mathbf{i} - 10.8\mathbf{j}) \text{ kg} \cdot \text{m/s}.}$

(b) Assuming no external influences, momentum is conserved, so we have

$\mathbf{p} = (-1.9\mathbf{i} - 10.8\mathbf{j}) \text{ kg} \cdot \text{m/s} = (2.4 \text{ kg})(2.5\mathbf{i} \text{ m/s}) + (1.6 \text{ kg})\mathbf{v}_2'$, which gives

$\mathbf{v}_2' = \boxed{(-4.9\mathbf{i} - 6.8\mathbf{j}) \text{ m/s}.}$

(c) Because mass is conserved, we have $m_2' = 2.4 \text{ kg} + 1.6 \text{ kg} - 2.1 \text{ kg} = 1.9 \text{ kg}$.

Because momentum is conserved, we have

$\mathbf{p} = (-1.9\mathbf{i} - 10.8\mathbf{j}) \text{ kg} \cdot \text{m/s} = (2.1 \text{ kg})\mathbf{v}_1' + (1.9 \text{ kg})[(-2.5\mathbf{j} + 1.3\mathbf{k}) \text{ m/s}]$, which gives

$\mathbf{v}_1' = \boxed{(-0.90\mathbf{i} - 2.9\mathbf{j} - 1.2\mathbf{k}) \text{ m/s}.}$

(d) The initial total kinetic energy is

$K_1 + K_2 = \tfrac{1}{2}(2.4 \text{ kg})[(-2.0 \text{ m/s})^2 + (-3.5 \text{ m/s})^2] + \tfrac{1}{2}(1.6 \text{ kg})[(1.8 \text{ m/s})^2 + (-1.5 \text{ m/s})^2]$

$= \boxed{24 \text{ J}.}$

For part (b) the final total kinetic energy is

$K_1' + K_2' = \tfrac{1}{2}(2.4 \text{ kg})(2.5 \text{ m/s})^2 + \tfrac{1}{2}(1.6 \text{ kg})[(-4.9 \text{ m/s})^2 + (-6.8 \text{ m/s})^2] = \boxed{64 \text{ J}.}$

Because this is greater than the initial kinetic energy, some internally stored energy was transformed into kinetic energy.

For part (c) the final total kinetic energy is

$K_1' + K_2' = \tfrac{1}{2}(2.1 \text{ kg})[(-0.90 \text{ m/s})^2 + (-2.9 \text{ m/s})^2 + (-1.2 \text{ m/s})^2] + \tfrac{1}{2}(1.9)[(-2.5 \text{ m/s})^2 + (1.3 \text{ m/s})^2]$

$= \boxed{19 \text{ J}.}$

Because this is less than the initial kinetic energy, some energy was stored as potential or lost due to internal friction.

13. For the vertical motion, with up positive, we find the speed at the ground from

$v^2 = v_0^2 + 2a\,\Delta y = 0 + 2(-9.8 \text{ m/s}^2)(-11 \text{ m})$, which gives $v = 15 \text{ m/s}$.

If the person does not bounce, the impulse on the person produces the change in momentum:

$J_{person} = \Delta p = (80 \text{ kg})[0 - (-15 \text{ m/s})] = 1.2 \times 10^3 \text{ kg} \cdot \text{m/s}$.

Then the impulse on the net is

$J_{net} = -J_{person} = \boxed{-1.2 \times 10^3 \text{ kg} \cdot \text{m/s, down}.}$

Once the person hits the net, we find the time to stop from

$\Delta y = v_{av}\,\Delta t;$ $-0.70 \text{ m} = \tfrac{1}{2}[0 + (-15 \text{ m/s})]\,\Delta t$, which gives $\Delta t = 0.093 \text{ s}$.

Thus, $F_{av} = J_{person}/\Delta t = (1.2 \times 10^3 \text{ kg} \cdot \text{m/s})/(0.093 \text{ s}) = \boxed{1.3 \times 10^4 \text{ N, up}.}$

19. (a) We find the assumed constant acceleration from

$v^2 = v_0^2 + 2a\,\Delta x;$

$0 = (45 \text{ m/s})^2 + 2a_{av}(0.25 \text{ m})$, which gives $a_{av} = 4.1 \times 10^3 \text{ m/s}^2$.

By applying Newton's second law, we get

$F_{av} = ma_{av} = (0.14 \text{ kg})(4.1 \times 10^3 \text{ m/s}^2) = \boxed{5.7 \times 10^2 \text{ N}.}$

(b) The work done is $W = F_{av}\,\Delta x = (5.7 \times 10^2 \text{ N})(0.25 \text{ m}) = \boxed{1.4 \times 10^2 \text{ J}.}$

(c) The impulse exerted changes the momentum of the ball:

$F_{av}\,\Delta t = \Delta p$, from which we get

$\Delta t = \Delta p/F_{av} = (0.14 \text{ kg})(0 + 45 \text{ m/s})/(5.7 \times 10^2 \text{ N}) = \boxed{0.011 \text{ s}.}$

21. For a perfectly elastic collision, the ball will rebound to the same height above the step; the speed as it leaves a step will be the same as when it hits the step. For the first step: $h_1 = 0.60$ m; for the second step: $h_2 = 0.80$ m; for the Nth step: $h_N = 0.60$ m $+ (N - 1)0.20$ m.

For the fall and rise we use energy conservation to get $v_N = \sqrt{2gh_N}$.

For the impulse at the Nth step we have

$$
\begin{aligned}
J_N &= \Delta p = m[v_N - (-v_N)] \\
&= 2mv_N = 2m\sqrt{2gh_N} \\
&= 2(0.150 \text{ kg})\sqrt{2g[0.60 \text{ m} + (N-1)(0.20\text{ m})]} \\
&= 0.30\sqrt{2g(0.40 + 0.20N)} \text{ kg}\cdot\text{m/s} = \boxed{0.60\sqrt{g(0.20 + 0.10N)} \text{ kg}\cdot\text{m/s}.}
\end{aligned}
$$

27. For this one-dimensional motion, we take the direction of the first object for the positive direction. For this perfectly inelastic collision, we use momentum conservation:

$$m_1v_1 + m_2v_2 = (m_1 + m_2)V;$$

$$m_1v + m_2(-v) = (m_1 + m_2)\tfrac{1}{2}v; \text{ which gives } \boxed{m_1/m_2 = 3.}$$

29. We let V be the speed of the block and bullet immediately after the collision and before the pendulum swings. This uses the fact that the collision takes place in such a short time that there is no motion of the pendulum during the collision.
For this perfectly inelastic collision, we use momentum conservation:

$mv + 0 = (M + m)V$, which gives $V/v = m/(M + m)$.

(a) The fractional change in the kinetic energy is

$$
\begin{aligned}
\Delta K/K_i &= [\tfrac{1}{2}(M + m)V^2 - \tfrac{1}{2}mv^2] / \tfrac{1}{2}mv^2 \\
&= [(M + m)/m](V/v)^2 - 1 = -M/(m + M),
\end{aligned}
$$

so the fraction lost $= \boxed{M/(m + M).}$

(b) For the pendulum motion after the collision we use energy conservation; the kinetic energy becomes gravitational potential energy:

$$\tfrac{1}{2}(M + m)V^2 = (m + M)gh.$$

We combine this with the result from momentum conservation to get

$$V = \sqrt{2gh} = mv/(m + M), \text{ which gives } \boxed{v = [(m + M)/m]\sqrt{2gh}.}$$

31. For the system of the two persons and the bobsled, all horizontal forces are internal, so momentum will be conserved. We use a coordinate system with the positive direction opposite to the direction the persons jump.

   (a) For the first jump, with velocities relative to the ice, we use momentum conservation of the two person-bobsled system to find the velocity of the remaining person and the bobsled:
   $$p_{1i} + p_{2i} + p_{si} = p_{1f} + p_2' + p_s';$$
   $$0 + 0 + 0 = m_1(-v) + (m_2 + M)v', \text{ from which we get } v' = m_1 v/(m_2 + M).$$
   For the second jump, with velocities relative to the ice, we use momentum conservation of the remaining person-bobsled system to find the velocity of the bobsled:
   $$p_2' + p_s' = p_{2f} + p_{sf};$$
   $$(m_2 + M)v' = m_2(v' - v) + Mv_f, \text{ from which we get}$$
   $$v_f = (Mv' + m_2 v)/M.$$
   Using the result from the first jump, we get $v_f = \boxed{v[m_2{}^2 + M(m_1 + m_2)]/M(m_2 + M) \text{ forward.}}$

   (b) For the first jump, with velocities relative to the ice, we have:
   $$p_{1i} + p_{2i} + p_{si} = p_{1f} + p_2' + p_s';$$
   $$0 + 0 + 0 = m_1 v' + m_2(-v) + Mv', \text{ from which we get } v' = m_2 v/(m_1 + M).$$
   For the second jump, with velocities relative to the ice, we have:
   $$p_1' + p_s' = p_{1f} + p_{sf};$$
   $$(m_1 + M)v' = m_1(v' - v) + Mv_f, \text{ from which we get}$$
   $$v_f = (Mv' + m_1 v)/M.$$
   Using the result from the first jump, we get $v_f = \boxed{v[m_1{}^2 + M(m_1 + m_2)]/M(m_1 + M) \text{ forward.}}$
   This is similar to the answer for part (a), with m_1 and m_2 interchanged.

   (c) When they jump together, we use momentum conservation:
   $$p_{1i} + p_{2i} + p_{si} = p_{1f} + p_2' + p_s';$$
   $$0 + 0 + 0 = (m_1 + m_2)(-v) + Mv_f, \text{ from which we get } v_f = \boxed{v\,(m_1 + m_2)/M \text{ forward.}}$$

35. When the large mass rebounds perfectly elastically from the wall, its kinetic energy does not change, so its velocity reverses direction with the same magnitude.
   For the elastic collision of the two masses, we use momentum conservation:
   $$mv_1 + Mv_2 = mv_3 + Mv_4;$$
   $$(0.126 \text{ kg})(0.875 \text{ m/s}) + (9.66 \text{ kg})(-0.875 \text{ m/s}) = (0.126 \text{ kg})v_3 + (9.66 \text{ kg})v_4.$$
   Because the collision is elastic, the relative speed does not change:
   $$v_1 - v_2 = -(v_3 - v_4), \quad \text{or} \quad 0.875 \text{ m/s} - (-0.875 \text{ m/s}) = -(v_3 - v_4).$$
   Combining these two equations for v_3 and v_4, we get
   $$v_4 = -0.84 \text{ m/s} \quad \text{and} \quad v_3 = -2.59 \text{ m/s}, \text{ so the return speed is } \boxed{2.59 \text{ m/s.}}$$

37. Because the tension does no work, when a pendulum of mass m swings through an angle θ, we use energy conservation to find the speed at the bottom:

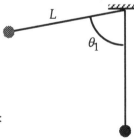

   $$v = \sqrt{2gh} = \sqrt{2g(L - L\cos\theta)}.$$
   The speed of the lighter mass just before the collision is
   $$v_1 = \sqrt{2gL(1 - \cos\theta_1)} = \sqrt{2(9.8 \text{ m/s}^2)(1 \text{ m})(1 - \cos 80°)} = 4.0 \text{ m/s.}$$
   For the elastic collision of the two masses, we use momentum conservation:
   $$mv_1 + Mv_2 = mv_3 + Mv_4;$$
   $$(0.400 \text{ kg})(4.0 \text{ m/s}) + (0.800 \text{ kg})(0) = (0.400 \text{ kg})v_3 + (0.800 \text{ kg})v_4.$$
   Because the collision is elastic, the relative speed does not change:
   $$v_1 - v_2 = -(v_3 - v_4), \quad \text{or} \quad 4.0 \text{ m/s} - 0 = -(v_3 - v_4).$$
   Combining these two equations for v_3 and v_4, we get
   $$v_3 = -0.80 \text{ m/s. The negative sign indicates a rebound.}$$
   We find the angle the lighter mass swings after the collision from
   $$v_3 = \sqrt{2gL(1 - \cos\theta_3)}; \quad 0.80 \text{ m/s} = \sqrt{2(9.8 \text{ m/s}^2)(1 \text{ m})(1 - \cos\theta_3)}, \text{ which gives } \theta_3 = \boxed{15°.}$$

39. (a) Because both objects fall the same distance h, we find their speed just before they hit the floor from

$v_1 = \boxed{\sqrt{2gh}\,.}$

(b) If the ball rebounds elastically, its velocity just reverses direction, so $v_2 = -v_1 = -\sqrt{2gh}$.

The speed of each object will be $\boxed{\sqrt{2gh}\,.}$

(c) For the elastic collision of the two objects, we use momentum conservation:

$mv_1 + Mv_2 = mv_3 + Mv_4;$

$m\sqrt{2gh} + M(-\sqrt{2gh}) = mv_3 + Mv_4.$

Because the collision is elastic, the relative speed does not change:

$v_1 - v_2 = -(v_3 - v_4),$ or $\sqrt{2gh} - (-\sqrt{2gh}) = -(v_3 - v_4).$

Combining these two equations, we get

$v_3 = -[(3M - m)/(m + M)]\sqrt{2gh} = \boxed{[(m - 3M)/(m + M)]\sqrt{2gh}\,,\text{ up.}}$

(d) We find the rebound height for the marble from

$|v_3| = \sqrt{2gh'}\,;$ $|(m - 3M)/(m + M)|\sqrt{2gh} = \sqrt{2gh'}$, which gives

$h' = \boxed{[(m - 3M)/(m + M)]^2 h.}$

(e) If $M \gg m$, we get $h' = \boxed{9h.}$

43. For the collision of the two masses, we use momentum conservation:

$m\mathbf{v} + 0 = m\mathbf{v}_1 + m\mathbf{v}_2;$

$(2.50 \text{ m/s}) \mathbf{i} + 0 = [(0.50 \text{ m/s})\mathbf{i} + (-1.00 \text{ m/s})\mathbf{j}] + \mathbf{v}_2 ,$ which gives

$\mathbf{v}_2 = \boxed{(2.00 \text{ m/s})\mathbf{i} + (1.00 \text{ m/s})\mathbf{j}.}$

To test whether the collision is elastic, we find the kinetic energy before and after:

$K_i = \tfrac{1}{2}m(2.50 \text{ m/s})^2 = 3.13\ m;$

$K_f = \tfrac{1}{2}m[(0.50 \text{ m/s})^2 + (-1.00 \text{m/s})^2 + (2.00 \text{ m/s})^2 + (-1.00 \text{ m/s})^2] = 3.13\ m.$

Because the kinetic energy is conserved, the collision is $\boxed{\text{elastic.}}$

Because the masses are the same, we could also test that the final velocities are perpendicular:

$\mathbf{v}_1 \cdot \mathbf{v}_2 = [(0.50 \text{ m/s})\mathbf{i} + (-1.00 \text{ m/s})\mathbf{j}] \cdot [(2.00 \text{ m/s})\mathbf{i} + (1.00 \text{ m/s})\mathbf{j}] = 0.$

47. We will take x to the right and y up. Because the speed of the bullet is so high, we ignore the effect of gravity and assume that it travels in a straight line until it hits the block. The time for the collision to take place is so small that we can ignore the effect of gravity on the system until after the collision. For the perfectly inelastic collision of the two masses, we use momentum conservation:

x-direction: $m_{block}v_{block,i} - m_{bullet}v_{bullet,i} \cos 60° = (m_{block} + m_{bullet})v_x ;$

$(0.80 \text{ kg})(10 \text{ m/s}) - (0.0050 \text{ kg})(550 \text{ m/s}) \cos 60° = (0.805 \text{ kg})v_x ,$ which gives $v_x = 8.2 \text{ m/s}.$

y-direction: $0 - m_{bullet}v_{bullet,i} \cos 60° = (m_{block} + m_{bullet})v_y ;$

$0 + (0.0050 \text{ kg})(550 \text{ m/s}) \sin 60° = (0.805 \text{ kg})v_y ,$ which gives $v_y = 3.0 \text{ m/s}.$

The velocity of the block immediately after the collision is

$\mathbf{v} = (8.2 \text{ m/s})\mathbf{i} + (3.0 \text{ m/s})\mathbf{j} = \boxed{8.7 \text{ m/s, } 20° \text{ above the horizontal.}}$

49. For the collision we use momentum conservation:

x-direction: $m_{car}v_{car,i} + 0 = m_{car}v_{car,f} \cos \theta_{car} + m_{truck}v_{truck,f} \cos \theta_{truck},$

$(1000 \text{ kg})v_{car,i} = (1000 \text{ kg})v_{car,f} \cos 60° + (1500 \text{ kg})(21.6 \text{ m/s}) \cos 33.7°;$

y-direction: $0 + 0 = -m_{car}v_{car,f} \sin \theta_{car} + m_{ruck}v_{truck,f} \sin \theta_{truck},$

$0 = -(1000 \text{ kg})v_{car,f} \sin 60° + (1500 \text{ kg})(21.6 \text{ m/s}) \sin 33.7°.$

When we solve these two equations for the two unknowns, we get

$v_{car,f} = 20.8 \text{ m/s},$ and $v_{car,i} = 37.4 \text{ m/s} = 84 \text{ mi/h}.$

The driver of the car was speeding!

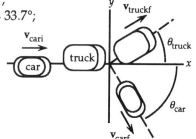

55. Because the system of the children and the ball is isolated, the total momentum remains constant. Because the initial momentum is zero, the initial speed of the center of mass is zero, so the final speed of the center of mass will be ☐ 0. ☐

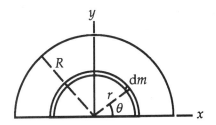

61. The uniform area mass density of the object is $\sigma = M/A = 2M/\pi R^2$. The shape of the object suggests the polar coordinate system r, θ shown in the diagram.
From the symmetry of the object, we have ☐ $X = 0$. ☐
For the y-position of the center of mass, we select a semi-circular arc at radius r of thickness dr. Within this arc we select a representative dm at the angle θ, which has area $r\,d\theta\,dr$ and mass $dm = \sigma r\,d\theta\,dr$. Then $y = r\sin\theta$. We find Y by integration over θ and r:

$$Y = \frac{1}{M}\int y\,\sigma\,dA = \frac{\sigma}{M}\int_0^R r\,dr\int_0^\pi r\sin\theta\,d\theta = \frac{\sigma}{M}\int_0^R r^2\,dr\int_0^\pi \sin\theta\,d\theta$$

$$= \frac{\sigma}{M}\left(\frac{R^3}{3}\right)(-\cos\theta)\Big|_0^\pi = -\frac{\sigma R^3}{3M}\big[(-1)-(+1)\big]$$

$$= \frac{2\sigma R^3}{3M} = \frac{2}{3}\left(\frac{2M}{\pi R^2}\right)R^3 = \boxed{\frac{4R}{3\pi}} \text{ from the center of the arc, along the bisector.}$$

67. Because the speed change is very small, we can use the differential form of the equation of motion:
$dv = -u_{ex}(dm/m)$;
$10\text{ m/s} = -(10^3\text{ m/s})(dm/m)$, which gives $dm/m = -10^{-2}$. ☐ 1% of mass must be discarded. ☐

71. (a) In the reference frame of the center of mass of the student, we use conservation of momentum, assuming he spits one seed of mass m:
$0 + 0 = Mv + m(-u)$, or $v = mu/M$;
which means that, because $M \gg m$ and does not change with each seed, each time he spits one seed his speed increases by mu/M with respect to the ice. After n seeds he will have a speed
$v_n = nmu/M$ and after 100 seeds he will have a speed of
$v_f = 100mu/M$.
If he spits n seeds with a speed of u/n, again in the reference frame of the center of mass of the student, from conservation of momentum we get
$0 + 0 = Mv + nm(-u/n)$,
which means that each time he spits n seeds his speed increases by mu/M with respect to the ice. After spitting 100 seeds in groups of n, his speed with respect to the ice will be
$v_f = (100/n)(mu/M) = (100mu/M)/n$, which is less than before.
Thus, it is better to ☐ spit out seeds one at a time. ☐
(b) $v_{max} = 100mu/M = (100)(1 \times 10^{-3}\text{ kg})(3\text{ m/s})/(50\text{ kg}) = \boxed{6 \times 10^{-3}\text{ m/s}}$.

75.   (a)

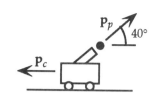

(b) In the horizontal direction momentum is conserved:

$0 = p_p \cos 35° - p_c$;

$0 = (5\ \text{kg})(8800\ \text{m/s}) \cos 35° - (800\ \text{kg})v_c$, which gives

$v_c = \boxed{4.1\ \text{m/s recoil.}}$

(c) The recoiling cannon's component of momentum perpendicular to the ground remains zero because of the upward impulse provided by the ground. During the firing, the upward impulse on the projectile, which gives it the vertical momentum, means there is a downward impulse on the cannon. The normal force from the ground will increase to counteract this.

77. We find the speed falling or rising through a height h from energy conservation:

$\frac{1}{2}mv^2 = mgh$,   or $v = \sqrt{2gh}$.

(a) The speed of each mass just before the collision is

$v_1 = \sqrt{2gh}$.

We choose the positive direction at the bottom in the direction of the motion of the larger mass. For the conservation of momentum during the collision we write

$Mv_1 + m(-v_1) = Mv_3 + mv_4$, with both final velocities assumed to be positive.

Because the collision is elastic, the relative speed does not change:

$v_1 - (-v_1) = -(v_3 - v_4)$ or $2v_1 = -v_3 + v_4$.

Combining these two equations, we get

$v_4 = [(3M - m)/(M + m)]v_1 = [(3M - m)/(M + m)]\sqrt{2gh}$.

We find the rebound height for the marble from

$v_4 = \sqrt{2gh'}$;

$[(3M - m)/(M + m)]\sqrt{2gh} = \sqrt{2gh'}$, which gives $h' = [(3M - m)/(M + m)]^2 h$.

The overshoot is

$h' - h = \boxed{[8M(M - m)/(M + m)^2]h.}$

(b) For the conservation of momentum during the perfectly inelastic collision, we write

$Mv_1 + m(-v_1) = (M + m)v$, which gives

$v = [(M - m)/(M + m)]\sqrt{2gh} = \sqrt{2gh''}$.

From this we get

$h'' = \boxed{[(M - m)/(M + m)]^2 h}$

and the "overshoot" is

$h'' - h = -[4mM/(M + m)^2]h.$ (The combined masses will not reach the lip.)

CHAPTER 9

When drawing a force diagram for the analysis of torques, care should be taken to draw the forces at the point of application. This allows the moment arm to be found from the diagram.

7. (a) From the definition of average angular acceleration, we have
$$\alpha_{av} = \Delta\omega/\Delta t = [0 - (10{,}000 \text{ rev/min})(2\pi \text{ rad/rev})/(60 \text{ s/min})]/(4.00 \text{ s}) = \boxed{-2.62 \times 10^2 \text{ rad/s}^2.}$$
    (b) For a constant acceleration, the angle turned through is
$$\begin{aligned}
\theta &= \omega_0 t + \tfrac{1}{2}\alpha t^2 \\
&= (10{,}000 \text{ rev/min})(2\pi \text{ rad/rev})/(60 \text{ s/min})]/(4.00 \text{ s}) + \tfrac{1}{2}(-2.62 \times 10^2 \text{ rad/s}^2)(4.00 \text{ s})^2 \\
&= 2.09 \times 10^3 \text{ rad.}
\end{aligned}$$
    The distance traveled by a point on the rim is
$$s = r\theta = (8 \times 10^{-2} \text{ m})(2.09 \times 10^3 \text{ rad}) = \boxed{1.67 \times 10^2 \text{ m.}}$$

9. The angular velocity after 5 s is
$$\omega_1 = \omega_0 + \alpha_1 t = 0 + (0.4 \text{ rad/s}^2)(5 \text{ s}) = 2.0 \text{ rad/s,}$$
    which is the constant angular velocity for the next 30 s.
    (a) Because it is moving at a constant angular velocity of 2.0 rad/s after 20 s, we have
$$\alpha_{av} = \Delta\omega/\Delta t = (2.0 \text{ rad/s} - 0)/(20 \text{ s}) = \boxed{0.10 \text{ rad/s}^2.}$$
    (b) During the first 5 s, the angle turned through is
$$\theta_1 = \omega_0 t + \tfrac{1}{2}\alpha_1 t^2 = 0 + \tfrac{1}{2}(0.4)(5)^2 = 5.0 \text{ rad.}$$
    During the next 30 s, the angle turned through is
$$\theta_2 = \omega_1 t_2 = 2.0(30) = 60 \text{ rad.}$$
    Because the slowing down is the reverse of the initial motion, the angle turned through is
$$\theta_3 = 5.0 \text{ rad.}$$
    The total number of revolutions is
$$\theta_{total} = \theta_1 + \theta_2 + \theta_3 = (5 \text{ rad} + 60 \text{ rad} + 5 \text{ rad})/(2\pi \text{ rad/rev}) = \boxed{11 \text{ rev.}}$$
    (c) The child will travel
$$s = r\theta = (3 \text{ m})(70 \text{ rad}) = \boxed{2.1 \times 10^2 \text{ m.}}$$

19. The constant tension will produce a constant acceleration, which we find from the string's motion:
$$x - x_0 = v_0 t + \tfrac{1}{2}at^2$$
$$0.8 \text{ m} = 0 + \tfrac{1}{2}a(1.5 \text{ s})^2, \text{ which gives } a = 0.71 \text{ m/s}^2.$$
    The final speed of the string is
$$v = v_0 + at = 0 + (0.71 \text{ m/s}^2)(1.5 \text{ s}) = 1.07 \text{ m/s.}$$
    Because this is the speed of a point on the rim of the spool, the final angular speed of the spool is
$$\omega = v/R = (1.07 \text{ m/s})/(1 \times 10^{-2} \text{ m}) = 107 \text{ rad/s.}$$
    The work done by the tension as the string unwinds produces the kinetic energy of the spool:
$$W = \Delta K = \tfrac{1}{2}I\omega_f^2 - \tfrac{1}{2}I\omega_i^2;$$
$$(20 \text{ N})(0.8 \text{ m}) = \tfrac{1}{2}I(107 \text{ rad/s})^2 - 0, \text{ which gives } \boxed{I = 2.8 \times 10^{-3} \text{ kg} \cdot \text{m}^2.}$$

21. The rotational inertia of the ball about the axis is
$$I = mr^2 = m(L \sin \theta)^2.$$
    The rotational kinetic energy is
$$K = \tfrac{1}{2}I\omega^2 = \tfrac{1}{2}mL^2\omega^2 \sin^2 \theta = \tfrac{1}{2}(0.75 \text{ kg})(1.5 \text{ m})^2(25 \text{ rad/s})^2 \sin^2 30° = \boxed{1.3 \times 10^2 \text{ J.}}$$
    For the same rotation at an angle of 60°, we have
$$K = \tfrac{1}{2}I\omega^2 = \tfrac{1}{2}mL^2\omega^2 \sin^2 \theta = \tfrac{1}{2}(0.75 \text{ kg})(1.5 \text{ m})^2(25 \text{ rad/s})^2 \sin^2 60° = \boxed{4.0 \times 10^2 \text{ J.}}$$

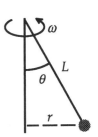

25. The rotational inertia of the earth, considered a uniform sphere, is
$$I_{earth} = (2/5)M_e R_e^2 = (2/5)(6 \times 10^{24} \text{ kg})(6.4 \times 10^6 \text{ m})^2 ;$$
$$I_{earth} = 9.8 \times 10^{37} \text{ kg} \cdot \text{m}^2.$$
For the neutron star we must find the radius. For a sphere, $M = \rho(4/3)\pi R^3$, so
$$R^3 = 3M/4\pi\rho = 3(5M_{Sun})/4\pi\rho = 3(5)(2 \times 10^{30} \text{ kg})/4\pi(6 \times 10^{17} \text{ kg/m}^3), \text{ from which we get}$$
$$R = 1.6 \times 10^4 \text{ m}.$$
The rotational inertia of the neutron star is
$$I_{neutron\ star} = (2/5)MR^2 = (2/5)(10 \times 10^{30} \text{ kg})(1.6 \times 10^4 \text{ m})^2;$$
$$\boxed{I_{neutron\ star} = 1.0 \times 10^{39} \text{ kg} \cdot \text{m}^2 \approx 10\, I_{earth}.}$$

31. We can consider the sphere to consist of three spheres:
a sphere of density ρ_1 and radius R_1;
a sphere of density ρ_2 and radius R;
a sphere of density $-\rho_2$ and radius R_1.
The total mass is
$$M = \rho_1(\tfrac{4}{3}\pi R_1^3) + \rho_2(\tfrac{4}{3}\pi R^3) + (-\rho_2)(\tfrac{4}{3}\pi R_1^3)$$
$$= \tfrac{4}{3}\pi[\rho_1 R_1^3 + \rho_2(R^3 - R_1^3)].$$
Each sphere has a rotational inertia about an axis through the center of $(2/5)MR^2$.
The total rotational inertia is
$$I = (2/5)\rho_1(\tfrac{4}{3}\pi R_1^3)R_1^2 + (2/5)\rho_2(\tfrac{4}{3}\pi R^3)R^2 + (2/5)(-\rho_2)(\tfrac{4}{3}\pi R_1^3)R_1^2$$
$$= (8/15)\pi[\rho_1 R_1^5 + \rho_2(R^5 - R_1^5)] = \boxed{(8/15)\pi[(\rho_1 - \rho_2)R_1^5 + \rho_2 R^5].}$$

33. We use the parallel-axis theorem for each sphere to find its rotational inertia about the center of mass.
Thus for the small sphere of density 4ρ, we have
$$I_1 = I_{1CM} + M d_1^2 = (2/5)M_1 R_1^2 + M_1 d_1^2$$
$$= (2/5)4\rho(4/3)\pi(R/2)^3 R^2 + 4\rho(4/3)\pi(R/2)^3[(R/2) - (R/6)]^2.$$
For the large sphere of density ρ, we have
$$I_2 = I_{2CM} + M d_2^2 = (2/5)M_2 R_2^2 + M_2 d_2^2$$
$$= (2/5)\rho(4/3)\pi R^3 R^2 + \rho(4/3)\pi R^3 (R/6)^2.$$

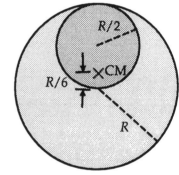

The total rotational inertia is
$$I = I_1 + I_2 = \rho(4/3)\pi R^3[(2/5)4(1/32) + 4(1/8)(1/9) + (2/5) + (1/36)]R^2,$$
which reduces to
$$I = \rho(4/3)\pi R^3(8/15)R^2.$$
The total mass of the object is
$$M = \rho(4/3)\pi R^3 + 4\rho(4/3)\pi(R/2)^3 = \rho(4/3)\pi R^3(3/2).$$
When we use this in the expression for I, we get
$$\boxed{I = (16/45)MR^2.}$$

39. Because each force is perpendicular to the rod and the torques are in the same direction, we have
$$\tau_{net} = F_{1y}(\tfrac{1}{2}L) + F_{2y}(\tfrac{1}{2}L) = \boxed{\tfrac{1}{2}L(F_{1y} + F_{2y}) \text{ perpendicular to the rod.}}$$

41. (a) We choose the clockwise direction as positive.
$$\tau_{pivot} = (7 \text{ N})(1.4 \text{ m}) - (3 \text{ N})(2.5 \text{ m}) = \boxed{2.3 \text{ N} \cdot \text{m.}}$$
  (b) We see that changing the 7-N force to 5.4 N will
  make the torque zero. An upward force of 8.4 N at the
  pivot will make the resultant force equal to zero.

43. The increased tension on one side of the belt is produced by the friction force. Because the pulley is rotating at constant speed, the torque of the motor must balance the torque from this increased tension:

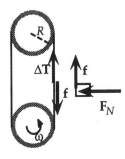

$\tau_{\text{driver}} = \tau_{\text{friction}} = \mu_k F_N R$
$= (1.2)(4 \text{ N})(0.008 \text{ m}) = \boxed{0.38 \text{ N} \cdot \text{m}.}$

47. We choose the upward direction as positive. For the system of student and bicycle wheel, the torque which turns the wheel is an internal torque. There are no external vertical torques, so the angular momentum of the student-bicycle wheel system in the direction of the vertical axis is conserved:

$I_w \omega_1 + 0 = I_w \omega_2 + I_s \omega_s;$
$(1.2 \text{ kg} \cdot \text{m}^2)(4\pi \text{ rad/s}) = (1.2 \text{ kg} \cdot \text{m}^2)(-4\pi \text{ rad/s}) + (8 \text{ kg} \cdot \text{m}^2)\omega_s$, which gives $\omega_s = \boxed{3.8 \text{ rad/s up.}}$
Thus the student will $\boxed{\text{rotate in the direction of the original rotation of the wheel.}}$

49. The rotational inertia of the object depends on the radius, $I = kR^2$, where the factor k will depend on the mass distribution. Because the spherical symmetry is preserved, as the radius decreases this factor will not change. Because there are no external torques, the angular momentum will be conserved:

$L = I_1 \omega_1 = I_2 \omega_2$, or $kR_1^2 \omega_1 = kR_2^2 \omega_2$, which gives $R_2/R_1 = (\omega_1/\omega_2)^{1/2}$.
The fractional change in radius is
$(R_2 - R_1)/R_1 = (R_2/R_1) - 1 = (\omega_1/\omega_2)^{1/2} - 1 = [(3.24592 \text{ rev/s})/(3.24608 \text{ rev/s})]^{1/2} - 1 = \boxed{-2.46 \times 10^{-5}.}$

53. We write $\Sigma F_x = ma_x$ from the force diagram for the cylinder:

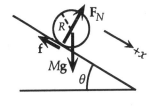

$Mg \sin\theta - f = Ma;$
We write $\Sigma\tau = I\alpha$ about the center of mass from the force diagram for the cylinder:
$fR = I\alpha.$
For the rolling motion we have $a = R\alpha.$
By successively eliminating α and f, we get
$a = g \sin\theta /[1 + (I/MR^2)].$
For the shell, $I = MR^2$; so $a = \frac{1}{2}g \sin\theta.$
For the linear motion we use
$x = v_0 t + \frac{1}{2}at^2 = 0 + \frac{1}{2}[\frac{1}{2}(9.8 \text{ m/s}^2) \sin 20°](4 \text{ s})^2 = \boxed{13.4 \text{ m.}}$
For the solid cylinder, $I = \frac{1}{2}MR^2$; so $a = \frac{2}{3}g \sin\theta.$
For the linear motion we use
$x = v_0 t + \frac{1}{2}at^2 = 0 + \frac{1}{2}[\frac{2}{3}(9.8 \text{ m/s}^2) \sin 20°](4 \text{ s})^2 = \boxed{17.9 \text{ m.}}$

55. The rotational inertia of the system about its axis is

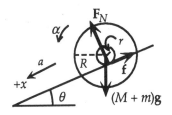

$I = \frac{1}{2}MR^2 + \frac{1}{2}mr^2$
$= \frac{1}{2}(0.80 \text{ kg})(0.12 \text{ m})^2 + \frac{1}{2}(0.10 \text{ kg})(0.02 \text{ m})^2 = 5.8 \times 10^{-3} \text{ kg} \cdot \text{m}^2.$
For the rolling motion, we write $\Sigma F_x = ma_x$ from the force diagram:
$(M + m)g \sin\theta - f = (M + m)a;$
We write $\Sigma\tau = I\alpha$ about the center of mass from the force diagram:
$fr = I\alpha.$
For the rolling motion without slipping, we have $a = r\alpha.$
By successively eliminating α and f, we get
$a = (M + m)gr^2 \sin\theta /[I + (M + m)r^2]$
$= (0.90 \text{ kg})(9.8 \text{ m/s}^2)(0.02 \text{ m})^2 \sin 5° /[(5.8 \times 10^{-3} \text{ kg} \cdot \text{m}^2) + (0.90 \text{ kg})(0.02 \text{ m})^2] = \boxed{0.050 \text{ m/s}^2.}$

61. When an object rolls without slipping, no work is done by the static friction force. As the cylinder rolls up the incline, energy is conserved; the kinetic energy decreases and the gravitational potential energy increases until the object comes to a momentary rest. For a uniform rolling object, the speed of the center of mass and the angular speed are related by $v = R\omega$.

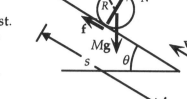

The kinetic energy of rotation is $K_{rot} = \frac{1}{2}I\omega^2 = \frac{1}{2}(\frac{1}{2}MR^2)(v/R)^2 = \frac{1}{4}Mv^2$.

The kinetic energy of translation is $K_{trans} = \frac{1}{2}Mv^2$.

For the conservation of energy, with the reference level for U at the starting point, we have

$K_i + U_i = K_f + U_f$;

$\frac{1}{4}Mv_0^2 + \frac{1}{2}Mv_0^2 + \ + 0 = 0 + Mg(s \sin\theta)$; which gives $s = \boxed{3v_0^2/4g \sin\theta.}$

(An alternate approach is to use the analysis of the solution to Problem 53.)

65. The yo-yo will roll to the left with an angular velocity out of the page. We write $\Sigma F_x = ma_x$ from the force diagram for the yo-yo:

$F - f_s = Ma$;

We choose out of the page as positive and write $\Sigma\tau = I\alpha$ about the center of mass from the force diagram for the yo-yo:

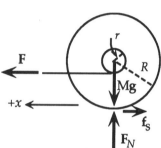

$f_s r - FR = I\alpha = \frac{1}{2}MR^2\alpha$.

With rolling without slipping, we have $a = R\alpha$.

By successively eliminating α and a, we get

$F = 3f_s/[1 + (2r/R)]$.

Thus the force F will be maximum when the friction force is maximum, which is

$f_{s\,max} = \mu F_N = \mu Mg$.

Therefore

$F_{max} = 3\mu Mg/[1 + (2r/R)] = \boxed{3\mu MgR/(R + 2r).}$

CHAPTER 10

7. For the projectile motion using the coordinate system shown, we find the position and velocity components:

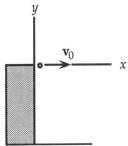

$x = v_{0x}t = v_x t;$

$y = v_{0y}t - \frac{1}{2}gt^2 = 0 - \frac{1}{2}gt^2;$

$v_x = $ constant;

$v_y = v_{0y} - gt = 0 - gt.$

When we express these as vectors, we have

$\mathbf{r} = v_x t\mathbf{i} - \frac{1}{2}gt^2\mathbf{j}; \quad \mathbf{p} = m\mathbf{v} = mv_x\mathbf{i} - mgt\mathbf{j}.$

The angular momentum about the origin is

$$\begin{aligned}\mathbf{L} &= \mathbf{r} \times \mathbf{p} = (v_x t\mathbf{i} - \tfrac{1}{2}gt^2\mathbf{j}) \times (mv_x\mathbf{i} - mgt\mathbf{j}) \\ &= -\tfrac{1}{2}mgv_x t^2\mathbf{k} \\ &= -\tfrac{1}{2}(0.060\text{ kg})(9.8\text{ m/s}^2)(25\text{ m/s})t^2\mathbf{k} \\ &= \boxed{-7.4t^2\mathbf{k}\text{ kg}\cdot\text{m}^2/\text{s (into the page).}}\end{aligned}$$

9. From the position, $\mathbf{r} = \frac{1}{2}at^2\mathbf{i} + vt\mathbf{j} + (\frac{1}{2}bt^2 - wt)\mathbf{k}$, we find the velocity:

$\mathbf{v} = d\mathbf{r}/dt = at\mathbf{i} + v\mathbf{j} + (bt - w)\mathbf{k}.$

The angular momentum is

$$\begin{aligned}\mathbf{L} &= \mathbf{r} \times \mathbf{p} = m[\tfrac{1}{2}at^2\mathbf{i} + vt\mathbf{j} + (\tfrac{1}{2}bt^2 - wt)\mathbf{k}] \times [at\mathbf{i} + v\mathbf{j} + (bt - w)\mathbf{k}] \\ &= m\{[vt(bt - w) - (\tfrac{1}{2}bt^2 - wt)v]\mathbf{i} + [(\tfrac{1}{2}bt^2 - wt)at - \tfrac{1}{2}at^2(bt - w)]\mathbf{j} + [(\tfrac{1}{2}at^2)v - vt(at)]\mathbf{k}\};\end{aligned}$$

$$\boxed{\mathbf{L} = \tfrac{1}{2}mbvt^2\,\mathbf{i} - \tfrac{1}{2}mawt^2\,\mathbf{j} - \tfrac{1}{2}mavt^2\,\mathbf{k}.}$$

13. (a) Because each mass is the same distance from the axis, the rotational inertia of the system is

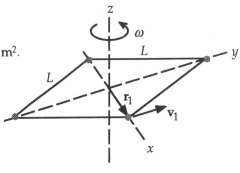

$I = 4m(\tfrac{1}{2}L\sqrt{2})^2 = 2mL^2 = 2(0.1\text{ kg})(0.20\text{ m})^2 = 8.0 \times 10^{-3}\text{ kg}\cdot\text{m}^2.$

The angular momentum of the system is

$\mathbf{L} = I\omega = (8.0 \times 10^{-3}\text{ kg}\cdot\text{m}^2.)(8\text{ rad/s})\mathbf{k}$

$= \boxed{6.4 \times 10^{-2}\,\mathbf{k}\text{ kg}\cdot\text{m}^2/\text{s (along }\omega\text{-direction).}}$

(b) With the coordinate system shown on the diagram, for the mass along the x-axis, we have

$\mathbf{r}_1 = \tfrac{1}{2}L\sqrt{2}\mathbf{i}$, and $\mathbf{v}_1 = \omega \times \mathbf{r}_1 = (\omega\mathbf{k}) \times (\tfrac{1}{2}L\sqrt{2}\mathbf{i}) = \tfrac{1}{2}\omega L\sqrt{2}\mathbf{j}.$

The angular momentum of this mass is

$$\begin{aligned}\mathbf{L}_1 &= \mathbf{r}_1 \times m\mathbf{v}_1 = (\tfrac{1}{2}L\sqrt{2})m(\tfrac{1}{2}\omega L\sqrt{2})(\mathbf{i} \times \mathbf{j}) = \tfrac{1}{2}mL^2\omega\mathbf{k} \\ &= \tfrac{1}{2}(0.1\text{ kg})(0.20\text{ m})^2(8\text{ rad/s})\mathbf{k} = 1.6 \times 10^{-2}\,\mathbf{k}\text{ kg}\cdot\text{m}^2.\end{aligned}$$

From the symmetry of the system, each of the other masses will have the same angular momentum, so the total is

$\mathbf{L} = 4\mathbf{L}_1 = 6.4 \times 10^{-2}\,\mathbf{k}\text{ kg}\cdot\text{m}^2$, which is the $\boxed{\text{same}}$ as in part (a).

15. In each case we have $L = I\omega$, so we find the rotational inertia.

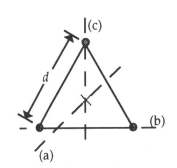

(a) All masses are equidistant from the axis, so we have

$I = 3m[d/(2\cos 30°)]^2 = md^2.$

$\boxed{L = m\omega d^2\text{ along the axis of rotation.}}$

(b) Only one mass contributes to the rotational inertia, so we have

$I = m(d\cos 30°)^2 = \tfrac{3}{4}md^2.$

$\boxed{L = \tfrac{3}{4}m\omega d^2\text{ along the axis of rotation.}}$

(c) Two masses contribute to the rotational inertia, so we have

$I = 2m(\tfrac{1}{2}d)^2 = \tfrac{1}{2}md^2;$

$\boxed{L = \tfrac{1}{2}m\omega d^2\text{ along the axis of rotation.}}$

23. The perpendicular distance from the axis A to the initial path of
the ball is
$$r_i = d\sin\{2[(\pi/2) - \theta]\} = d\sin(\pi - 2\theta) = d\sin(2\theta).$$
The initial angular momentum about A is
$$L_i = r_i mv = \boxed{mvd\sin(2\theta) \text{ up.}}$$
Because the final velocity passes through the axis, we have
$$L_f = \boxed{0.}$$
The wall exerts an impulsive force perpendicular to the wall on the
ball and thus an impulsive torque that changes the angular
momentum about point A.

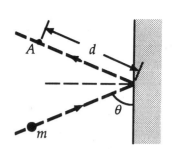

27. Before the children land on the merry-go-round, they have zero angular momentum about the axis. From
the conservation of angular momentum of the system of merry-go-round and children, we have
$$L = I_1\omega_1 = I_2\omega_2 = (I_1 + 2m_{child}R^2)\omega_2 ;$$
$(120\text{ kg}\cdot\text{m}^2)(2.5\text{ rad/s}) = [(120\text{ kg}\cdot\text{m}^2) + 2(25\text{ kg})(1.7\text{ m})^2]\omega_2$, which gives $\omega_2 = \boxed{1.1\text{ rad/s.}}$

35. The (negative) work decreases the kinetic energy of the flywheel:
$$W = \Delta K = \tfrac{1}{2}I(\omega_f^2 - \omega_i^2)$$
$- 1200\text{ J} = \tfrac{1}{2}(0.033\text{ kg}\cdot\text{m}^2)[\omega_f^2 - (490\text{ rad/s})^2]$, which gives $\boxed{\omega_f = 409\text{ rad/s.}}$

39. For Newton's second law, both force and centripetal acceleration are central, so we have
$$F = -kr = -mv^2/r, \text{ which gives } v = \sqrt{k/m}\ r.$$
The Bohr quantization condition is
$$L = mvr = n\hbar, n = 1, 2, \dots .$$
We combine these to get $mvr = m\sqrt{k/m}\ r^2 = n\hbar$, which gives $\boxed{r_n = (n^2\hbar^2/mk)^{1/4}, n = 1, 2, \dots .}$
For v we get $\boxed{v_n = (n^2\hbar^2 k/m^3)^{1/4}, n = 1, 2, \dots .}$
From this we can get the kinetic energy:
$$K = \tfrac{1}{2}mv^2 = \tfrac{1}{2}m(n^2\hbar^2 k/m^3)^{1/2};$$
$$\boxed{K_n = \sqrt{k/m}\ n\hbar/2, n = 1, 2, \dots .}$$

41. The permissible energies of the hydrogen atom are
$$E_n = -(13.6\text{ eV})/n^2; \quad E_1 = -13.6\text{ eV}, E_2 = -3.4\text{ eV}, E_3 = -1.5\text{ eV}, E_4 = -0.85\text{ eV}, \text{ etc.}$$
$\boxed{\text{There is no energy level 2.0 eV above the lowest state,}}$ i. e., at -11.6 eV, so an energy of 2.0 eV cannot
be absorbed.
Possible excitation energies are $E_n - E_1 = [13.6 - (13.6/n^2)]\text{ eV}, n = 2, 3, \dots.$
From the above values, $\boxed{\text{possible excitation energies are 10.2 eV, 12.1 eV, 12.75 eV,} \dots .}$

45. We find the center of mass relative to the pivot:
$$\begin{aligned} R_{CM} &= (m_1 r_1 + m_2 r_2)/(m_1 + m_2) \\ &= [(0.8\text{ kg})(-0.16\text{ m}) + (1\text{ kg})(0.10\text{ m})]/(0.8\text{ kg} + 1\text{ kg}) \\ &= -0.016\text{ m.} \end{aligned}$$
The rotational inertia of the mass about the spin axis is zero, so
the rotational inertia of the system about the spin axis is $m_2 R^2$.
We can treat this as a top, with a precessional rate of
$$\begin{aligned} \omega_p &= MgR_{CM}/I\omega = MgR_{CM}/m_2 R^2\omega \\ &= (1.8\text{ kg})(9.8\text{ m/s}^2)(0.016\text{ m})/(1\text{ kg})(0.10\text{ m})^2(10\text{ rad/s}) = \boxed{2.8\text{ rad/s.}} \end{aligned}$$

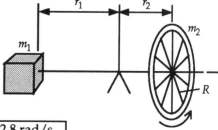

Rotation

47. We choose the reference level for gravitational potential energy at
    the initial position. The kinetic energy will be the translational
    energy of the center of mass and the rotational energy about the
    center of mass. Because there is no work done by friction while the
    cylinder is rolling, for the work–energy theorem we have

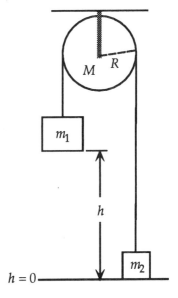

   $$W_{net} = \Delta K + \Delta U;$$
   $$0 = (\tfrac{1}{2}Mv^2 + \tfrac{1}{2}I\omega^2 - 0) + Mg(0 - \ell \sin\theta).$$
   Because the cylinder is rolling, $v = R\omega$.
   The rotational inertia is $\tfrac{1}{2}MR^2$. Thus we get

   $$\tfrac{1}{2}M(R\omega)^2 + \tfrac{1}{2}(\tfrac{1}{2}MR^2)\omega^2 = Mg\ell \sin\theta, \text{ which we write as}$$
   $$\omega^2 = 4g\ell(\sin\theta)/3R^2 = 4(9.8 \text{ m/s}^2)(1.5 \text{ m})(\sin 28°)/3(0.042 \text{ m})^2, \text{ which gives } \omega = \boxed{72 \text{ rad/s.}}$$

49. We treat the child as a point mass moving on a radial line of the platform.
    For the system of child and platform, angular momentum is conserved:

   $$L = I_{platform}\omega_0 = (I_{platform} + m_{child}R^2)\omega_1;$$
   $$(450 \text{ kg}\cdot\text{m}^2)(0.8 \text{ rad/s}) = [450 \text{ kg}\cdot\text{m}^2 + (32 \text{ kg})(2 \text{ m})^2]\omega_1, \text{ which gives } \omega_1 = \boxed{0.62 \text{ rad/s.}}$$
   The change in energy is

   $$\Delta K = \tfrac{1}{2}(I_{platform} + m_{child}R^2)\omega_1{}^2 - \tfrac{1}{2}I_{platform}\omega_0{}^2$$
   $$= \tfrac{1}{2}[450 \text{ kg}\cdot\text{m}^2 + (32 \text{ kg})(2 \text{ m})^2](0.62 \text{ rad/s})^2 - \tfrac{1}{2}(450 \text{ kg}\cdot\text{m}^2)(0.8 \text{ rad/s})^2 = \boxed{-33 \text{ J.}}$$
   The work was done by the force of friction between the child and the platform, which is necessary to
   enable the child to walk.

51. (a) For the system of the two blocks and pulley, with no
        slipping of the rope over the pulley, no work will be done
        by nonconservative forces. The rope ensures that each
        block has the same speed v and the angular speed of the
        pulley is $\omega = v/R$. We choose the reference level for
        gravitational potential energy at the floor.
        The rotational inertia of the pulley is

   $$I = \tfrac{1}{2}MR^2 = \tfrac{1}{2}(2 \text{ kg})(0.1 \text{ m})^2 = 0.10 \text{ kg}\cdot\text{m}^2.$$
   For the work–energy theorem we have

   $$W_{net} = \Delta K + \Delta U;$$
   $$0 = (\tfrac{1}{2}m_1v^2 + \tfrac{1}{2}m_2v^2 + \tfrac{1}{2}I\omega^2 - 0) + m_1g(0 - h) + m_2g(h - 0);$$
   $$\tfrac{1}{2}(4 \text{ kg})v^2 + \tfrac{1}{2}(1 \text{ kg})v^2 + \tfrac{1}{2}(0.10 \text{ kg}\cdot\text{m}^2)(v/0.1 \text{ m})^2 =$$
   $$(4 \text{ kg} - 1 \text{ kg})(9.8 \text{ m/s}^2)(1.5 \text{ m}), \text{ which gives}$$
   $$v = \boxed{3.8 \text{ m/s.}}$$

    (b) Because the motion has constant acceleration, we have

   $$y = \tfrac{1}{2}(v + v_0)t;$$
   $$1.5 \text{ m} - (3.8 \text{ m/s} - 0)t, \text{ which gives } t = \boxed{0.78 \text{ s.}}$$

57. Because mass M is stationary, the tension in the string, which
    provides the centripetal acceleration of the mass m, is $T = Mg$.
    Thus, for $\Sigma F_r = ma_r$ for the mass m, we have
    $Mg = mv_1^2/r_1 = mr_1\omega_1^2$;
    $M(9.8 \text{ m/s}^2) = (0.2 \text{ kg})(0.8 \text{ m})(40 \text{ rad/s})^2$, which gives
    $M = \boxed{26 \text{ kg.}}$

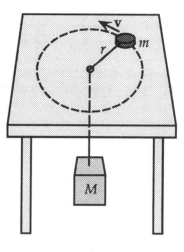

Because the mass M is increased slowly, the acceleration of the
mass M will be very small and we can assume that $T = Mg$ at all
times. For the mass m, the tension has no torque acting
about the center of rotation, so we have conservation of
angular momentum:
$mv_1r_1 = mv_2r_2$, or $r_1^2\omega_1 = r_2^2\omega_2$;
$(0.8 \text{ m})^2(40 \text{ rad/s}) = (0.7 \text{ m})^2\omega_2$, which gives $\omega_2 = \boxed{52 \text{ rad/s.}}$
For $\Sigma F_r = ma_r$ for the mass m, we have
$M'g = mr_2\omega_2^2$;
$M'(9.8 \text{ m/s}^2) = (0.2 \text{ kg})(0.7 \text{ m})(52 \text{ rad/s})^2$, which gives $M' = 38 \text{ kg}$.
The increase in M is $38 \text{ kg} - 26 \text{ kg} = \boxed{12 \text{ kg.}}$

63. If we call the length of the string L, the angle that the top has turned when the string comes off is
    $\theta = L/R$. With a constant force (and thus constant torque), the accelerations will be constant.
    We write $\Sigma\tau = I\alpha$ about the center of mass of the top:
    $FR = \frac{1}{2}MR^2\alpha$, which gives $\alpha = 2F/MR$.
    For the angular motion we find the time of the accelerated motion from
    $\theta = \theta_0 + \frac{1}{2}\alpha t^2$;
    $L/R = \frac{1}{2}(2F/MR)t^2$, which gives $t = \sqrt{LM/F}$.
    (a) We write $\Sigma F = ma$ for the center of mass:
        $F = Ma$, which gives $a = F/M$.
        The speed of the center of mass when the string drops off is
        $$v = v_0 + at = 0 + (F/M)\sqrt{LM/F} = \sqrt{FL/M}$$
        $$= \sqrt{(0.6 \text{ N})(1 \text{ m})/(0.10 \text{ kg})} = \boxed{2.4 \text{ m/s.}}$$
    (b) We find the angular speed from
        $\omega^2 = \omega^2 + 2\alpha\theta$
        $= 0 + 2(2F/MR)(L/R) = 4FL/MR^2$
        $= 4(0.6 \text{ N})(1 \text{ m})/(0.10 \text{ kg})(0.02 \text{ m})^2$, which gives $\omega = \boxed{2.4 \times 10^2 \text{ rad/s.}}$
Note that $v \neq R\omega$, because v is the speed of the center of mass, not the relative speed of the surface of the
cylinder.

CHAPTER 11

The major step in solving static equilibrium problems is to draw the force diagram. Forces should be placed at their points of application and distance, angles (if necessary), and possible locations for calculating torques shown symbolically. You may also want to show the positive direction for torques on the diagram. Referral should be made to the diagram as each torque is found.

3. We choose the coordinate system shown, with positive torques clockwise. We write $\Sigma\tau = I\alpha$ about the point A from the force diagram for the board:

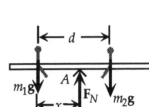

$\Sigma\tau_A = MgD - F_{N2}L = 0$, which gives
$F_{N2} = Mg(D/L) = (24 \text{ kg})(9.8 \text{ m/s}^2)(0.9 \text{ m})/(2.2 \text{ m}) = 96 \text{ N}.$
We write $\Sigma F_y = ma_y$ from the force diagram for the board and worker:
$F_{N1} + F_{N2} - Mg = 0$, which gives
$F_{N1} = Mg - F_{N2}$
$= (24 \text{ kg})(9.8 \text{ m/s}^2) - 96 \text{ N} = 139 \text{ N}.$
The forces on the workmen are the reactions to these normal forces:

| 139 N down   and   96 N down. |

7. We choose the coordinate system shown, with positive torques clockwise. We write $\Sigma\tau = I\alpha$ about the pivot point A (note that this eliminates the unknown force) from the force diagram for the seesaw and children:

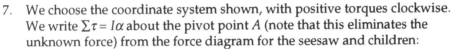

$\Sigma\tau_A = m_2g(d - x) - m_1gx = 0$, which gives
$x = m_2d/(m_1 + m_2) = (40 \text{ kg})(2.8 \text{ m})/(25 \text{ kg} + 40 \text{ kg})$
$= \boxed{1.7 \text{ m.}}$

9. We choose the coordinate system shown, with positive torques clockwise. As the plank is moved out from the roof, the effective normal force acts at a point closer to the edge. When the normal force reaches the edge, the plank is on the verge of tipping. We write $\Sigma\tau = I\alpha$ about the edge A (note that this eliminates the unknown normal force) from the force diagram for the plank, the load and the concrete:

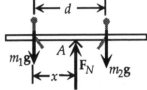

$\Sigma\tau_A = M_1gx - M_2g(L - x) = 0$, which gives
$x = M_2L/(M_1 + M_2) = (15 \text{ kg})(2.4 \text{ m})/(30 \text{ kg} + 15 \text{ kg}) = \boxed{0.8 \text{ m.}}$

13. The maximum distance for the top book to remain on the bottom book will be reached when its center of mass is over the edge of the bottom book. The maximum distance for the bottom book to remain on the table will be reached when the center of mass of the combination is over the edge of the table. If we take the edge of the table as the origin, for the location of the center of mass we have

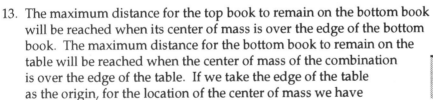

$R = m(x - \tfrac{1}{2}L) + mx = 0$, which gives $x = \tfrac{1}{4}L.$
The distance to the extreme edge of the top book is
$D = \tfrac{1}{4}L + \tfrac{1}{2}L = \boxed{\tfrac{3}{4}L.}$
If we add a third book, the maximum distance for the top two books to remain on the bottom book will be reached when their composite center of mass is over the edge of the bottom book. The maximum distance for the bottom book to remain on the table will be reached when the center of mass of the three books is over the edge of the table. If we take the edge of the table as the origin, we have

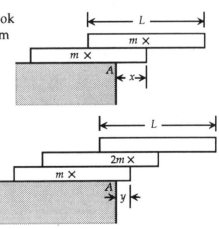

$R = m(y - \tfrac{1}{2}L) + 2my = 0$, which gives $y = L/6.$
The distance to the extreme edge of the top book is
$D = (L/6) + (3L/4) = \boxed{11L/12.}$

19. We choose the coordinate system shown, with positive torques
counterclockwise.

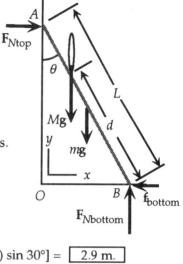

(a) We write $\sum \tau = I\alpha$ about the point B from the force diagram
for the ladder and man:
$$\sum \tau_B = mg(\tfrac{1}{2}L)\sin\theta + Mgd\sin\theta - F_{Ntop}L\cos\theta = 0.$$
We write $\sum F_x = ma_x$ from the force diagram for the ladder and man:
$$F_{Ntop} - f_{bottom} = 0.$$
We write $\sum F_y = ma_y$ from the force diagram for the ladder and man:
$$F_{Nbottom} - (m + M)g = 0.$$
As the man climbs the ladder, the friction force at the bottom increases.
At the point where slipping begins,
$$f_{bottom} = f_{bottom,max} = \mu_s F_{Nbottom} = \mu_s(m + M)g.$$
Then we have
$$F_{Ntop} = \mu_s(m + M)g.$$
Using these in the torque equation, we get
$$d = [\mu_s(m + M)L\cos\theta - m(L/2)\sin\theta]/(M\sin\theta)$$
$$= [0.4(10\text{ kg} + 80\text{ kg})(4\text{ m})\cos 30° - 10(4\text{ m}/2)\sin 30°]/[(80\text{ kg})\sin 30°] = \boxed{2.9\text{ m.}}$$

(b) We write $\sum \tau = I\alpha$ about the point A from the force diagram for the ladder and man:
$$\sum \tau_A = F_{Nbottom}(L\sin\theta) - Mg(L - d)\sin\theta - mg(\tfrac{1}{2}L)\sin\theta - f_{bottom}L\cos\theta = 0.$$
We write $\sum \tau = I\alpha$ about the point O from the force diagram for the ladder and man:
$$\sum \tau_O = F_{Nbottom}(L\sin\theta) - Mg(L - d)\sin\theta - mg(\tfrac{1}{2}L)\sin\theta - F_{Ntop}L\cos\theta = 0.$$
We can combine these with the torque about the point B to obtain the force equations.
If we add the equations for points A and O, we get
$$F_{Ntop} = f_{bottom}, \text{ which is the } x\text{-equation.}$$
If we subtract the equations for points B and O, we get
$$F_{Nbottom} = \mu_s(m + M)g, \text{ which is the } y\text{-equation.}$$

21. We choose the coordinate system shown, with positive torques clockwise.

(a) The force exerted by the person on the rope is equal to the tension in the rope. Because the rope
is smooth, the tension is the same on both sides of the pulley. From the equilibrium of the engine,
we have
$$T = Mg = (30\text{ kg})(9.8\text{ m/s}^2) = 2.9 \times 10^2 \text{ N.}$$
We write $\sum \tau = I\alpha$ about the point A from the force diagram for the strut and engine:
$$\sum \tau_A = -T\cos\theta\, L\sin 45° - T\sin\theta\, L\cos 45° + mg(\tfrac{1}{2}L)\cos 45° + MgL\cos 45° = 0.$$
Note that we found the torque from the tension by finding the torque from each of its components.
Because $\sin 45° = \cos 45°$ and the factor L is in each term, when we use $T = Mg$ we get
$$-M\cos\theta - M\sin\theta + \tfrac{1}{2}m + Mg = 0, \text{ or}$$
$$\cos\theta + \sin\theta = (M + \tfrac{1}{2}m)/M = [(30\text{ kg}) + \tfrac{1}{2}(12.5\text{ kg})]/(30\text{ kg}) = 1.21.$$
If we use $\sin^2\theta + \cos^2\theta = 1$, we get a quadratic equation for $\cos\theta$:
$$\cos^2\theta - 1.21\cos\theta + 0.46, \text{ which has two solutions: } \theta = 13.5° \text{ and } 76°.$$
We choose the smaller angle as more practical.
The person exerts a force of $\boxed{2.9 \times 10^2 \text{ N } 13.5° \text{ above the horizontal.}}$

(b) We write $\sum F_x = ma_x$ from the force diagram for the strut and engine:
$$F_H - T\cos\theta = 0, \text{ which gives}$$
$$F_H = (2.9 \times 10^2 \text{ N})\cos 13.5° = 2.8 \times 10^2 \text{ N.}$$
We write $\sum F_y = ma_y$ from the force diagram for the strut and engine:
$$F_V + T\sin\theta - (m + M)g = 0, \text{ which gives}$$
$$F_V = -(2.9 \times 10^2 \text{ N})\sin 13.5° + (12.5\text{ kg} + 30\text{ kg})(9.8\text{ m/s}^2) = 3.5 \times 10^2 \text{ N.}$$
The forces exerted by the strut on the ground are the reactions to these:
$$\boxed{F_V = \ 3.5 \times 10^2 \text{ N down \quad and \quad } F_H = \ 2.8 \times 10^2 \text{ N.}}$$

27. We choose the coordinate system shown, with positive torques clockwise. On the force diagram we have
added the two forces on each leg at A and on each leg at B.

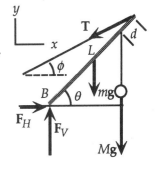

(a) We write $\Sigma\tau = I\alpha$ about the point B from the force diagram for the seat:

$\Sigma\tau_B = -mg\ell_2 + 2F_{NA}\ell_1 = 0$, which gives

$F_{NA} = \frac{1}{2}mg\ell_2/\ell_1$.

We write $\Sigma\tau = I\alpha$ about the point A from the force diagram
for the seat:

$\Sigma\tau_A = -mg(\ell_2 - \ell_1) + 2F_{NB}\ell_1 = 0$, which gives

$F_{NB} = \frac{1}{2}mg(\ell_2 - \ell_1)/\ell_1$.

The forces on the table will be the reactions to these forces:

$\boxed{F_{NA} = \frac{1}{2}mg\ell_2/\ell_1 \text{ down}}$ and $\boxed{F_{NB} = \frac{1}{2}mg(\ell_2 - \ell_1)/\ell_1 \text{ up.}}$

(b) As $\ell_2 \to 0$; $F_{NA} \to 0$, $F_{NB} \to -mg/2$. Because the normal forces
cannot become negative, the seat will lose contact at B and turn clockwise.
As $\ell_1 \to 0$; $F_{NA} \to \infty$, $F_{NB} \to \infty$. The normal forces will have no net torque about the contact point,
and the seat will turn counterclockwise.

(c) From the above expressions, we get

$F_{NA} = \frac{1}{2}mg\ell_2/\ell_1 = \frac{1}{2}(10 \text{ kg})(9.8 \text{ m/s}^2)(0.30 \text{ m})/(0.20 \text{ m}) = \boxed{74 \text{ N}}$ and

$F_{NB} = \frac{1}{2}mg(\ell_2 - \ell_1)/\ell_1 = \frac{1}{2}(10 \text{ kg})(9.8 \text{ m/s}^2)[(0.30 \text{ m}) - (0.20 \text{ m})]/(0.20 \text{ m}) = \boxed{25 \text{ N.}}$

31. We choose the coordinate system shown, with positive torques clockwise.
The three unknowns are F_H, F_V, and Mg. We write $\Sigma\tau = I\alpha$ about the
point B (which eliminates F_H and F_V) from the force diagram for the beam:

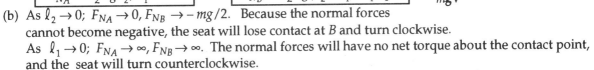

$\Sigma\tau_B = mg(\frac{1}{2}L \cos\theta) + Mg[(L - d) \sin\theta] - T \cos\phi (L \sin\theta) - T \sin\phi (L \cos\theta) = 0$,
which gives

$M = \{[TL(\cos\phi \sin\theta - \sin\phi \cos\theta)/g] - \frac{1}{2}mL \cos\theta\}/[(L - d) \cos\theta]$.

We see that maximum M corresponds to maximum T, so we have

$M_{max} = \{[(10 \times 10^3 \text{ N})(3 \text{ m})(\cos 30° \sin 45° - \sin 30° \cos 45°)/(9.8 \text{ m/s}^2)] -$
$\frac{1}{2}(100 \text{ kg})(3 \text{ m}) \cos 45°\}/[(3 \text{ m} - 0.25 \text{ m}) \cos 45°]$

$= \boxed{353 \text{ kg.}}$

35. We choose the coordinate system shown, with positive torques clockwise.
We use the point A at the ground as the axis so the force exerted by the
ground will produce no torque. Because the wind force is variable, we
determine its torque by finding the torque on a small element dh at the
height h and then adding (integrating) the torques from all of the
differential elements.
We write $\Sigma\tau = 0$ about the point A from the force diagram for the beam:

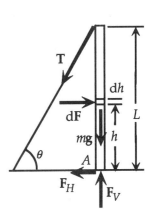

$\Sigma\tau_A = \displaystyle\int_0^L h \, dF - T \cos\theta \, L = 0$, which we can write as

$T \cos\theta \, L = \displaystyle\int_0^L h\alpha h \, dh = \frac{1}{3}\alpha h^3 \Big|_0^L = \frac{1}{3}\alpha L^3$, which gives

$T = \alpha L^2/(3 \cos\theta) = (50 \text{ N/m}^2)(20 \text{ m})/(3 \cos 60°)$

$= \boxed{1.33 \times 10^4 \text{ N.}}$

37. We choose the coordinate system shown, with positive torques
   clockwise. From the symmetry of the placement, the tensions
   in the two cables will be equal.
   We write $\Sigma\tau = I\alpha$ about the axis A (which eliminates the forces
   from the brackets) from the force diagram for the shelf and sack:

   $\Sigma\tau_A = Mg(\frac{1}{2}L) - (2T\sin\theta)L = 0$, which gives

   $T = \frac{1}{4}Mg/\sin\theta$
   $= \frac{1}{4}(20\text{ kg})(9.8\text{ m/s}^2)/\sin 45° = \boxed{69\text{ N.}}$

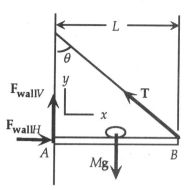

41. We choose the coordinate system shown, with positive torques clockwise.
   The center of mass of a triangle is $\frac{1}{3}$ the distance from its base.
   We write $\Sigma\tau = I\alpha$ about the point A from the force diagram for the sign:

   $\Sigma\tau_A = Mg(\frac{1}{3}L) - T(L\tan\theta) = 0$, which gives

   $T = \frac{1}{3}Mg/(\tan\theta)$
   $= \frac{1}{3}(15\text{ kg})(9.8\text{ m/s}^2)/(\tan 30°) = \boxed{85\text{ N.}}$

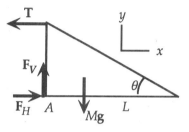

43. To climb the step, the roller will roll about the contact point A.
   We choose the coordinate system shown, with positive torques
   clockwise. We find the angle θ from

   $\sin\theta = (R - h)/R$
   $= (30\text{ cm} - 15\text{ cm})/(30\text{ cm}) = 0.50$, which gives $\theta = 30°$.
   When the roller goes over the curb, contact with
   the ground is lost and $F_{N_1} = 0$.
   We write $\Sigma\tau = I\alpha$ about the point A:
   $F(R - h) - MgR\cos\theta = I_A\alpha$.
   The minimum force occurs when $\alpha = 0$:
   $F_{min} = (MgR\cos\theta)/(R - h)$.
   $= (80\text{ kg})(9.8\text{ m/s}^2)(0.30\text{ m})(\cos 30°)/(0.30\text{ m} - 0.15\text{ m}) = \boxed{1.36\times 10^3\text{ N.}}$

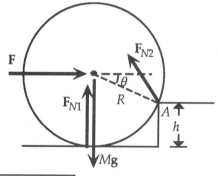

45. We choose the coordinate system shown, with positive torques clockwise. On the force diagram we have added the two forces on each near leg and on each far leg.

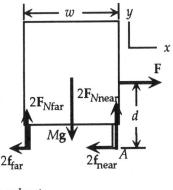

(a) We write $\Sigma F_x = ma_x$ from the force diagram for the chest:

$$F - 2f_{far} - 2f_{near} = 0.$$

We write $\Sigma F_y = ma_y$ from the force diagram for the chest:

$$2F_{Nfar} + 2F_{Nnear} - Mg = 0, \text{ which gives}$$
$$F_{Nfar} + F_{Nnear} = \tfrac{1}{2}Mg.$$

When this is used in the x-equation, we get

$$\begin{aligned} F &= 2(f_{far} + f_{near}) \\ &= 2\mu_k(F_{Nfar} + F_{Nnear}) = \mu_k Mg. \end{aligned}$$

Because three of the forces will have no torque about the point A, we choose it as the axis and write $\Sigma\tau = I\alpha$ from the force diagram for the chest:

$$\Sigma\tau_A = Fd - Mg(\tfrac{1}{2}w) + (2F_{Nfar})w = 0, \text{ which gives}$$
$$\begin{aligned} F_{Nfar} &= (\tfrac{1}{2}Mgw - Fd)/2w \\ &= \tfrac{1}{4}Mg - \tfrac{1}{2}\mu_k Mg(d/w) \\ &= Mg[\tfrac{1}{4} - \tfrac{1}{2}\mu_k(58 \text{ cm}/58 \text{ cm})] = Mg(\tfrac{1}{4} - \tfrac{1}{2}\mu_k). \end{aligned}$$

From the y-equation, we get

$$F_{Nnear} = \tfrac{1}{2}Mg - Mg(\tfrac{1}{4} - \tfrac{1}{2}\mu_k) = Mg(\tfrac{1}{4} + \tfrac{1}{2}\mu_k).$$

When we include the friction forces, we have

near legs:	$Mg(\tfrac{1}{4} + \tfrac{1}{2}\mu_k)$ up and $\mu_k Mg(\tfrac{1}{4} + \tfrac{1}{2}\mu_k)$ to the left;
far legs:	$Mg(\tfrac{1}{4} - \tfrac{1}{2}\mu_k)$ up and $\mu_k Mg(\tfrac{1}{4} - \tfrac{1}{2}\mu_k)$ to the left.

(b) The chest will topple if the normal forces on the far legs become zero:

$$F_{Nfar} = Mg(\tfrac{1}{4} - \tfrac{1}{2}\mu_k) = 0, \text{ which gives} \quad \boxed{\mu_k = 0.5.}$$

47. We choose the coordinate system shown, with positive torques clockwise.

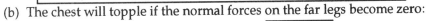

We write $\Sigma\tau = I\alpha$ about the point A from the force diagram for the cylinder:

$$\Sigma\tau_A = F_N R\sin\theta - f_s(R + R\cos\theta) = 0, \text{ which gives}$$
$$f_s = [(\sin\theta)/(1 + \cos\theta)]F_N = \tan(\tfrac{1}{2}\theta)\,F_N.$$

For static friction, $f_s \leq \mu_s F_N$; the minimum value of μ_s is when $f_s = \mu_s F_N$.

Thus we have

$$\mu_{s\,min}F_N = \tan(\tfrac{1}{2}\theta)\,F_N, \text{ which gives}$$
$$\mu_{s\,min} = \boxed{\tan(\tfrac{1}{2}\theta).}$$

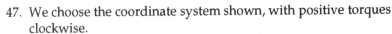

CHAPTER 12

3. (a) We write the dimensions of $F(r) = h/r^3$ as
$$[F] = [h]\,[r^{-3}] \quad \text{or} \quad [MLT^{-2}] = [h][L^{-3}], \text{ which gives } [h] = \boxed{[ML^4T^{-2}].}$$
    (b) For the circular motion, the central force provides the centripetal acceleration:
$$F_r = ma_r;$$
$$h/R^3 = mv^2/R, \text{ which gives } v = \sqrt{h/m}\,/R.$$
    The angular momentum is
$$L = mvR = m(\sqrt{h/m}\,/R\,)R = \boxed{\sqrt{mh}\,.}$$
    (c) The period of the circular motion is
$$T = 2\pi R/v = 2\pi R/(\sqrt{h/m}\,/R\,) = 2\pi\sqrt{m/h}\ R^2, \text{ which we write as}$$
$$\boxed{T/R^2 = 2\pi\sqrt{m/h}\ = \text{a constant.}}$$

15. With the reference level for potential energy at the ground, we use energy conservation to relate the maximum height to the initial speed:
$$K_i + U_i = K_f + U_f\,;$$
$$\tfrac{1}{2}mv_0^2 + 0 = 0 + mgh, \text{ which gives } v_0^2 = 2gh.$$
Because we assume that the initial speed is constant, with g_{Mars} from Problem 8, we have
$$g_{\text{Mars}}h_{\text{Mars}} = g_e h_e, \text{ or}$$
$$h_{\text{Mars}} = (g_e/g_{\text{Mars}})h_e = [(9.8 \text{ m/s}^2)/(3.70 \text{ m/s}^2)](1.85 \text{ m}) = \boxed{4.9 \text{ m.}}$$

19. Because the escape speed is the speed at the surface necessary to get far away with no final speed, we use energy conservation, with the reference level for potential energy at infinity:
$$K_i + U_i = K_f + U_f\,;$$
$$\tfrac{1}{2}mv_{\text{esc}}^2 - (GM_s m/R_s) = 0 + 0, \text{ which gives}$$
$$v_{\text{esc}} = \sqrt{2GM/R}$$
$$= \sqrt{2(6.67\times10^{-11} \text{ N·m}^2/\text{kg}^2)(1.99\times10^{30} \text{ kg})/(6.96\times10^8 \text{ m})} = \boxed{6.18\times10^5 \text{ m/s.}}$$

23. For the material to just barely stay on the surface, the upward normal force from the ground will be zero so the gravitational attraction must provide the centripetal acceleration for the circular motion:
$$GMm/R^2 = mv^2/R = mR\omega^2, \text{ so we have}$$
$$\omega^2 = GM/R^3 = G\rho(\tfrac{4}{3}\pi R^3)/R^3 = \tfrac{4}{3}G\rho\pi$$
$$= \tfrac{4}{3}(6.67\times10^{-11} \text{ N·m}^2/\text{kg}^2)(4800 \text{ kg/m}^3)\pi, \text{ which gives}$$
$$\omega = \boxed{1.6\times10^{-3} \text{ rad/s.}}$$

31. From Kepler's third law, we have
$$T^2 = 4\pi^2 R^3/GM$$
$$= 4\pi^2[(3393 + 95)\times10^3 \text{ m}]^3/(6.67\times10^{-11} \text{ N·m}^2/\text{kg}^2)(6.42\times10^{23} \text{ kg}), \text{ which gives}$$
$$T = 6.25\times10^3 \text{ s} = \boxed{1.74 \text{ h.}}$$

35. For the conservation of angular momentum, we have
    $L = mv (\cos 45°) R = mv_f r$, which gives $v = (r/R)v_f / \cos 45°$.
    For the conservation of energy, we have
    $K_i + U_i = K_f + U_f$;
    $\frac{1}{2}mv^2 - (GMm/R) = \frac{1}{2}mv_f^2 - (GMm/r)$;
    We see that the mass m cancels. If we multiply by 2 and use the escape speed, given by $v_{esc}^2 = 2GM/R$,
    and the result from angular momentum conservation, with $r = 2R$, we get
    $[(r/R)v_f/(\cos 45°)]^2 - v_{esc}^2 = v_f^2 - v_{esc}^2(R/r)$;
    $[2v_f/(\cos 45°)]^2 - v_{esc}^2 = v_f^2 - \frac{1}{2}v_{esc}^2$, which reduces to
    $v_f^2 = v_{esc}^2/14 = (11.2 \text{ km/s})^2/14$, which gives
    $v_f = \boxed{2.99 \text{ km/s.}}$

37. (b) At the nearest approach, the potential energy   (a)
    will be minimum and the speed will be maximum.
    For the conservation of angular momentum, we have
    $L = mv_0 d = mv_{max} r_{min}$, which gives $r_{min} = v_0 d/v_{max}$.
    For the conservation of energy, we have
    $K_0 + U_0 = K + U$;
    $\frac{1}{2}mv_0^2 + 0 = \frac{1}{2}mv_{max}^2 - (GMm/r_{min})$.

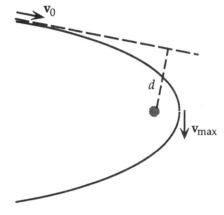

    Using the result from angular momentum conservation, we get
    $v_{max}^2 - (2GMv_{max}/v_0 d) - v_0^2 = 0$;
    $v_{max}^2 - [2(6.67 \times 10^{-11} \text{ N} \cdot \text{m}^2/\text{kg}^2)(1.99 \times 10^{30} \text{ kg})v_{max}/$
    $\qquad (12 \times 10^3 \text{ m/s})(3 \times 10^{11} \text{ m})] - (12 \times 10^3 \text{ m/s})^2 = 0$.
    Solving this quadratic equation for v_{max}, we get
    $v_{max} = -1.92 \times 10^3 \text{ m/s}$, and $+7.56 \times 10^4 \text{ m/s}$.
    The positive answer is the physical result: $v_{max} = \boxed{75.6 \text{ km/s perpendicular to the radius.}}$
    (c) From the conservation of angular momentum, we get
    $r_{min} = v_0 d/v_{max} = (12 \text{ km/s})(3 \times 10^8 \text{ km})/(75.6 \text{ km/s}) = \boxed{4.76 \times 10^7 \text{ km.}}$

39. (a) In the circular orbit, for Newton's second law, we have
$GMm/r_0{}^2 = mv_0{}^2/r_0$,   or
$v_0{}^2 = GM/r_0$
$= (6.67 \times 10^{-11} \, \text{N} \cdot \text{m}^2/\text{kg}^2)(6 \times 10^{24} \, \text{kg})/[(6.37 + 2.00) \times 10^6 \, \text{m}]$, which gives
$v_0 = \boxed{6.91 \times 10^3 \, \text{m/s}.}$

(b) The angular momentum is
$L = mv_0 r_0 = (300 \, \text{kg})(6.91 \times 10^3 \, \text{m/s})(8.37 \times 10^6 \, \text{m}) = \boxed{1.74 \times 10^{13} \, \text{kg} \cdot \text{m}^2/\text{s}.}$

(c) If the direction of motion immediately after the firing is unchanged and the speed is $v_1 = \frac{1}{2}v_0$, the angular momentum is
$L = mv_1 r_0 = \frac{1}{2}mv_0 r_0 = \frac{1}{2}(1.74 \times 10^{13} \, \text{kg} \cdot \text{m}^2/\text{s}^2) = \boxed{8.68 \times 10^{12} \, \text{kg} \cdot \text{m}^2/\text{s}.}$

(d) The orbit will be elliptical. For the conservation of angular momentum for the new orbit, at the lowest point of the orbit we have
$mvr = L = mv_1 r_0 = \frac{1}{2}mv_0 r_0$, which gives $v = \frac{1}{2}(r_0/r)v_0$.
For the conservation of energy, we have
$K_1 + U_1 = K + U$;
$\frac{1}{2}mv_1{}^2 - (GMm/r_0) = \frac{1}{2}mv^2 - (GMm/r)$;
$\frac{1}{2}(\frac{1}{2}v_0)^2 - (GM/r_0) = \frac{1}{2}(\frac{1}{2}v_0)^2(r_0/r)^2 - (GM/r)$.
Using the expression from part (a), we have
$(1/8)(GM/r_0) - (GM/r_0) = (1/8)(GM/r_0)(r_0/r)^2 - (GM/r)$, which reduces to
$(r_0/r)^2 - 8(r_0/r) + 7 = 0$.
When we solve this quadratic equation, we get
$r_0/r = 1$ and 7.
The value 1 is the initial position of the satellite. Thus the perigee is
$r = r_0/7 = (8.37 \times 10^6 \, \text{m})/7 = 1.20 \times 10^6 \, \text{m}$.
Because this is less than the radius of the Earth, $\boxed{\text{the satellite crashes.}}$

41. (a) In the circular orbit, for Newton's second law, we have
$GMm/r_0{}^2 = mv_0{}^2/r_0$,   or
$v_0{}^2 = GM/r_0$
$= (6.67 \times 10^{-11} \, \text{N} \cdot \text{m}^2/\text{kg}^2)(5.98 \times 10^{24} \, \text{kg})/[(6.37 + 0.30) \times 10^6 \, \text{m}]$, which gives
$v_0 = \boxed{7.73 \times 10^3 \, \text{m/s}.}$

(b) The angular momentum is
$L = mv_0 r_0 = (2000 \, \text{kg})(7.73 \times 10^3 \, \text{m/s})(6.67 \times 10^6 \, \text{m}) = \boxed{1.03 \times 10^{14} \, \text{kg} \cdot \text{m}^2/\text{s}.}$

(c) The total energy is
$E = \frac{1}{2}mv_0{}^2 - (GMm/r_0) = -\frac{1}{2}GMm/r_0$
$= -\frac{1}{2}(6.67 \times 10^{-11} \, \text{N} \cdot \text{m}^2/\text{kg}^2)(5.98 \times 10^{24} \, \text{kg})(2000 \, \text{kg})/(6.67 \times 10^6 \, \text{m}) = \boxed{-5.98 \times 10^{10} \, \text{J}.}$
Remember that the potential energy is zero at infinity.

(d) Because the force is radial, the torque is $\boxed{\text{zero.}}$

(e) Because the torque is zero, the angular momentum will be conserved:
$mvr = mv_0 r_0$.
This shows that the speed will decrease as r increases, but vr will remain constant. From part (a) we see that for circular orbits, $v^2 r$ is constant. If r changes, both of these cannot be true.
$\boxed{\text{The orbit cannot be circular.}}$

45. From Example 12–5, we have

$F = -\tfrac{4}{3}\pi G m \rho r$.

The spring force, $F = -kx$, has a potential energy: $\tfrac{1}{2}kx^2$. Because the gravitational force in the tunnel has the same form, it also has a potential energy, given by

$U = \tfrac{1}{2}(\tfrac{4}{3}\pi G m \rho)r^2$, with $U = 0$ at the center.

Because the density is

$\rho = M/(\tfrac{4}{3}\pi R^3)$, this becomes

$U = \tfrac{1}{2}(GMm/R^3)r^2$.

For conservation of energy as the mass falls from the surface to the center, we have

$K_i + U_i = K_f + U_f$;

$0 + \tfrac{1}{2}(GMm/R^3)R^2 = \tfrac{1}{2}mv^2 + 0$;

$(6.67 \times 10^{-11}\ \text{N} \cdot \text{m}^2/\text{kg}^2)(5.98 \times 10^{24}\ \text{kg})/(6.37 \times 10^6\ \text{m}) = v^2$, which gives

$v = \boxed{7.9 \times 10^3\ \text{m/s.}}$

Note that the result is independent of the mass m.

55. (a) For a "weightless" circular orbit, the gravitational force provides the centripetal acceleration:

$GM/R^2 = \tfrac{4}{3}\pi G m \rho R = mv_0^2/R$, or

$R^2 = 3v_0^2/4\pi G\rho = 3(2.0\ \text{m/s})^2/4\pi(6.67 \times 10^{-11}\ \text{N} \cdot \text{m}^2/\text{kg}^2)(5.2 \times 10^3\ \text{kg/m}^3)$, which gives

$\boxed{R = 1.66 \times 10^3\ \text{m.}}$

(b) The escape speed is

$v_{\text{esc}} = (2GM/R)^{1/2} = v_0\sqrt{2} = (2.0\ \text{m/s})\sqrt{2} = \boxed{2.8\ \text{m/s.}}$

(c) The surface speed at the equator is

$v = 2\pi R/T = 2\pi(1.66 \times 10^3\ \text{m})/(12\ \text{h})(3600\ \text{s/h}) = 0.24\ \text{m/s}$.

By walking in the direction of the rotation, he would need a speed relative to the surface of 1.76 m/s to orbit the asteroid.

59. For a circular orbit, the attractive gravitational force provides the centripetal acceleration:

$GMm/r^2 = mv^2/r$, which gives $v^2 = GM/r$.

The energies are

$K = \tfrac{1}{2}mv^2 = GMm/2r$,

$U = -GMm/r$, and

$E = K + U = (GMm/2r) - (GMm/r) = -GMm/2r$.

If the radius changes slowly, we can approximate the change in energy by differentiating:

$\Delta E \approx dE = -(-GMm/2r^2)\,dr \approx (GMm/2r^2)\,\Delta r$.

For the work-energy theorem, for one revolution, we have

$W_f = \Delta E$;

$-f(2\pi r) = (GMm/2r^2)\,\Delta r$, which gives

$\boxed{\Delta r = -4\pi f r^3/GMm.}$ The radius decreases.

The changes in the energies are

$\boxed{\begin{array}{l} \Delta U \approx dU = -(-GMm/r^2)\,dr \approx (GMm\,\Delta r)/r^2 = -4\pi f r. \\ \Delta E = -2\pi f r. \\ \Delta K \approx dK = (-GMm/2r^2)\,dr \approx -(GMm\,\Delta r)/r^2 = 2\pi f r. \end{array}}$

The total energy decreases, but the kinetic energy increases.

61. (a) While the astronaut throws the wrench, linear momentum is conserved. Using the initial reference frame of the astronaut, in which he wants to acquire a speed back toward the ship, we have
$$(m + m_W)(0) = m(-v_i) + m_W v_W;$$
$$0 = (115 \text{ kg})(-0.05 \text{ m/s}) + (3 \text{ kg})v_W, \text{ which gives}$$
$$v_W = \boxed{1.92 \text{ m/s relative to himself, which is } 1.97 \text{ m/s relative to the ship.}}$$
   (b) We find his average acceleration toward the ship produced by the gravitational attraction from
$$F_{av} = GMm/r_{av}^2 = ma_{av};$$
$$a_{av} = (6.67 \times 10^{-11} \text{ N} \cdot \text{m}^2/\text{kg}^2)(10^5 \text{ kg})/(12 \text{ m} + 0.5 \text{ m})^2 = 4.3 \times 10^{-8} \text{ m/s}^2.$$
   To find the time, we use
$$x = \tfrac{1}{2}a_{av}t^2;$$
$$1 \text{ m} = \tfrac{1}{2}(4.3 \times 10^{-8} \text{ m/s}^2)t^2, \text{ which gives}$$
$$t = 6.8 \times 10^3 \text{ s} = \boxed{1.9 \text{ h.}}$$

63. (a) From the symmetry of the arrangement, we see that the net force on a star is toward the center of the triangle. Thus the net force can provide the necessary centripetal acceleration, so the arrangement is $\boxed{\text{possible.}}$
   (b) If D is the separation of a pair of stars, for Newton's second law for one star we have
$$F_{net} = 2F \cos 30° = 2(GM^2/D^2) \cos 30° = ma = mR\omega^2 = MR(2\pi/T)^2.$$
   The distance R from the center to a star is $\tfrac{1}{2}D/\cos 30°$, so we have
$$T = \pi(D^3/GM)^{1/2}/\cos 30°$$
$$= \pi[(1.5 \times 10^{11} \text{ m})^3/(6.67 \times 10^{-11} \text{ N} \cdot \text{m}^2/\text{kg}^2)(1.99 \times 10^{30} \text{ kg})]^{1/2}/\cos 30°$$
$$= \boxed{1.83 \times 10^7 \text{ s}} \quad (212 \text{ days}).$$
   (c) To test the stability, we consider small changes in D. When D decreases, the force increases, but the required centripetal force decreases; thus the star system will collapse. When D increases, the force decreases, but the required centripetal force increases; thus the star system will separate. The system is $\boxed{\text{unstable.}}$

65. With the Earth pictured as an onion, we find the energy required to remove each layer and take it to infinity. A layer of radius r will have a mass $dm = \rho 4\pi r^2\, dr$, where the density is $\rho = M_e/(4/3)\pi R_e^3$. Because the layer being removed is outside the remaining mass m, the potential energy of the layer is
$$dU = -(Gm\, dm)/r, \text{ with the reference level at infinity.}$$
Thus the energy required to remove this layer is
$$dE = 0 - [(-Gm\, dm)/r] = (Gm\, dm)/r$$
$$= [G\rho(\tfrac{4}{3}\pi r^3)\rho 4\pi r^2\, dr]/r = \tfrac{1}{3}[G\rho^2(4\pi)^2 r^4\, dr].$$
We find the total energy required by integrating over all of the layers:
$$E = \int_0^{R_e} \frac{G\rho^2(4\pi)^2 r^4\, dr}{3} = \frac{G\rho^2(4\pi)^2}{3}\left(\frac{r^5}{5}\right)\Big|_0^{R_e}$$
$$= \frac{G\rho^2(4\pi)^2}{3^2}\left(\frac{3R_e^5}{5}\right) = G\left(\frac{\rho 4\pi R_e^3}{3}\right)^2\left(\frac{3}{5R_e}\right) = \boxed{\tfrac{3}{5}GM_e^2/R_e.}$$

CHAPTER 13

3. The general expression for x is
    $x = A \cos(\omega t + \delta)$, from which we get
    $v = dx/dt = -A\omega \sin(\omega t + \delta)$.
    Comparing this to $v = 0.4 \sin(\omega t + \pi)$ m/s, we see that
    $A\omega = 0.4$ m/s, so $A = (0.4 \text{ m/s})/(2.00 \text{ rad/s}) = 0.2$ m.
    Because $\sin(\theta + \pi) = -\sin \theta$, we see that $\delta = 0$. Thus
    $x = 0.2 \cos(\omega t)$ m.
    We obtain a from
    $a = dv/dt = 0.4\omega \cos(\omega t + \pi) = -0.8 \cos(\omega t)$ m/s^2.

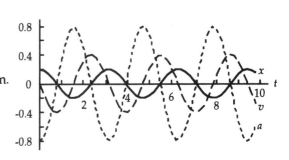

13. (a) We find the angular frequency from
        $\omega = \sqrt{k/m} = \sqrt{(0.5 \text{ N/m})/(0.2 \text{ kg})} = \boxed{1.58 \text{ rad/s.}}$
        The period is $T = 2\pi/\omega = 2\pi/(1.58 \text{ rad/s}) = \boxed{3.97 \text{ s.}}$
    (b) From $v_{max} = A\omega$, we get
        $A = v_{max}/\omega = (2.0 \text{ m/s})/(1.58 \text{ rad/s}) = \boxed{1.26 \text{ m.}}$

15. The period is the time for one cycle: $T = 1.71$ s.
    Because the total distance traveled is twice that from one extreme to the other, we have
        $A = D/4 = (6.98 \text{ cm})/4 = 1.75$ cm.
    (a) We define the average speed as total distance traveled/time:
        average speed $= D/T = (6.98 \text{ cm})/(1.71 \text{ s}) = \boxed{4.08 \text{ cm/s.}}$
    (b) The angular frequency is $\omega = 2\pi/T = 2\pi/(1.71 \text{ s}) = 3.67$ rad/s. Thus
        $v_{max} = A\omega = (1.75 \text{ cm})(3.67 \text{ rad/s}) = \boxed{6.41 \text{ cm/s}}$ and
        $a_{max} = A\omega^2 = (1.75 \text{ cm})(3.67 \text{ rad/s})^2 = \boxed{23.6 \text{ cm/s}^2.}$

17. (a) If we write $x = A \sin(\omega t + \delta)$, then $v = dx/dt = A\omega \cos(\omega t + \delta)$.
        From the initial conditions at $t = 0$, for these two equations we have
            $3 \text{ cm} = A \sin \delta$, and $-5 \text{ cm/s} = A(3.0 \text{ rad/s}) \cos \delta$, which give us two equations for A and δ.
        Dividing these two equations, we have
            $\tan \delta = -1.8$, which gives $\delta = -1.06$ rad and $+2.08$ rad. To correspond to the initial conditions, we
        choose the one that gives a positive value for the sine and a negative value for the cosine: $\delta = 2.08$ rad.
        From this value, we obtain $A = 3.44$ cm. The displacement is
            $\boxed{x = (3.44 \text{ cm}) \sin[(3.0 \text{ rad/s})t + 2.08 \text{ rad}].}$
    (b) We find the times when the position is 3 cm from
            $3 \text{ cm} = (3.44 \text{ cm}) \sin[(3.0 \text{ rad/s})t + 2.08 \text{ rad}]$, which gives
            $(3.0 \text{ rad/s})t + 2.08 \text{ rad} = 1.06 \text{ rad} + n2\pi, \ n = 0. \pm 1, \pm 2, \ldots$, and
            $(3.0 \text{ rad/s})t + 2.08 \text{ rad} = 2.08 \text{ rad} + n2\pi, \ n = 0. \pm 1, \pm 2, \ldots$.
        Because 2.08 rad corresponds to $t = 0$, the next time the position is 3 cm occurs when
            $(3.0 \text{ rad/s})t + 2.08 \text{ rad} = 1.06 \text{ rad} + 2\pi$, which gives $\boxed{t = 1.75 \text{ s.}}$
    (c) We find the times when the speed is 5 cm/s from   (d)
            $\pm 5 \text{ cm/s} = (3.44 \text{ cm})(3.0 \text{ rad/s}) \cos[(3.0 \text{ rad/s})t + 2.08 \text{ rad}]$.
        These two equations have four solutions:
            $(3.0 \text{ rad/s})t + 2.08 \text{ rad} = 1.06 \text{ rad} + n2\pi, \ n = 0. \pm 1, \pm 2, \ldots$,
            $(3.0 \text{ rad/s})t + 2.08 \text{ rad} = 2.08 \text{ rad} + n2\pi, \ n = 0. \pm 1, \pm 2, \ldots$,
            $(3.0 \text{ rad/s})t + 2.08 \text{ rad} = 4.20 \text{ rad} + n2\pi, \ n = 0. \pm 1, \pm 2, \ldots$, and
            $(3.0 \text{ rad/s})t + 2.08 \text{ rad} = 5.22 \text{ rad} + n2\pi, \ n = 0. \pm 1, \pm 2, \ldots$.
        Beginning with $t = 0$, the four times are
            $\boxed{0, \ 0.71 \text{ s}, \ 1.05 \text{ s}, \ 1.75 \text{ s.}}$
        Additional times occur when the period of the motion, $T = 2.1$ s,
        is added to these times.

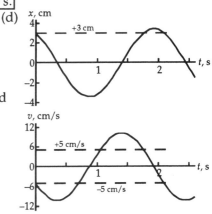

29. The angular frequency is $\omega = 2\pi f = 2\pi(20 \text{ s}^{-1}) = 40\pi \text{ rad/s}$.
The maximum force is the force that provides the maximum acceleration:
$$F_{max} = mA\omega^2 = (1.20 \text{ kg})(0.8 \times 10^{-2} \text{ m})(40\pi \text{ rad/s})^2 = \boxed{1.5 \times 10^2 \text{ N.}}$$

31. We use a coordinate system with up positive.
At the equilibrium position of the 25-g mass, we have $\Sigma F = k\,\Delta y - mg = 0$. Thus
$$k = mg/\Delta y = (25 \times 10^{-3} \text{ kg})(9.8 \text{ m/s}^2)/(12 \times 10^{-2} \text{ m}) = 2.0 \text{ N/m}.$$
For the 75-g mass, the angular frequency is
$$\omega = \sqrt{k/m} = \sqrt{(2.0 \text{ N/m})/(75 \times 10^{-3} \text{ kg})} = 5.2 \text{ rad/s}.$$
Thus the period is $T = 2\pi/\omega = 2\pi/(5.2 \text{ rad/s}) = \boxed{1.2 \text{ s.}}$

33. In the equilibrium position, we have
$$F_{net} = F_{20} - F_{10} = 0, \quad \text{or} \quad F_{10} = F_{20}.$$
When the object is moved to the right a distance x, we have
$$F_{net} = F_{20} - k_2x - (F_{10} + k_1x) = -(k_1 + k_2)x.$$
The effective spring constant is $k_{eff} = k_1 + k_2$, so the angular frequency is
$$\omega = \sqrt{k_{eff}/m} = \sqrt{[(100 \text{ N/m}) + (200 \text{ N/m})]/(0.06 \text{ kg})} = 71 \text{ rad/s}.$$
If we write the displacement as $x = A\sin(\omega t + \delta)$, at $t = 0$ we have
$A = 0.01 \text{ m} = A\sin(0 + \delta)$, which gives $\delta = \pi/2$, so
$$\boxed{x = (0.01 \text{ m}) \sin[(71 \text{ rad/s})t + \pi/2].}$$

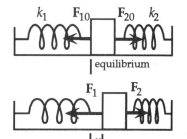

39. Because the tension is central, it does no work. The total energy, which is conserved, is
$$E = \tfrac{1}{2}mv_{top}^2 + U_{top} = \tfrac{1}{2}mv_{bottom}^2 + U_{bottom}.$$
Each speed is $R\omega$. We choose the reference level for U at the bottom, so we have
$$\tfrac{1}{2}mR^2\omega_{top}^2 + mg(2R) = \tfrac{1}{2}R^2\omega_{bottom}^2 + 0.$$
After canceling common factors, we have
$$R\omega_{top}^2 + 4g = R\omega_{bottom}^2;$$
$(1.2 \text{ m})(2.2 \text{ rad/s})^2 + 4(9.8 \text{ m/s}^2) = (1.2 \text{ m})\omega_{bottom}^2$, which gives $\boxed{\omega_{bottom} = 6.1 \text{ rad/s.}}$

43. The angular frequency is $\omega = \sqrt{k/m} = \sqrt{(1 \text{ N/m})/(0.2 \text{ kg})} = 2.24 \text{ rad/s}$.
Because $x = x_{max}$ at $t = 0$, we can write
$x = x_{max}\sin[\omega t + (\pi/2)]$, from which we get
$v = x_{max}\omega\cos[\omega t + (\pi/2)] = v_{max}\cos[\omega t + (\pi/2)]$.
Using the data at $t = 0.5$ s, we get
$1.5 \text{ m/s} = v_{max}\cos[(2.24 \text{ rad/s})(0.5 \text{ s}) + (\pi/2)]$, which gives $v_{max} = \boxed{1.67 \text{ m/s.}}$
Then $x_{max} = v_{max}/\omega = (1.67 \text{ m/s})/(2.24 \text{ rad/s}) = \boxed{0.74 \text{ m.}}$
The total energy is $E = \tfrac{1}{2}kx_{max}^2 = \tfrac{1}{2}(1 \text{ N/m})(0.74 \text{ m})^2 = \boxed{0.27 \text{ J.}}$ This is also $\tfrac{1}{2}mv_{max}^2$.

49. (a) The period of a simple pendulum depends only on
its length. Without the peg, we have
$$T_1 = 2\pi\sqrt{L/g} = 2\pi\sqrt{(1.0 \text{ m})/(9.8 \text{ m/s}^2)} = \quad 2.0 \text{ s}.$$
When the string is caught by the peg, we have
$$T_2 = 2\pi\sqrt{L/g} = 2\pi\sqrt{(0.5 \text{ m})/(9.8 \text{ m/s}^2)} = \quad 1.4 \text{ s}.$$
Because there is half of each of these swings in
each basic period, we have
$$T = \tfrac{1}{2}T_1 + \tfrac{1}{2}T_2 = \tfrac{1}{2}(2.0 \text{ s}) + \tfrac{1}{2}(1.4 \text{ s}) = \boxed{1.7 \text{ s.}}$$
(b) The tension in the string is always perpendicular
to the displacement and thus does no work, so the
total energy, $E = \tfrac{1}{2}mv^2 + mgh$, is conserved and
$v = 0$ at the maximum height on each side;
the ball must rise to the same height: $\boxed{0.05 \text{ m.}}$

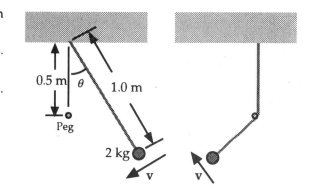

53. We use the parallel-axis theorem to get the rotational inertia about the pin:
$$I = I_{CM} + Md^2 = (1/12)ML^2 + M(\tfrac{1}{2}L - y)^2,$$
where $d = \tfrac{1}{2}L - y$ is the distance from the center to the pin. For the physical pendulum we have

$$T = 2\pi\sqrt{\frac{I}{Mgr}} = 2\pi\sqrt{\frac{M\left[\frac{1}{12}L^2 + \left(\frac{1}{2}L - y\right)^2\right]}{Mg\left(\frac{1}{2}L - y\right)}} = \boxed{2\pi\sqrt{\frac{2\left(L^2 - 3Ly + 3y^2\right)}{3(L - 2y)g}}}.$$

55. We can treat the wire as a physical pendulum. If we call the length of each side L, the rotational inertia about the pivot is $I = \tfrac{1}{3}(\tfrac{1}{2}M)L^2 + \tfrac{1}{3}(\tfrac{1}{2}M)L^2 = \tfrac{1}{3}ML^2$. In the equilibrium position the center of mass is directly below the pivot and midway between the center of mass of each side, so its distance from the pivot is $r = \tfrac{1}{2}L \cos 45°$. Thus

$$f = \frac{1}{T} = \frac{1}{2\pi}\sqrt{\frac{Mgr}{I}} = \frac{1}{2\pi}\sqrt{\frac{Mg\left(\frac{L}{2}\cos\theta\right)}{\frac{1}{3}ML^2}} = \frac{1}{2\pi}\sqrt{\frac{3g\cos\theta}{2L}}$$

$$= \frac{1}{2\pi}\sqrt{\frac{3(9.8 \text{ m/s}^2)\cos 45°}{2(0.40 \text{ m})}} = \boxed{0.81 \text{ Hz.}}$$

57. (a) We find the rotational inertia of the disk about the nail from the parallel-axis theorem:
$$I = I_{CM} + Md^2 = \boxed{\tfrac{1}{2}MR^2 + M\ell^2.}$$

(b) Gravity provides the restoring torque. From Newton's second law for rotation, $\tau = I\alpha$, we have
$-Mg\ell \sin\theta = I\, d^2\theta/dt^2$, which for small angles becomes
$-Mg\ell\,\theta = M(\tfrac{1}{2}R^2 + \ell^2)\, d^2\theta/dt^2$,   or
$$\boxed{-g\ell\,\theta = (\tfrac{1}{2}R^2 + \ell^2)\, d^2\theta/dt^2.}$$

(c) The result for part (b) shows that the angular motion is simple harmonic with $\omega^2 = 2g\ell/(R^2 + 2\ell^2)$, so
$$T = \frac{2\pi}{\omega} = \boxed{2\pi\sqrt{\frac{R^2 + 2\ell^2}{2g\ell}}.}$$

(d) For small ℓ, we have $T = \dfrac{2\pi R}{\sqrt{2g\ell}}$, so as $\ell \to 0$, $\boxed{T \to \infty; \text{ no torque,}}$ which means uniform rotation.

59. (a) The rotational inertia of the dumbbell about the center is
$$I = 2[m(\tfrac{1}{2}\ell)^2] = \tfrac{1}{2}m\ell^2.$$
Because γ corresponds to the rotational force constant, we find the period from

$$T = 2\pi\sqrt{\frac{I}{\gamma}} = 2\pi\sqrt{\frac{m\ell^2}{2\gamma}} = 2\pi\sqrt{\frac{(80 \text{ g})(30 \text{ cm})^2}{2(2\times10^5 \text{ g·cm}^2/\text{s}^2)}} = \boxed{2.7 \text{ s.}}$$

(b) The constant total energy is the maximum potential energy:
$$E = \tfrac{1}{2}\gamma\theta_{max}^2 = \tfrac{1}{2}(2\times10^5 \text{ g·cm}^2/\text{s}^2)(10^{-3} \text{ kg/g})(10^{-2} \text{ m/cm})^2(0.1 \text{ rad})^2 = \boxed{1.0\times10^{-4} \text{ J.}}$$

(c) The constant total energy is the maximum kinetic energy:
$$E = \tfrac{1}{2}I(d\theta/dt)_{max}^2 = \tfrac{1}{2}(\tfrac{1}{2}m\ell^2)(d\theta/dt)_{max}^2;$$
$1.0\times10^{-4} \text{ J} = \tfrac{1}{2}[\tfrac{1}{2}(0.080 \text{ kg})(0.30 \text{ m})^2](d\theta/dt)_{max}^2$, which gives $(d\theta/dt)_{max} = 0.24$ rad/s.
Because the speed is tangential, we have
$$v_{max} = \tfrac{1}{2}\ell(d\theta/dt)_{max} = \tfrac{1}{2}(0.30 \text{ m})(0.24 \text{ rad/s}) = \boxed{3.6\times10^{-2} \text{ m/s.}}$$

65. For the undamped angular frequency we have $\omega_0^2 = k/m$, or $m = k/\omega_0^2$.
   For critical damping we have
   $$b_c = (4mk)^{1/2} = 2k/\omega_0 = 2(184 \text{ N/m})/(3880 \text{ rad/s}) = \boxed{0.095 \text{ kg/s.}}$$

69. If we write $x = Ae^{-bt/2m}\sin(\omega't + \delta)$, the maxima will occur approximately when $\sin(\omega't + \delta) = 1$.
   From the given data we have
   $$6.0 \text{ cm} = Ae^{-b(1.5 \text{ s})/2m} \quad \text{and} \quad 5.6 \text{ cm} = Ae^{-b(2.5 \text{ s})/2m}.$$
   We have two equations for $b/2m$ and A, which give $b/2m = 0.0690 \text{ s}^{-1}$ and $A = 6.65 \text{ cm}$.
   We find the angular frequency from the time between adjacent maxima:
   $$\omega' = 2\pi/T' = 2\pi/(2.5 \text{ s} - 1.5 \text{ s}) = 2\pi \text{ rad/s.}$$
   Because the first maximum occurs at 1.5 s, we can find δ from
   $$\sin(\omega't + \delta) = \sin[(2\pi \text{ rad/s})(1.5 \text{ s}) + \delta] = 1, \text{ or } 3\pi + \delta = \pi/2, \text{ which gives } \delta = -5\pi/2 \text{ rad, or } -\pi/2.$$
   The positions are found from $x = Ae^{-bt/2m}\sin(\omega't + \delta)$:
   $$x_{3.0} = (6.65 \text{ cm})e^{-(0.0690 /s)(3.0 \text{ s})}\sin[(2\pi \text{ rad/s})(3.0 \text{ s}) - (\pi/2)] = \boxed{-5.40 \text{ cm.}}$$
   $$x_{4.8} = (6.65 \text{ cm})e^{-(0.0690 /s)(4.8 \text{ s})}\sin[(2\pi \text{ rad/s})(4.8 \text{ s}) - (\pi/2)] = \boxed{-1.48 \text{ cm.}}$$
   $$x_0 = (6.65 \text{ cm})e^{-(0.0690 /s)(0 \text{ s})}\sin[(2\pi \text{ rad/s})(0 \text{ s}) - (\pi/2)] = \boxed{-6.65 \text{ cm.}}$$

75. (a) The natural frequency of the system is $\omega_0 = \sqrt{k/m} = \sqrt{(86 \text{ N/m})/(0.548 \text{ kg})} = 12.5 \text{ rad/s}$.
   For the damped harmonic motion we have
   $$\omega^2 = \omega_0^2 - (b^2/2m^2) = (k/m) - (b^2/2m^2);$$
   $$(12.2 \text{ rad/s})^2 = [(86 \text{ N/m})(0,548 \text{ kg})] - [b^2/2(0.548 \text{ kg})^2)], \text{ which gives } b = \boxed{2.2 \text{ N·s/m.}}$$
   (b) The lifetime is defined as $\tau = m/b = (0.548 \text{ kg})/(2.8 \text{ N·s/m}) = \boxed{0.25 \text{ s.}}$
   (c) The sharpness can be expressed as $\Delta\omega = 2b/m = 2(2.2 \text{ N·s/m})/(0.548 \text{ kg}) = \boxed{8 \text{ rad/s,}}$
   or with the Q factor: $Q = \omega_0\tau = (12.5 \text{ rad/s})(0.25 \text{ s}) = \boxed{3.1.}$

79. If Δy is the stretch of the spring at equilibrium, from the initial condition $k\,\Delta y_1 = mg$, or $k = mg/\Delta y_1$.
   The force must have the same magnitude at any location in the spring, while the stretch increases linearly from the top. For the same hanging mass, when the spring is cut in two, the tension in the spring will be the same as before, so the stretch at equilibrium will be $\Delta y_2 = \frac{1}{2}\Delta y_1$.
   Thus $k_2 = mg/\Delta y_2 = 2(mg/\Delta y_1) = 2k$ and $\omega_2 = \sqrt{k_2/m} = \sqrt{2k/m} = \boxed{\sqrt{2}\,\omega.}$

81. We call the length of the unstretched cord L.
   (a) If we consider the energies, at the bridge there is gravitational potential energy only, with $h = 0$ at the river level. When the cord is at its maximum extension at the water surface, there is elastic potential energy only. Calling the maximum extension of the cord ΔL, we use energy conservation:
   $$K_i + U_{gi} + U_{cordi} = K_f + U_{gf} + U_{cordf}:$$
   $$0 + mgH + 0 = 0 + 0 + \tfrac{1}{2}k(\Delta L)^2, \text{ or } mgH = \tfrac{1}{2}(10\,mg/H)(\Delta L)^2, \text{ which gives } \Delta L = H/\sqrt{5}.$$
   From $L + \Delta L = H$, we have $L = H - \Delta L = H[1 - (1/\sqrt{5})] = \boxed{0.553\,H.}$
   (b) At the final equilibrium position, the stretch is $\Delta L_0 = mg/k = mg/(10\,mg/H) = H/10$.
   The final height above the water is $h = H - L - \Delta L_0 = H(1 - 0.553 - 0.1) = \boxed{0.347H.}$

85. We find the rotational inertia about the pivot, with the twins as point masses and the bar as two rods:

$$I = 2[\tfrac{1}{3}(\tfrac{1}{2}M_{Bar})(\tfrac{1}{2}L)^2] + 2M_T(\tfrac{1}{2}L)^2 = (\tfrac{1}{3}M_{Bar} + 2M_T)(\tfrac{1}{2}L)^2$$
$$= [\tfrac{1}{3}(12 \text{ kg}) + 2(32\text{kg})][\tfrac{1}{2}(5.0 \text{ m})]^2 = 425 \text{ kg} \cdot \text{m}^2.$$

Before oscillating, the center of mass is directly below the pivot at a distance of

$$r = [2M_T(\tfrac{1}{2}L \sin 7°) + 2(\tfrac{1}{2}M_{Bar})(\tfrac{1}{2})(\tfrac{1}{2}L \sin 7°)]/(2M_T + M_{Bar})$$
$$= \{2(32 \text{ kg})[\tfrac{1}{2}(5.0 \text{ m}) \sin 7°] + 2[\tfrac{1}{2}(12 \text{ kg})](\tfrac{1}{2})[\tfrac{1}{2}(5.0 \text{ m}) \sin 7°]\}/[2(32 \text{ kg}) + 12 \text{ kg}] = 0.28 \text{ m}.$$

We find the period of the physical pendulum from

$$T = 2\pi\sqrt{\frac{I}{mgr}} = 2\pi\sqrt{\frac{425 \text{ kg} \cdot \text{m}^2}{[2(32 \text{ kg}) + 12 \text{ kg}](9.8 \text{ m}/s^2)(0.28 \text{ m})}} = \boxed{9.0 \text{ s.}}$$

87. (a) The equilibrium separation occurs where the potential energy is a minimum, which we find from the derivative of $U = (A/r^2) - (e^2/r)$:

$$dU/dr = (-2A/r^3) + (e^2/r^2) = (-2A + e^2r)/r^3 = 0, \text{ which gives } r_0 = \boxed{2A/e^2.}$$

(b) We want to find the expression for U in terms of the displacement from equilibrium, r_0x:

$U = [A/r_0^2(1 + x)^2] - [e^2/r_0(1 + x)]$. When we use the expansions for small x, we get
$U = (A/r_0^2)(1 - 2x + 3x^2 - ...) - (e^2/r_0)(1 - x + x^2 - ...)$. We collect the powers of x up to x^2:
$U = (A/r_0^2) - (e^2/r_0) + [(-2A/r_0^2) + (e^2/r_0)]x + [(3A/r_0^2) - (e^2/r_0)]x^2.$

If we use the result of part (a), the coefficient of the x term is 0 and we get
$U = (-e^4/4A) + (e^4/4A)x^2.$

To find the effective force constant, we express this in terms of the displacement from equilibrium r_0x:
$U = (-e^4/4A) + (e^4/4A)(e^2/2A)^2(r_0x)^2 = (-e^4/4A) + \tfrac{1}{2}(e^8/8A^3)(r_0x)^2.$

This has the form of the elastic potential energy $\tfrac{1}{2}k(\Delta r)^2$, so the motion will be simple harmonic, with $k = e^8/8A^3$. The kinetic energy is $K = \tfrac{1}{2}\mu v^2$, with the reduced mass $\mu = mm/(m + m) = m/2$. The angular frequency is

$$\omega = \sqrt{k/\mu} = \sqrt{(e^8/8A^3)/(m/2)} = \boxed{e^4/\sqrt{4A^3m}.}$$

91. Because the cylinder is rolling, the contact point A is instantaneously at rest, so we can use Newton's law for rotation: $\Sigma\tau_A = I_A\alpha$.
We use the parallel-axis theorem to find the rotational inertia about A:

$$I_A = I_{CM} + Md^2 = \tfrac{1}{2}mR^2 + mR^2 = \tfrac{3}{2}mR^2.$$

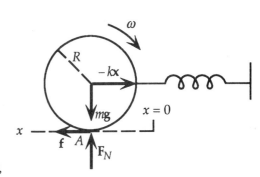

The only torque is from the restoring spring force with a moment arm of R. From Newton's second law, we have
$$-kxR = (\tfrac{3}{2}mR^2)\alpha.$$
The stretch of the spring x is also the distance rolled, $x = R\theta$, so we have
$$-kR^2 \theta = \tfrac{3}{2}mR^2\, d^2\theta/dt^2,$$
which corresponds to angular simple harmonic motion.
We see that the $k_{effective} = kR^2$, so
$$\omega = \sqrt{k_{effective}/I} = \sqrt{kR^2/\tfrac{3}{2}mR^2} = \sqrt{2k/3m}.$$
The frequency is
$$f = \frac{\omega}{2\pi} = \boxed{\frac{1}{2\pi}\sqrt{2k/3m}.}$$

CHAPTER 14

The best way to keep the signs correct in problems involving the Doppler effect is to use two steps. Find the wavelength produced by the source; remember that the wavelength decreases in front of a moving source, while the wavelength increases behind a moving source. The frequency that this wavelength produces at a receiver is determined by the relative velocity of the sound wave.

3. For the wavelength and frequency we have
   $$\lambda = L/N, \text{ and } f = 1/\tau.$$
   We find the tension from
   $$f = (1/\lambda)\sqrt{T/\mu} = (N/L)\sqrt{T/\mu};$$
   $$1/(0.10 \text{ s}) = [3.5/(2.7 \text{ m})]\sqrt{T/(220 \times 10^{-3} \text{ kg/m})}, \text{ which gives } \boxed{T = 13 \text{ N.}}$$

9. The speed of a wave in the rope is
   $$v = \sqrt{T/\mu} = \sqrt{(200 \text{ N})/(40 \text{ g/cm})(10^{-3} \text{ kg/g})(100 \text{ cm/m})} = 7.07 \text{ m/s}.$$
   We assume that the disturbance at the end is so slight that it is equivalent to a fixed point. There are nodes at each end, so the possible wavelengths are $\lambda_n = 2L/n$, and the possible frequencies are
   $$f_n = v/\lambda_n = nv/2L \text{ and } \omega_n = 2\pi f_n; \ \tau_n = 2\pi/\omega_n.$$
   The values for the three lowest frequencies are
   $$f_1 = 1(7.07 \text{ m/s})/2(1 \text{ m}) = \boxed{3.54 \text{ Hz,}} \ \omega_1 = 2\pi(3.54 \text{ Hz}) = \boxed{22.2 \text{ rad/s,}} \ \tau_1 = 2\pi/(22.2 \text{ rad/s}) = \boxed{0.28 \text{ s.}}$$
   $$f_2 = 2f_1 = \boxed{7.07 \text{ Hz,}} \ \omega_2 = 2\omega_1 = \boxed{44.4 \text{ rad/s,}} \ \tau_2 = \tau_1/2 = \boxed{0.14 \text{ s.}}$$
   $$f_2 = 3f_1 = \boxed{10.6 \text{ Hz,}} \ \omega_3 = 3\omega_1 = \boxed{66.6 \text{ rad/s,}} \ \tau_3 = \tau_1/3 = \boxed{0.094 \text{ s.}}$$

13.

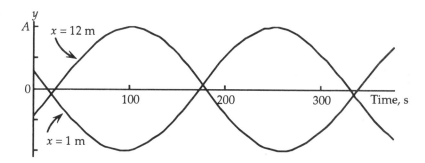

17. For the traveling wave, $z(x, t) = f(x - vt)$, we have the following derivatives:
   $$\frac{\partial z}{\partial x} = \frac{df}{d(x-vt)}\frac{\partial(x-vt)}{\partial x} = \frac{df}{d(x-vt)};$$

   $$\frac{\partial^2 z}{\partial x^2} = \frac{\partial\left[\dfrac{df}{d(x-vt)}\right]}{\partial x} = \frac{d^2 f}{d(x-vt)^2}\frac{\partial(x-vt)}{\partial x} = \frac{d^2 f}{d(x-vt)^2};$$

   $$\frac{\partial z}{\partial t} = \frac{df}{d(x-vt)}\frac{\partial(x-vt)}{\partial t} = \frac{df}{d(x-vt)}(-v);$$

   $$\frac{\partial^2 z}{\partial t^2} = (-v)\frac{\partial\left[\dfrac{df}{d(x-vt)}\right]}{\partial t} = (-v)\frac{d^2 f}{d(x-vt)^2}\frac{\partial(x-vt)}{\partial t} = (-v)^2\frac{d^2 f}{d(x-vt)^2}.$$

   Comparing the second derivatives, we see that
   $$\frac{\partial^2 z}{\partial t^2} = v^2\frac{\partial^2 z}{\partial x^2}, \quad \text{which is a form of the wave equation.}$$

19. If we consider a length L of the rug which is much greater than the length of the hump, the hump will travel the length L in a time $\Delta t = L/v$. In this time the rug, with mass $m = \mu L$, has moved a distance of $D = 0.1$ m, so the average speed of the rug is $v_{av} = D/(L/v)$.
    The average momentum carried by the pulse is
    $$p_{av} = m v_{av} = \mu L[D/(L/v)] = \mu D v = (1.5 \text{ kg/m})(0.1 \text{ m})(2 \text{ m/s}) = \boxed{0.3 \text{ kg} \cdot \text{m/s.}}$$

29. The speed of the wave depends on k, ℓ, and m. We write the relation as $v = k^\alpha \ell^\beta m^\gamma$. When we substitute the dimensions, we get
    $$[LT^{-2}] = [MT^{-2}]^\alpha [L]^\beta [M]^\gamma.$$
    By equating the exponents for each dimension, we obtain $\alpha + \gamma = 0$, $\beta = 1$, and $-2\alpha = -1$.
    These equations are satisfied by $\alpha = \frac{1}{2}$, $\beta = 1$, $\gamma = -\frac{1}{2}$, so $\boxed{v = \ell(k/m)^{1/2}.}$

31. The speed of the wave is
    $$v = (T/\mu)^{1/2} = [(800 \text{ N})/(12 \text{ g/cm})(10\text{–}3 \text{ kg/g})(10^2 \text{ cm/m})]^{1/2} = 26 \text{ m/s}.$$
    The angular frequency is
    $$\omega = 2\pi f = 2\pi v/\lambda = 2\pi(26 \text{ m/s})/(0.7 \text{ m}) = 230 \text{ rad/s}.$$
    The average power is
    $$P_{av} = \tfrac{1}{2}\mu z_0^2 \omega^2 v = \tfrac{1}{2}(1.2 \text{ kg/m})(5 \times 10^{-2} \text{ m})^2 (230 \text{ rad/s})^2 (26 \text{ m/s})$$
    $$= \boxed{2.1 \times 10^3 \text{ W.}}$$
    Because the speed depends on the medium and thus does not change, we have
    $$\omega' = 2\pi v/\lambda' = 2\pi v/2\lambda = \tfrac{1}{2}\omega.$$
    Because there is no change in amplitude or speed, the only change in the power is from the change in ω, so we have
    $$P_{av}' = \tfrac{1}{4}P_{av} = \tfrac{1}{4}(2.1 \times 10^3 \text{ W}) = \boxed{5.3 \times 10^2 \text{ W.}}$$

35. For the left-moving traveling wave, with wave speed $v = \omega/k$, we have
    $$z(x, t) = z_0 \sin(kx + \omega t),$$
    $$\partial z/\partial x = z_0 k \cos(kx + \omega t), \text{ and}$$
    $$\partial z/\partial t = z_0 \omega \cos(kx + \omega t).$$
    The transverse component of the force acting on the portion of string to the left of x acts upward:
    $F = + T \,\partial z/\partial x$, so the power delivered to the string to the left is
    $$P = \mathbf{F} \cdot \mathbf{v}_{transverse} = + T (\partial z/\partial x)(\partial z/\partial t)$$
    $$= Tk\omega z_0^2 \cos^2(kx + \omega t) = \mu v^2 k \omega z_0^2 \cos^2(kx + \omega t) = \mu v \omega^2 z_0^2 \cos^2(kx + \omega t).$$
    Thus the power delivered is positive.

37. For a right-moving wave, we have
    $$z = f(x - vt),$$
    $$\frac{\partial z}{\partial x} = \frac{df}{d(x - vt)}\frac{\partial(x - vt)}{\partial x} = \frac{df}{d(x - vt)};$$
    $$\frac{\partial z}{\partial t} = \frac{df}{d(x - vt)}\frac{\partial(x - vt)}{\partial t} = \frac{df}{d(x - vt)}(-v).$$
    For the power delivered, we have
    $$P = \mathbf{F} \cdot \mathbf{v}_{transverse} = -T\frac{\partial z}{\partial x}\frac{\partial z}{\partial t} = -T(-v)\left[\frac{df}{d(x - vt)}\right]^2 = Tv\left[\frac{df}{d(x - vt)}\right]^2 = \mu v\left(\frac{\omega}{k}\right)^2\left[\frac{df}{d(x - vt)}\right]^2,$$
    where we have used $v = \omega/k$. The power delivered is positive.

43. We find the intensity of the sound 5 m from the source from
$$\beta = 10 \log_{10}(I/I_0);$$
$$83 \text{ dB} = 10 \log(I/10^{-12} \text{ W/m}^2), \text{ which gives } I = 2.0 \times 10^{-4} \text{ W/m}^2.$$
Because the sound has the same intensity in all directions, the power output is the total rate at which energy passes through a sphere:
$$P = IA = I(4\pi r^2) = (2.0 \times 10^{-4} \text{ W/m}^2)4\pi(5 \text{ m})^2 = \boxed{6.3 \times 10^{-2} \text{ W.}}$$

49. The speed of the train is $(140 \text{ km/h})/(3.6 \text{ ks/h}) = 38.9 \text{ m/s}.$
  (a) Because the wavelength in front of a moving source decreases, the wavelength traveling toward the observer is
$$\lambda = (v - v_T)/f_s = (330 \text{ m/s} - 39 \text{ m/s})/(333 \text{ Hz}) = 0.874 \text{ m}.$$
This wavelength approaches the observer at a speed of v. He hears a frequency
$$f_a = v/\lambda = (330 \text{ m/s})/(0.874 \text{ m}) = \boxed{378 \text{ Hz.}}$$
  (b) The wavelength traveling toward the observer is
$$\lambda_b = v/f_T = (330 \text{ m/s})/(333 \text{ Hz}) = 0.991 \text{ m}.$$
This wavelength approaches the observer at a relative speed of $v + v_O$. He hears a frequency
$$f_b = (v + v_O)/\lambda_b = (330 \text{ m/s} + 39 \text{ m/s})/(0.991 \text{ m}) = \boxed{372 \text{ Hz.}}$$

53. The speed of the ambulance is $(95 \text{ km/h})/(3.6 \text{ ks/h}) = 26 \text{ m/s}.$
Because the wavelength in front of a moving source decreases, the wavelength approaching the cliff is
$$\lambda_1 = (v - v_A)/f_s = (330 \text{ m/s} - 26 \text{ m/s})/(1600 \text{ Hz}) = 0.190 \text{ m}.$$
The frequency of the sound reaching the cliff is also the frequency of the sound that leaves the cliff. Reflection from the stationary cliff does not change the wavelength. This wavelength approaches the ambulance at a relative speed of $v + v_A$, so the frequency heard at the ambulance is
$$f_2 = (v + v_A)/\lambda_1 = (330 \text{ m/s} + 26 \text{ m/s})/(0.190 \text{ m}) = \boxed{1.88 \times 10^3 \text{ Hz.}}$$

55. We choose a coordinate system with origin at the top of the tower, down positive, and $t = 0$ when the tuning fork is dropped. If we call t_f the time of fall for the tuning fork, we have
$$v_f = v_{0f} + gt_f = 0 + 10t_f, \quad \text{and} \quad y_f = y_{0f} + v_{0f} + gt_f^2 = 0 + 0 + 5.0t_f^2.$$
Because the wavelength behind a moving source increases, the wavelength leaving the tuning fork at t_f is
$$\lambda = (v + v_f)/f_s = (v + 10t_f)/f_s.$$
When this wavelength, traveling at speed v, arrives at the top of the tower, the student will hear a frequency
$$f = v/\lambda = vf_s/(v + 10t_f).$$
Because the wave leaves the position at y_f at time t_f and travels up at a speed v, we find the time at which this frequency is heard (that is, arrives at $y = 0$) from
$$y = y_f - v(t - t_f);$$
$$0 = 5.0t_f^2 - 330(t - t_f).$$
This is a quadratic equation for t_f in terms of t, which has the positive solution
$$t_f = 33\left(\sqrt{1 + 0.060t} - 1\right).$$
We substitute this into the frequency result to get
$$f = \frac{(330 \text{ m/s})(440 \text{ Hz})}{330 \text{ m/s} + 330\left(\sqrt{1 + 0.060t} - 1\right)} = \boxed{\frac{440 \text{ Hz}}{\sqrt{1 + 0.060t}}.}$$

61. From the relation between level and intensity, $\beta = 10 \log_{10}(I/I_0)$, we have
$$\beta_2 - \beta_1 = 10 \log_{10}(I_2/I_0) - 10 \log_{10}(I_1/I_0) = 10 \log_{10}(I_2/I_1);$$
$$120 \text{ dB} - 55 \text{ dB} = 10 \log_{10}(I_2/I_1), \text{ which gives}$$
$$I_2/I_1 = 3.2 \times 10^6.$$
Because the power delivered is $P = I(\text{area})$ and the area is the same, this is also the ratio of power:
$$\boxed{P_2/P_1 = 3.2 \times 10^6.}$$

65. The tangential speed of the tuning fork is
    $$v_t = R\omega = (1.0 \text{ m})2\pi(3 \text{ rev/s}) = 9.4 \text{ m/s}.$$

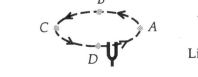

Listener

(a) At points A and C, the velocity of the fork
     is perpendicular to the direction of the
     sound travel, so the second person would
     hear a frequency of 512 Hz.
   At point B, the wavelength moving toward
     the receiver is increased:
    $$\begin{aligned}\lambda_B &= (v + v_t)/f \\ &= (330 \text{ m/s} + 9.4 \text{ m/s})/(512 \text{ Hz}) \\ &= 0.662 \text{ m}.\end{aligned}$$
   The frequency received is
    $$\begin{aligned}f_B &= v/\lambda_B \\ &= (330 \text{ m/s})/(0.662 \text{ m}) \\ &= \boxed{484 \text{ Hz, which is the minimum frequency.}}\end{aligned}$$
   At point D, the wavelength moving toward the receiver is decreased:
    $$\lambda_D = (v - v_t)/f_F = (330 \text{ m/s} - 9.4 \text{ m/s})/(512 \text{ Hz}) = 0.626 \text{ m}.$$
   The frequency received is
    $$f_D = v/\lambda_D = (330 \text{ m/s})/(0.626 \text{ m}) = \boxed{543 \text{ Hz, which is the maximum frequency.}}$$
(b) The highest frequency corresponds to the tuning fork moving toward the listener (point D).
   The lowest frequency corresponds to the tuning fork moving away from the listener (point B).

67. The enhancement of the sound intensity occurs when a standing wave is produced in the air column.
    Because the water vibration is much smaller than the air vibration, the top of the water column can be
    considered a fixed point for the standing wave in the air column. Adjacent points of enhanced intensity
    are separated by $\lambda/2$, so the wavelength is
    $$\lambda = 2(1.0 \text{ m} - 0.6 \text{ m}) = 0.8 \text{ m}.$$
    The speed of sound in the air is
    $$v = f\lambda = (440 \text{ Hz})(0.8 \text{ m}) = \boxed{3.5 \times 10^2 \text{ m/s}.}$$

69. For the standing modes of the wire, there are fixed points at $x = 0$ and $x = L$.
    The lowest mode has the longest wavelength, $\lambda_1 = 2L$. This corresponds to the wave number
    $$k_1 = 2\pi/\lambda_1 = \pi/L, \text{ so } \omega_1^2 = (T/\mu)k_1^2 + \alpha^2 k_1^4 = (T/\mu)(\pi/L)^2 + \alpha^2(\pi/L)^4.$$
    The frequency is
    $$\boxed{f_1 = \frac{\omega_1}{2\pi} = \frac{1}{2L}\sqrt{\frac{T}{\mu} + \alpha\frac{\pi^2}{L^2}}.}$$
    The second mode has the wavelength, $\lambda_2 = L$. This corresponds to the wave number
    $$k_2 = 2\pi/\lambda_2 = 2\pi/L, \text{ so } \omega_2^2 = (T/\mu)k_2^2 + \alpha^2 k_2^4 = (T/\mu)(2\pi/L)^2 + \alpha^2(2\pi/L)^4.$$
    The frequency is
    $$\boxed{f_2 = \frac{\omega_2}{2\pi} = \frac{1}{L}\sqrt{\frac{T}{\mu} + 4\alpha\frac{\pi^2}{L^2}}.}$$
    We see that the second term inside the square root, which is a measure of the stiffness, is greater, relative
    to the first term, for the second mode. The stiffness is a more important effect for $\boxed{\text{the higher frequency.}}$

CHAPTER 15

5. Because the waves travel in the same medium and have the same frequency, they will also have the same speed and wavelength. We can write the combined wave, with $A = 1$ cm, as

$$\psi(x, t) = 3A \sin(kx - \omega t) + 2A \sin[kx - \omega t - (\pi/2)] = 3A \sin(kx - \omega t) + 2A \cos(kx - \omega t).$$

We want to express this as

$$\psi(x, t) = B \sin(kx - \omega t + \alpha) = B \sin(kx - \omega t) \cos \alpha + B \cos(kx - \omega t) \sin \alpha.$$

Comparing coefficients of the trigonometric functions, we have

$3A = B \cos \alpha$ and $2A = B \sin \alpha$, which gives

$\tan \alpha = \frac{2}{3}$, or $\alpha = 33.7°$;   and $B = 3A/\cos \alpha = 3(1 \text{ cm})/\cos 33.7° = \boxed{3.6 \text{ cm.}}$

7. We want

$$A \sin(kx - \omega t) + A \sin(kx + \omega t + \delta) = A \sin(kx + \omega t + \delta').$$

We use the trigonometric identity for the sum of two sine functions to get

$$2A \sin(kx + \omega t + \tfrac{1}{2}\delta) \cos(\tfrac{1}{2}\delta) = A \sin(kx + \omega t + \delta').$$

Comparing the two sides, we see that

$\delta' = \tfrac{1}{2}\delta$, and $2A \cos(\tfrac{1}{2}\delta) = A$, or $\cos(\tfrac{1}{2}\delta) = \tfrac{1}{2}$, which gives $\boxed{\delta = 2\pi/3 \text{ and } -2\pi/3.}$

Physically, these are the same solution, with the roles of the two waves reversed.

13. We write the wave as

$$\psi(x, t) = A \sin(kx - \omega t) + A \sin(kx + \omega t + \theta).$$

Travel in opposite directions produces the difference in the sign of ωt.

With the definitions $u = kx + (\theta/2)$, and $v = \omega t + (\theta/2)$, we have

$\psi(x, t)$ $= A \sin(u - v) + A \sin(u + v)$, which expands to

$\psi(x, t)$ $= A(\sin u \cos v - \cos u \sin v + \sin u \cos v + \cos u \sin v) = A(2 \sin u \cos v)$

$= 2A \sin[kx + (\theta/2)] \cos[\omega t + (\theta/2)]$, which represents a standing wave.

15. For a standing wave, the distance between nodes must be a multiple of $\lambda/2$. The largest possible wavelength is $\lambda/2 = 6$ m $- 3$ m, which gives $\boxed{\lambda = 6 \text{ m.}}$

There is a node at $x = 0$. Because the first wave has a maximum here at $t = 0$, the second wave must have a minimum here at $t = 0$.

21. Because the 4 beats/s are heard, this must be the pulse frequency. With the small fractional change of the frequency, we approximate the changes by differentials:

$f = v/\lambda = (1/\lambda)(T/\mu)^{1/2}$;

$df = (1/\lambda)(\tfrac{1}{2})(1/T\mu)^{1/2}\, dT.$

Dividing the two equations, we get

$df/f = \tfrac{1}{2}(dT/T);$

$(-4 \text{ Hz})/(512 \text{ Hz}) = \tfrac{1}{2}dT/T$, which gives $dT/T = -0.016 = \boxed{-1.6\%.}$

27. We know that the path-length difference from two sources separated by d to the point an angle θ from the centerline is

    $\Delta L = d \sin \theta$.

    (a) For the first non-central maximum, this path-length difference must be λ:

    $\sin \theta = \Delta L / d = \lambda / d = (3 \text{ cm})/(5 \text{ cm}) = 0.6$, which gives $\boxed{\theta_a = 37°.}$

    (b) If the sources have opposite phases, this is equivalent to an additional path difference of $\frac{1}{2}\lambda$, so the path-length difference for the first non-central maximum must be $\frac{1}{2}\lambda$:

    $\sin \theta = \Delta L / d = \frac{1}{2}\lambda / d = \frac{1}{2}(3 \text{ cm})/(5 \text{ cm}) = 0.3$, which gives $\boxed{\theta_b = 17.5°.}$

    (c) With the sources in phase, we have

    $\sin \theta = \Delta L / d = \lambda / d = (6 \text{ cm})/(5 \text{ cm}) > 1$, which means $\boxed{\text{there is no non-central maximum.}}$

    With the sources having opposite phases, we have

    $\sin \theta = \Delta L / d = \frac{1}{2}\lambda / d = \frac{1}{2}(6 \text{ cm})/(5 \text{ cm}) = 0.6$, which gives $\boxed{\theta_c = 37°.}$

31. With $\psi_1 = (3 \text{ cm}) \sin(\pi r_1 - \pi v t)$ and $\psi_2 = (3 \text{ cm}) \sin(\pi r_2 - \pi v t)$, we see that the two waves are coherent and have the same wavelength, frequency and speed. If r is in cm and t is in s, the wavelength is

    $\lambda = 2\pi/\pi = 2 \text{ cm}$.

    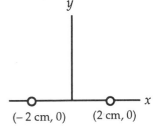

    (a) From the diagram, we see that on the y-axis, the path-length difference will be zero. Because the two sources have the same amplitude, all points on the y-axis will have a maximum. On the x-axis between the sources, the maxima are located where the path-length difference is a multiple of the wavelength. From the diagram, the path-length difference is

    $\Delta L = (x + 2 \text{ cm}) - (2 \text{ cm} - x) = n\lambda = n(2 \text{ cm})$, where $n = 0, \pm 1, \pm 2, \ldots$.

    Between the sources we have $\boxed{x = -1 \text{ cm}, 0, +1 \text{ cm}.}$

    (b) Because the two waves are in phase along $x = 0$ and the distance from either source is $r = (y^2 + 2^2)^{1/2}$, the sum of the two waves is

    $\psi = \psi_1 + \psi_2 = (3 \text{ cm}) \sin(\pi r - \pi v t) + (3 \text{ cm}) \sin(\pi r - \pi v t) = \boxed{(6 \text{ cm}) \sin(\pi r - \pi v t), r = (y^2 + 4)^{1/2}.}$

    (This ignores the decrease in amplitude as the wave spreads.)

33. (a) Where the constructive interference is maximum, the amplitude is twice the amplitude of one wave. Because the intensity depends on the square of the amplitude, the intensity is four times the intensity of one wave.

    (b) Where an interference minimum occurs, the amplitude is zero; thus the intensity is zero.

    (c) If there is no interference, the rate at which energy is received is the sum of that from each wave; thus the intensity is twice the intensity from one wave.

    (d) For parts (a) and (b), there are regions where the intensity is $4I$ of one source but other regions where $I = 0$. Average intensity over a circle will be $2I$ of one source. For part (c), a doubling of intensity I means $2I$ of one source. $\boxed{\text{Total energy is conserved.}}$

35. Because each part of the pulse moves at the same speed, the shape does not change. For motion in the $+x$-direction, we replace x by $x - vt$:

    $z(x, 0) = 0$ for $x < -a$ and $x > 0$,
    $z(x, 0) = k(x + a)$ for $-a < x < 0$.

    $\boxed{\begin{array}{l} z(x, t) = 0 \text{ for } x < (vt - a) \text{ and } x > vt, \\ z(x, t) = k(x - vt + a) \text{ for } (vt - a) < x < vt. \end{array}}$

    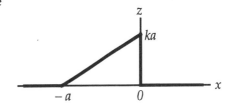

37. We are given the wave

$z(x, t) = z_0 e^{-(x - vt)^2/\alpha^2}$, so we have

$$v_t = \frac{\partial z}{\partial t} = z_0 \left[-2(x - vt) \left(\frac{-v}{\alpha^2} \right) \right] e^{-(x - vt)^2/\alpha^2}$$

$$= z_0 \left[2(x - vt) \left(\frac{v}{\alpha^2} \right) \right] e^{-(x - vt)^2/\alpha^2};$$

$$a_t = \frac{\partial^2 z}{\partial t^2} = z_0 \left[-\frac{2v^2}{\alpha^2} + 4 (x - vt)^2 \left(\frac{v^2}{\alpha^4} \right) \right] e^{-(x - vt)^2/\alpha^2}.$$

45. The tension and frequency will not change across the boundary. The wave number will change:

$$k = \frac{\omega}{v} = \omega \sqrt{\frac{\mu}{T}}, \text{ so we have } k_i = k_r = k_1 = \frac{\omega}{\sqrt{T}} \sqrt{\mu_1}, \; k_t = k_2 = \frac{\omega}{\sqrt{T}} \sqrt{\mu_2}.$$

The equations for the waves are

$z_i = A \cos(k_1 x - \omega t);$
$z_r = B \cos(k_1 x + \omega t + \delta_r);$
$z_t = C \cos(k_2 x - \omega t + \delta_t).$

The slopes of the waves are

$\partial z_i / \partial x = -Ak_1 \sin(k_1 x - \omega t);$
$\partial z_r / \partial x = -Bk_1 \sin(k_1 x + \omega t + \delta_r);$
$\partial z_t / \partial x = -Bk_2 \sin(k_2 x - \omega t + \delta_t).$

The string is unbroken at $x = 0$, so we have

$A \cos(-\omega t) + B \cos(\omega t + \delta_r) = C \cos(-\omega t + \delta_t) = C \cos(\omega t - \delta_t).$

Expanding the second and third terms, we get

$A \cos(\omega t) + B[\cos(\omega t) \cos \delta_r - \sin(\omega t) \sin \delta_r] = C[\cos(\omega t) \cos \delta_t + \sin(\omega t) \sin \delta_t].$

For this to be satisfied at any time, the coefficients of $\cos(\omega t)$ and $\sin(\omega t)$ must separately be equal:

$A + B \cos \delta_r = C \cos \delta_t;$ (1)
$-B \sin \delta_r = C \sin \delta_t.$ (2)

For there to be no kinks at $x = 0$, we have

$Ak_1 \sin(-\omega t) + Bk_1 \sin(\omega t + \delta_r) = Ck_2 \sin(-\omega t + \delta_t) = -Ck_2 \sin(\omega t - \delta_t)$, which expands to
$-Ak_1 \sin(\omega t) + Bk_1[\sin(\omega t) \cos \delta_r + \cos(\omega t) \sin \delta_r] = -Ck_2[\sin(\omega t) \cos \delta_t - \cos(\omega t) \sin \delta_t].$

For this to be satisfied at any time, the coefficients of $\cos(\omega t)$ and $\sin(\omega t)$ must separately be equal:

$(-A + B \cos \delta_r)k_1 = -Ck_2 \cos \delta_t;$ (3)
$Bk_1 \sin \delta_r = Ck_2 \sin \delta_t.$ (4)

These four equations can be written to make evident the four unknowns, $B/A, C/A, \delta_r, \delta_t$:

$1 + (B/A) \cos \delta_r = (C/A) \cos \delta_t;$ (1′)
$-(B/A) \sin \delta_r = (C/A) \sin \delta_t;$ (2′)
$1 - (B/A) \cos \delta_r = (C/A)(k_2/k_1) \cos \delta_t;$ (3′)
$(B/A) \sin \delta_r = (C/A)(k_2/k_1) \sin \delta_t.$ (4′)

We assume that $k_2/k_1 > 1$. By adding Eq. (2′) and Eq. (4′), we see that $\sin \delta_t = 0$.
By adding Eq. (1′) and Eq. (3′), we see that $\cos \delta_t > 0$, which gives $\delta_t = 0$.
Then Eq. (2′) gives $\sin \delta_r = 0$, and Eq. (3′) gives $\cos \delta_r < 0$, which gives $\delta_r = \pi$.
When we substitute these results in the equations, we can solve for the amplitude ratios to get

$$\frac{B}{A} = \frac{k_2 - k_1}{k_2 + k_1} = \frac{\sqrt{\mu_2} - \sqrt{\mu_1}}{\sqrt{\mu_2} + \sqrt{\mu_1}};$$
$$\frac{C}{A} = \frac{2k_2}{k_2 + k_1} = \frac{2\sqrt{\mu_1}}{\sqrt{\mu_2} + \sqrt{\mu_1}}.$$

Then the wave equations are

$z_i = A \cos(k_1 x - \omega t);$
$z_r = B \cos(k_1 x + \omega t + \pi);$
$z_t = C \cos(k_2 x - \omega t).$

If we had assumed $k_2/k_1 < 1$, we would find $\delta_r = 0$ and get the same results for the amplitude ratios, with k_1 and k_2 interchanged.

53. The angular frequency of the turntable is $\omega = (120 \text{ rev}/\text{min})(2\pi \text{ rad}/\text{rev})(1 \text{ min}/60 \text{ s}) = 4\pi \text{ rad}/\text{s}$.
    The tangential speed of the sources is
    $$v_t = R\omega = (0.50 \text{ m})(4\pi \text{ rad}/\text{s}) = 2\pi \text{ m}/\text{s}.$$
    If we take $t = 0$ when a source has a tangential velocity directly toward the detector, the wavelength from the approaching source is
    $$\lambda_1 = [v - v_t \cos(\omega t)]/f_s \text{ and the frequency detected will be}$$
    $$f_1 = v/\lambda = vf_s/[v - v_t \cos(\omega t)].$$
    The frequency from the second source on the opposite side of the turntable can be obtained by adding π rad to the angle:
    $$f_2 = vf_s/[v - v_t \cos(\omega t + \pi)] = vf_s/[v + v_t \cos(\omega t)].$$
    The pulse frequency is
    $$f_{\text{pulse}} = f_1 - f_2 = \{vf_s/[v - v_t \cos(\omega t)]\} - \{vf_s/[v + v_t \cos(\omega t)]\}$$
    $$= f_s(\{1/[1 - (v_t/v) \cos(\omega t)]\} - \{1/[1 + (v_t/v) \cos(\omega t)]\}).$$
    Because $v_t \ll v$, we can use the approximation $1/(1 - x) \approx 1 + x$:
    $$f_{\text{pulse}} \approx f_s[1 + (v_t/v) \cos(\omega t) - 1 + (v_t/v) \cos(\omega t)]$$
    $$= 2f_s(v_t/v) \cos(\omega t) = 2(1800 \text{ Hz})[(2\pi \text{ m}/\text{s})/(330 \text{ m}/\text{s})] \cos(4\pi t) = \boxed{69 \cos(4\pi t) \text{ Hz.}}$$

55. The direct wave travels $L_1 = 10$ m, while the reflected wave travels a distance
    $$L_2 = 2[(20 \text{ m})^2 + (5 \text{ m})^2]^{1/2} = 41.2 \text{ m}.$$
    The path-length difference is $\Delta L = L_2 - L_1 = 31.2$ m.
    The reflection from the wall causes a π rad phase shift that is equivalent to $\lambda/2$. Consequently, for a maximum interference, the path-length difference must be
    $$\Delta L = \lambda/2, 3\lambda/2, 5\lambda/2, \ldots, \text{ which gives } \lambda_n = \Delta L/(n + \tfrac{1}{2}), \text{ where } n = 0, 1, 2, \ldots.$$
    The frequencies are
    $$f_n = v/\lambda_n = (n + \tfrac{1}{2})v/\Delta L = (n + \tfrac{1}{2})(330 \text{ m}/\text{s})/(31.2 \text{ m}) = (n + \tfrac{1}{2})10.6 \text{ Hz}, \text{ where } n = 0, 1, 2, \ldots.$$
    The three frequencies above 80 Hz are given by $n = 8, 9,$ and 10:
    $$f_8 = \boxed{90 \text{ Hz,}} \quad f_9 = \boxed{101 \text{ Hz,}} \quad f_{10} = \boxed{111 \text{ Hz.}}$$

61. For the sum of the two waves, we have
    $$z = z_1 + z_2$$
    $$= z_0 \sin(k_1 x - \omega_1 t) + z_0 \cos(k_2 x - \omega_2 t) = z_0 \{\sin(k_1 x - \omega_1 t) - \sin[k_2 x - \omega_2 t - (\pi/2)]\}.$$
    We use a trigonometric identity to add the two sine functions:
    $$z = 2z_0 \sin[\tfrac{1}{2}(k_1 - k_2)x - \tfrac{1}{2}(\omega_1 - \omega_2)t + (\pi/4)] \cos[\tfrac{1}{2}(k_1 + k_2)x - \tfrac{1}{2}(\omega_1 + \omega_2)t - (\pi/4)].$$
    If $\Delta f = f_1 - f_2$ is small, $\Delta\omega = \omega_1 - \omega_2$ is small, and we can say $\tfrac{1}{2}(\omega_1 + \omega_2) = \omega_1$.
    Because $k = \omega/v$ and v is the same in a nondispersive medium,
    $$\Delta k = k_1 - k_2 \text{ is small, and we can say } \tfrac{1}{2}(k_1 + k_2) = k_1.$$
    With these approximations, we have
    $$z \approx 2z_0 \{\sin[\tfrac{1}{2}\Delta k\, x - \tfrac{1}{2}\Delta\omega\, t + (\pi/4)] \cos[k_1 x - \omega_1 t - (\pi/4)]\}.$$
    The cosine function is the average traveling wave. The sine function is the envelope, with a long wavelength and low frequency, so we have
    $$\boxed{z(\text{envelope}) \approx \sin[\Delta k(x/2) - \Delta\omega(t/2) + (\pi/4)].}$$
    For the frequency of the envelope, we are interested in the absolute value of z, so the frequency is
    $$\boxed{f = 2(\Delta\omega/2)/2\pi = (\omega_1 - \omega_2)/2\pi.}$$
    The wavelength is $\boxed{\lambda = \tfrac{1}{2}2\pi/(\tfrac{1}{2}\Delta k) = 2\pi/(k_1 - k_2).}$

63. As suggested in the problem statement, we take the phase of each wave to be zero, $\delta = 0$.
    (a) The direction of motion determines the relative sign of the kx and ωt terms:

    $$y_1(x, t) = y_0 \sin(k_1 x + \omega_1 t);$$
    $$y_2(x, t) = y_0 \sin(k_2 x - \omega_2 t).$$

    (b) In a nondispersive medium, we have

    $$\omega_2 = v k_2 = k_2 \omega_1 / k_1.$$

    (c) We have $\omega_2 = \omega_1 + \delta\omega = \omega_1 k_2 / k_1 = \omega_1(k_1 + \delta k)/k_1 = \omega_1 + (\omega_1/k_1)\delta k$, which gives

    $$\delta\omega = (\omega_1/k_1)\delta k = v\,\delta k.$$

    (d) For the sum of the two waves, we have

    $$y = y_1 + y_2$$
    $$= y_0[\sin(k_1 x + \omega_1 t) + \sin(k_2 x - \omega_2 t)].$$

    We use a trigonometric identity to add the two sine functions:

    $$y = 2y_0 \sin[\tfrac{1}{2}(k_1 + k_2)x + \tfrac{1}{2}(\omega_1 - \omega_2)t] \cos[\tfrac{1}{2}(k_1 - k_2)x + \tfrac{1}{2}(\omega_1 + \omega_2)t].$$
    $$= 2y_0 \sin(k_{av} x - \tfrac{1}{2}\delta\omega\, t) \cos(-\tfrac{1}{2}\delta k\, x + \omega_{av} t).$$

    With $\delta k \to 0$ and $\delta\omega \to 0$, we have $k_{av} = k_1$ and $\omega_{av} = \omega_1$. The wave is

    $$y = 2y_0 \sin(k_1 x) \cos(\omega_1 t),$$ which is a standing wave with $\lambda = 2\pi/k_1$ and $\omega = \omega_1$.

    (e) From the expansion in part (d), we have

    $$y = 2y_0 \sin(k_{av} x - \tfrac{1}{2}\delta\omega\, t) \cos(-\tfrac{1}{2}\delta k\, x + \omega_{av} t).$$

    The first wave has

    $$v_\ell = \tfrac{1}{2}\delta\omega/k_{av} = v\,\delta k/2k_{av} \quad \text{and} \quad \lambda_\ell = 2\pi/k_{av}.$$

    The second wave has

    $$v_r = \omega_{av}/\tfrac{1}{2}\delta k = 2vk_{av}/\delta k \quad \text{and} \quad \lambda_r = 2\pi/\tfrac{1}{2}\delta k = 4\pi/\delta k.$$

CHAPTER 16

11. The normal force on the pyramid must equal the weight of the pyramid. The reaction to this force is the force exerted on the table by the pyramid, so for the pressure we have

$$p = N/A = mg/A$$
$$= (1.8 \text{ kg})(9.8 \text{ m/s}^2)/(15 \times 10^{-2} \text{ m})^2 = \boxed{7.8 \times 10^2 \text{ N/m}^2.}$$

Expansion will increase A, but mg will not change; thus $\boxed{p \text{ will decrease.}}$

19. The force in the large piston must equal the weight of the car. The pressures in the two pistons will be the same, so we have

$$p = F/A_1 = mg/A_2;$$
$$F/\tfrac{1}{4}\pi D_1^2 = mg/\tfrac{1}{4}\pi D_2^2, \quad \text{or} \quad F = mg(D_1/D_2)^2 = (1200 \text{ kg})(9.8 \text{ m/s}^2)[(30 \text{ cm})/(2 \text{ cm})]^2 = \boxed{52 \text{ N.}}$$

With no frictional losses, the work done by **F** must increase the potential energy of the car:

$$Fh_1 = mgh_2;$$
$$(52 \text{ N})(0.5 \text{ m}) = (1200 \text{ kg})(9.8 \text{ m/s}^2)h_2 \text{ , which gives } h_2 = 2.2 \times 10^{-3} \text{ m} = \boxed{2.2 \text{ mm.}}$$

Note that, for an incompressible fluid, this can also be obtained by equating the volume changes. The decrease in fluid volume in the smaller piston must be the increase in the fluid volume in the larger piston.

23. Two forces act on the box: the buoyant force, which depends on the submerged volume, up and the force of gravity down. When the box floats, the net force is zero. If the fraction of the box that is above the surface is f, we have

$$F_{\text{net}} = 0 = \rho_{\text{water}}g(1-f)V - \rho_{\text{box}}gV;$$
$$(1000 \text{ kg/m}^3)g(1-0.16)V = \rho_{\text{box}}gV, \text{ which gives } \rho_{\text{box}} = \boxed{8.4 \times 10^2 \text{ kg/m}^3.}$$

25. We take the raft to be made of N logs, each of length L and radius r. We assume that for the minimum area, the logs are completely in the water and the boys and equipment are not. Because the net force is zero, we have

$$F_{\text{buoy}} = m_{\text{total}}g;$$
$$\rho_{\text{water}}g(\pi r^2)NL = (m_{\text{boys}} + m_{\text{equipment}})g + \rho_{\text{tree}}g(\pi r^2)NL.$$

Because the specific gravity is the ratio of the density to the density of water, $\rho_{\text{tree}} = 0.8\rho_{\text{water}}$, so we have

$$(1-0.8)\rho_{\text{water}}\pi r^2 NL = m_{\text{boys}} + m_{\text{equipment}};$$
$$(1-0.8)(1.00 \times 10^3 \text{ kg/m}^3)\pi(0.10 \text{ m})^2 NL = 400 \text{ kg, which gives } NL = 63.6 \text{ m.}$$

The minimum area of the raft is $L(Nd) = NLd = (63.6 \text{ m})(0.20 \text{ m}) = \boxed{12.7 \text{ m}^2.}$

27. The weight of the object is the tension in the spring. Because the net force is zero, we have

$$T + F_{\text{buoy}} = mg.$$

Because the density of air is much less than the density of alcohol and the density of the object (which must be greater than the density of alcohol, otherwise the object would float), we ignore the buoyant force of the air to get

$$T_{\text{air}} = mg, \quad \text{and} \quad T_{\text{alcohol}} + F_{\text{buoyant,alcohol}} = mg.$$

If we subtract these equations, we get

$$T_{\text{air}} - T_{\text{alcohol}} = F_{\text{buoyant,alcohol}} = \rho_{\text{alcohol}}gV, \quad \text{or} \quad V = (T_{\text{air}} - T_{\text{alcohol}})/\rho_{\text{alcohol}}g.$$

The density of the object is

$$\rho = m/V = mg/Vg = T_{\text{air}}/Vg = T_{\text{air}}\rho_{\text{alcohol}}/(T_{\text{air}} - T_{\text{alcohol}})$$
$$= (0.45 \text{ N})(0.79 \times 10^3 \text{ kg/m}^3)/(0.45 \text{ N} - 0.089 \text{ N}) = \boxed{0.96 \times 10^3 \text{ kg/m}^3.}$$

29. When the cube floats, the net force is zero. Each buoyant force depends on the volume in the fluid, so we have

$F_{buoy1} + F_{buoy2} = mg$;

$[\rho_1(H - D) + \rho_2 D]gA = \rho_3 gHA$, which reduces to

$\rho_3 = \rho_1 + (\rho_2 - \rho_1)(D/H)$.

Because $(\rho_2 - \rho_1) > 0$ and $D/H > 0$, we see that $\rho_3 > \rho_1$.

If we rewrite the result as

$\rho_3 = \{\rho_2 - [1 + (H/D)\rho_1]\}(D/H)$,

we see that $\rho_3 < \rho_2$, so the condition is

$\boxed{\rho_1 < \rho_3 < \rho_2.}$

The fraction of the cube in the water is

$\boxed{D/H = (\rho_3 - \rho_1)/(\rho_2 - \rho_1).}$

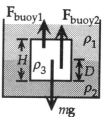

31. For the minimum number of balloons N, the net force will be zero. If we ignore the buoyant force on the child, we have

$F_{buoy} = m_{child}g + Nm_{balloon}g + Nm_{He}g$;

$N\rho_{air}g\frac{4}{3}\pi R^3 = m_{child}g + Nm_{balloon}g + N\rho_{He}g\frac{4}{3}\pi R^3$, which can be rearranged to

$N[\frac{4}{3}\pi R^3(\rho_{air} - \rho_{He}) - m_{balloon}] = m_{child}$;

$N\{\frac{4}{3}\pi(0.21 \text{ m})^3[(1.29 \text{ kg/m}^3) - (0.18 \text{ kg/m}^3)] - 0.0035 \text{ kg}\} = 32 \text{ kg}$, which gives $\boxed{N = 8.1 \times 10^2.}$

37. If we ignore air resistance, the water has projectile motion. The maximum range is achieved with an initial angle of 45° and is given by

$R = 2v_0^2/g$, or $v_0 = (gR/2)^{1/2}$.

From the mass conservation equation of continuity at the nozzle, with constant density, we have

$v_1 A_1 = v_2 A_2$;

$(gR_1/2)^{1/2}\frac{1}{4}\pi d_1^2 = (gR_2/2)^{1/2}\frac{1}{4}\pi d_2^2$.

After canceling the common factors, we have

$(1.5 \text{ m})^{1/2}(1.2 \text{ cm})^2 = (18 \text{ m})^{1/2}d_2^2$, which gives $d_2 = \boxed{0.81 \text{ cm.}}$

45. (a) We ignore the change in water depth, which varies from 15 cm to 0 cm as the basement empties. If we use Bernoulli's equation between the basement and the end of the hose, we have

$p_1 + \frac{1}{2}\rho v_1^2 + \rho g h_1 = p_2 + \frac{1}{2}\rho v_2^2 + \rho g h_2$;

$p_{atm} + 0 + (1.0 \times 10^3 \text{ kg/m}^3)(9.8 \text{ m/s}^2)(3 \text{ m}) = p_{atm} + \frac{1}{2}(1.0 \times 10^3 \text{ kg/m}^3)v_2^2 + 0$, which gives

$\boxed{v_2 = 7.7 \text{ m/s.}}$

(b) From the equation of continuity (mass conservation), we have

$dV/dt = v_2 A_2$, so the time to empty the basement is

$t = V/v_2 A_2$

$= (7.5 \text{ m})(12 \text{ m})(15 \times 10^{-2} \text{ m})/(7.7 \text{ m/s})\pi(0.6 \times 10^{-2} \text{ m})^2 = 1.6 \times 10^4 \text{ s} = \boxed{4.3 \text{ h.}}$

47. (a) Because all of the pipelets are equivalent, the speeds must be the same in each. From the equation of continuity, we have

$v_0 A_0 = v_1 A_1 + v_2 A_2 + v_3 A_3 + v_4 A_4 = 4v_1 A_1$;

$(0.3 \text{ m/s})\frac{1}{4}\pi(10 \text{ cm})^2 = 4v_1\frac{1}{4}\pi(2 \text{ cm})^2$, which gives

$\boxed{v_1 = v_2 = v_3 = v_4 = 1.9 \text{ m/s.}}$

(b) If we use Bernoulli's equation between the pipe and a pipelet, we have

$p_1 + \frac{1}{2}\rho v_1^2 + \rho g h_1 = p_2 + \frac{1}{2}\rho v_2^2 + \rho g h_2$;

$(2.5 \text{ atm})(1.01 \times 10^5 \text{ Pa}) + \frac{1}{2}(1.0 \times 10^3 \text{ kg/m}^3)(0.3 \text{ m/s})^2 + 0 =$

$p_2 + \frac{1}{2}(1.0 \times 10^3 \text{ kg/m}^3)(1.9 \text{ m/s})^2 + (1.0 \times 10^3 \text{ kg/m}^3)(9.8 \text{ m/s}^2)(3.5 \text{ m})$, which gives

$\boxed{p_2 = 2.16 \times 10^5 \text{ Pa} \ (2.14 \text{ atm}).}$

53. The volume of water is constant. If we call L the length of the wall,
    D the width of the original state when the depth is H_0 , and the wall
    moves a distance x with the final depth H, we have
    $$H_0 LD = HL(D + x), \text{ or } H = H_0 D/(D + x).$$
    At the maximum compression, the wall will momentarily come to rest; the
    initial and final kinetic energies are zero. The change in gravitational
    potential energy of the water, determined by the change in height of the
    center of mass, will increase the elastic potential energy of the spring:
    $$\tfrac{1}{2}kx^2 = mg\,\Delta h = mg\tfrac{1}{2}(H_0 - H) = \tfrac{1}{2}mgH_0\{1 - [D/(D + x)]\} = \tfrac{1}{2}mgH_0[x/(D + x)];$$
    $$\tfrac{1}{2}(180 \text{ N/m})x^2 = \tfrac{1}{2}(2 \text{ kg})(9.8 \text{ m/s}^2)(0.10 \text{ m})[x/(0.15 \text{ m} + x)].$$
    This is a cubic equation for x. A numerical solution gives the real positive
    value: $x = 0.054 \text{ m} = \boxed{5.4 \text{ cm.}}$
    Note that, with no energy dissipated, the wall will oscillate.

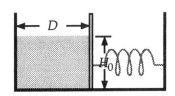

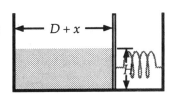

57. (a) If we use Bernoulli's equation for the time after the water leaves the fountain until it reaches its
        highest elevation, we have
        $$p_1 + \tfrac{1}{2}\rho v_1{}^2 + \rho g h_1 = p_2 + \tfrac{1}{2}\rho v_2{}^2 + \rho g h_2;$$
        $$p_{\text{atm}} + \tfrac{1}{2}\rho v_1{}^2 + 0 = p_{\text{atm}} + 0 + \rho g h_2, \text{ which gives } \underline{\text{a familiar result:}}$$
        $$v_1{}^2 = 2gh_2 = 2(9.8 \text{ m/s}^2)(37 \text{ m}), \text{ which gives } \boxed{v_1 = 27 \text{ m/s.}}$$
    (b) If we estimate the speed from a 1-cm diameter hose to be 3 m/s, we have
        $$vA \approx (3 \text{ m/s})\pi(0.5 \times 10^{-2} \text{ m})^2 \approx 2 \times 10^{-4} \text{ m}^3/\text{s}.$$
        The rate of flow of the fountain is 0.051 m³/s, which is $\boxed{\approx 200(\text{the rate of flow of the hose}).}$
    (c) The rate of flow in the pipe must equal the rate coming out of the orifice:
        $$dV/dt = vA;$$
        $$0.051 \text{ m}^3/\text{s} = (27 \text{ m/s})A, \text{ which gives } A = \boxed{1.9 \times 10^{-3} \text{ m}^2.}$$
    (d) If we use Bernoulli's equation across the nozzle, we have
        $$p_1 + \tfrac{1}{2}\rho v_1{}^2 + \rho g h_1 = p_2 + \tfrac{1}{2}\rho v_2{}^2 + \rho g h_2;$$
        $$p + 0 + 0 = p_{\text{atm}} + \tfrac{1}{2}(1.0 \times 10^3 \text{ kg/m}^3)(27 \text{ m/s})^2 + 0, \text{ which gives}$$
        $$p - p_{\text{atm}} = \boxed{3.65 \times 10^5 \text{ Pa (gauge).}}$$

59. Consider a cycle in which a block starts at the bottom, is inserted, rises, and then falls down to the bottom
    outside the tank. When the block is inserted, the surface of the water must rise. If the block has volume
    V and mass M, a volume V of water has been lifted to the top, so we have
    $$W_{\text{insertion}} = \rho_w Vgh.$$
    This represents work that must be done on the system. As the block rises the height h, the work done by
    the buoyant force and the force of gravity is
    $$W = F_{\text{buoy}}h - Mgh = \rho_w Vgh - Mgh.$$
    When the block falls back to the bottom, the work done by gravity is
    $$W_g = + Mgh.$$
    Thus the net work extracted from the system is
    $$W + Mgh - W_{\text{insertion}} = 0.$$
    Dissipation of energy from viscous effects during the rise will lead to a negative work extracted from the
    system or a net loss of energy. Thus we see that the work of insertion is greater than the energy acquired
    from the buoyant force by at least the potential energy gain during the rise.

61. We choose down as positive.
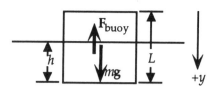
   (a) In the equilibrium position, the net force is zero, so we have
   $F_{buoy} = mg;$
   $\rho_{water}gL^2h = \rho gL^3$, which gives
   $\rho = \rho_{water}(h/L) = (1.0 \times 10^3 \text{ kg/m}^3)(12 \text{ m})/(20 \text{ m})$
   $= \boxed{6.0 \times 10^2 \text{ kg/m}^3.}$
   (b) When the block is pushed down a distance Δ, the net force is
   $F_{net} = -\rho_{water}gL^2(h + \Delta) + \rho gL^3.$
   When we use the equilibrium condition, we have
   $F_{net} = -\rho_{water}gL^2\Delta = -(1.0 \times 10^3 \text{ kg/m}^3)(9.8 \text{ m/s}^2)(0.20 \text{ m})^2 \Delta = \boxed{-0.39 \times 10^3 \Delta \quad \text{N} \quad \text{(up)}.}$
   (c) When the block is a distance Δ above the equilibrium position, we have
   $F_{net} = -\rho_{water}gL^2(h - \Delta) + \rho gL^3$
   $= +\rho_{water}gL^2\Delta = \boxed{0.39 \times 10^3 \Delta \quad \text{N} \quad \text{(down)}.}$
   (d) From parts (b) and (c), we see that the effective force constant is $\rho_{water}gL^2$. The frequency is
   $f = (k/4\pi^2 m)^{1/2} = (\rho_{water}gL^2/4\pi^2\rho L^3)^{1/2} = (\rho_{water}g/4\pi^2\rho L)^{1/2}$
   $= [(1.0 \times 10^3 \text{ kg/m}^3)(9.8 \text{ m/s}^2)/4\pi^2(6.0 \times 10^2 \text{ kg/m}^3)(0.20 \text{ m})]^{1/2} = \boxed{1.44 \text{ Hz.}}$

63. (a) We find the static pressure at the bottom of the tank from
   $p_1 = p_0 + \rho gh_1$
   $= (1.01 \times 10^5 \text{ Pa}) + (1.0 \times 10^3 \text{ kg/m}^3)(9.8 \text{ m/s}^2)(8 \text{ m}) = \boxed{1.8 \times 10^5 \text{ Pa.}}$
   (b) We find the static pressure in the horizontal pipe from
   $p_2 = p_0 + \rho gh_2$
   $= (1.01 \times 10^5 \text{ Pa}) + (1.0 \times 10^3 \text{ kg/m}^3)(9.8 \text{ m/s}^2)(39 \text{ m}) = \boxed{4.8 \times 10^5 \text{ Pa.}}$
   (c) From the equation of continuity, we have
   $dV/dt = v_0A_0 = v_1A_1;$
   $v_0\pi(6 \text{ m})^2 = (80 \text{ L/s})(10^{-3} \text{ m}^3/\text{L})$, which gives $v_0 = \boxed{7.0 \times 10^{-4} \text{ m/s.}}$
   (d) If we use Bernoulli's equation from the top, with $v_0 \approx 0$, to the opening, we have
   $p_0 + \tfrac{1}{2}\rho v_0^2 + \rho gh_0 = p_3 + \tfrac{1}{2}\rho v_3^2 + \rho gh_3;$
   $p_{atm} + 0 + \rho gh_0 = p_{atm} + \tfrac{1}{2}\rho v_3^2 + \rho gh_3,$ or
   $v_3^2 = 2g(h_0 - h_3) = 2(9.8 \text{ m/s}^2)(8 \text{ m})$, which gives $v_3 = \boxed{13 \text{ m/s.}}$
   (e) The volume flow is
   $dV/dt = v_3A_3 = (13 \text{ m/s})\pi(3 \times 10^{-3} \text{ m})^2 = 0.37 \times 10^{-3} \text{ m}^3/\text{s} = \boxed{0.37 \text{ L/s.}}$

67. The pressure at the bottom of the cone depends only on the height of the ice cream:

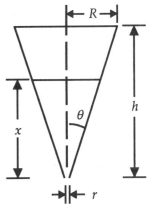

$p_{bottom} = p_0 + \rho g h$, which gives

$$p_{bottom} - p_0 = (1.2 \times 10^3 \text{ kg/m}^3)(9.8 \text{ m/s}^2)(0.10 \text{ m})$$

$$= \boxed{1.2 \times 10^3 \text{ Pa (gauge)}.}$$

To determine the speed of the ice cream at the opening when the height of the ice cream is x, we use Bernoulli's equation between the top of the ice cream and the hole:

$$p_0 + \tfrac{1}{2}\rho v_0{}^2 + \rho g h_0 = p + \tfrac{1}{2}\rho v^2 + \rho g h;$$

$$p_{atm} + 0 + \rho g x = p_{atm} + \tfrac{1}{2}\rho v_{bottom}{}^2 + 0, \text{ which gives}$$

$$v_{bottom} = \sqrt{2gx}.$$

The volume of ice cream in the cone is

$V = \tfrac{1}{3}\pi x^3 \tan^2 \theta$, so the rate at which the volume changes when x changes is

$$dV/dt = d(\tfrac{1}{3}\pi x^3 \tan^2 \theta)/dt = (\pi x^2 \tan^2 \theta)(dx/dt).$$

The flow of ice cream from the hole is

$$dV/dt = v_{bottom}A_{bottom} = \sqrt{2gx}\,\pi r^2 = -(\pi x^2 \tan^2 \theta)(dx/dt),$$

where the minus sign is needed, because a positive v means a decrease in volume.
This is a differential equation for x, which we integrate after rearranging to separate the variables:

$$-\int_0^T \frac{r^2\sqrt{2g}}{\tan^2 \theta}\, dt = \int_h^0 x^{3/2}\, dx;$$

$$-\frac{r^2\sqrt{2g}}{\tan^2 \theta}\, T = \frac{2}{5}\left(0 - h^{5/2}\right), \text{ which gives}$$

$$T = \frac{2}{5}\frac{h^2\tan^2 \theta}{r^2}\sqrt{\frac{h}{2g}} = \frac{2}{5}\left(\frac{R}{r}\right)^2\sqrt{\frac{h}{2g}} = \frac{2}{5}\left(\frac{3 \text{ cm}}{0.05 \text{ cm}}\right)^2\sqrt{\frac{10 \times 10^{-2} \text{ m}}{2(9.8 \text{ m/s}^2)}}$$

$$= 103 \text{ s} = \boxed{8.6 \text{ min.}}$$

CHAPTER 17

9. Because the volume of the gas is constant, the pressure and temperature are linearly related, so we have
$p_2/p_1 = T_2/T_1$, or $\Delta p/p = \Delta T/T$;
$\Delta p/(1 \text{ Pa}) = (349 \text{ K} - 271\text{K})/(271\text{K})$, which gives $\boxed{\Delta p = 0.29 \text{ Pa.}}$

13. Because 0°R in the Reaumur scale corresponds to 273 K and 80°R corresponds to 373 K, the conversion between the two linear scales is
$T - 273 = (373 - 273)t_R/(80-0)$, or $\boxed{T = 1.25t_R + 273.}$
Because 0°R in the Reaumur scale corresponds to 32°F and 80°R corresponds to 212°F, the conversion between the two linear scales is
$t_F - 32 = (212 - 32)t_R/(80-0)$, or $\boxed{t_F = 2.25t_R + 32.}$

15. We can expand the definition of the Rankine scale:
$t_R = (9/5)T = (9/5)(t_C + 273.15) = (9/5)[(5/9)(t_F - 32) + 273.15] = t_F + 459.67.$
Using the appropriate conversion, we have
$t_R = (9/5)(t_C + 273.15) = (9/5)(100 + 273.15) = \boxed{672°\text{R.}}$
$t_R = (9/5)T = (9/5)(4.2) = \boxed{7.6°\text{R.}}$
$t_R = (9/5)T = (9/5)(6000) = \boxed{10,800°\text{R.}}$
$t_R = t_F + 459.67 = 32 + 459.67 = \boxed{492°\text{R.}}$
$t_R = t_F + 459.67 = -30 + 459.67 = \boxed{430°\text{R.}}$
$t_R = (9/5)(t_C + 273.15) = (9/5)(-25 + 273.15) = \boxed{447°\text{R.}}$

17. The air in the tires has constant mass and volume. With temperature defined by
$T = kpV/m$, we have
$T_2/T_1 = p_2/p_1$;
$T_2 = T_1 p_2/p_1$
$= (288 \text{ K})(38 \text{ psi} + 14.5 \text{ psi})/(32 \text{ psi} + 14.5 \text{ psi}) = 325 \text{ K} = \boxed{52°\text{C} \ (126°\text{F}).}$

19. Both the water and the copper will expand. The expansion of the copper will increase the volume of the bowl as if it were copper. For the excess that spills we have
$\Delta V_w - \Delta V_c = V_0(\beta_w - \beta_c) \Delta T$
$= (1500 \text{ cm}^3)[(2.07 \times 10^{-4} \text{ /K}) - (5.1 \times 10^{-5} \text{ /K})](50°\text{C} - 20°\text{C}) = \boxed{7.0 \text{ cm}^3.}$
Note that a temperature change has the same magnitude in °C and K.

21. For the two states of the gas we can write
$p_1V_1 = nRT_1$ and $p_2V_2 = nRT_2$, which can be combined to give
$(p_2/p_1)(V_2/V_1) = T_2/T_1$;
$(1.50\, p_1/p_1)(V_2/2500 \text{ cm}^3) = (0.85\, T_1/T_1)$, which gives $V_2 = \boxed{1.4 \times 10^3 \text{ cm}^3.}$

31. If each molecule occupies a cube of side d, the number of molecules is
$N = V/d^3.$
We can also find the number from the mass:
$N = (m/M)N_A.$
When we equate these two, we can find d, which is also the separation:
$[(1 \text{ g})/(18 \text{ g/mole})](6.02 \times 10^{23} \text{ molecules/mol}) = (10^{-6} \text{ m}^3)/d^3$, which gives $d = \boxed{3.1 \times 10^{-10} \text{ m.}}$

33. (a) The pressure is
$$p = (0.6 \text{ atm})(1.01 \times 10^5 \text{ Pa/atm}) = \boxed{6.06 \times 10^4 \text{ Pa.}}$$
The temperature is
$$T = 35°C + 273 = \boxed{308 \text{ K.}}$$
(b) The volume of the container is the volume of the gas, which we find from the ideal gas law:
$$pV = NkT;$$
$$(6.06 \times 10^4 \text{ Pa})V = (5 \times 10^{22} \text{ molecules})(1.38 \times 10^{-23} \text{ J/K})(308 \text{ K}), \text{ which gives } V = \boxed{3.5 \times 10^{-3} \text{ m}^3.}$$
(c) Because the volume is constant, we have
$$p_2/p_1 = T_2/T_1;$$
$$p_2/(0.6 \text{ atm}) = (120 + 273)\text{K}/(308 \text{ K}), \text{ which gives } p_2 = \boxed{0.77 \text{ atm.}}$$

37. From $pV = nRT = (m/M)RT$, we have
$$\begin{aligned} \rho &= m/V = pM/RT \\ &= (1.01 \times 10^5 \text{ Pa})(4 \text{ g})(10^{-3} \text{ kg/g})/(8.314 \text{ J/mol} \cdot \text{K})(8 \text{ K}) = \boxed{6.07 \text{ kg/m}^3.} \end{aligned}$$

41. (a) If we consider the gas in the tank, before and after the release, we can use the ideal gas law to get
$$p_2V_2/p_1V_1 = n_2T_2/n_1T_1, \text{ or } (p_2/p_1)(V_2/V_1) = (n_2/n_1)(T_2/T_1);$$
$$[p_2/(120 \text{ atm})](1) = (1/2)(313 \text{ K}/263 \text{ K}), \text{ which gives } p_2 = \boxed{71 \text{ atm.}}$$
(b) If we consider the gas in the tank, before and after the release, we have the same analysis. The order of the processes does not affect the result:
$$p_2 = \boxed{71 \text{ atm.}}$$

43. The initial full volume of the pump is
$$V_0 = \pi r^2 L = \pi(1.25 \text{ cm})^2(35 \text{ cm}) = 172 \text{ cm}^3.$$
The volume remaining in the pump at the end of the downward push is
$$V_f = 0.05V_0 = 0.05(172 \text{ cm}^3) = 8.6 \text{ cm}^3.$$
As the piston is pushed down, the pressure goes up until the tire pressure is reached. At constant temperature, we find the volume at which air starts to enter the tire from

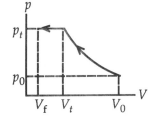

$$p_0V_0 = p_tV_t;$$
$$(1 \text{ atm})(172 \text{ cm}^3) = (1 \text{ atm} + 2.5 \text{ atm})V_t, \text{ which gives } V_t = 49.1 \text{ cm}^3.$$
We assume that the change in pressure in the tire as the air is pumped into it can be neglected. The volume transferred to the tire at a gauge pressure of 2.5 atm is
$$\Delta V = V_t - V_f = 49.1 \text{ cm}^3 - 8.6 \text{ cm}^3 = \boxed{40.5 \text{ cm}^3.}$$
This corresponds to a volume at 1 atm of
$$\Delta V' = \Delta V(p_t/p_0) = (40.5 \text{ cm}^3)(3.5 \text{ atm}/1 \text{ atm}) = \boxed{142 \text{ cm}^3.}$$

47. Because the mass of the gas does not change, we compare two
    states of the gas:
    $(p_a/p_b)(V_a/V_b) = (T_a/T_b)$.
    We are given $p_1 = 10^5$ Pa, $V_1 = 300$ cm³, $T_1 = 293$ K.

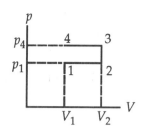

(a) For state 2, we have $\boxed{p_2 = p_1 = 10^5 \text{ Pa}, \ V_2 = 2V_1 = 600 \text{ cm}^3.}$
    We compare state 2 with state 1:
    $(p_2/p_1)(V_2/V_1) = (T_2/T_1)$;
    $(1)(2) = T_2/293$ K, which gives $\boxed{T_2 = 568 \text{ K} = 313°\text{C}.}$

(b) For state 3, we have $\boxed{p_3 = 1.30p_2 = 1.30 \times 10^5 \text{ Pa}, \ V_3 = V_2 = 600 \text{ cm}^3.}$
    We compare state 3 with state 2:
    $(p_3/p_2)(V_3/V_2) = (T_3/T_2)$;
    $(1.30)(1) = T_3/568$ K, which gives $\boxed{T_3 = 762 \text{ K} = 489°\text{C}.}$

(c) For state 4, we have $\boxed{p_4 = p_3 = 1.30 \times 10^5 \text{ Pa}, \ V_4 = V_1 = 300 \text{ cm}^3.}$
    We compare state 4 with state 3:
    $(p_4/p_3)(V_4/V_3) = (T_4/T_3)$;
    $(1)(0.5) = T_4/762$ K, which gives $\boxed{T_4 = 381 \text{ K} = 108°\text{C}.}$

53. We assume that the temperature-pressure relation is linear and extrapolate to $p = 0$:
    $(T_2 - T_0) = (T_2 - T_1)(p_2 - p_0)/(p_2 - p_1)$;
    $(373.35 \text{ K} - T_0) = (373.35 \text{ K} - 373.25 \text{ K})(0.4 \text{ atm} - 0)/(0.4 \text{ atm} - 0.2 \text{ atm})$, which gives $T_0 = \boxed{373.15 \text{ K.}}$

55. (a) If we expand the expression for a van der Waals gas, we get
    $(pV/n) + (an/V) - pb - (abn^2/V^2) = RT$.
    If the ideal gas law, $pV = nRT$, is valid also, we have
    $(an/V) - pb - (abn^2/V^2) = 0$, or $(pVb/n) = a - (abn/V)$.
    We substitute the ideal gas law again to get
    $RT_Bb = a - (abn/V)$, which gives $\boxed{T_B = (a/bR)[1 - (bn/V)].}$

(b) For dilute argon, we have
    $T_B = (a/bR) = (0.140 \text{ m}^6 \cdot \text{Pa/mol}^2)/(4.00 \times 10^{-5} \text{ m}^3/\text{mol})(8.314 \text{ J/mol} \cdot \text{K}) = \boxed{421 \text{ K (148°C).}}$

63. From the Stefan–Boltzmann formula for the radiated power per unit area, $E = \sigma T^4$, we find the total
    power radiated from
    $P = EA = \sigma AT^4$
    $= (5.68 \times 10^{-8} \text{ W/m}^2 \cdot \text{K}^4)4\pi(6.95 \times 10^8 \text{ m})^2(6000 \text{ K})^4 = \boxed{4.47 \times 10^{26} \text{ W.}}$

69. We find the densities of the gases from
    $\rho = m/V = pM/RT$;
    $\rho_{air} = (1.01 \times 10^5 \text{ Pa})(28.9 \text{ g/mol})(10^{-3} \text{ kg/g})/(8.314 \text{ J/mol} \cdot \text{K})(283 \text{ K}) = 1.24 \text{ kg/m}^3$;
    $\rho_{He} = (1.01 \times 10^5 \text{ Pa})(4 \text{ g/mol})(10^{-3} \text{ kg/g})/(8.314 \text{ J/mol} \cdot \text{K})(283 \text{ K}) = 0.17 \text{ kg/m}^3$.
    In equilibrium, we have $F_{net} = 0$, so the buoyant force equals the total weight:
    $F_{buoy} = m_{He}g + Mg$;
    $\rho_{air}gV_{balloon} = \rho_{He}gV_{balloon} + Mg$;
    $(1.24 \text{ kg/m}^3)gV_{balloon} = (0.17 \text{ kg/m}^3)gV_{balloon} + (230 \text{ kg})g$, which gives
    $V_{balloon} = 215 \text{ m}^3$.
    The mass of helium needed is
    $m = \rho_{He}V_{balloon} = (0.17 \text{ kg/m}^3)(215 \text{ m}^3) = \boxed{37 \text{ kg.}}$

71. The volume and mass of the gas do not change, and the pressure difference is
    $p_2 - p_1 = mg/A$, where m is the additional mass added to the piston.
    The two states of the gas are
    $p_1 V = nRT_1$ and $p_2 V = nRT_2$.
    If we subtract these equations, we get
    $(p_2 - p_1)V = nR(T_2 - T_1)$, or
    $V/n = R(T_2 - T_1)/(p_2 - p_1) = R(T_2 - T_1)A/mg$
    $= (8.314 \text{ J/mol} \cdot \text{K})(80°C - 20°C)(70 \text{ cm}^2)(10^{-4} \text{ m}^2/\text{cm}^2)/(0.5 \text{ kg})(9.8 \text{ m/s}^2)$
    $= 0.71 \text{ m}^3/\text{mol}$.
    The volume of 0.2 mol is
    $V = (0.71 \text{ m}^3/\text{mol})(0.2 \text{ mol}) = \boxed{0.14 \text{ m}^3.}$

73. From the ideal gas law, we have
    $\rho_0 = m/V = pm/nRT = pM/RT$, where we have used the definition of molecular weight.
    For the speed of sound, we have
    $$v_{sound} = \sqrt{\frac{\gamma p}{\rho_0}} = \sqrt{\frac{\gamma RT}{M}} .$$
    If we assume the temperature ranges from 0°C to 45°C, we have

    $$v_1 = \sqrt{\frac{1.4 \left(8.314 \text{ J/mol} \cdot \text{K}\right)\left(273 \text{ K}\right)}{\left(28.9 \times 10^{-3} \text{ kg/mol}\right)}} = 332 \text{ m/s};$$

    $$v_2 = \sqrt{\frac{1.4 \left(8.314 \text{ J/mol} \cdot \text{K}\right)\left(318 \text{ K}\right)}{\left(28.9 \times 10^{-3} \text{ kg/mol}\right)}} = 358 \text{ m/s}.$$

    The speed change is $\boxed{26 \text{ m/s.}}$

77. (a) We use the approximation $1/(1-x) \approx 1 + x$, when $x \ll 1$:
    $[p + a(n/V)^2]/RT = (n/V)[1 - (b\,n/V)]^{-1} \approx (n/V)[1 + (b\,n/V)]$;
    $[p + a(n/V)^2]/RT = (n/V) + b(n/V)^2$, which can be rearranged to
    $p/RT = (n/V) + b(n/V)^2 - (a/RT)(n/V)^2$ with $bn/V \ll 1$.
    (b) If the result of part (a) is expressed as
    $p/RT = (n/V) + B_2(T)(n/V)^2$, then $B_2(T) = b - (a/RT) = b[1 - (a/bRT)]$.
    If $T < a/bR$, $(a/bRT) > 1$, and we see that $B_2(T) < 0$;
    if $T = a/bR$, $(a/bRT) = 1$, so $B_2(T) = 0$;
    if $T > a/bR$, $(a/bRT) < 1$, so $B_2(T) > 0$.
    The value of $T = a/bR$ is the result in Problem 55, when $bn/V \ll 1$.

CHAPTER 18

In earlier problems involving the dynamics of rigid bodies, it was necessary to specify the object in order to determine the forces acting on it. Similarly, the system must be specified for applications of the first law of thermodynamics. Then the appropriate signs can be interpreted for the work and heat flow terms. In heat flow problems, careful attention must be paid to units.

7. We neglect the heat capacity of the glass.
    (a) Because the amount of ice is relatively large, we assume that not all of the ice will melt and the final temperature will be 0°C. The total heat flow into the system is zero. If we let m represent the mass in grams of the ice that melts, we have
    $$\Delta Q_{sys} = \Delta Q_w + \Delta Q_i = 0, \quad \text{or} \quad m_w c_w \Delta T_w + m L_i = 0;$$
    $(200 \text{ g})(1.00 \text{ cal/g} \cdot \text{K})(0°C - 25°C) + [m/(18 \text{ g/mol})](6.0 \times 10^3 \text{ J/mol})/(4.185 \text{ J/cal}) = 0$, which gives $m = 62.8$ g.
    Because this is less than 100 g, our assumption is valid. 37.2 g of ice remains.

    The glass contains 262.8 g of water and 37.2 g of ice at 0°C.

    (b) Because the amount of ice is relatively small, we assume that all of the ice will melt and only water will be present at a temperature above 0°C. The total heat flow into the system is zero, so we have
    $$\Delta Q_{sys} = \Delta Q_w + \Delta Q_i = 0, \quad \text{or} \quad m_w c_w \Delta T_w + m_i L_i + m_i c_w \Delta T_i = 0;$$
    $(250 \text{ g})(1.00 \text{ cal/g} \cdot \text{K})(T - 25°C) + [(50 \text{ g})/(18 \text{ g/mol})](6.0 \times 10^3 \text{ J/mol})/(4.185 \text{ J/cal}) +$
    $\qquad (50 \text{ g})(1.00 \text{ cal/g} \cdot \text{K})(T - 0°C) = 0$, which gives
    $T = 7.6°C.$
    Because this is greater than 0°C, our assumption is valid. All of the ice has melted.

    The glass contains 300 g of water at 7.6°C.

11. The total heat flow into the system is zero, so we have
    $$\Delta Q_{sys} = \Delta Q_w + \Delta Q_{Fe} + \Delta Q_{Ag} = 0, \quad \text{or} \quad m_w c_w \Delta T_w + m_{Fe} c_{Fe} \Delta T_{Fe} + m_{Ag} c_{Ag} \Delta T_{Ag} = 0;$$
    $(200 \text{ g})(1.00 \text{ cal/g} \cdot \text{K})(T - 293 \text{ K}) + (200 \text{ g})(0.112 \text{ cal/g} \cdot \text{K})(T - 400 \text{ K}) +$
    $\qquad (200 \text{ g})(0.0557 \text{ cal/g} \cdot \text{K})(T - 400 \text{ K}) = 0$, which gives $T = 308$ K.
    Note that we need do only one step.

13. The total heat flow into the system is zero, so we have
    $$n_g c'_g \Delta T + n_s c'_s \Delta T = 0$$
    $(1 \text{ mol})(3.00 \text{ cal/mol} \cdot \text{K})(4.185 \text{ J/cal})(T - 100 \text{ K}) + (2 \text{ mol})\tfrac{3}{2}(8.314 \text{ J/mol} \cdot \text{K})(T - 200 \text{ K}) = 0$,
    which gives $T =$ 167 K.

17. We assume that all of the kinetic energy loss in the collision heats the lead. From conservation of energy during the fall, the kinetic energy of the bag just before hitting the ground is equal to the initial potential energy, so for N drops, we have
    $$\Delta Q = W = -\Delta U, \quad \text{or} \quad mc \Delta T = -(0 - Nmgh), \quad \text{which becomes}$$
    $N = c \Delta T/gh = (128 \text{ J/kg} \cdot \text{K})(5.0 \text{ K})/(9.8 \text{ m/s}^2)(2.0 \text{ m}) = 32.7 =$ 33 times.

21. (a) We assume that all of the kinetic energy, which comes from the decrease in potential energy, goes into the heat flow into the water from the turbulent motion at the base of the falls:
$$\Delta Q = K = U, \quad \text{or} \quad mc\,\Delta T = mgh.$$
We have to be careful with the units. If m is the mass in kg, we have
$$m(10^3 \text{ g/kg})(1.0 \text{ cal/g} \cdot \text{K})(4.185 \text{ J/cal})\,\Delta T = m(9.8 \text{ m/s}^2)(2648 \text{ ft})(0.3048 \text{ m/ft}), \text{ which gives}$$
$\Delta T = 1.9°\text{C}$, so we have $T = \boxed{11.9°\text{C.}}$

  (b) For Niagara Falls, the only change is the height, so we have
$$m(10^3 \text{ g/kg})(1.0 \text{ cal/g} \cdot \text{K})(4.185 \text{ J/cal})\,\Delta T = m(9.8 \text{ m/s}^2)(50 \text{ m}), \text{ which gives}$$
$\Delta T = 0.12°\text{C}$, so we have $T = \boxed{10.1°\text{C.}}$

  (c) The total potential energy change is the same but the kinetic energy is less at bottom, because of the negative work done by the frictional force during the fall. This negative work generates a heat flow, which raises the temperature of the water and the air. Because some of the heat flow is into the air, ΔT of the water is less than in part (a).

  (d) Most of the energy required to provide the heat of vaporization comes from the water, which causes a negative ΔT. Because the heat of vaporization is so large, this effect is greater than the positive ΔT from the fall, which produces a net negative temperature change.

29. (a) Because βV must have the dimensions of p, we can write
$$[\beta] = [p]/[V] = [ML^{-1}T^{-2}]/[L^3] = \boxed{[ML^{-4}T^{-2}],}$$
which has units of N/m^5.

  (b) The integral to determine the work is

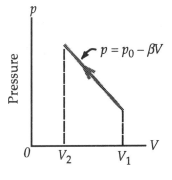

$$W = \int_{V_1}^{V_2} p\,dV = \int_{V_1}^{V_2} \left(p_0 - \beta V\right) dV$$
$$= \left(p_0 V - \tfrac{1}{2}\beta V^2\right)\Big|_{V_1}^{V_2}$$
$$= p_0\left(V_2 - V_1\right) - \tfrac{1}{2}\beta\left(V_2^2 - V_1^2\right)$$
$$= \boxed{\left(V_2 - V_1\right)\left[p_0 - \tfrac{1}{2}\beta\left(V_2 + V_1\right)\right].}$$

  (c) From the direction of the path on the p–V diagram, the area under the curve is negative.
The evaluation of the area under the p–V curve to determine the work is
$$W = -\tfrac{1}{2}(p_2 + p_1)(V_1 - V_2)$$
$$= -\tfrac{1}{2}[(p_0 - \beta V_1) + (p_0 - \beta V_2)](V_1 - V_2)$$
$$= \boxed{(V_2 - V_1)[p_0 - \tfrac{1}{2}\beta(V_2 + V_1)],} \text{ which is the result from part (b).}$$

31. We find the temperature at C from the ideal gas law for the two states of the gas:
$$p_A V_A = nRT_A \quad \text{and} \quad p_C V_C = nRT_C.$$

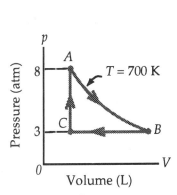

Because the number of moles and the volume are constant, we have
$$T_C = T_A(p_C/p_A) = (700 \text{ K})[(3 \text{ atm})/(8 \text{ atm})] = 263 \text{ K}.$$
For the work done in the isothermal expansion, we have
$$W_{A \to B} = nRT_A \ln(V_B/V_A) = nRT_A \ln(p_A/p_B)$$
$$= (2 \text{ mol})(8.314 \text{ J/mol} \cdot \text{K})(700 \text{ K}) \ln[(8 \text{ atm})/(3 \text{ atm})]$$
$$= 1.14 \times 10^4 \text{ J}.$$
For the isobaric compression, we have
$$W_{B \to C} = \int p\,dV = p(V_C - V_B) = nR(T_C - T_B)$$
$$= (2 \text{ mol})(8.314 \text{ J/mol} \cdot \text{K})(263 \text{ K} - 700 \text{ K}) = -0.73 \times 10^4 \text{ J}.$$
Because there is no volume change from C to A, we have
$$W_{C \to A} = \int p\,dV = 0.$$
The total work done by the gas in one cycle is
$$W_{\text{cycle}} = W_{A \to B} + W_{B \to C} + W_{C \to A} = (+1.14 - 0.73) \times 10^4 \text{ J} = \boxed{4.1 \times 10^3 \text{ J.}}$$

35. We can find the internal energy change from the path $A \rightarrow C \rightarrow B$:
$$\Delta U = U_B - U_A = -W_{A \rightarrow C \rightarrow B} + Q_{A \rightarrow C \rightarrow B}$$
$$= -20,000 \text{ cal} + 40,000 \text{ cal} = +20,000 \text{ cal}.$$

Because internal energy depends only on the state, this is the change for any path between A and B.

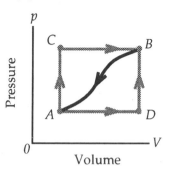

(a) For the path $A \rightarrow D \rightarrow B$, the first law of thermodynamics gives us
$$U_B - U_A = -W_{A \rightarrow D \rightarrow B} + Q_{A \rightarrow D \rightarrow B} ;$$
$$+20,000 \text{ cal} = -7,000 \text{ cal} + Q_{A \rightarrow D \rightarrow B} , \text{ which gives}$$
$$\boxed{Q_{A \rightarrow D \rightarrow B} = 2.7 \times 10^4 \text{ cal.}}$$

(b) For the return path $B \rightarrow A$, the work done by the gas is negative, and the first law of thermodynamics gives us
$$U_A - U_B = -W_{B \rightarrow A} + Q_{B \rightarrow A} ;$$
$$-20,000 \text{ cal} = -(-15,000 \text{ cal}) + Q_{B \rightarrow A} , \text{ which gives}$$
$$\boxed{Q_{B \rightarrow A} = -3.5 \times 10^4 \text{ cal.}}$$
The negative value means the heat is $\boxed{\text{liberated.}}$

37. (a) The internal energy of an ideal gas depends only on the temperature:
$$U_A = nc'_V T_A = nc'_V (p_A V_A / nR) = \boxed{(c'_V / R) p_A V_A;}$$
$$U_B = nc'_V T_B = nc'_V (p_B V_B / nR) = \boxed{(c'_V / R) p_B V_B;}$$
$$U_C = nc'_V T_C = nc'_V (p_C V_C / nR) = \boxed{(c'_V / R) p_B V_A.}$$

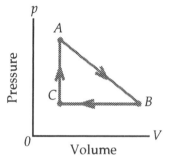

(b) The work done by the gas in the cycle is the area of the triangle enclosed by the path:
$$W_{\text{net}} = \tfrac{1}{2}(p_A - p_B)(V_B - V_C) = \boxed{\tfrac{1}{2}(p_A - p_B)(V_B - V_A).}$$

39. (a) We use the ideal gas law to find the temperatures:
$$T_i = p_i V_i / nR = (1.2 \times 10^5 \text{ N/m}^2)(3.5 \times 10^{-3} \text{ m}^3)/[(5 \text{ g})/(28 \text{ g/mol})](8.314 \text{ J/mol} \cdot \text{K}) = \boxed{2.8 \times 10^2 \text{ K.}}$$
$$T_f = p_f V_f / nR = (1.8 \times 10^5 \text{ N/m}^2)(5.0 \times 10^{-3} \text{ m}^3)/[(5 \text{ g})/(28 \text{ g/mol})](8.314 \text{ J/mol} \cdot \text{K}) = \boxed{6.1 \times 10^2 \text{ K.}}$$

(b) The internal energy of an ideal gas depends only on the temperature:
$$\Delta U = nc'_V \Delta T$$
$$= [(5 \text{ g})/(28 \text{ g/mol})](4.94 \text{ cal/mol} \cdot \text{K})(4.185 \text{ J/cal})(6.1 \times 10^2 \text{ K} - 2.8 \times 10^2 \text{ K}) = \boxed{1.2 \times 10^3 \text{ J.}}$$

(c) The work done by the gas is the area under the p–V curve:
$$W = \tfrac{1}{2}(p_i + p_f)(V_f - V_i)$$
$$= \tfrac{1}{2}[(1.2 + 1.8) \times 10^5 \text{ N/m}^2][(5.0 - 3.5) \times 10^{-3} \text{ m}^3] = \boxed{2.3 \times 10^2 \text{ J.}}$$

(d) We use the first law of thermodynamics to find the heat flow:
$$\Delta U = -W + Q ;$$
$$1.2 \times 10^3 \text{ J} = -2.3 \times 10^2 \text{ J} + Q, \text{ which gives } Q = \boxed{1.4 \times 10^3 \text{ J (into the gas).}}$$

41. We find the internal energy change from the first law of thermodynamics:
$$\Delta U = -W + Q = -8 \text{ J} + 5 \text{ J} = -3 \text{ J}.$$
For an ideal gas, we have
$$\Delta U = nc'_V \Delta T;$$
$$-3 \text{ J} = (0.2 \text{ mol})(20.8 \text{ J/mol} \cdot \text{K}) \Delta T, \text{ which gives } \Delta T = \boxed{-0.72 \text{ K (cooled).}}$$
Note that we use $\Delta U = nc'_V \Delta T$, even though the process is not at constant volume. The internal energy is a state function, so the change depends only on the initial and final states, not the process.

51. The relation between p and V for an adiabatic process is $pV^\gamma =$ constant, so we have
$$p_2/p_1 = (V_1/V_2)^\gamma = (1/2)^{1.4} = 0.38.$$
The pressure change for the adiabatic process is $\Delta p_{adiabatic} = 0.62p_1$.
The relation between p and V for an isothermal process is $pV =$ constant, so we have
$$p_2/p_1 = V_1/V_2 = 1/2 = 0.50.$$
The pressure change for the isothermal process is $\Delta p_{isothermal} = 0.50p_1$.
Thus we have
$$\boxed{\Delta p_{adiabatic}/\Delta p_{isothermal} = 0.62/0.50 = 1.24.}$$

53. The relation between p and V for an adiabatic process is $pV^\gamma =$ constant, so we have
$$p_2/p_1 = (V_1/V_2)^\gamma;$$
$$5 = 3^\gamma, \text{ which gives } \gamma = C_p/C_V = \ln 5/\ln 3 = 1.465; \quad \boxed{C_V/C_p = 0.683.}$$
For an ideal gas, $C_p - C_V = nR$. Using the above ratio, we find
$$C_V = nR/0.465 = 2.15nR, \text{ and } C_p = 3.15nR.$$
We use the first law of thermodynamics for the adiabatic process to find the work done by the gas:
$$\Delta U = -W + Q, \text{ which gives}$$
$$W = -\Delta U = -C_V(T_2 - T_1) = -C_V(p_2V_2 - p_1V_1)/nR$$
$$= -(2.15)](5p_1)(V_1/3) - p_1V_1] = -1.43\,p_1V_1.$$
The work done on the gas is
$$\boxed{W_{on} = +1.43\,p_1V_1.}$$
The internal energy change is
$$\Delta U = -W = -(-1.43\,p_1V_1) = \boxed{+1.43\,p_1V_1.}$$

55. From Problem 17–43 we have the initial full volume of the pump: $V_0 = 172$ cm^3, and the volume remaining in the pump at the end of the downward push: $V_f = 8.6$ cm^3. As the piston is pushed down, the pressure goes up until the tire pressure is reached.
For an adiabatic process, we find the volume at which air starts to enter the tire from
$$p_t/p_0 = (V_0/V_t)^\gamma;$$
$$(1 \text{ atm})/(3.5 \text{ atm}) = [(172 \text{ cm}^3)/V_t]^{1.4}, \text{ which gives } V_t = 70.3 \text{ cm}^3.$$
We find the temperature at the time air starts to enter the tire from
$$p_tV_t/p_0V_0 = T_t/T_0;$$
$$(3.5 \text{ atm})(70.3 \text{ cm}^3)/(1 \text{ atm})(172 \text{ cm}^3) = T_t/(295 \text{ K}), \text{ which gives } T_t = 422 \text{ K} = 149°\text{C}.$$
We assume that the change in pressure in the tire as the air is pumped into it can be neglected.
The volume transferred to the tire at a gauge pressure of 2.5 atm and 422 K is
$$\Delta V = V_t - V_f = 70.3 \text{ cm}^3 - 8.6 \text{ cm}^3 = \boxed{61.7 \text{ cm}^3.}$$
We find the corresponding volume at 1 atm and 295 K from
$$p_t\Delta V/p_0\Delta V' = T_t/T_0;$$
$$(3.5 \text{ atm})(61.7 \text{ cm}^3)/(1 \text{ atm})\Delta V' = (422 \text{ K})/(295 \text{ K}), \text{ which gives } \boxed{\Delta V' = 151 \text{ cm}^3.}$$
This is more than the volume transferred in the isothermal process: 142 cm^3.

61. The molar heat capacity of oxygen at constant volume is
$$c'_V = (4.97 \text{ cal/mol} \cdot \text{K})(4.185 \text{ J/cal}) = 20.8 \text{ J/mol} \cdot \text{K}.$$
The molar heat capacity of oxygen at constant pressure is
$$c'_p = c'_V + R = (20.8 \text{ J/mol} \cdot \text{K}) + (8.314 \text{ J/mol} \cdot \text{K}) = 29.1 \text{ J/mol} \cdot \text{K}.$$
We find the number of moles from the ideal gas law:
$$pV = nRT;$$
$$(1.01 \times 10^5 \text{ Pa})(10 \times 10^{-3} \text{ m}^3) = n(8.314 \text{ J/mol} \cdot \text{K})(273 \text{ K}). \text{ which gives } n = 0.445 \text{ mol}.$$
(a) From the ideal gas law, we know that tripling the volume at constant pressure will also triple the temperature. The heat flow is
$$Q = nc'_p\Delta T = (0.445 \text{ mol})(29.1 \text{ J/mol} \cdot \text{K})(3 - 1)(273 \text{ K}) = \boxed{+7.1 \times 10^3 \text{ J}.}$$
(b) From the ideal gas law, we know that doubling the pressure at constant volume will also double the temperature. The heat flow is
$$Q = nc'_V\Delta T = (0.445 \text{ mol})(20.8 \text{ J/mol} \cdot \text{K})(2 - 1)(273 \text{ K}) = \boxed{+2.5 \times 10^3 \text{ J}.}$$

69. (a) Using the ideal gas law, we have

$$p_A V_A = nRT_A = nRT_B = p_B V_B;$$
$$p_B = p_A(V_A/V_B) = (1 \text{ atm})(10) = \boxed{10 \text{ atm.}}$$

(b) For the work done by the gas during the isothermal compression, we have

$$W_{A \to B} = nRT_A \ln(V_B/V_A)$$
$$= (1 \text{ mol})(8.314 \text{ J/mol} \cdot \text{K})(300 \text{ K}) \ln(0.1) = \boxed{-5.7 \times 10^3 \text{ J.}}$$

The negative value means that work is done on the gas.

(c) Because there is no change in internal energy of an ideal gas for an isothermal process, the first law of thermodynamics gives us

$$\Delta U = 0 = -W + Q;$$
$$Q_{A \to B} = W_{A \to B} = \boxed{-5.7 \times 10^3 \text{ J (from the gas).}}$$

71. We assume that all of the ice melts. The ice must absorb heat as the temperature rises to 32°F, then it melts and the resulting water absorbs heat as the temperature rises above 32°F. The net heat flow is zero, so we find the final temperature from

$$Q_{\text{net}} = Q_{\text{ice}} + Q_{\text{melt}} + Q_{\text{icew}} + Q_{\text{water}} = 0;$$
$$m_{\text{ice}} c_{\text{ice}} \Delta T_{\text{ice}} + m_{\text{ice}} L_f + m_{\text{ice}} c_{\text{water}} \Delta T_{\text{icew}} + m_{\text{water}} c_{\text{water}} \Delta T_{\text{water}} = 0;$$
$$4(25 \text{ g})(2.04 \text{ J/g} \cdot \text{K})(32°F - 10°F)[(5 \text{ K})/(9°F)] + 4(25 \text{ g})[(1 \text{ mol})/18 \text{ g})](6.0 \times 10^3 \text{ J/mol}) +$$
$$4(25 \text{ g})(1.00 \text{ cal/g} \cdot \text{K})(T_f - 32°F)[(5 \text{ K})/(9°F)](4.185 \text{ J/cal}) +$$
$$(150 \text{ cm}^3)(1.00 \text{ g/cm}^3)(1.00 \text{ cal/g} \cdot \text{K})(T_f - 70°F)[(5 \text{ K})/(9°F)](4.185 \text{ J/cal}) = 0, \text{ which gives}$$
$$T_f = -6.8°F.$$

We had assumed that there was only water left, but we cannot have water below 32°F. The final temperature must be $\boxed{32°F,}$ and there will be a mixture of ice and water. We can find the amount of ice that melts from

$$Q_{\text{net}} = Q_{\text{ice}} + Q_{\text{melt}} + Q_{\text{water}} = 0;$$
$$m_{\text{ice}} c_{\text{ice}} \Delta T_{\text{ice}} + x_{\text{ice}} L_f + m_{\text{water}} c_{\text{water}} \Delta T_{\text{water}} = 0;$$
$$4(25 \text{ g})(2.04 \text{ J/g} \cdot \text{K})(32°F - 10°F)[(5 \text{ K})/(9°F)] + x_{\text{ice}}[(1 \text{ mol})/18 \text{ g})](6.0 \times 10^3 \text{ J/mol}) +$$
$$(150 \text{ cm}^3)(1.00 \text{ g/cm}^3)(1.00 \text{ cal/g} \cdot \text{K})(32°F - 70°F)[(5 \text{ K})/(9°F)](4.185 \text{ J/cal}) = 0, \text{ which gives}$$
$$x_{\text{ice}} = 32 \text{ g of ice melts.}$$

73. Using the ideal gas law, we have

$$p_A V_A = nRT_A;$$
$$(5 \text{ atm})(1.01 \times 10^5 \text{ Pa/atm})V_A = (1 \text{ mol})(8.314 \text{ J/mol} \cdot \text{K})(673 \text{ K}),$$

which gives $V_A = 1.11 \times 10^{-2} \text{ m}^3$.

Because $T_A = T_B = 673 \text{ K}$ and $V_B = 2V_A = 2.22 \times 10^{-2} \text{ m}^3$, we have

$$p_B = p_A(V_A/V_B) = (5 \text{ atm})(1/2) = 2.5 \text{ atm.}$$

Because $V_B = V_C$ and $p_C = \frac{1}{2} p_B = 1.25 \text{ atm}$, we have

$$T_C = T_B(p_C/p_B) = (673 \text{ K})(1/2) = 336 \text{ K.}$$

Because $V_D = V_A$ and $p_A = 4p_D$, we have

$$T_D = T_A(p_D/p_A) = (673 \text{ K})(1/4) = 168 \text{ K.}$$

These values are summarized in the following table:

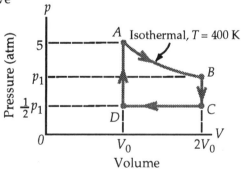

Point	p	T	V
A	5.0 atm	673 K	$1.11 \times 10^{-2} \text{ m}^3$
B	2.5 atm	673 K	$2.22 \times 10^{-2} \text{ m}^3$
C	1.25 atm	336 K	$2.22 \times 10^{-2} \text{ m}^3$
D	1.25 atm	168 K	$1.11 \times 10^{-2} \text{ m}^3$

There is work done by the gas only for the isothermal expansion and the isobaric compression, so we have

$$W_{\text{net}} = W_{A \to B} + W_{C \to D}$$
$$= nRT_A \ln(V_B/V_A) + p_C(V_D - V_C)$$
$$= (1 \text{ mol})(8.314 \text{ J/mol} \cdot \text{K})(673 \text{ K}) \ln(2/1) +$$
$$(1.25 \text{ atm})(1.01 \times 10^5 \text{ Pa/atm})[(1.11 - 2.22) \times 10^{-2} \text{ m}^3]$$
$$= \boxed{+2.48 \times 10^3 \text{ J.}}$$

75. We are given the following values:

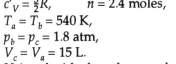

$$c'_V = \tfrac{5}{2}R, \qquad n = 2.4 \text{ moles},$$
$$T_a = T_b = 540 \text{ K},$$
$$p_b = p_c = 1.8 \text{ atm},$$
$$V_c = V_a = 15 \text{ L}.$$

(a) Using the ideal gas law, we have
$$p_a V_a = nRT_a;$$
$$p_a(15 \text{ L})(10^{-3} \text{ m}^3/\text{L}) = (2.4 \text{ mol})(8.314 \text{ J/mol} \cdot \text{K})(540 \text{ K}),$$
which gives $\boxed{p_a = 7.18 \times 10^5 \text{ Pa} \ (7.1 \text{ atm}).}$

(b) Because $T_a = T_b = 540$ K, we have
$$V_b = V_a(p_a/p_b) = (15 \text{ L})[(7.1 \text{ atm})/(1.8 \text{ atm})] = 59 \text{ L} = \boxed{5.9 \times 10^{-2} \text{ m}^3.}$$

(c) We find the work done from the area under the path:
$$\begin{aligned} W_{a \to b} &= \int p \, dV = \tfrac{1}{2}(p_a + p_b)(v_b - v_a) \\ &= \tfrac{1}{2}(7.1 \text{ atm} + 1.8 \text{ atm})(1.01 \times 10^5 \text{ Pa/atm})(59 \text{ L} - 15 \text{ L})(10^{-3} \text{ m}^3/\text{L}) = \boxed{2.0 \times 10^4 \text{ J.}} \end{aligned}$$

(d) Because $p_b = p_c$, we have
$$T_c = T_b(V_c/V_b) = (540 \text{ K})[(15 \text{ L})/(59 \text{ L})] = \boxed{137 \text{ K.}}$$

(e) For any process, we find the internal energy change from
$$\begin{aligned} \Delta U_{c \to a} &= nc'_V(T_a - T_c) \\ &= (2.4 \text{ mol})\tfrac{5}{2}(8.314 \text{ J/mol} \cdot \text{K})(540 \text{ K} - 137 \text{ K}) = \boxed{+2.0 \times 10^4 \text{ J.}} \end{aligned}$$

(f) The path from a to b is not isothermal. We find the net work from the area enclosed by the cycle:
$$\begin{aligned} W_{\text{net}} &= \text{area of triangle} = \tfrac{1}{2}(p_a - p_c)(V_b - V_c) \\ &= \tfrac{1}{2}(7.1 \text{ atm} - 1.8 \text{ atm})(1.01 \times 10^5 \text{ Pa/atm})(59 \text{ L} - 15 \text{ L})(10^{-3} \text{ m}^3/\text{L}) = \boxed{+1.2 \times 10^4 \text{ J.}} \end{aligned}$$

CHAPTER 19

3. We find the number of components from the molecular weight of CH_2:
$$N = \rho V / (\text{mass}/\text{molecule})$$
$$= (0.9 \text{ g/cm}^3)(0.1 \text{ L})(10^3 \text{ cm}^3/\text{L})/[(14 \text{ g/mol})/(6.02 \times 10^{23} \text{ molecules/mol})]$$
$$= \boxed{3.9 \times 10^{24} \text{ components.}}$$
The area of the layer is
$$A = N(10^{-15} \text{ cm}^2) = (3.9 \times 10^{24})(10^{-15} \text{ cm}^2) = 3.9 \times 10^9 \text{ cm}^2 = \boxed{3.9 \times 10^5 \text{ m}^2.}$$

7. (a) The internal energy of an ideal gas depends only on the temperature:
$$U = nc'_V T = \tfrac{3}{2}nRT = \tfrac{3}{2}pV$$
$$= \tfrac{3}{2}(5 \text{ atm})(1.01 \times 10^5 \text{ Pa/atm})(5000 \text{ cm}^3)(10^{-6} \text{ m}^3/\text{cm}^3) = \boxed{3.8 \times 10^3 \text{ J.}}$$
    (b) $T = pV/nR$
$$= (5 \text{ atm})(1.01 \times 10^5 \text{ Pa/atm})(5000 \text{ cm}^3)(10^{-6} \text{ m}^3/\text{cm}^3)/(1 \text{ mol})(8.314 \text{ J/mol} \cdot \text{K}) = \boxed{304 \text{ K.}}$$
    (c) On the molecular level, the internal energy is the kinetic energy of the molecules:
$$U = \tfrac{1}{2}Nm\langle \mathbf{v}^2 \rangle;$$
$$304 \text{ J} = \tfrac{1}{2}(6.02 \times 10^{23} \text{ molecules})(3.36 \times 10^{-26} \text{ kg/molecule})\langle \mathbf{v}^2 \rangle, \text{ which gives } \boxed{\langle \mathbf{v}^2 \rangle = 3.8 \times 10^5 \text{ m}^2/\text{s}^2.}$$
    (d) From part (c), we have
$$v_{\text{rms}} = \langle \mathbf{v}^2 \rangle^{1/2} = (3.8 \times 10^5 \text{ m}^2/\text{s}^2)^{1/2} = \boxed{6.1 \times 10^2 \text{ m/s.}}$$

11. The relation between pressure and rms speed is
$$pV = \tfrac{2}{3}U = \tfrac{2}{3}(\tfrac{1}{2}Nm\langle v^2 \rangle)), \text{ which gives}$$
$$p = \tfrac{1}{3}(N/V)m\langle v^2 \rangle$$
$$= \tfrac{1}{3}[(3 \times 10^{-20} /\text{ly}^3)/(9.5 \times 10^{15} \text{ m/ly})^3](3 \times 10^{41} \text{ kg})(100 \times 10^3 \text{ m/s})^2 = \boxed{3 \times 10^{-17} \text{ Pa.}}$$

23. The mass of a water droplet is
$$m_{\text{water}} = \rho \tfrac{4}{3}\pi R^3 = (1.0 \times 10^3 \text{ kg/m}^3)\tfrac{4}{3}\pi(0.5 \times 10^{-6} \text{ m})^3 = 5.2 \times 10^{-16} \text{ kg.}$$
Equilibrium means the temperature of the water droplets is 300 K, so we have
$$v_{\text{rms}}^2 = 3kT/m_{\text{water}}$$
$$= 3(1.38 \times 10^{-23} \text{ J/K})(300 \text{ K})/(5.2 \times 10^{-16} \text{ kg}), \text{ which gives } \boxed{v_{\text{rms}} = 4.9 \times 10^{-3} \text{ m/s.}}$$

29. (a) For each die, the probability of getting a six is 1/6. Because the dice throws are independent, the probability of getting four sixes is
$$P = (1/6)(1/6)(1/6)(1/6) = \boxed{1/1296.}$$
    (b) If we wanted the four numbers to be on particular dice, we would have the same answer as part (a). Because we do not care which number appears on which die, we need to find the number of ways that we can arrange 6, 6, 6, and 5, without regard to order. If we write these out, 6, 6, 6, 5; 6, 6, 5, 6; 6, 5, 6, 6; 5, 6, 6, 6; we find there are 4 ways to arrange the numbers. Thus the probability is
$$P = 4(1/1296) = \boxed{1/324.}$$
    [Note: The general expression for the number of ways that n things can be arranged is $n!$. If k of the things are identical, the number of ways is reduced by a factor of $k!$. Thus we have $(n!/k!)(1/1296) = (4!/3!)(1/1296) = 1/324.$]

33. (a) Using the value at the midpoint of each range, we find the average grade from
$$\langle g \rangle = (1/N)(\Sigma g N_g)$$
$$= (1/87)[(-40)(3) + (-20)(18) + (0)(29) + (20)(22) + (40)(15)] = \boxed{+6.4.}$$
   (b) $\langle D^2 \rangle = \Sigma(D^2 N_D)/N = [\Sigma(g - \langle g \rangle)^2 N_g]/N$
$$= (1/87)[(-46.4)^2(3) + (-26.4)^2(18) + (-6.4)^2(29) + (13.6)^2(22) + (33.6)^2(15)] = \boxed{474.}$$
   (c) Because the midpoint of the range is $+20$, we have
$$\langle g \rangle = (1/N)(\Sigma g N_g) = (1/87)(+20)(87) = \boxed{+20.}$$

35. We call the angle of the needle from the x-axis θ, which has the range $0 < \theta < \pi$. Because any angle is equally likely, we find the average value of the y-component of the needle's length from

$$\langle L_y \rangle = \langle L \sin \theta \rangle = L \langle \sin \theta \rangle = L \frac{\int_0^\pi \sin \theta \, d\theta}{\int_0^\pi d\theta} = \frac{L}{\pi}\left(- \cos \theta\right)\Big|_0^\pi = \frac{2L}{\pi}.$$

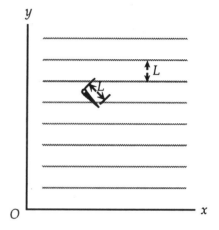

The needle will hit a line if the center falls within L/π on either side of a line. Because the center of the needle has an equal probability of falling anywhere, we have
$$P(\text{falling on a line}) = [(L/\pi) + (L/\pi)]/L = 2/\pi.$$

39. We find the average speed from

$$\langle v \rangle = \int_0^\infty v \, F(\mathbf{v}) \, d^3v.$$

d^3v is a differential volume element in a 3-dimensional velocity space. Because the function $F(\mathbf{v})$ depends only on the speed, which is the "distance" from the origin, we can choose a spherical shell of radius v and thickness dv for the differential volume element: $d^3v \rightarrow 4\pi v^2 \, dv$.

Using the Maxwell distribution, the integral becomes

$$\langle v \rangle = \int_0^\infty 4\pi \frac{m}{2\pi kT}^{3/2} v^3 \, e^{-\frac{mv^2}{2kT}} \, dv.$$

Both kT and $\frac{1}{2}mv^2$ have the dimensions of energy; thus $m/2kT$ has the dimensions of v^{-2}. We change to the dimensionless variable $z = (m/2kT)^{1/2}v$, which means
$dv = (2kT/m)^{1/2} \, dz$, so the integral becomes

$$\langle v \rangle = \frac{4\pi}{\pi^{3/2}}\left(\frac{2kT}{m}\right)^{1/2} \int_0^\infty z^3 \, e^{-z^2} \, dz = \frac{4}{\sqrt{\pi}}\left(\frac{2kT}{m}\right)^{1/2}\left(\frac{1}{2}\right), \text{ which gives } \boxed{\langle v \rangle = v_{\text{av}} = \left(2/\sqrt{\pi}\right)\sqrt{2kT/m}.}$$

41. (a) We let $\mathbf{v} - \mathbf{u} = \mathbf{w}$. Because $\mathbf{u}$ is a constant vector, the differential velocity volume is $d^3v = dv_x\,dv_y\,dv_z = dw_x\,dw_y\,dw_z = d^3w$. To find Z, we normalize the distribution:

$$\int F(\mathbf{v})\,d^3v = \frac{1}{Z}\int e^{-\frac{m(\mathbf{v}-\mathbf{u})^2}{2kT}}\,d^3v = \frac{1}{Z}\int e^{-\frac{mw^2}{2kT}}\,d^3w.$$

Because the components for $\mathbf{v}$ and $\mathbf{w}$ range from $-\infty$ to $+\infty$, the integral is the same as for the usual distribution; thus Z will be the same.

(b) We use the same substitution to find the averages:

$$\langle \mathbf{v}\rangle = \frac{1}{Z}\int \mathbf{v}e^{-\frac{m(\mathbf{v}-\mathbf{u})^2}{2kT}}\,d^3v = \frac{1}{Z}\int (\mathbf{w}+\mathbf{u})e^{-\frac{mw^2}{2kT}}\,d^3w$$

$$= \frac{1}{Z}\int \mathbf{w}e^{-\frac{mw^2}{2kT}}\,d^3w + \frac{1}{Z}\int \mathbf{u}e^{-\frac{mw^2}{2kT}}\,d^3w\,.$$

Because the vector $\mathbf{w}$ has no preferred direction, the first term is 0. Because the vector $\mathbf{u}$ is constant, it can be brought outside the integral and the second term is $\mathbf{u}$, so we have

$$\boxed{\langle \mathbf{v}\rangle = \mathbf{u}.}$$

$$\langle \mathbf{v}^2\rangle = \frac{1}{Z}\int \mathbf{v}^2 e^{-\frac{m(\mathbf{v}-\mathbf{u})^2}{2kT}}\,d^3v = \frac{1}{Z}\int (\mathbf{w}+\mathbf{u})^2 e^{-\frac{mw^2}{2kT}}\,d^3w$$

$$= \frac{1}{Z}\int \mathbf{w}^2 e^{-\frac{mw^2}{2kT}}\,d^3w + \frac{1}{Z}\int 2\mathbf{u}\cdot\mathbf{w}\,e^{-\frac{mw^2}{2kT}}\,d^3w + \frac{1}{Z}\int \mathbf{u}^2 e^{-\frac{mw^2}{2kT}}\,d^3w\,.$$

The first term is the integral for v_{rms}^2 with no wind. The second term is the same as the first term in part (b) and is 0. The third term is u^2, so we have

$$\boxed{\langle \mathbf{v}^2\rangle = v_{rms,0}^2 + u^2.}$$

These results are consistent with a random velocity distribution superimposed on a uniform velocity.

45. From the equipartition of energy, each degree of freedom contributes an energy of $\frac{1}{2}RT$ for each mole; and thus $\frac{1}{2}R$ to the molar specific heat. For s degrees of freedom, we have

$$c'_V = s(\tfrac{1}{2}R).$$

If we treat the gases as ideal, we have $c'_p - c'_V = R$ and

$$\gamma = c_p/c_V = c'_p/c'_V = 1 + (R/c'_V) = 1 + (2/s).$$

For the gases we have

argon: $\gamma = (0.124)/(0.074) = 1.67 \approx 1 + (2/3)$, or $s = 3$.

   This corresponds to 3 translational degrees of freedom, as expected for a monatomic gas.

oxygen: $\gamma = (0.219)/(0.157) = 1.39 \approx 1 + (2/5)$, or $s = 5$.

   This corresponds to 3 translational and 2 rotational degrees of freedom, as expected for a diatomic gas with a dumbbell shape.

water vapor: $\gamma = (0.445)/(0.335) = 1.33 \approx 1 + (2/6)$, or $s = 6$.

   This corresponds to 3 translational and 3 rotational degrees of freedom, as expected for a triatomic gas with the atoms not linear.

carbon dioxide: $\gamma = (0.201)/(0.156) = 1.29 \approx 1 + (2/7)$, or $s = 7$.

   This corresponds to 3 translational degrees of freedom, 2 rotational degrees of freedom for a linear arrangement, and 2 vibrational degrees of freedom.

47. For the normalized probability distribution, to find the probability that a molecule has speed greater than $0.90\,v_{rms}$, we integrate the speed distribution:

$$P(v > 0.90v_{rms}) = \int_{0.90v_{rms}}^{\infty} 4\pi\left(\frac{m}{2\pi kT}\right)^{3/2} v^2\, e^{-\frac{mv^2}{2kT}}\, dv.$$

We change to the dimensionless variable $u = (m/2kT)^{1/2}v$, which means
$$dv = (2kT/m)^{1/2}\,du,$$
and the lower limit becomes

$$0.90v_{rms}\sqrt{\frac{m}{2kT}} = 0.90\sqrt{\frac{3kT}{m}}\sqrt{\frac{m}{2kT}} = 0.90\sqrt{\frac{3}{2}} = 1.102,$$

so the integral becomes

$$P(v > 0.90v_{rms}) = \frac{4}{\sqrt{\pi}}\int_{1.102}^{\infty} u^2\, e^{-u^2}\, du\ .$$

Because the integral does not contain T in either the limits or the integrand, the result is independent of temperature.

If the limit is a fixed speed v, after the change of variable the lower limit is $v\sqrt{\frac{m}{2kT}}$. Because the limit depends on T, the result of the integration will also depend on the temperature.

51. The cross-section presented by an air molecule is
$$\sigma = \pi D^2 = \pi(3.0 \times 10^{-10}\ \text{m})^2 = 28 \times 10^{-20}\ \text{m}^2.$$
The mean free path is
$$\lambda = 1/n\sigma\sqrt{2},\ \text{ or }\ n = 1/\lambda\sigma\sqrt{2}.$$
For an ideal gas, $p = nkT$, where n is the molecular density.
For a mean free path of 0.01 cm, we have
$$n = 1/(0.01 \times 10^{-2}\ \text{m})(28 \times 10^{-20}\ \text{m}^2)\sqrt{2} = 2.5 \times 10^{22}\ /\text{m}^3,\ \text{ and}$$
$$p = (2.5 \times 10^{22}\ /\text{m}^3)(1.38 \times 10^{-23}\ \text{J/K})(300\ \text{K}) = 1.0 \times 10^2\ \text{N/m}^2 \simeq \boxed{10^{-3}\ \text{atm.}}$$
For a mean free path of 100 cm, we have
$$n = 1/(100 \times 10^{-2}\ \text{m})(28 \times 10^{-20}\ \text{m}^2)\sqrt{2} = 2.5 \times 10^{18}\ /\text{m}^3,\ \text{ and}$$
$$p = (2.5 \times 10^{18}\ /\text{m}^3)(1.38 \times 10^{-23}\ \text{J/K})(300\ \text{K}) = 1.0 \times 10^{-2}\ \text{N/m}^2 \simeq \boxed{10^{-7}\ \text{atm.}}$$

55. The collision cross section is $\sigma = \pi D^2$, and the mean free path is
$$\lambda = 1/n\sigma\sqrt{2},$$
so the number of atoms is
$$\begin{aligned} N\ &= nV = V/\lambda\pi D^2\sqrt{2}\\ &= (0.10\ \text{m})^3/(4 \times 10^{-4}\ \text{m})\pi(10^{-10}\ \text{m})^2\sqrt{2} = \boxed{5.6 \times 10^{19}\ \text{atoms.}}\end{aligned}$$
For a mean free path of 40 cm, we have
$$\begin{aligned} N\ &= V/\lambda\pi D^2\sqrt{2}\\ &= (0.10\ \text{m})^3/(0.40\ \text{m})\pi(10^{-10}\ \text{m})^2\sqrt{2} = \boxed{5.6 \times 10^{16}\ \text{atoms.}}\end{aligned}$$
Because the pressure is directly proportional to the density, which is inversely proportional to the mean free path, we have
$$p_2 = p_1(\lambda_1/\lambda_2) = (3 \times 10^4\ \text{Pa})[(4 \times 10^{-4}\ \text{m})/(0.40\ \text{m})] = \boxed{30\ \text{Pa.}}$$

59. (a) Because the potential energy is zero at ∞, we find the escape speed by setting the total energy to 0:

$\frac{1}{2}mv^2 - (GMm/R) = 0$, or

$v_{esc}^2 = 2gR$

$= 2(3.5 \text{ m/s}^2)(2.4 \times 10^6 \text{ m})$, which gives $\boxed{v_{esc} = 4.1 \times 10^3 \text{ m/s.}}$

(b) The average kinetic energy of the H_2 molecule is

$\frac{1}{2}mv_{rms}^2 = \frac{3}{2}kT$.

If the rms speed equals the escape speed, we have

$T = \frac{1}{3}mv_{esc}^2/k$

$= \frac{1}{3}[(2 \times 10^{-3} \text{ kg/mol})/(6.02 \times 10^{23} /\text{mol})](4.1 \times 10^3 \text{ m/s})^2/(1.38 \times 10^{-23} \text{ J/K}) = \boxed{1350 \text{ K.}}$

$\boxed{\text{As faster molecules escape, } v_{rms} \text{ and thus } T \text{ decrease.}}$

(c) At a lower temperature, there would be a similar effect, because for the Maxwell distribution the speed of some H_2 molecules is greater than the escape speed, but the fraction would be less; so H_2 molecules would escape more slowly.

(d) The lighter element will have a higher rms speed. Over time, the lighter component will escape more rapidly and will make up a smaller fraction of the remaining atmosphere.

61. The pressure is

$p = [(10^{-9} \text{ torr})/(760 \text{ torr/atm})](1.01 \times 10^5 \text{ Pa/atm}) = 1.3 \times 10^{-7} \text{ Pa.}$

(a) The mean free path is

$\lambda = 1/n\sigma\sqrt{2} = kT/p\pi D^2\sqrt{2}$

$= (1.38 \times 10^{-23} \text{ J/K})(300 \text{ K})/\pi(10^{-10} \text{ m})^2(1.3 \times 10^{-7} \text{ Pa})\sqrt{2} = \boxed{7.0 \times 10^5 \text{ m.}}$

(b) Because the mean free path is inversely proportional to the pressure, we have

$\lambda_2/\lambda_1 = p_1/p_2 = (10^{-9} \text{ torr})/(10^{-6} \text{ torr}) = 10^{-3}.$

$\boxed{\text{The mean free path decreases by a factor of 1000 to } 7.0 \times 10^2 \text{ m.}}$

65. We will use $\mathbf{P}$ for the momentum, to distinguish it from pressure. If we consider a wall perpendicular to the x-axis, the momentum change from a rebounding collision of a particle with momentum component P_x with the wall is

$\Delta P_x = 2P_x.$

In a time Δt, one-half of the particles with a speed component of c_x in a cylinder of area A and height $c_x \Delta t$ will strike the wall. (The other half are moving away from the wall.) If the density of particles is n, the total change in momentum in time Δt is

$\Delta P_{x,total} = 2P_x(\frac{1}{2}nc_xA \Delta t).$

The rate of change of momentum is the force on the particles and, by Newton's third law, the force on the wall. The pressure is the force per unit area, so we have

$p = (1/A)(\Delta P_{x,total}/\Delta t) = nP_xc_x.$

The pressure from all of the particles is

$p = n\langle P_xc_x \rangle.$

Because there is no preferred direction for the speeds, we have

$\langle P_xc_x \rangle = \langle P_yc_y \rangle = \langle P_zc_z \rangle = \frac{1}{3}\langle \mathbf{P} \cdot \mathbf{c} \rangle = \frac{1}{3}\langle Pc \rangle.$

The pressure is

$\boxed{p = \frac{1}{3}n\langle Pc \rangle = \frac{1}{3}n\langle E \rangle.}$

The average energy times the number of particles is the internal energy of the gas, so we have

$pV = \frac{1}{3}N\langle E \rangle = \frac{1}{3}U.$

67. When the gases mix and equilibrium is reached, the average energy of each molecule is the same. The energy of a monatomic gas, like helium, is $\frac{3}{2}RT$ per mole and the energy of a diatomic gas, like oxygen, is $\frac{5}{2}RT$ per mole. Because the total energy has not changed, when the two gases reach the equilibrium temperature T, we have

$n_1(\frac{3}{2}RT_1) + n_2(\frac{5}{2}RT_2) = (n_1\frac{3}{2}R + n_2\frac{5}{2}R)T;$

$(2.5 \text{ mol})(\frac{3}{2}R)(293 \text{ K}) + (1 \text{ mol})\frac{5}{2}R(273 \text{ K}) = [(2.5 \text{ mol})\frac{3}{2}R + (1 \text{ mol})\frac{5}{2}R]T$, which gives

$T = \boxed{285 \text{ K } (12°C).}$

We have 3.5 mol of an ideal gas at this temperature, so we have

$p = nRT/V$

$= (3.5 \text{ mol})(8.314 \text{ J/mol} \cdot \text{K})(285 \text{ K})/(0.3 \text{ m}^3) = \boxed{2.8 \times 10^4 \text{ Pa.}}$

69. (a) We find the density from the ideal gas law:

$n = N/V = p/kT$

$= [(10^{-11} \text{ torr})/(760 \text{ torr/atm})](1.01 \times 10^5 \text{ Pa/atm})/(1.38 \times 10^{-23} \text{ J/K})(300 \text{ K})$

$= \boxed{3.2 \times 10^{11} \text{ atoms/m}^3.}$

(b) The mass of a atom is

$m = M/N_A = (4 \times 10^{-3} \text{ kg/mol})/(6.02 \times 10^{23} /\text{mol}) = 6.64 \times 10^{-27} \text{ kg.}$

The average speed of the molecules is

$v_{av} = (8kT/\pi m)^{1/2}$

$= [8(1.38 \times 10^{-23} \text{ J/K})(300 \text{ K})/\pi(6.64 \times 10^{-27} \text{ kg})]^{1/2} = 1.26 \times 10^3 \text{ m/s.}$

We find the mean collision time from

$\tau = 1/n\sigma v_{av}\sqrt{2}$

$= 1/(3.2 \times 10^{11} \text{ molecules/m}^3)\pi(10^{-10} \text{ m})^2(1.26 \times 10^3 \text{ m/s})\sqrt{2} = \boxed{5.6 \times 10^4 \text{ s.}}$

(c) The mean free path is

$\lambda = v_{av}\tau$

$= (1.26 \times 10^3 \text{ m/s})(5.6 \times 10^4 \text{ s}) = \boxed{7.1 \times 10^7 \text{ m.}}$

CHAPTER 20

3. (a) The possible outcomes are

1 2 3 4	1 2 3 5	1 2 3 6	1 2 4 5	1 2 4 6
1 2 5 6	1 3 4 5	1 3 4 6	1 3 5 6	1 4 5 6
2 3 4 5	2 3 4 6	2 3 5 6	2 4 5 6	3 4 5 6

   (b) There are 15 equally likely outcomes, so the probability of drawing any one like 1 2 3 4 is $\boxed{1/15.}$

   (c) Drawing two white and two black tokens means two even numbers and two odd numbers. There are nine such combinations, so the probability is $9/15 = \boxed{3/5.}$

9. (a) The maximum efficiency is
   $$\eta = W/Q_h = 1 - (T_c/T_h);$$
   $$0.55 = 1 - [T_c/(723 \text{ K})], \text{ which gives } \boxed{T_c = 325 \text{ K} = 52°C.}$$

   (b) We find the heat flow into the system from the hot reservoir from
   $$Q_h = W/\eta = (5 \text{ kWh})/0.55 = \boxed{9.1 \text{ kWh.}}$$
   We find the heat flow rejected to the cold reservoir from energy conservation:
   $$W = Q_h - Q_c;$$
   $$5 \text{ kWh} = 9.1 \text{ kWh} - Q_c, \text{ which gives } \boxed{Q_c = 4.1 \text{ kWh.}}$$

11. For the operating temperatures, the maximum efficiency is
    $$\eta = 1 - (T_c/T_h) = 1 - [(250 \text{ K})/(450 \text{ K})] = 0.444.$$
    The maximum work possible from the heat flow at the high temperature is
    $$W = \eta Q_h = (0.444)(3.0 \times 10^8 \text{ J}) = 1.33 \times 10^8 \text{ J}, \text{ which is greater than the claimed work delivered.}$$
    Thus there is no violation of the second law of thermodynamics.
    If we look at energy conservation, we have
    $$W = Q_h - Q_c;$$
    $$1.0 \times 10^8 \text{ J} \neq (3.0 \times 10^8 \text{ J}) - (1.4 \times 10^8 \text{ J});$$
    $\boxed{\text{there is a violation of the first law of thermodynamics.}}$

15. We will take all Qs and Ws as positive magnitudes. Engine A drives engine B. We adjust the values so the heat flow rejected by engine A has the same magnitude as the heat flow absorbed by engine B from the cold reservoir. From the definition of efficiency, we have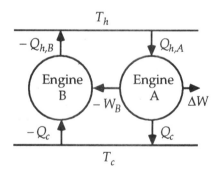
    $$\eta = W/Q_h = (Q_h - Q_c)/Q_h = 1 - (Q_c/Q_h).$$
    We use this for the two engines (both Qs have negative signs for engine B) and subtract:
    $$\eta_A - \eta_B = Q_c[(1/Q_{h,B}) - (1/Q_{h,A})].$$
    If engine A is more efficient, this is positive, so $Q_{h,B} < Q_{h,A}$; there is a net heat flow from T_h.
    We apply the first law of thermodynamics to each engine:
    $$W_A = Q_{h,A} - Q_c;$$
    $$-W_B = -Q_{h,B} + Q_c.$$
    If we add these two equations, we get
    $$-W_B + W_A = Q_{h,A} - Q_{h,B} > 0$$
    We have $W_A > W_B$, which means there is net work available from engine A, beyond what is needed to run engine B. The net effect is to have a net heat flow from T_h with work done, which violates the Kelvin statement.
    The argument cannot be reversed, because engine A is not reversible. The heat flows will be different, and the efficiency of engine A will be different.

25. First we check to see if the first law of thermodynamics is satisfied. The net heat transfer is
$(dQ/dt)_h - (dQ/dt)_c = 94 \text{ cal/s} - 80 \text{ cal/s} = 14 \text{ cal/s}.$
The rate at which work is done by the compressor is
$P = dW/dt = (60 \text{ W})/(4.185 \text{ J/cal}) = 14 \text{ cal/s},$ so the law is satisfied.
We check to see if the second law of thermodynamics is satisfied.
The coefficient of performance of the refrigerator is
$K_{\text{ref}} = (dQ/dt)_h/(dW/dt) = (94 \text{ cal/s})/(14 \text{ cal/s}) = 6.7.$
The ideal coefficient of performance for these temperatures is
$K_{\text{ref}} = T_c/(T_h - T_c) = (253 \text{ K})/(298 \text{ K} - 253 \text{ K}) = 5.6.$
Because this is less than the specifications and actual coefficients of performance will be significantly less than ideal, the statement ⎢ cannot be trusted. ⎢

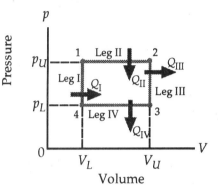

29. (a) We find the temperatures from the ideal gas law:
$p_L V_U = nRT_{34},$ which gives ⎢ $T_3 = p_L V_U/nR;$ ⎢
$p_L V_L = nRT_{41},$ which gives ⎢ $T_4 = p_L V_L/nR.$ ⎢
(b) Even though leg IV is not at constant volume, we find the internal energy change for an ideal gas from
$\Delta U_{IV} = C_V(T_4 - T_3)$
$= C_V p_L(V_L - V_U)/nR =$ ⎢ $-C_V p_L(V_U - V_L)/nR.$ ⎢
(c) For the first law of thermodynamics, we have
$\Delta U_{IV} = -W_{IV} - Q_{IV};$
$-C_V p_L(V_U - V_L)/nR = -p_L(V_L - V_U) - Q_{IV},$ which gives
⎢ $Q_{IV} = +[1 + (C_V/nR)]p_L(V_U - V_L).$ ⎢
(d) To find the efficiency, we must find the absorbed heat flow and the net work done. The net work done is the enclosed rectangular area:
$W = (p_U - p_L)(V_U - V_L).$
Because heat is absorbed on legs I and II, we find the heat flows for these legs.
For leg I, we have
$Q_I = C_V(T_1 - T_4) = +C_V V_L(p_U - p_L)/nR.$
· For leg II, we use the analysis of part (c):
$Q_{II} = W_{II} + \Delta U_{II} = p_U(V_U - V_L) + [C_V p_U(V_U - V_L)/nR] = [1 + (C_V/nR)]p_U(V_U - V_L).$
The net heat absorbed is
$Q_{\text{abs}} = Q_I + Q_{II} = [C_V V_L(p_U - p_L)/nR] + [1 + (C_V/nR)]p_U(V_U - V_L)$
$= (C_V/nR)(p_U V_U - p_L V_L) + p_U(V_U - V_L).$
The efficiency is
$\eta = W/Q_{\text{abs}} = (p_U - p_L)(V_U - V_L)/[(C_V/nR)(p_U V_U - p_L V_L) + p_U(V_U - V_L)]$
$=$ ⎢ $nR(p_U - p_L)(V_U - V_L)/[C_V(p_U V_U - p_L V_L) + nRp_U(V_U - V_L)].$ ⎢

31. (a) We must find the heat flow absorbed and the net work
    done. Because there is no heat flow for an adiabatic
    process, there are two heat flows. These occur for the
    constant volume processes, for which no work is done,
    so we have
    $$Q_{23} = \Delta U_{23} = C_V(T_3 - T_2) \quad \text{(liberated, because } T_3 < T_2)$$
    and $Q_{41} = \Delta U_{41} = C_V(T_1 - T_4)$ (absorbed).
    There is no internal energy change for the cycle,
    so we have
    $$\Delta U = -W + Q = 0, \text{ which gives } W = Q_{41} + Q_{23}.$$
    The efficiency is
    $$\eta_{\text{Otto}} = W/Q_{41}$$
    $$= C_V[(T_1 - T_4) + (T_3 - T_2)]/C_V(T_1 - T_4)$$
    $$= 1 - [(T_2 - T_3)/(T_1 - T_4)].$$

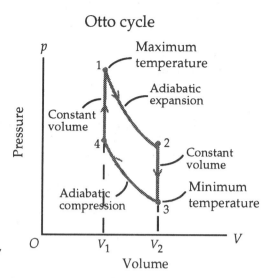

Otto cycle

Maximum temperature

1

Adiabatic expansion

Constant volume

4   2

Constant volume

Adiabatic compression

Minimum temperature

3

Pressure

Volume

V_1 V_2

(b) For an adiabatic process of an ideal gas, we have
    $pV^\gamma = $ a constant, or, with the use of the ideal gas law,
    $TV^{\gamma-1} = $ a constant.
    If we use this for the two adiabatic processes, we have
    $$T_2 V_2^{\gamma-1} = T_1 V_1^{\gamma-1} \quad \text{and} \quad T_3 V_2^{\gamma-1} = T_4 V_1^{\gamma-1}.$$
    The efficiency can be written
    $$\eta_{\text{Otto}} = 1 - [(T_2 - T_3)/(T_1 - T_4)]$$
    $$= 1 - \{[T_1(V_1/V_2)^{\gamma-1} - T_4(V_1/V_2)^{\gamma-1}]/(T_1 - T_4)\} = 1 - (V_1/V_2)^{\gamma-1}.$$

33. We call the volume at point 2 V_a and the pressure at
    point 3 p_b. We must find the heat flow absorbed and
    the net work done. We find these for each leg.
    Leg (1 → 2):
    $$Q_{12} = C_p(T_2 - T_1) \quad \text{absorbed;}$$
    $$\Delta U_{12} = C_V(T_2 - T_1);$$
    $$W_{12} = Q_{12} - \Delta U_{12} = (C_p - C_V)(T_2 - T_1);$$
    Leg (2 → 3):
    $$Q_{23} = 0;$$
    $$\Delta U_{23} = C_V(T_3 - T_2);$$
    $$W_{23} = -\Delta U_{23} = -C_V(T_3 - T_2).$$
    Leg (3 → 4):
    $$W_{34} = 0;$$
    $$\Delta U_{34} = C_V(T_4 - T_3);$$
    $$Q_{34} = \Delta U_{34} + W_{34} = C_V(T_4 - T_3) \quad \text{liberated.}$$
    Leg (4 → 1):
    $$Q_{41} = 0;$$
    $$\Delta U_{41} = C_V(T_1 - T_4);$$
    $$W_{41} = -\Delta U_{41} = -C_V(T_1 - T_4).$$
    The net work is
    $$W = W_{12} + W_{23} + W_{34} + W_{41}$$
    $$= (C_p - C_V)(T_2 - T_1) - C_V(T_3 - T_2) + 0 - C_V(T_1 - T_4)$$
    $$= C_p(T_2 - T_1) - C_V(T_3 - T_4).$$
    The efficiency is
    $$\eta_{\text{Diesel}} = W/Q_{\text{abs}} = W/Q_{12}$$
    $$= [C_p(T_2 - T_1) - C_V(T_3 - T_4)]/C_p(T_2 - T_1) = \boxed{1 - [(T_3 - T_4)/\gamma(T_2 - T_1)].}$$

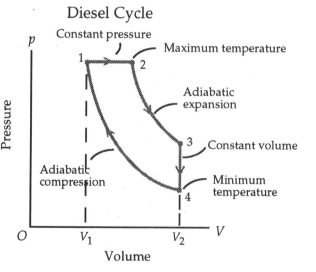

Diesel Cycle

Constant pressure

Maximum temperature

1   2

Adiabatic expansion

3 Constant volume

Adiabatic compression

Minimum temperature

4

Pressure

Volume

V_1 V_2

37. For the system of iron and water, the net heat flow is zero:

$Q_{iron} + Q_{water} = m_{iron}c_{iron}\Delta T_{iron} + m_{water}c_{water}\Delta T_{water} = 0;$

$(10^3 \text{ g})(0.107 \text{ cal/g}\cdot\text{K})(T - 80°C) + (0.5 \times 10^3 \text{ cm}^3)(1 \text{ g/cm}^3)(1 \text{ cal/g}\cdot\text{K})(T - 20°C) = 0$, which gives

$\boxed{T = 31°C.}$

We calculate the entropy change for each component from a reversible path, where it is in contact with an infinite series of reservoirs:

$\Delta S = \int dQ/T = \int mc \, dT/T = mc \ln(T_f/T_i);$

$\Delta S_{iron} = (10^3 \text{ g})(0.107 \text{ cal/g}\cdot\text{K})(4.185 \text{ J/cal}) \ln[(304 \text{ K})/(353 \text{ K})] = -67 \text{ J/K};$

$\Delta S_{water} = (0.5 \times 10^3 \text{ cm}^3)(1 \text{ g/cm}^3)(1 \text{ cal/g}\cdot\text{K})(4.185 \text{ J/cal}) \ln[(304 \text{ K})/(293 \text{ K})] = +77 \text{ J/K}.$

The total entropy change is

$\Delta S = \Delta S_{iron} + \Delta S_{water} = \boxed{+10 \text{ J/K,}}$ positive, as expected for an irreversible process.

39. For the system of the two quantities of water, the net heat flow is zero:

$Q_1 + Q_2 = m_1 c_{water}\Delta T_1 + m_2 c_{water}\Delta T_2 = 0;$

$(0.3 \times 10^3 \text{ g})(1 \text{ cal/g}\cdot\text{K})(T - 70°C) + (0.2 \times 10^3 \text{ g})(1 \text{ cal/g}\cdot\text{K})(T - 15°C) = 0$, which gives

$T = 48°C.$

We calculate the entropy change for each component from a reversible path, where it is in contact with an infinite series of reservoirs:

$\Delta S = \int dQ/T = \int mc \, dT/T = mc \ln(T_f/T_i);$

$\Delta S_1 = (0.3 \times 10^3 \text{ g})(1 \text{ cal/g}\cdot\text{K})(4.185 \text{ J/cal}) \ln[(321 \text{ K})/(343 \text{ K})] = -83 \text{ J/K};$

$\Delta S_2 = (0.2 \times 10^3 \text{ g})(1 \text{ cal/g}\cdot\text{K})(4.185 \text{ J/cal}) \ln[(321 \text{ K})/(288 \text{ K})] = +91 \text{ J/K}.$

Because there is no interaction with the environment, the change in entropy of the universe is

$\Delta S = \Delta S_1 + \Delta S_2 = \boxed{+8 \text{ J/K,}}$ positive, as expected for an irreversible process.

45. In an isothermal compression of an ideal gas there is no change in internal energy but there is work done on the gas, which is a negative work done by the gas. From the first law of thermodynamics, there must be a heat flow from the gas: $Q < 0$. The entropy change is

$\Delta S = Q/T < 0.$

$\boxed{\text{The entropy of the gas decreases.}}$ Note that the entropy of the surroundings increases.

47. For a rapid (adiabatic) free expansion of an ideal gas, we have

$Q = 0, \quad W = 0.$

From the first law of thermodynamics, we have

$\Delta U = -W + Q = 0.$

Because the internal energy depends only on the temperature, this means $\boxed{\Delta T = 0.}$

We find the entropy change of the ideal gas from

$\Delta S = C_V \ln(T_0/T_0) + nR \ln(V/V_0)$

$= 0 + (1 \text{ mol})(8.314 \text{ J/mol}\cdot\text{K}) \ln(50) = \boxed{+32.5 \text{ J/K.}}$

53. Because the internal energy depends only on the temperature, for the isothermal compression of the gas we have

$\Delta U = 0, W < 0,$ and therefore $Q = W < 0.$

The entropy change of the gas is

$\boxed{\Delta S_{gas} = Q/T < 0 \text{ (decreases).}}$

The entropy change of the reservoir is

$\boxed{\Delta S_{reservoir} = -Q/T > 0 \text{ (increases).}}$

For the universe, we have $\Delta S_{universe} = \Delta S_{gas} + \Delta S_{reservoir} = 0.$

There is no conflict with the second law of thermodynamics, because this is a reversible process.

55. (a) For the free expansion, there is no work done by the gas and no heat flow. From the first law of thermodynamics, we have

$\Delta U = - W + Q = 0$; therefore there is no temperature change: $\Delta T = 0$.

$\boxed{T = T_0.}$

(b) We find the entropy change from

$\Delta S_{gas} = C_V \ln(T/T_0) + nR \ln(V/V_0)$

$= 0 + nR \ln(3V_0/V_0) = \boxed{+ nR \ln(3).}$

(c) Because there has been no change in the rest of the universe, we have

$\Delta S_{universe} = \Delta S_{gas} = \boxed{+ nR \ln(3).}$

(d) We have a reversible process in which work is done at the expense of internal energy; the temperature decreases. There is no heat flow, so the entropy change is

$\boxed{\Delta S = \int dQ/T = 0.}$

59. (a) The initial volume of the helium is

$V_1 = nRT/p_1 = (2 \text{ mol})(8.314 \text{ J/mol} \cdot \text{K})(273 \text{ K})/(1.01 \times 10^5 \text{ Pa}) = 4.50 \times 10^{-2} \text{ m}^3$.

From the ideal gas law, for an isothermal process, we have

$V_2 = (p_1/p_2)V_1 = \tfrac{1}{2}V_1 = \boxed{2.25 \times 10^{-2} \text{ m}^3.}$

(b) Because the process is isothermal, we have $T_2 = T_1 = \boxed{273 \text{ K.}}$

(c) We find the entropy change from

$\Delta S = C_V \ln(T_2/T_1) + nR \ln(V/V_0)$

$= 0 + (2 \text{ mol})(8.314 \text{ J/mol} \cdot \text{K}) \ln(\tfrac{1}{2}) = \boxed{-11.5 \text{ J/K.}}$

63. (a) For the ideal operation of the engine we have

$\eta_{engine} = 1 - (T_c/T_h) = 1 - [(300\ K)/(400\ K)] = 0.25.$

The heat absorbed from the hot reservoir is

$Q_{h,engine} = W/\eta_{engine} = (100\ J)/0.25 = \boxed{400\ J.}$

The heat rejected to the cold reservoir is

$Q_{c,engine} = Q_{h,engine} - W = 400\ J - 100\ J = \boxed{300\ J.}$

For the ideal operation of the refrigerator we have

$K_{refrigerator} = T_c/(T_h - T_c) = (300\ K)/[(400\ K) - (300\ K)] = 3.00.$

The heat rejected to the cold reservoir is

$Q_{c,refrigerator} = K_{refrigerator}W = 3.00(100\ J) = \boxed{300\ J.}$

The heat absorbed from the hot reservoir is

$Q_{h,refrigerator} = W + Q_{c,refrigerator} = 100\ J + 300\ J = \boxed{400\ J.}$

For the combined system we have

$Q_{h,net} = Q_{h,refrigerator} - Q_{h,engine} = 400\ J - 400\ J = \boxed{0;}$

$Q_{c,net} = Q_{c,engine} - Q_{c,refrigerator} = 300\ J - 300\ J = \boxed{0.}$

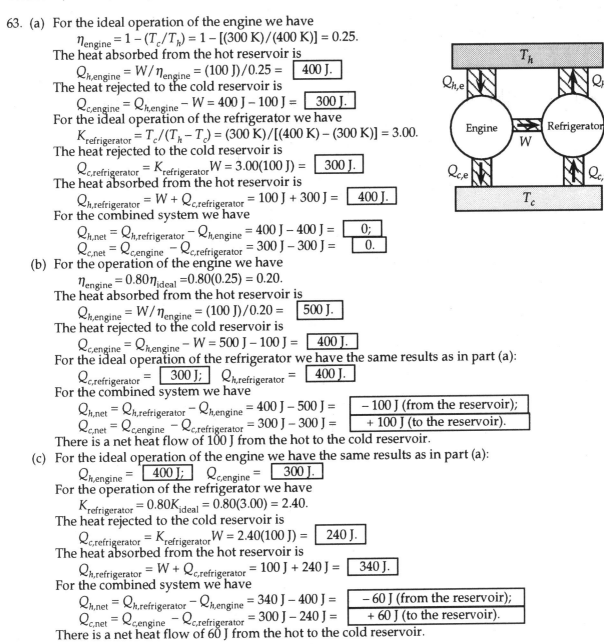

(b) For the operation of the engine we have

$\eta_{engine} = 0.80\eta_{ideal} = 0.80(0.25) = 0.20.$

The heat absorbed from the hot reservoir is

$Q_{h,engine} = W/\eta_{engine} = (100\ J)/0.20 = \boxed{500\ J.}$

The heat rejected to the cold reservoir is

$Q_{c,engine} = Q_{h,engine} - W = 500\ J - 100\ J = \boxed{400\ J.}$

For the ideal operation of the refrigerator we have the same results as in part (a):

$Q_{c,refrigerator} = \boxed{300\ J;} \quad Q_{h,refrigerator} = \boxed{400\ J.}$

For the combined system we have

$Q_{h,net} = Q_{h,refrigerator} - Q_{h,engine} = 400\ J - 500\ J = \boxed{-100\ J\ \text{(from the reservoir);}}$

$Q_{c,net} = Q_{c,engine} - Q_{c,refrigerator} = 300\ J - 300\ J = \boxed{+100\ J\ \text{(to the reservoir).}}$

There is a net heat flow of 100 J from the hot to the cold reservoir.

(c) For the ideal operation of the engine we have the same results as in part (a):

$Q_{h,engine} = \boxed{400\ J;} \quad Q_{c,engine} = \boxed{300\ J.}$

For the operation of the refrigerator we have

$K_{refrigerator} = 0.80K_{ideal} = 0.80(3.00) = 2.40.$

The heat rejected to the cold reservoir is

$Q_{c,refrigerator} = K_{refrigerator}W = 2.40(100\ J) = \boxed{240\ J.}$

The heat absorbed from the hot reservoir is

$Q_{h,refrigerator} = W + Q_{c,refrigerator} = 100\ J + 240\ J = \boxed{340\ J.}$

For the combined system we have

$Q_{h,net} = Q_{h,refrigerator} - Q_{h,engine} = 340\ J - 400\ J = \boxed{-60\ J\ \text{(from the reservoir);}}$

$Q_{c,net} = Q_{c,engine} - Q_{c,refrigerator} = 300\ J - 240\ J = \boxed{+60\ J\ \text{(to the reservoir).}}$

There is a net heat flow of 60 J from the hot to the cold reservoir.

67. We are given $\gamma = 1.4 = c'_p/c'_V$. Because $c'_p - c'_V = R$, we have
$\qquad \gamma - 1 = R/c'_V$.
For the adiabatic expansion, we have
$\qquad T_h V_{min}{}^{\gamma-1} = T_c V_{max}{}^{\gamma-1}$, or
$\qquad V_{min}/V_{max} = (T_c/T_h)^{1/(\gamma-1)}$.
For leg (i), the adiabatic expansion, we have
$\qquad Q_i = 0$;
$\qquad W_i = -\Delta U_i = -nc'_V(T_c - T_h) = nc'_V(T_h - T_c)$.
For leg (ii), the isothermal compression, we have
$\qquad \Delta U = 0$;
$\qquad Q_{ii} = W_{ii} = nRT_c \ln(V_{min}/V_{max})$
$\qquad\qquad = nRT_c[1/(\gamma-1)] \ln(T_c/T_h) = nc'_V T_c \ln(T_c/T_h)$, which is < 0.
For leg (iii), at constant volume, we have
$\qquad W_{iii} = 0$;
$\qquad Q_{iii} = \Delta U_{iii} = nc'_V(T_h - T_c)$, which is > 0.
Heat is absorbed only in step (iii), so the efficiency of the cycle is
$\qquad \eta = W/Q_{iii} = (W_i + W_{ii})/Q_{iii} = [nc'_V(T_h - T_c) + nc'_V T_c \ln(T_c/T_h)]/nc'_V(T_h - T_c)$
$\qquad\quad = 1 - [T_c/(T_h - T_c)] \ln(T_h/T_c)$
$\qquad\quad = 1 - [(303 \text{ K}/(603 \text{ K} - 303 \text{ K})] \ln[(603 \text{ K})/(303 \text{ K})] = \boxed{0.30.}$
This is $\boxed{60\% \text{ of the Carnot efficiency:}}$
$\qquad \eta_{Carnot} = 1 - (T_c/T_h) = 1 - [(303 \text{ K})/(603 \text{ K})] = 0.50.$

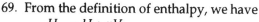

69. From the definition of enthalpy, we have
$\qquad H = U + pV$
$\qquad\quad = nc_V T + nRT = (nc_V + nR)T = \boxed{nc_p T.}$
The enthalpy values are useful in constant pressure processes, when $Q = \Delta H$.

CHAPTER 21

3. To determine the filling ratio we find the length of the basic cube edge a in
    terms of the atomic diameter d and the number of atoms in the basic cube.
    Simple cubic cell: There is one atom at each corner. From the figure we see that
    $a = d$.
    The atom at each of the 8 corners has $1/8$ of its volume in the cell, so
    there is 1 atom/basic cube.
    Filling ratio = volume of atoms/volume of cube
    $$= (1)\tfrac{4}{3}\pi(\tfrac{1}{2}d)^3/d^3 = \boxed{0.524.}$$

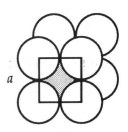

    Face centered cubic cell: There is one atom at each corner and one atom at the
    center of each of the six faces. From the figure we see that the face
    diagonal contains 2 atomic diameters:
    $a\sqrt{2} = 2d$, or $a = d\sqrt{2}$.
    The atom at each of the 8 corners has $1/8$ of its volume in the cell and the
    atom at each of the 6 faces has $1/2$ of its volume in the cell, so there are
    $8(1/8) + 6(1/2) = 4$ atoms/basic cube.
    Filling ratio = $(4)\tfrac{4}{3}\pi(\tfrac{1}{2}d)^3/(d\sqrt{2})^3 = \boxed{0.740.}$

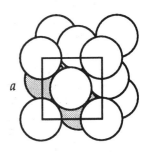

11. If we take the separation of the atoms when unstretched to be $y = 10^{-10}$ m, we have
    $e = \Delta L/L = \Delta y/y$;
    $(2 \text{ m})/(3 \times 10^3 \text{ m}) \approx \Delta y/(10^{-10} \text{ m})$, which gives $\boxed{\Delta y \approx 10^{-13} \text{ m.}}$

13. We call the length of the beam L_{beam}. The initial length of the cable is
    $L_0 = L_{\text{beam}}/\cos 30° = 2L_{\text{beam}}/\sqrt{3}$ and the distance along the wall from the
    beam to the cable is $D = L_{\text{beam}} \tan 30° = L_{\text{beam}}/\sqrt{3}$. When the load is
    hung at the end of the beam, there must be an additional tension in
    the cable to maintain equilibrium:
    $\sum \tau_A = (\Delta T \sin 30°)L_{\text{beam}} - Mg \, L_{\text{beam}} = 0$, or
    $\Delta T = Mg/\sin 30° = (30 \text{ kg})(9.8 \text{ m/s}^2)/\sin 30° = 588$ N,
    which is an additional stress in the cable:
    $\Delta T/\pi r^2 = (588 \text{ N})/\pi(1 \times 10^{-3} \text{ m})^2 = 1.87 \times 10^8 \text{ N/m}^2$.
    (Note that even when the initial tension is added, this is less
    than the tensile strength, so the cable does not break.)
    The additional stress produces an elongation of the cable:
    $\Delta L/L_0 = \Delta T/AY$.
    This elongation causes the beam to drop below the horizontal
    (exaggerated in the diagram). We find the angle θ from a geometrical
    formula for a triangle:
    $D^2 + L_{\text{beam}}^2 - 2DL_{\text{beam}} \cos \theta = (L_0 + \Delta L)^2 = L_0^2[1 + (\Delta L/L_0)]^2 \approx L_0^2[1 + (2 \Delta L/L_0)]$,
    where we have used the fact that $\Delta L \ll L_0$. Because $D^2 + L_{\text{beam}}^2 = L_0^2$, this becomes
    $- 2(L_{\text{beam}}/\sqrt{3})L_{\text{beam}} \cos \theta = (2L_{\text{beam}}/\sqrt{3})^2 2 \Delta L/L_0$, which reduces to
    $\cos \theta = -(4/\sqrt{3})(\Delta L/L_0) = -(4/\sqrt{3})(\Delta T/AY)$
    $= (4/\sqrt{3})(588 \text{ N})/[\pi(1 \times 10^{-3} \text{ m})2(2.1 \times 10^5 \text{ MN/m}^2)] = 0.00206$, which gives $\theta = 90.12°$.
    Thus the beam is $\boxed{0.12° \text{ below the horizontal.}}$

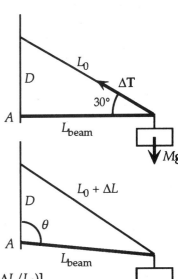

15. From Example 21–1, the compressional strain is
$$e = \Delta L/L = 1.2 \times 10^{-3}.$$
We find the fractional change in the diameter from Poisson's ratio:
$$\Delta d/d = -\sigma e = -0.3(1.2 \times 10^{-3}) = -3.6 \times 10^{-4}.$$
The change in the diameter is
$$\Delta d = (-3.6 \times 10^{-4})(2.0 \text{ cm}) = -7.2 \times 10^{-4} \text{ cm, so the new diameter is}$$
$$d' = d + \Delta d = 2.0 \text{ cm} - (7.2 \times 10^{-4} \text{ cm}) = \boxed{1.99928 \text{ cm.}}$$

17. (a) We have an induced transverse strain, given by $e_{tr} = -\sigma e_1$. The total strain on a side of the cube will be due to the direct strain from the applied pressure and the induced transverse strain from the other two sides:
$$e = \Delta L/L = e_1 + 2e_{tr} = e_1(1 - 2\sigma).$$
The volume of the cube is $V = L^3$, so a small change in L means a small change in V, given by
$$\Delta V = 3L^2 \Delta L, \text{ which gives a volume strain:}$$
$$\Delta V/V = 3 \Delta L/L = 3e_1(1 - 2\sigma).$$
If we use the relation between stress and strain, we have
$$\Delta V/V = 3[(F/A)/Y](1 - 2\sigma).$$
All three sides are under pressure, which is the stress, so we have for the magnitude
$$\Delta V/V = 3p(1 - 2\sigma)/Y.$$
(b) The bulk modulus is defined as
$$B = -p/(\Delta V/V).$$
If we substitute the given volume change, which must be negative for a positive pressure, we get
$$B = -p/[-3p(1 - 2\sigma)/Y] = Y/3(1 - 2\sigma).$$

27. We find the moduli from
$$Y = \rho v_p^2 = (5 \times 10^3 \text{ kg/m}^3)(8 \times 10^3 \text{ m/s})^2 = \boxed{3 \times 10^5 \text{ MN/m}^2;}$$
$$G = \rho v_s^2 = (5 \times 10^3 \text{ kg/m}^3)(4 \times 10^3 \text{ m/s})^2 = \boxed{8 \times 10^4 \text{ MN/m}^2.}$$
The time for each wave to travel a distance L is
$$t_P = L/v_P \quad \text{and} \quad t_S = L/v_S, \quad \text{so the difference in arrival times is}$$
$$\Delta t = L[(1/v_S) - (1/v_P)];$$
$$(27 \text{ min})(60 \text{ s/min}) = L\{[1/(4 \times 10^3 \text{ m/s})] - [1/(8 \times 10^3 \text{ m/s})]\},$$
which gives $\boxed{L = 1.3 \times 10^7 \text{ m} = 1.3 \times 10^4 \text{ km.}}$

31. The increase in area means an increase in radius and circumference. We treat the circumference as a length, with a change $\Delta C = C\alpha \Delta t$. Because $C = 2\pi R$, the fractional change in radius is
$$\Delta R/R = \Delta C/C = \alpha \Delta T.$$
The area is $A = \pi R^2$, so a small change in R causes a small change in A:
$$\Delta A = 2\pi R \Delta R, \quad \text{or} \quad \Delta A/A = 2 \Delta R/R = 2\alpha \Delta T;$$
$$0.016 = 2\alpha(10°C), \text{ which gives } \boxed{\alpha = 8.0 \times 10^{-4} \text{ K}^{-1}.}$$

35. (a) The best material is the one that has the smallest change in length when the temperature changes. The material with the smallest coefficient of thermal expansion is $\boxed{\text{steel.}}$

(b) Because the changes in length, and thus period, will be small, we approximate them with differentials:

$f = 1/\tau = (1/2\pi)(g/L)^{1/2}$; $\Delta f \approx (1/2\pi)(-\frac{1}{2})(g/L^3)^{1/2}\,\Delta L$, which gives

$\Delta f/f \approx -\frac{1}{2}(\Delta L/L) = -\frac{1}{2}\alpha\,\Delta T$,   or $\Delta f \approx -\frac{1}{2}f\alpha\,\Delta T$.

Aluminum:

$\Delta f \approx -\frac{1}{2}(1.00000\ \text{Hz})(2.30\times 10^{-5}\ /\text{K})(\pm 25^\circ\text{C}) = \pm 29\times 10^{-5}\ \text{Hz}$, so the range is

$\boxed{0.99971\ \text{Hz} \le f_{\text{aluminum}} \le 1.00029\ \text{Hz}.}$

Copper:

$\Delta f \approx -\frac{1}{2}(1.00000\ \text{Hz})(1.67\times 10^{-5}\ /\text{K})(\pm 25^\circ\text{C}) = \pm 21\times 10^{-5}\ \text{Hz}$, so the range is

$\boxed{0.99979\ \text{Hz} \le f_{\text{copper}} \le 1.00021\ \text{Hz}.}$

Steel:

$\Delta f \approx -\frac{1}{2}(1.00000\ \text{Hz})(1.05\times 10^{-5}\ /\text{K})(\pm 25^\circ\text{C}) = \pm 13\times 10^{-5}\ \text{Hz}$, so the range is

$\boxed{0.99987\ \text{Hz} \le f_{\text{steel}} \le 1.00013\ \text{Hz}.}$

37. The expansion of the copper causes the enclosed volume to increase as if it were copper. (Consider the expansion of a solid sphere of copper. The volume increase means that the volume enclosed by the outer spherical shell must increase as if it were copper.) The volume of ethyl alcohol that spills is

$\Delta V = \Delta V_{\text{alcohol}} - \Delta V_{\text{copper}} = V_0\beta_{\text{alcohol}}\,\Delta T - V_0(3\alpha_{\text{copper}})\Delta T$.

We find the percentage of the ethyl alcohol that spills from

$\Delta V/V_0 = (\beta_{\text{alcohol}} - 3\alpha_{\text{copper}})\,\Delta T$

$= [(1.12\times 10^{-3}\ \text{K}^{-1}) - 3(1.67\times 10^{-5}\ \text{K}^{-1})](25^\circ\text{C} - 5^\circ\text{C})(100) = \boxed{2.1\%.}$

45. We find the rate of thermal energy loss from

$dQ/dt = A\,\Delta T/R$;

Pine: $dQ/dt = (4\ \text{ft})(8\ \text{ft})(35^\circ\text{F})/6(1.3\ \text{ft}^2\cdot\text{h}\cdot{}^\circ\text{F/Btu}) = \boxed{144\ \text{Btu/h.}}$

Fiberglas: $dQ/dT = (4\ \text{ft})(8\ \text{ft})(35^\circ\text{F})/6(3.0\ \text{ft}^2\cdot\text{h}\cdot{}^\circ\text{F/Btu}) = \boxed{62\ \text{Btu/h.}}$

For the two thermal resistances in parallel, we have

$A/R_{\text{eff}} = (A_1/R_1) + (A_2/R_2) = [(0.10\,A)/6(1.3\ \text{ft}^2\cdot\text{h}\cdot{}^\circ\text{F/Btu})] + [(0.90\,A)/6(3.0\ \text{ft}^2\cdot\text{h}\cdot{}^\circ\text{F/Btu})]$,

which gives $\boxed{R_{\text{eff}} = 16\ \text{ft}^2\cdot\text{h}\cdot{}^\circ\text{F/Btu.}}$

47. (a) We call the temperature at the interface T_3. In the steady state, the rate of thermal energy flow is the same for each slab:

$dQ/dt = dQ_1/dt = dQ_2/dt$;

$\kappa_{\text{eff}}A(T_1 - T_2)/(L_1 + L_2) = \kappa_1 A(T_1 - T_3)/L_1 = \kappa_2 A(T_3 - T_2)/L_2$.

We treat this as two equations:

$T_1 - T_3 = (L_1/\kappa_1)[\kappa_{\text{eff}}/(L_1 + L_2)](T_1 - T_2)$;

$T_3 - T_2 = (L_2/\kappa_2)[\kappa_{\text{eff}}/(L_1 + L_2)](T_1 - T_2)$.

If we add these two equations, we get

$T_1 - T_2 = [(L_1/\kappa_1) + (L_2/\kappa_2)][\kappa_{\text{eff}}/(L_1 + L_2)](T_1 - T_2)$, which gives

$(L_1 + L_2)/\kappa_{\text{eff}} = (L_1/\kappa_1) + (L_2/\kappa_2)$, which gives

$\boxed{\kappa_{\text{eff}} = \kappa_1\kappa_2(L_1 + L_2)/(\kappa_2 L_1 + \kappa_1 L_2).}$

(b) To find T_3 we substitute κ_{eff} into one of the equations:

$T_3 = T_2 + (L_2/\kappa_2)[\kappa_1\kappa_2/(\kappa_2 L_1 + \kappa_1 L_2)](T_1 - T_2) = \boxed{(\kappa_1 L_2 T_1 + \kappa_2 L_1 T_2)/(\kappa_1 L_2 + \kappa_2 L_1).}$

53. We consider the cylinder to be made up of cylindrical shells. A representative shell has radius r, thickness dr, and area $2\pi rL$, where L is a length of the pipe. The rate of thermal energy flow through each shell, dQ/dt, is the same:

$$dQ/dt = \kappa A\, dT/dr = \kappa 2\pi rL\, dT/dr.$$

We rewrite this and then integrate:

$$dr/2\pi \kappa r = L\, dT/(dQ/dt),$$

$$\int_{R_1}^{R_1 + \alpha} \frac{dr}{2\pi \kappa r} = \int_{T_1}^{T_2} \frac{L}{dQ/dt}\, dT = \frac{L}{dQ/dt}\int_{T_1}^{T_2} dT\,;$$

$$\frac{1}{2\pi\kappa}\ln\left(\frac{R_1 + \alpha}{R_1}\right) = \frac{L}{dQ/dt}\left(T_2 - T_1\right), \text{ which gives} \quad \boxed{\frac{1}{L}\frac{dQ}{dt} = \frac{2\pi\kappa\left(T_2 - T_1\right)}{\ln\left[1 + (\alpha/R_1)\right]}.}$$

55. The power output of the stove must equal the rate of thermal energy loss through the wall:

$$P = dQ/dt = kA\,\Delta T/L$$
$$= (0.35\ \text{W/m}\cdot\text{K})(2.5\ \text{m})(2.0\ \text{m})[8°C - (-25°C)]/(6\times 10^{-2}\ \text{m}) = \boxed{9.6\times 10^2\ \text{W.}}$$

57. The heat loss through the crack is

$$dQ/dt = kA\,\Delta T/L$$
$$= (25\times 10^{-2}\ \text{W/m}\cdot\text{K})(6\times 10^{-4}\ \text{m}^2)(30°C)/(5\times 10^{-3}\ \text{m}) = \boxed{0.90\ \text{W.}}$$

59. The thermal changes in the components of the reinforced concrete are

$$\Delta L_{rod} = L_0 \alpha_{rod}\,\Delta T \quad \text{and} \quad \Delta L_{concrete} = L_0 \alpha_{concrete}\,\Delta T.$$

If we assume that the new length is the length of the iron rods, there must be a stress in the concrete to produce an additional strain to bring the concrete to the same length:

$$\Delta L_{net}/L_0 = (\alpha_{rod} - \alpha_{concrete})\,\Delta T = \text{stress}/Y_{concrete};$$
$$\text{stress} = Y_{concrete}(\alpha_{rod} - \alpha_{concrete})\,\Delta T$$
$$= (20\times 10^9\ \text{N/m}^2)[(12.5\times 10^{-6}\ \text{K}^{-1}) - (12.0\times 10^{-6}\ \text{K}^{-1})](30°C - 10°C) = \boxed{2.0\times 10^5\ \text{N/m}^2.}$$

Note that this is a tensile stress, which is not well supported by concrete.

CHAPTER 22

We have a new force which acts at a distance. All of the previous techniques for analyzing forces, such as drawing a force diagram and selecting a coordinate system, are used.

5. From symmetry considerations, each time two identical cork balls touch, the charge is shared evenly. At the first touch, the first cork ball (and the second cork ball) will have $\frac{1}{2}$ of the original charge:

$q_1 = \frac{1}{2}q_0 = \frac{1}{2}(-4 \times 10^{-10}\,\text{C}) =$ $\boxed{-2 \times 10^{-10}\,\text{C,}}$ and the number of electrons gained is

$N_1 = (2 \times 10^{-10}\,\text{C})/(1.602 \times 10^{-19}\,\text{C/electron}) =$ $\boxed{1.25 \times 10^9\,\text{electrons.}}$

At the second touch, the second cork ball (and the third cork ball) will have $\frac{1}{2}q_1$:

$q_2 = \frac{1}{2}q_1 = \frac{1}{2}(-2 \times 10^{-10}\,\text{C}) = -1 \times 10^{-10}\,\text{C,}$ and the number of electrons gained is

$N_2 = (1 \times 10^{-10}\,\text{C})/(1.602 \times 10^{-19}\,\text{C/electron}) = 6.2 \times 10^8\,\text{electrons.}$

$q_3 = q_2 = \boxed{-1 \times 10^{-10}\,\text{C,}\ 6.2 \times 10^8\,\text{electrons.}}$

15. The two up quarks will repel each other with a force

$F_{\text{up-up}} = kq_1q_2/r_{12}^2$

$= (9 \times 10^9\,\text{N} \cdot \text{m}^2/\text{C}^2)\frac{2}{3}(1.6 \times 10^{-19}\,\text{C})\frac{2}{3}(1.6 \times 10^{-19}\,\text{C})/(1.5 \times 10^{-15}\,\text{m})^2 =$ $\boxed{46\,\text{N repulsion.}}$

The up and down quarks will attract each other with a force

$F_{\text{up-down}} = kq_1q_3/r_{13}^2$

$= (9 \times 10^9\,\text{N} \cdot \text{m}^2/\text{C}^2)\frac{2}{3}(1.6 \times 10^{-19}\,\text{C})\frac{1}{3}(1.6 \times 10^{-19}\,\text{C})/(1.5 \times 10^{-15}\,\text{m})^2 =$ $\boxed{23\,\text{N attraction.}}$

23. The Coulomb force is

$F = kq_1q_2/r^2 = kq_1(q - q_1)/r^2 = (qq_1 - q_1^2)k/r^2$, with q_1 as the variable.

To find q_1 that maximizes the force, we set $dF/dq_1 = 0$:

$dF/dq_1 = (q - 2q_1)k/r^2 = 0$, which gives $\boxed{q_1/q = \frac{1}{2}.}$

This means that $\boxed{q_2/q = \frac{1}{2},}$ which we would expect from the symmetry of the force law.

25. (a) The opposite charges attract. We find the magnitude of the Coulomb force from

$F = ke^2/r^2 = (9 \times 10^9\,\text{N} \cdot \text{m}^2/\text{C}^2)(1.6 \times 10^{-19}\,\text{C})^2/(3 \times 10^{-10}\,\text{m})^2$

$= \boxed{2.6 \times 10^{-9}\,\text{N toward the proton (centripetal).}}$

(b) The attractive Coulomb force provides the centripetal acceleration of the circular motion:

$F = mv^2/r$;

$(2.6 \times 10^{-9}\,\text{N}) = (9.11 \times 10^{-31}\,\text{kg})v^2/(3 \times 10^{-10}\,\text{m})$, which gives $\boxed{v = 9.2 \times 10^5\,\text{m/s.}}$

(c) We find the frequency from

$f = v/2\pi r = (9.2 \times 10^5\,\text{m/s})/2\pi(3 \times 10^{-10}\,\text{m}) = \boxed{4.9 \times 10^{14}\,\text{Hz.}}$

(d) We find the spring constant from

$k = (2\pi f)^2 m = [2\pi(4.9 \times 10^{14}\,\text{Hz})]^2(9.11 \times 10^{-31}\,\text{kg}) = \boxed{8.6\,\text{N/m.}}$

27. The charges are separated by $r = 2L \sin\theta$.

From the force diagram, we apply $\Sigma\mathbf{F} = 0$, to get

horizontal: $T \sin\theta = F = kqq/r^2$;

vertical: $T \cos\theta = mg$.

If we divide the two equations, we get

$\tan\theta = F/mg = kq^2/r^2mg = kq^2/(2L \sin\theta)^2mg$

$\tan 10° = (9 \times 10^9\,\text{N} \cdot \text{m}^2/\text{C}^2)q^2/[2(0.20\,\text{m}) \sin 10°]^2(0.20 \times 10^{-3}\,\text{kg})(9.8\,\text{m/s}^2)$,

which gives $\boxed{q = 1.4 \times 10^{-8}\,\text{C.}}$

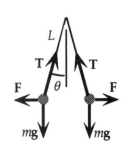

29. Because q_1 and q_2 attract each other, they must have opposite signs and their product will be negative. We can take this into account by giving the force a negative value:

$F_{12} = kq_1q_2/r_{12}^2$;

$-1.4 \times 10^{-2}\,\text{N} = (9 \times 10^9\,\text{N} \cdot \text{m}^2/\text{C}^2)q_1q_2/(15.0 \times 10^{-2}\,\text{m})^2$, which gives $q_1q_2 = -3.5 \times 10^{-14}\,\text{C}^2$.

Because q_2 and q_3 attract each other, they must have opposite signs and their product will be negative. We can take this into account by giving the force a negative value:

$F_{23} = kq_2q_3/r_{23}^2$;

$-3.8 \times 10^{-3}\,\text{N} = (9 \times 10^9\,\text{N} \cdot \text{m}^2/\text{C}^2)q_2q_3/(20.0 \times 10^{-2}\,\text{m})^2$, which gives $q_2q_3 = -1.7 \times 10^{-13}\,\text{C}^2$.

Because q_1 and q_3 repel each other, they must have the same sign and their product will be positive. We can take this into account by giving the force a positive value:

$F_{13} = kq_1q_3/r_{13}^2$;

$+5.2 \times 10^{-3}\,\text{N} = (9 \times 10^9\,\text{N} \cdot \text{m}^2/\text{C}^2)q_1q_3/(10.0 \times 10^{-2}\,\text{m})^2$, which gives $q_1q_3 = +5.8 \times 10^{-14}\,\text{C}^2$.

We have three equations for three unknowns, q_1, q_2, and q_3. If we assume that q_1 is positive, when we combine the equations we get

$$\boxed{q_1 = +1.1 \times 10^{-7}\,\text{C}, \quad q_2 = -3.2 \times 10^{-7}\,\text{C}, \quad q_3 = +5.3 \times 10^{-7}\,\text{C}.}$$

If we took q_1 to be negative, we would get the same magnitudes, with q_1 and q_3 negative and q_2 positive.

33. Because the two charges have opposite signs, the charge Q must be on the x-axis outside the two, where the two forces on Q will be in opposite directions. The net force will be zero when the two magnitudes are equal:

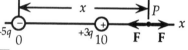

$k5qQ/r_1^2 = k3qQ/r_2^2$, or, when we cancel common factors,

$5/x^2 = 3/(x-10)^2$, which gives $x = 5.6$, and 44.4.

The point outside the two charges where the force is zero is at $\boxed{x = 44.4.}$

37. The forces on each quark are shown in the diagram.
For the positive "up" quark on the right, we have

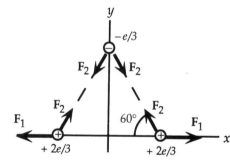

$\Sigma F_x = F_1 - F_2 \cos 60°$
$= (46\,\text{N}) - (23\,\text{N}) \cos 60°$
$= 34\,\text{N}$;
$\Sigma F_y = F_2 \sin 60°$
$= (23\,\text{N}) \sin 60° = +20\,\text{N}$.

When we combine these components, we get

$\boxed{F_+ = 39\,\text{N} \ \ 30° \text{above the x-axis, the line joining the two "up" quarks.}}$

From symmetry, the force on the left "up" quark will be 39 N 30° above the $-x$-axis.
For the negative "down" quark at the top, we have

$\Sigma F_x = F_2 \cos 60° - F_2 \cos 60° = 0$;
$\Sigma F_y = -F_2 \sin 60° - F_2 \sin 60°$
$= -2(23\,\text{N}) \sin 60° = -40\,\text{N}$.

The force on the "down" quark is

$\boxed{F_- = 40\,\text{N} \ \ \text{toward the center of the line joining the two "up" quarks.}}$

Note that the sum of the three forces is zero, within the limitation of significant figures.

39. (a) From the symmetry of the charges and the distances, we have
$F_1 = F_2 = F_3 = F_4$, so
$\boxed{\sum F = 0,}$ the negative charge is in equilibrium.
    (b) If the negative charge is moved slightly toward one of the positive charges, the attractive force toward that charge will increase, while the attractive force toward the opposite corner will decrease. The net force will be away from the equilibrium point, so the equilibrium will be $\boxed{\text{unstable.}}$
    (c) If the negative charge is moved perpendicular to the plane a small distance, each of the four attractive forces will have a component pointing back toward the plane. From symmetry, the net force, the sum of these four forces, will be toward the equilibrium point, so equilibrium will be $\boxed{\text{stable.}}$

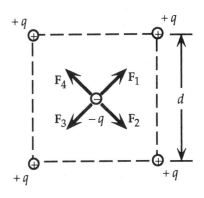

41. (a) The three forces acting on q are shown in the figure. Their magnitudes are
$$F_1 = F_2 = k2qq/(2L)^2 = \tfrac{1}{2}kq^2/L^2;$$
$$F_3 = k4qq/(2L\sqrt{2})^2 = \tfrac{1}{2}kq^2/L^2.$$
The net force acting on q is
$$\mathbf{F}_{net} = \mathbf{F}_1 + \mathbf{F}_2 + \mathbf{F}_3$$
$$= (-\tfrac{1}{2}kq^2/L^2)\mathbf{i} + (\tfrac{1}{2}kq^2/L^2)\mathbf{j} -$$
$$\{[(\tfrac{1}{2}kq^2/L^2)\cos 45°]\,\mathbf{i} + [(\tfrac{1}{2}kq^2/L^2)\sin 45°]\,\mathbf{j}\}$$
$$= (\tfrac{1}{2}kq^2/L^2)\{[-(2+\sqrt{2})/2]\mathbf{i} + [(2-\sqrt{2})/2]\mathbf{j}\}$$
$$= \boxed{(\sqrt{3})kq^2/2L^2,\ 9.7° \text{ above the } -x\text{-axis.}}$$

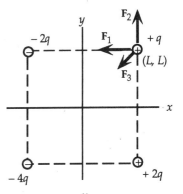

    (b) The four forces acting on Q are shown in the figure. Their magnitudes are
$$F_1 = F_3 = k2qQ/(L\sqrt{2})^2 = kqQ/L^2;$$
$$F_2 = kqQ/(L\sqrt{2})^2 = kqQ/2L^2;$$
$$F_4 = k4qQ/(L\sqrt{2})^2 = 2kqQ/L^2.$$
The vector addition to find the net force is simplified if we use a rotated $x'y'$-coordinate system, as shown on the diagram. The net force acting on q is
$$\mathbf{F}_{net} = \mathbf{F}_1 + \mathbf{F}_2 + \mathbf{F}_3 + \mathbf{F}_4$$
$$= (kqQ/L^2)\mathbf{j}' - (kqQ/2L^2)\mathbf{i}' + (kqQ/L^2)\mathbf{j}' - (2kqQ/L^2)\mathbf{i}'$$
$$= (kqQ/L^2)[-2.5\mathbf{i}' + 2\mathbf{j}']$$
$$= 3.2kqQ/L^2,\ 38.7° \text{ above the } -x'\text{-axis, or} \boxed{3.2kqQ/L^2,\ 6.3° \text{ below the } -x\text{-axis.}}$$

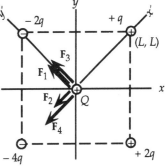

45. We align the rod along the x-axis with one end at the origin, as shown in the figure. The linear charge density is $\lambda = Q/L$, so the charge on the element dx is $dQ = (Q/L)\,dx$. Each differential element produces a differential force in the $+x$-direction. The total force is

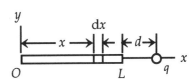

$$\mathbf{F} = \int \mathbf{i}\, dF_x = \int_0^L \frac{kq\lambda}{r^2}\mathbf{i}\, dx = \frac{kqQ}{L}\mathbf{i} \int_0^L \frac{dx}{(L-x+d)^2}$$

$$= \frac{kqQ}{L}\mathbf{i}\left(\frac{1}{L-x+d}\right)\Big|_0^L = \frac{kqQ}{L}\mathbf{i}\left(\frac{1}{d} - \frac{1}{L+d}\right) = \frac{kqQ}{d(L+d)}\mathbf{i}.$$

$\boxed{\text{The force on } q \text{ is } kqQ/d(L+d) \text{ away from the rod.}}$

47. If we consider the plate to be a series of concentric rings, each ring will produce a force away from the plate, as shown in the figure. We choose a representative ring of radius r and thickness dr. The area charge density of the plate is

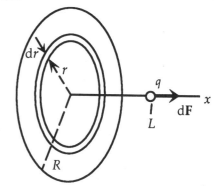

$\sigma = Q/\pi R^2$, so the charge on the ring is

$dq = \sigma\, 2\pi r\, dr = (2Q/R^2)r\, dr.$

We use the result for a ring from Example 22–6 to find the total force by summing (integrating) the forces from all of the rings:

$$\mathbf{F} = \int d\mathbf{F} = \int_0^R \frac{2kqQL}{R^2}\frac{r\,dr}{\left(r^2 + L^2\right)^{3/2}}\mathbf{i} = \frac{2kqQL}{R^2}\mathbf{i}\left[\frac{-1}{\left(r^2 + L^2\right)^{1/2}}\right]\Bigg|_0^R$$

$$= \frac{2kqQL}{R^2}\left[\frac{-1}{\left(R^2 + L^2\right)^{1/2}} - \frac{-1}{L}\right]\mathbf{i} = \frac{2kqQ}{R^2}\left[1 - \frac{L}{\left(R^2 + L^2\right)^{1/2}}\right]\mathbf{i}.$$

For the given data, we get

$$F = \frac{2\left(9\times10^9\ \text{N·m}^2/\text{C}^2\right)\left(0.65\times10^{-6}\ \text{C}\right)\left(1.6\times10^{-6}\ \text{C}\right)}{\left(8\times10^{-2}\ \text{m}\right)^2}\left\{1 - \frac{5\ \text{cm}}{\left[\left(8\ \text{cm}\right)^2 + \left(5\ \text{cm}\right)^2\right]^{1/2}}\right\}\mathbf{i}$$

$= \boxed{1.4\ \text{N away from the center.}}$

51. We place the wire in a vertical plane, as shown. From the symmetry of the charge distribution , we know that the force on q will be down. (The horizontal force from each differential element on one side of the ring will be balanced by the horizontal force of an element at the same height on the other side.) The linear charge density of the wire is $\lambda = Q/\pi R$. We use an element $dQ = \lambda R\, d\theta$ at an angle θ from the horizontal. We find the net force by summing (integrating) the vertical components:

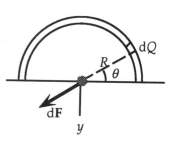

$$F = \int dF_y = \int_0^\pi \frac{kq}{R^2}\sin\theta\, dQ = \int_0^\pi \frac{kqQ}{\pi R^3}\left(\sin\theta\right) R\, d\theta$$

$$= \frac{kqQ}{\pi R^2}\left(-\cos\theta\right)\Bigg|_0^\pi = \frac{2kqQ}{\pi R^2}.$$

Using the given data, we get

$F = 2(9\times10^9\ \text{N·m}^2/\text{C}^2)(0.30\times10^{-6}\ \text{C})(0.75\times10^{-6}\ \text{C})/\pi(0.050\ \text{m})^2 = \boxed{0.52\ \text{N.}}$

55. In the equilibrium position, the net force is zero. From the force diagram, we have

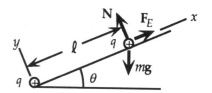

$\Sigma F_x = F_E - mg\sin\theta = 0;$

$kqq/\ell^2 = mg\sin\theta;$

$(9\times10^9\ \text{N·m}^2/\text{C}^2)(2\times10^{-8}\ \text{C})^2/(0.08\ \text{m})^2 =$

$(0.5\times10^{-3}\ \text{kg})(9.8\ \text{m/s}^2)\sin\theta,$

which gives

$\sin\theta = 0.115,$ $\boxed{\theta = 6.6°.}$

59. (a) The middle charge is repelled by each of the other charges.
         The net force is

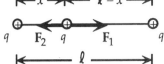

$$F_{net} = F_1 - F_2 ;$$
$$F_{net} = kq^2\{(1/x^2) - [1/(\ell - x)^2]\} \text{ away from the closer charge.}$$

We reduce this to

$$F_{net} = kq^2[\ell(\ell - 2x)/x^2(\ell - x)^2].$$

For the net force to be zero, we have

$$\ell - 2x = 0, \quad \text{or} \quad \boxed{x = \ell/2,} \quad \text{which we expect from symmetry considerations.}$$

   (b) We call the displacement from equilibrium $\Delta x = x - (\ell/2)$, so $x = (\ell/2) + \Delta x$. When we substitute this into the expression for the net force, we get

$$F_{net} = kq^2\{\ell(-2\,\Delta x)/[(\ell/2) + \Delta x]^2[(\ell/2) - \Delta x]^2\}, \quad \text{or}$$
$$\boxed{\mathbf{F}_{net} = -2kq^2\ell\,\Delta x/[(\ell/2)^2 - (\Delta x)^2]^2.}$$

   (c) When $\Delta x \ll \ell$, we can drop the Δx^2 term in the denominator:

$$\mathbf{F}_{net} = -kq^2(32\,\Delta x/\ell^3)\mathbf{i}, \text{ which has the form of a restoring spring force.}$$

The oscillation frequency for small displacements is

$$f = (1/2\pi)(k_{eff}/m)^{1/2} = \boxed{(1/2\pi)(32kq^2/\ell^3m)^{1/2}.}$$

61. From the analogy with the gravitational force, we know that if we replace distribution 2 with a point charge q_2, the force exerted on q_2 by distribution 1 is the same as if distribution 1 were a point charge. From Newton's third law, the force on distribution 1 by q_2 is the reaction to the force on q_2, thus distribution 1 can be treated as a point charge when there is an external point charge.

    Similarly, if we replace distribution 1 with a point charge q_1, the force exerted on q_1 by distribution 2 is the same as if distribution 2 were a point charge. From Newton's third law, the force on distribution 2 by q_1 is the reaction to the force on q_1, thus distribution 2 can be treated as a point charge when there is an external point charge.

    Thus we can simultaneously treat both spherically symmetric distributions as point charges to find the force between them.

CHAPTER 23

The electric field is a vector, so a diagram much like a force diagram should be used to analyze the addition of electric fields.

5.   For a regular hexagon, we have the angles shown. The edge is $L = 10$ cm.
For the distances from the charges we have
$$r_1 = r_5 = L = 10 \text{ cm}; \quad r_2 = r_4 = 2L \cos 30° = 17.3 \text{ cm}; \quad r_3 = 2L = 20 \text{ cm}.$$
We take advantage of the symmetry of the charges to simplify
the vector addition of the individual fields. Because $q_1 = -q_5$,
and $r_1 = r_5$, the magnitudes of E_1 and E_5 will be equal.
Their resultant will be in the y-direction:
$$\begin{aligned}
E_1 + E_5 &= 2[(1/4\pi\varepsilon_0)q_1/r_1^2] \sin 60° \, \mathbf{j} \\
&= 2[(9 \times 10^9 \text{ N} \cdot \text{m}^2/\text{C}^2)(2 \times 10^{-6} \text{ C})/(0.10 \text{ m})^2] \sin 60° \, \mathbf{j} \\
&= (3.12 \times 10^6 \text{ N/C}) \, \mathbf{j}.
\end{aligned}$$
Because $q_2 = q_4$, and $r_2 = r_4$, the magnitudes of E_2 and E_4 will
be equal. Their resultant will be in the x-direction:
$$\begin{aligned}
E_2 + E_4 &= 2[(1/4\pi\varepsilon_0)q_2/r_2^2] \cos 30° \, \mathbf{i} \\
&= 2[(9 \times 10^9 \text{ N} \cdot \text{m}^2/\text{C}^2)(3 \times 10^{-6} \text{ C})/(0.173 \text{ m})^2] \cos 30° \, \mathbf{i} \\
&= (1.56 \times 10^6 \text{ N/C}) \, \mathbf{i}.
\end{aligned}$$
For E_3 we have
$$\begin{aligned}
E_3 &= -[(1/4\pi\varepsilon_0)q_3/r_3^2] \, \mathbf{i} \\
&= -[(9 \times 10^9 \text{ N} \cdot \text{m}^2/\text{C}^2)(4 \times 10^{-6} \text{ C})/(0.20 \text{ m})^2] \, \mathbf{i} \\
&= -(0.90 \times 10^6 \text{ N/C}) \, \mathbf{i}.
\end{aligned}$$
The resultant electric field is
$$\begin{aligned}
E &= E_1 + E_2 + E_3 + E_4 + E_5 \\
&= [(0.66 \times 10^6 \, \mathbf{i}) + 3.12 \times 10^6 \, \mathbf{j})] \text{ N/C} = \boxed{3.19 \times 10^6 \text{ N/C}, 78° \text{ above the } +x\text{-axis.}}
\end{aligned}$$

7.   (a) With the charges on the x-axis, the electric fields produced by the charges will have the same
magnitude and point in the $-x$-direction. The resultant field will be
$$\begin{aligned}
E &= 2(1/4\pi\varepsilon_0)[q/(\ell/2)]^2 \, (-\mathbf{i}) \\
&= \boxed{-(1/4\pi\varepsilon_0)(8q/\ell^2) \, \mathbf{i}.}
\end{aligned}$$

   (b) The electric fields produced by the charges will have the same magnitude and point in opposite
directions. The resultant field will be
$$\boxed{E = 0.}$$

   (c) We take a representative point on the y-axis. From the diagram,
we see that the electric fields produced by the charges will have
the same magnitude, and the resultant field will point away
from the origin.
If we call the distance from the origin d, we have
$$\begin{aligned}
E &= 2(1/4\pi\varepsilon_0)(q/r^2) \cos \theta = (1/4\pi\varepsilon_0)(2q/r^2)(d/r) \\
&= (1/4\pi\varepsilon_0)\{2qd/[d^2 + (\ell/2)^2]^{3/2}\}.
\end{aligned}$$
From the symmetry in the yz-plane, at a point d from the origin
we have
$$\boxed{E = (1/4\pi\varepsilon_0)\{2qd/[d^2 + (\ell/2)^2]^{3/2}\}} \text{ away from the origin.}$$
Note that, as expected, this reduces to 0 at the origin.

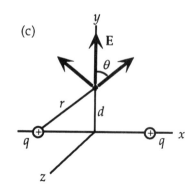

9. From the diagram, we see that the resultant electric field is

$$E = E_+ + E_- = \frac{1}{4\pi\varepsilon_0}\frac{q}{[r+(L/2)]^2}\mathbf{i} - \frac{1}{4\pi\varepsilon_0}\frac{q}{[r-(L/2)]^2}\mathbf{i}$$

$$= \frac{q}{4\pi\varepsilon_0}\left\{\frac{1}{[r+(L/2)]^2} - \frac{1}{[r-(L/2)]^2}\right\}\mathbf{i} = \frac{q}{4\pi\varepsilon_0}\left\{\frac{[r-(L/2)]^2 - [r+(L/2)]^2}{[r+(L/2)]^2[r-(L/2)]^2}\right\}\mathbf{i}$$

$$= \frac{q}{4\pi\varepsilon_0}\left\{\frac{-2rL}{[r+(L/2)]^2[r-(L/2)]^2}\right\}\mathbf{i} = -\frac{2qL}{4\pi\varepsilon_0 r^3}\left\{\frac{1}{[1+(L/2r)]^2[1-(L/2r)]^2}\right\}\mathbf{i}.$$

We express this in terms of the dipole moment:

$$\boxed{E = \frac{2p}{4\pi\varepsilon_0 r^3}\left\{\frac{1}{[1+(L/2r)]^2[1-(L/2r)]^2}\right\}.}$$

When $r \gg L$, we can approximate each of the terms in the denominator as 1, so the electric field along the axis of the dipole far from the dipole is

$$\boxed{E = \frac{2p}{4\pi\varepsilon_0 r^3}.}$$

27. We assume that the plates are large enough that they may be considered infinite plates. Each plate produces an electric field perpendicular to and away from the plate with a magnitude
$E = \sigma/2\varepsilon_0$.
Outside the two plates, the fields from the two plates are in the same direction, so we have

$$\boxed{E = (\sigma/2\varepsilon_0) + (\sigma/2\varepsilon_0) = \sigma/\varepsilon_0 \text{ perpendicular to the plates and away from them.}}$$

Between the plates, the fields from the two plates are in opposite directions, so we have

$$\boxed{E = (\sigma/2\varepsilon_0) - (\sigma/2\varepsilon_0) = 0.}$$

29. (a) From the symmetry of the charge distribution, we know that the electric field on the z-axis is along the z-axis. For a differential element we choose a ring of radius r and thickness dr. The charge on the ring is $dq = (Q/\pi R^2)2\pi r\, dr = (2Qr\, dr)/R^2$. Using the result for the field of a hoop of charge, we integrate over the disk:

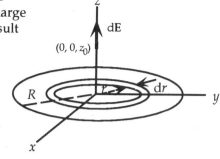

$$\mathbf{E} = \frac{1}{4\pi\varepsilon_0}\int \frac{z_0\, dq}{\left(r^2 + z_0^2\right)^{3/2}}\mathbf{k} = \frac{z_0 2Q}{4\pi\varepsilon_0 R^2}\int_0^R \frac{r\, dr}{\left(r^2 + z_0^2\right)^{3/2}}\mathbf{k}$$

$$= \frac{z_0 Q}{2\pi\varepsilon_0 R^2}\left[\frac{-1}{\left(r^2 + z_0^2\right)^{1/2}}\right]\Bigg|_0^R \mathbf{k}$$

$$= \boxed{\frac{z_0 Q}{2\pi\varepsilon_0 R^2}\left[\frac{1}{z_0} - \frac{1}{\left(R^2 + z_0^2\right)^{1/2}}\right]\mathbf{k}.}$$

(b) To find the field in the limit $z_0 \to \infty$, we rearrange the expression and then use the approximation $(1 + x)^{-1/2} \approx 1 - (x/2)$:

$$\mathbf{E} = \frac{Q}{2\pi\varepsilon_0 R^2}\left\{1 - \left[1 + \left(\frac{R}{z_0}\right)^2\right]^{-1/2}\right\}\mathbf{k} = \frac{Q}{2\pi\varepsilon_0 R^2}\left[1 - 1 + \frac{1}{2}\left(\frac{R}{z_0}\right)^2\right]\mathbf{k}$$

$$= \boxed{\frac{Q}{4\pi\varepsilon_0 z_0^2}\mathbf{k}.}$$

As we expect, the field is that of a point charge.

(c) To find the field in the limit $R \to \infty$, we consider the result from part (a). The second term will go to zero, so we have

$\mathbf{E} = Q/2\pi\varepsilon_0 R^2\mathbf{k}$.

The charge density of the disk is $\sigma = Q/\pi R^2$, so we can write

$\boxed{\mathbf{E} = (\sigma/2\varepsilon_0)\mathbf{k}.}$

As we expect, the field is that of an infinite plane.

The limits of parts (b) and (c) are not the same. Part (b) is equivalent to the disk being a point charge, while part (c) is equivalent to being very close to the disk.

31. We find the electric field from the vector sum of the field of a rod $\mathbf{E}_1$ and the field of a point charge $\mathbf{E}_2$:

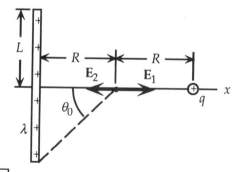

$\mathbf{E} = \mathbf{E}_1 + \mathbf{E}_2 = (\lambda/2\pi\varepsilon_0 R)\sin\theta_0\,\mathbf{i} - (q/4\pi\varepsilon_0 R^2)\mathbf{i}$

$= (1/4\pi\varepsilon_0)\{[(2\lambda\sin\theta_0)/R] - (q/R^2)\}\mathbf{i}$.

The angle for the endpoint of the rod is

$\theta_0 = \tan^{-1}(L/R) = \tan^{-1}(1) = 45°$.

The magnitude of the field is

$E = (9\times10^9\,\text{N}\cdot\text{m}^2/\text{C}^2)\{[2(15\times10^{-6}\,\text{C/m})(\sin 45°)/(0.15\,\text{m})] -$
$\qquad [(3\times10^{-6}\,\text{C})/(0.15\,\text{m})^2]\}$

$= 7.3\times10^4\,\text{N/C}$.

The resultant field is $\boxed{E = 7.3\times10^4\,\text{N/C toward the point charge.}}$

39. The force produced by the electric field of the wire on the negative charge is toward the wire and provides the centripetal force for the circular motion:

$F = mv^2/r$;

$q(\lambda/2\pi\varepsilon_0 r) = mv^2/r$, which gives a speed $\boxed{v = (q\lambda/2\pi\varepsilon_0 m)^{1/2},}$ which does not depend on r.

41. The force produced by the electric field of the wire on the negative charge is toward the wire and provides the centripetal force for the circular motion:

$F = mv^2/r$;

$q(\lambda/2\pi\varepsilon_0 r) = mv^2/r$, which gives a speed $v = (q\lambda/2\pi\varepsilon_0 m)^{1/2}$.

The period of the orbit is the time to traverse the circle:

$T = 2\pi r/v = [2\pi/(q\lambda/2\pi\varepsilon_0 m)^{1/2}]r$.

If the centripetal force is provided by a point charge, we have

$(1/4\pi\varepsilon_0)(qQ/r^2) = mv^2/r$, which gives

$v = (qQ/4\pi\varepsilon_0 mr)^{1/2}$.

The period of the orbit is

$T_{\text{point charge}} = 2\pi r/v = 2\pi(4\pi\varepsilon_0 m/qQ)^{1/2}r^{3/2}$, which has a different r dependence.

43. We use the coordinate system from Example 23–8. The initial horizontal component of the velocity is

$v_{0x} = (v_0{}^2 - v_{0y}{}^2)^{1/2} = [(5.0 \times 10^6 \text{ m/s})^2 - (2.0 \times 10^5 \text{ m/s})^2]^{1/2} = 5.0 \times 10^6 \text{ m/s}$.

We find the time for the electron to travel between the plates from the horizontal motion:

$t_1 = L_1/v_{0x} = (3 \times 10^{-2} \text{ m})/(5.0 \times 10^6 \text{ m/s}) = 6.0 \times 10^{-9} \text{ s}$.

The deflection at this time is

$y_1 = v_{0y}t_1 + \tfrac{1}{2}at_1{}^2 = v_{0y}t_1 + \tfrac{1}{2}(qE/m)t_1{}^2$

$= (2.0 \times 10^5 \text{ m/s})(6.0 \times 10^{-9} \text{ s}) + \tfrac{1}{2}[(1.6 \times 10^{-19} \text{ C})(10^3 \text{ N/C})/(9.1 \times 10^{-31} \text{ kg})](6.0 \times 10^{-9} \text{ s})^2$

$= 4.4 \times 10^{-3} \text{ m}$.

The vertical component of the velocity as the electron leaves the plates is

$v_{1y} = v_{0y} + at_1 = v_{0y} + (qE/m)t_1$

$= (2.0 \times 10^5 \text{ m/s}) + [(1.6 \times 10^{-19} \text{ C})(10^3 \text{ N/C})/(9.1 \times 10^{-31} \text{ kg})](6.0 \times 10^{-9} \text{ s})$

$= 1.25 \times 10^6 \text{ m/s}$.

After it leaves the plates, the electron travels in a straight line with a direction given by

$\tan \theta = v_{1y}/v_{0x} = (1.25 \times 10^6 \text{ m/s})/(5.0 \times 10^6 \text{ m/s}) = 0.25$.

The deflection while the electron travels this straight line is

$y_2 = L_2 \tan \theta = (12 \times 10^{-2} \text{ m})(0.25) = 3.0 \times 10^{-2} \text{ m}$.

The total deflection is

$y = y_1 + y_2 = (0.44 \times 10^{-2} \text{ m}) + (3.0 \times 10^{-2} \text{ m}) = 3.4 \times 10^{-2} \text{ m} = \boxed{3.4 \text{ cm.}}$

45. The electric field produced by the large plate is

$E = \sigma/2\varepsilon_0$

$= (10^{-6} \text{ C/m}^2)/2(8.85 \times 10^{-12} \text{ C}^2/\text{N} \cdot \text{m}^2) = 5.6 \times 10^4 \text{ N/C}$,

which attracts the negative charge. Using the force diagram, we find the equation of motion for the tangential direction:

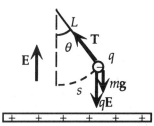

$-(mg + qE) \sin \theta = m \, d^2s/dt^2$.

If the angle is small, we have

$\sin \theta \approx \theta = s/L$, and the equation of motion for s becomes

$-[(mg + qE)/L]s = m \, d^2s/dt^2$.

This is the equation for simple harmonic motion. The effective force constant is

$k_{\text{eff}} = (mg + qE)/L$.

The angular frequency of the motion is

$\omega = (k_{\text{eff}}/m)^{1/2} = [(mg + qE)/Lm]^{1/2}$

$= \{[(5 \times 10^{-3} \text{ kg})(9.8 \text{ m/s}^2) + (2 \times 10^{-6} \text{ C})(5.6 \times 10^4 \text{ N/C})]/(1 \text{ m})(5 \times 10^{-3} \text{ kg})\}^{1/2} = \boxed{5.7 \text{ s}^{-1}.}$

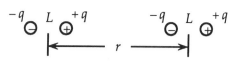

51. Each dipole is a pair of charges q separated by a distance L, such that $p = qL$. To find the force on the dipole on the right, we find the electric field at each of the charges produced by the charges of the other dipole. The field at the positive charge is

$$E_+ = \frac{1}{4\pi\varepsilon_0}\left[\frac{q}{r^2} - \frac{q}{(r+L)^2}\right] = \frac{q}{4\pi\varepsilon_0 r^2}\left\{1 - \frac{1}{\left[1+(L/r)\right]^2}\right\} \text{ to the right.}$$

The field at the negative charge is

$$E_- = \frac{1}{4\pi\varepsilon_0}\left[\frac{q}{(r-L)^2} - \frac{q}{r^2}\right] = \frac{q}{4\pi\varepsilon_0 r^2}\left\{\frac{1}{\left[1-(L/r)\right]^2} - 1\right\} \text{ to the right.}$$

The force on the dipole is

$$F = qE_+ + (-q)E_- = \frac{q^2}{4\pi\varepsilon_0 r^2}\left\{1 - \frac{1}{\left[1+(L/r)\right]^2} - \frac{1}{\left[1-(L/r)\right]^2} + 1\right\} \text{ to the right.}$$

Because $L \ll r$, we make use of the approximation $(1 \pm x)^{-2} \simeq 1 \mp 2x + 3x^2 \mp \dots$ and expand the terms:

$$F \simeq \frac{q^2}{4\pi\varepsilon_0 r^2}\left\{1 - \left[1 - 2\tfrac{L}{r} + 3\left(\tfrac{L}{r}\right)^2\right] - \left[1 + 2\tfrac{L}{r} + 3\left(\tfrac{L}{r}\right)^2\right] + 1\right\} = \frac{q^2}{4\pi\varepsilon_0 r^2}\left[-6\left(\tfrac{L}{r}\right)^2\right] = -\frac{6q^2 L^2}{4\pi\varepsilon_0 r^4} = \boxed{-\frac{6p^2}{4\pi\varepsilon_0 r^4}\text{(attraction).}}$$

57. We choose the coordinate system shown in the diagram, with the rods aligned parallel to the z-axis.
   (a) At points on the y-axis, we have
   $$\mathbf{E} = \mathbf{E}_+ + \mathbf{E}_- = (\lambda/2\pi\varepsilon_0)(\{1/[y - (R/2)]\} - \{1/[y + (R/2)]\})\mathbf{j}$$
   $$= \boxed{\{(\lambda R/2\pi\varepsilon_0)/[y^2 - (R/2)^2]\}\mathbf{j}.}$$
   (b) We see from the diagram that the symmetry along the x-axis means that the resultant field will have only a y-component. We find the field by doubling the y-component from one rod:
   $$\mathbf{E} = -2E_-\sin\theta\,\mathbf{j} = -(2\lambda/2\pi\varepsilon_0 r)[(R/2)/r]\mathbf{j}$$
   $$= -(\lambda R/2\pi\varepsilon_0 r^2)\mathbf{j} = \boxed{-\{(\lambda R/2\pi\varepsilon_0)/[x^2 + (R/2)^2]\}\mathbf{j}.}$$

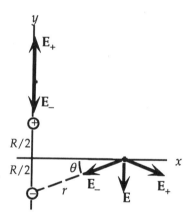

59. Each infinite plate produces a constant field perpendicular to the plate. The total electric field is
   $$\mathbf{E} = (\sigma_1/2\varepsilon_0)\mathbf{j} + (\sigma_2/2\varepsilon_0)\mathbf{i}.$$
   The constant force produced by this field on the particle causes a constant acceleration:
   $$\mathbf{a} = q\mathbf{E}/m = (q/2\varepsilon_0 m)(\sigma_1\mathbf{j} + \sigma_2\mathbf{i}).$$
   If the particle starts from rest, its position is
   $$\mathbf{r} = \mathbf{r}_0 + \mathbf{v}_0 t + \tfrac{1}{2}\mathbf{a}t^2 = (1\text{ m})\mathbf{i} + (1\text{ m})\mathbf{j} + 0 + (q/4\varepsilon_0 m)(\sigma_2\mathbf{i} + \sigma_1\mathbf{j})t^2$$
   $$= \{1 + [(1\times10^{-7}\text{ C})/4(8.85\times10^{-12}\text{ C}^2/\text{N}\cdot\text{m}^2)(1\times10^{-3}\text{ kg})](+3\times10^{-6}\text{ C/m}^2)t^2\}\mathbf{i} +$$
   $$\{1 + [(1\times10^{-7}\text{ C})/4(8.85\times10^{-12}\text{ C}^2/\text{N}\cdot\text{m}^2)(1\times10^{-3}\text{ kg})](-5\times10^{-6}\text{ C/m}^2)t^2\}\mathbf{j}$$
   $$= \boxed{(1 + 8.5t^2)\mathbf{i} + (1 - 14t^2)\mathbf{j}\text{ m, with }t\text{ in s.}}$$

61. (a) The electric field of the plate is perpendicular to and away from the plate. The force on the positive charge is away from the plate:

$$F = qE = q\sigma/2\varepsilon_0$$
$$= (1.6 \times 10^{-19}\,\text{C})(8.0 \times 10^{-6}\,\text{C/m}^2)/2(8.85 \times 10^{-12}\,\text{C}^2/\text{N} \cdot \text{m}^2)$$
$$= \boxed{7.2 \times 10^{-14}\,\text{N away from the plate.}}$$

  (b) The proton is fired toward the plate. We find the work from the work-energy theorem:

$$W = \Delta K$$
$$= 0 - (2 \times 10^6\,\text{eV})(1.6 \times 10^{-19}\,\text{J/eV}) = \boxed{-3.2 \times 10^{-13}\,\text{J.}}$$

  (c) Because the work is done by the electric field, we have

$$W = -Fd$$
$$-3.2 \times 10^{-13}\,\text{J} = -(7.2 \times 10^{-14}\,\text{N})d, \text{ which gives } \boxed{d = 4.4\,\text{m.}}$$

65. We are given the force

$$\mathbf{F} = \frac{q\lambda_0}{2\pi\varepsilon_0 L}\left\{\ln\left[\frac{R-(L/2)}{R+(L/2)}\right] + R\left[\frac{1}{R-(L/2)} - \frac{1}{R+(L/2)}\right]\right\}\mathbf{i}.$$

If we change variable to $x = L/2R$, the magnitude of the force becomes

$$F = \frac{q\lambda_0}{2\pi\varepsilon_0 L}\left[\ln\left(\frac{1-x}{1+x}\right) + \left(\frac{1}{1-x} - \frac{1}{1+x}\right)\right] = \frac{q\lambda_0}{2\pi\varepsilon_0 L}\left[\ln(1-x) - \ln(1+x) + \left(\frac{1}{1-x} - \frac{1}{1+x}\right)\right].$$

Using the approximate expansions for small x, we get

$$F = \frac{q\lambda_0}{2\pi\varepsilon_0 L}\left[\left(-x - \frac{x^2}{2} - \frac{x^3}{3} - \ldots\right) - \left(x - \frac{x^2}{2} + \frac{x^3}{3} - \ldots\right) + \left(1 + x + x^2 + x^3 + \ldots\right) - \left(1 - x + x^2 - x^3 + \ldots\right)\right]$$

$$= \frac{q\lambda_0}{2\pi\varepsilon_0 L}\left[\left(-2x - \frac{2x^3}{3} - \ldots\right) + \left(2x + 2x^3 + \ldots\right)\right] \simeq \frac{q\lambda_0}{2\pi\varepsilon_0 L}\left(\frac{4x^3}{3}\right).$$

In terms of the distance R, the force is

$$F = (q\lambda_0/2\pi\varepsilon_0 L)(4/3)(L/2R)^3 = q\lambda_0 L^2/12\pi\varepsilon_0 R^3.$$

The field of a dipole on the axis is $E = p/2\pi\varepsilon_0 R^3$, so the dipole moment is

$$\boxed{p = \lambda_0 L^2/6.}$$

CHAPTER 24

The integration for Gauss' law is performed by choosing an appropriate Gaussian surface based on the symmetry of the charge distribution. For parts of the surface where **E** is zero or perpendicular to **A**, the contribution to the integral is zero. For parts of the surface where **E** is parallel to **A** and constant, the contribution to the integral is EA.

3. On the ends of the cylinder the electric field is not constant, but it is always perpendicular to the area vector of the surface. On the sides of the cylinder the electric field is constant and parallel to the area vector. The flux through the cylinder is

$$\Phi = \oint\!\!\!\oint \mathbf{E}\cdot d\mathbf{A} = \iint_{end}\mathbf{E}\cdot d\mathbf{A} + \iint_{end}\mathbf{E}\cdot d\mathbf{A} + \iint_{side}\mathbf{E}\cdot d\mathbf{A}$$
$$= 0 + 0 + EA_{side} = \frac{\lambda}{2\pi\varepsilon_0 R}(2\pi Rh) = \boxed{\frac{\lambda h}{\varepsilon_0}.}$$

We see that the result is independent of R, so we get the same flux through a cylinder of radius $2R$.

7. The electric field and the area vector are parallel over the circle. Because the electric field varies over the surface, we find the flux by integrating. The field depends only on r, so we choose a circular ring of radius r and thickness dr as the differential area element:

$$\Phi = \iint \mathbf{E}\cdot d\mathbf{A} = \int_0^R E_0\left(1-\frac{r}{R}\right)(2\pi r\,dr) = 2\pi E_0\int_0^R\left(r-\frac{r^2}{R}\right)dr = 2\pi E_0\left(\frac{r^2}{2}-\frac{r^3}{3R}\right)\Big|_0^R = \boxed{\frac{\pi E_0 R^2}{3}.}$$

9. Because the angle between the electric field of the point charge and the area vector of the cylinder varies over the surface of the hemisphere, we find the flux by integration. We choose a band at an elevation z, which corresponds to an angle θ such that $z = R\tan\theta$. If we differentiate, we see that the band has thickness $dz = R\sec^2\theta\,d\theta$ so the area of this band is
$dA = 2\pi R\,dz = 2\pi R^2\sec^2\theta\,d\theta.$
From the symmetry we see that the flux will be the same for the upper and lower halves of the surface, so we double the result of the integration over the top half. The angle θ ranges from 0 to θ_0, with $\sin\theta_0 = h/R$. From the diagram, we see that θ is the angle between **E** and $d\mathbf{A}$ for all elements of the band, so we have

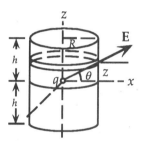

$$\Phi = \iint \mathbf{E}\cdot d\mathbf{A} = 2\int_0^{\theta_0}\frac{1}{4\pi\varepsilon_0}\frac{q}{(R/\cos\theta)^2}2\pi R^2\sec^2\theta\,(\cos\theta)\,d\theta = \frac{q}{\varepsilon_0}\int_0^{\theta_0}\cos\theta\,d\theta$$

$$= \frac{q}{\varepsilon_0}(\sin\theta)\Big|_0^{\theta_0} = \frac{q}{\varepsilon_0}\sin\theta_0 = \frac{q}{\varepsilon_0}\frac{h}{\sqrt{h^2+R^2}} = \boxed{\frac{q}{\varepsilon_0}\frac{1}{\sqrt{1+(R^2/h^2)}}.}$$

13. (a) We use the spherical surface within the charged surface as a Gaussian surface. Because there is no enclosed charge the total electric flux through the surface is $\boxed{\text{zero.}}$
    (b) We use the spherical surface outside the charged surface as a Gaussian surface. Because all of the charge is enclosed, the total electric flux through the surface is
$$\Phi = \oint\!\!\!\oint \mathbf{E}\cdot d\mathbf{A} = Q/\varepsilon_0$$
$$= (10^{-3}\,\text{C})/(8.85\times10^{-12}\,\text{C}^2/\text{N}\cdot\text{m}^2) = \boxed{1.13\times10^8\,\text{N}\cdot\text{m}^2/\text{C}.}$$

19. Because the charge at the origin is at the center of the cube, we know from symmetry that it will produce a flux out of each side that is 1/6 of the total flux it produces:

$$\Phi_1 = (1/6)\Phi_{charge} = (1/6)(Q/\varepsilon_0)$$
$$= (1/6)(5 \times 10^{-8}\ C)/(8.85 \times 10^{-12}\ C^2/N \cdot m^2) = 9.4 \times 10^2\ N \cdot m^2/C.$$

Because the uniform field is parallel to the x-axis, it produces no flux through the sides parallel to the x-axis. Through the sides parallel to the yz-plane, the uniform field produces a flux

$$\Phi_2 = EA = (3000\ N/C)(0.20\ m)^2 = 1.2 \times 10^2\ N \cdot m^2/C.$$

Because this flux enters the cube from the $+x$-axis and leaves the cube toward the $-x$-axis, we have

$9.4 \times 10^2\ N \cdot m^2/C$ out of the sides parallel to the xy- or yz-planes,
$10.6 \times 10^2\ N \cdot m^2/C$ out of the side perpendicular to the $-x$-axis,
$8.2 \times 10^2\ N \cdot m^2/C$ out of the side perpendicular to the $+x$-axis.

25. We choose a Gaussian surface with a top surface just above the ground and a bottom surface below the ground, each of area 40 acres, and the sides perpendicular to the ground. There is no flux through the sides, because $\mathbf{E} \cdot d\mathbf{A} = 0$. There is no flux through the bottom, because $\mathbf{E} = 0$. When we apply Gauss' law to this surface, we get

$$\oint \mathbf{E} \cdot d\mathbf{A} = \iint_{top} \mathbf{E} \cdot d\mathbf{A} + \iint_{side} \mathbf{E} \cdot d\mathbf{A} + \iint_{bottom} \mathbf{E} \cdot d\mathbf{A}$$
$$= -EA_{top} + 0 + 0 = Q/\varepsilon_0 \text{, which gives}$$

$$Q = -\varepsilon_0 EA$$
$$= -(8.85 \times 10^{-12}\ C^2/N \cdot m^2)(110\ N/C)(60\ acres)(4 \times 10^3\ m^2/acre) = \boxed{-2.3 \times 10^{-4}\ C.}$$

29. From the symmetry of the charge distribution, we know that the electric field must be radial, away from the axis of the cylinder, with a magnitude independent of the direction. For a Gaussian surface we choose a cylinder of length L and radius r, centered on the axis. On the ends of this surface, the electric field is not constant but $\mathbf{E}$ and $d\mathbf{A}$ are perpendicular, so we have $\mathbf{E} \cdot d\mathbf{A} = 0$. On the curved side, the field has a constant magnitude and $\mathbf{E}$ and $d\mathbf{A}$ are parallel, so we have $\mathbf{E} \cdot d\mathbf{A} = E\, dA$.

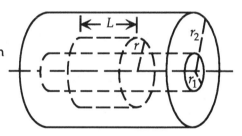

For the region where $r < r_1$, we apply Gauss' law:

$$\oint \mathbf{E} \cdot d\mathbf{A} = \iint_{ends} \mathbf{E} \cdot d\mathbf{A} + \iint_{side} \mathbf{E} \cdot d\mathbf{A} = 0 + EA_{side} = Q/\varepsilon_0;$$

$E2\pi rL = 0$, because there is no charge inside the Gaussian surface, which gives

$$\boxed{E = 0 \text{ for } r < r_1.}$$

For the region where $r_1 < r < r_2$, we apply Gauss' law:

$$\oint \mathbf{E} \cdot d\mathbf{A} = \iint_{ends} \mathbf{E} \cdot d\mathbf{A} + \iint_{side} \mathbf{E} \cdot d\mathbf{A} = 0 + EA_{side} = Q/\varepsilon_0;$$

$E2\pi rL = \rho(\pi r^2 - \pi r_1^2)L/\varepsilon_0$, which gives

$$\boxed{\mathbf{E} = [\rho(r^2 - r_1^2)/2\varepsilon_0 r]\hat{\mathbf{r}} \text{ for } r_1 < r < r_2.}$$

For the region where $r_2 < r$, we apply Gauss' law:

$$\oint \mathbf{E} \cdot d\mathbf{A} = \iint_{ends} \mathbf{E} \cdot d\mathbf{A} + \iint_{side} \mathbf{E} \cdot d\mathbf{A} = 0 + EA_{side} = Q/\varepsilon_0;$$

$E2\pi rL = \rho(\pi r_2^2 - \pi r_1^2)L/\varepsilon_0$, which gives

$$\boxed{\mathbf{E} = [\rho(r_2^2 - r_1^2)/2\varepsilon_0 r]\hat{\mathbf{r}} \text{ for } r_2 < r.}$$

33. The positive sheet produces an electric field directed away
from the sheet with a magnitude

$E_+ = \sigma_+/2\varepsilon_0$
$= (5 \times 10^{-6}\,C/m^2)/2(8.85 \times 10^{-12}\,C^2/N \cdot m^2)$
$= 2.8 \times 10^5\,N/C$.

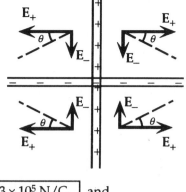

The negative sheet produces an electric field directed toward
the sheet with a magnitude

$E_- = \sigma_-/2\varepsilon_0$
$= (3 \times 10^{-6}\,C/m^2)/2(8.85 \times 10^{-12}\,C^2/N \cdot m^2)$
$= 1.7 \times 10^5\,N/C$.

If we consider the 1st quadrant, because the fields are perpendicular,
we have

$E = (E_+^2 + E_-^2)^{1/2} = [(2.8 \times 10^5\,N/C)^2 + (1.7 \times 10^5\,N/C)^2]^{1/2} = \boxed{3.3 \times 10^5\,N/C,}$ and

$\tan \theta = E_-/E_+ = (1.7 \times 10^5\,N/C)/(2.8 \times 10^5\,N/C) = 0.61$, which gives $\boxed{\theta = 31°.}$

From the diagram, we see that the fields in the other quadrants are mirror images of the field in the 1st
quadrant:

1st quadrant:	E at $-\theta$;
2nd quadrant:	E at $180° + \theta$;
3rd quadrant:	E at $180° - \theta$;
4th quadrant:	E at θ.

35. From the symmetry of the charge distribution, we know that
the electric field must be radial, with a magnitude independent
of the direction. For a Gaussian surface we choose a sphere of
radius r. On this surface, the field has a constant magnitude and
E and dA are parallel, so we have $\mathbf{E} \cdot d\mathbf{A} = E\,dA$. The inner charge
density is

$\rho_1 = Q_1/(\tfrac{4}{3}\pi R_1^3) = 3Q_1/4\pi R_1^3$
$= 3(-2 \times 10^{-6}\,C)/4\pi(0.03\,m)^3 = -1.77 \times 10^{-2}\,C/m^3$.

The outer surface charge density is

$\sigma_2 = Q_2/4\pi R_2^2$
$= (5 \times 10^{-6}\,C)/4\pi(0.08\,m)^2 = 6.2 \times 10^{-5}\,C/m^2$.

For the region where $r < R_1$, part of the charge on the inner sphere is enclosed; we apply Gauss' law:

$\oiint \mathbf{E} \cdot d\mathbf{A} = EA = Q/\varepsilon_0;$

$E4\pi r^2 = \rho_1(\tfrac{4}{3}\pi r^3)/\varepsilon_0$, which gives

$E = \rho_1 r/3\varepsilon_0$
$= [(-1.77 \times 10^{-2}\,C/m^3)/3(8.85 \times 10^{-12}\,C^2/N \cdot m^2)]r$
$= (-6.7 \times 10^8\,)r\,N/C$ with r in m;

$\boxed{\mathbf{E} = (-6.7 \times 10^8\,)r\,\hat{\mathbf{r}}\,N/C \text{ with } r \text{ in m, where } r < 3\,cm.}$

For the region where $R_1 < r < R_2$, all of the charge on the inner sphere is enclosed; we apply Gauss' law:

$\oiint \mathbf{E} \cdot d\mathbf{A} = EA = Q/\varepsilon_0;$

$E4\pi r^2 = Q_1/\varepsilon_0$, which gives

$E = (1/4\pi\varepsilon_0)Q_1/r^2$
$= (9 \times 10^9\,N \cdot m^2/C^2)(-2 \times 10^{-6}\,C)/r^2$
$= (-1.8 \times 10^4)/r^2\,N/C$ with r in m;

$\boxed{\mathbf{E} = [(-1.8 \times 10^4)/r^2]\hat{\mathbf{r}}\,N/C \text{ with } r \text{ in m, where } 3\,cm < r < 8\,cm.}$

For the region where $R_2 < r$, both charges are enclosed; we apply Gauss' law:

$\oiint \mathbf{E} \cdot d\mathbf{A} = EA = Q/\varepsilon_0;$

$E4\pi r^2 = (Q_1 + Q_2)/\varepsilon_0$, which gives

$E = (Q_1 + Q_2)/4\pi\varepsilon_0 r^2$
$= (9 \times 10^9\,N \cdot m^2/C^2)[(-2 \times 10^{-6}\,C) + (5 \times 10^{-6}\,C)]/r^2)\,N/C$
$= (2.7 \times 10^4)/r^2\,N/C$ with r in m;

$\boxed{\mathbf{E} = [(2.7 \times 10^4)/r^2]\hat{\mathbf{r}}\,N/C, \text{ with } r \text{ in m, where } 8\,cm < r.}$

45. From Gauss' law, we know that the flux through a Gaussian surface depends on the enclosed charge:
    $\Phi = Q_{enclosed}/\varepsilon_0$.
    For the region where $a < r < b$, we have $\Phi_1 = Q/\varepsilon_0$, so the enclosed charge is Q. This must be the total charge on the inner sphere. Because the sphere is conducting, the charge is located uniformly on the surface, with the charge density
    $$\boxed{\sigma_{inner\ sphere} = Q/4\pi a^2.}$$
    For the region within the conducting shell, where $b < r < R$, the electric flux is 0, so the net enclosed charge is 0. Because there is a charge Q on the inner sphere, there must be a charge $-Q$ on the inner surface of the shell. Because the shell is conducting, the charge is located uniformly on the surface, with the charge density
    $$\boxed{\sigma_{shell,\ inside} = -Q/4\pi b^2.}$$
    For the region outside the shell, where $R < r$, we have $\Phi_2 = 2Q/\varepsilon_0$, so the net enclosed charge is $2Q$. Because there is a charge Q on the inner sphere, there must be a net charge Q on the shell. Because there is a charge $-Q$ on the inner surface, there must be a charge $+2Q$ on the outer surface, located uniformly on the surface, with the charge density
    $$\boxed{\sigma_{shell,\ outside} = +2Q/4\pi R^2 = +Q/2\pi R^2.}$$

49. We find the flux through a side from
    $\Phi = \iint \mathbf{E} \cdot d\mathbf{A}.$
    For the sides perpendicular to the x-axis, we have
    $\Phi_{x=0} = \iint \mathbf{E} \cdot d\mathbf{A} = \iint (bx^2\mathbf{i}) \cdot (-dA\mathbf{i}) = -b(0)^2 a^2 = \boxed{0;}$
    $\Phi_{x=a} = \iint \mathbf{E} \cdot d\mathbf{A} = \iint (bx^2\mathbf{i}) \cdot (+dA\mathbf{i}) = b(a)^2 a^2 = \boxed{ba^4.}$
    For the sides parallel to the x-axis, we have
    $\Phi_{y=0} = \iint \mathbf{E} \cdot d\mathbf{A} = \iint (bx^2\mathbf{i}) \cdot (-dA\mathbf{j}) = \boxed{0;}$
    $\Phi_{y=a} = \iint \mathbf{E} \cdot d\mathbf{A} = \iint (bx^2\mathbf{i}) \cdot (+dA\mathbf{j}) = \boxed{0;}$
    $\Phi_{z=0} = \iint \mathbf{E} \cdot d\mathbf{A} = \iint (bx^2\mathbf{i}) \cdot (-dA\mathbf{k}) = \boxed{0;}$
    $\Phi_{z=a} = \iint \mathbf{E} \cdot d\mathbf{A} = \iint (bx^2\mathbf{i}) \cdot (+dA\mathbf{k}) = \boxed{0.}$

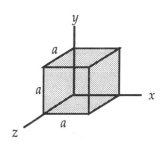

    We use Gauss' law to find the enclosed charge:
    $\Phi = \oiint \mathbf{E} \cdot d\mathbf{A} = q/\varepsilon_0;$
    $ba^4 = q/\varepsilon_0$, which gives $\boxed{q = \varepsilon_0 ba^4.}$

53. Because there is no charge enclosed by the tetrahedron, the net flux through all sides is 0:
    $\Phi_{net} = \Phi_{upper\ sides} + \Phi_{bottom}.$
    Thus we find the flux through the three upper sides from
    $\Phi_{upper\ sides} = -\Phi_{bottom} = -(E\mathbf{k}) \cdot A(-\mathbf{k})$
    $= +E(\tfrac{1}{2}L)(L\sin 60°) = \boxed{0.433EL^2.}$

55. From the symmetry of the field we construct a Gaussian surface which is a cylinder of length L and radius a with its axis along the axis of the field. Because the field is parallel to the ends of the cylinder, we have

$$\oiint \mathbf{E} \cdot d\mathbf{A} = E2\pi aL = Q_{enclosed}/\varepsilon_0 .$$

From the cylindrical symmetry of the field, the charge distribution within the cylinder must depend only on the distance from the axis, $\rho(r)$, so we have

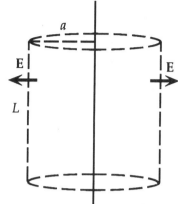

$$Q_{enclosed} = \int_0^a 2\pi rL\rho(r)\, dr . \quad \text{Thus}$$

$$\varepsilon_0 E2\pi aL = \int_0^a 2\pi rL\rho(r)\, dr \ , \ \text{or} \ \varepsilon_0 Ea = \int_0^a r\rho(r)\, dr .$$

We can write the left-hand side as an integral to get

$$\varepsilon_0 E \int_0^a dr = \int_0^a r\rho(r)\, dr .$$

Comparing the two integrands, we see that

$$\boxed{\rho(r) = \varepsilon_0 E/r .}$$

Note that this function diverges when $r \to 0$. The required field can be set up only beginning at some distance r_0 from the axis. Within r_0 only the total charge has to correspond to the required field:
$q/L = \varepsilon_0 E2\pi r_0$, which is finite.

57. We can express the linear electric field as $\mathbf{E}(x) = bx\,\mathbf{i}$. At $x = 0.5$ m, we have
$\mathbf{E}(0.5 \text{ m}) = (3000 \text{ N/C})\mathbf{i} = b(0.5 \text{ m})\mathbf{i}$, which gives $b = 6000 \text{ N/C}\cdot\text{m}$.
We call the area of the surface oriented in the yz-plane A. We choose a Gaussian surface consisting of the boundary of the region parallel to the x-axis and ends of area A at $x = 0$ and $x = x$. Because there is no flux through the sides and the field is constant at each end, we have

$$\Phi = \oiint \mathbf{E} \cdot d\mathbf{A} = Q_{enclosed}/\varepsilon_0;$$
$$E(x)\mathbf{i} \cdot A\mathbf{i} + E(0)\mathbf{i} \cdot A(-\mathbf{i}) = (1/\varepsilon_0) \int \rho A\, dx, \text{ which gives}$$
$$\int \rho\, dx = \varepsilon_0(bx) = +b\varepsilon_0 x.$$

Comparing the two sides, we see that

$$\boxed{\rho = + b\varepsilon_0 = +6000\varepsilon_0 \text{ C/m}^3 \text{ (constant).}}$$

Note that for the field to be 0 at $x = 0$, there must be external charges at $x < 0$.

59. We assume that the positive test charge is at a stable equilibrium point. The electric field there from the other charges is 0. A short distance from the stable equilibrium point, the electric field must be directed toward the point. We choose a small Gaussian surface around the equilibrium point:

$$\oiint \mathbf{E} \cdot d\mathbf{A} = Q_{enclosed}/\varepsilon_0 .$$

Because the field is directed into the surface, we have

$$\oiint \mathbf{E} \cdot d\mathbf{A} < 0,$$

which means that there is a negative charge at the equilibrium point. This is a contradiction, because the only charge there is the positive test charge. Thus the test charge cannot be in stable equilibrium.

CHAPTER 25

The electric potential is a scalar. As with potential energy, a reference level must be chosen. Because there are positive and negative charges, it is not possible to choose a reference level that makes all potentials positive. It is necessary to use the sign of the charge in the expression for the potential.

5. (a) Because there is no other charge present, no force is required to bring the charge from infinity:
$\boxed{W = 0.}$

(b) There is now a potential energy of the two charges, with the reference level at infinity. We find the work done by the non constant electric field from
$$W = -\Delta U = (-1/4\pi\varepsilon_0)(q_1 q_2 / r_1)$$
$$= -(9 \times 10^9\ \mathrm{C^2/N \cdot m^2})(3 \times 10^{-6}\ \mathrm{C})(5 \times 10^{-6}\ \mathrm{C})/(0.10\ \mathrm{m}) = \boxed{-1.35\ \mathrm{J.}}$$

(c) The work done by the external agent is the negative of the work done by the electric field:
$$W_F = -W = \boxed{+1.35\ \mathrm{J.}}$$

7. The potential energy is a scalar that depends only on the distance. The distances of the third charge from each of the others are
$$r_{a1} = [(\Delta x_1)^2 + (\Delta y_1)^2 + (\Delta z_1)^2]^{1/2} = [(30\ \mathrm{cm})^2 + 0 + (50\ \mathrm{cm} - 5\ \mathrm{cm})^2]^{1/2} = 54.1\ \mathrm{cm};$$
$$r_{a2} = [(\Delta x_2)^2 + (\Delta y_2)^2 + (\Delta z_2)^2]^{1/2} = [(30\ \mathrm{cm})^2 + 0 + (15\ \mathrm{cm} + 5\ \mathrm{cm})^2]^{1/2} = 62.6\ \mathrm{cm}.$$
The potential energy is
$$U_a = (1/4\pi\varepsilon_0)[(q_1 q_3 / r_{a1}) + (q_2 q_3 / r_{a2})] = (1/4\pi\varepsilon_0)q_3[(q_1/r_{a1}) + (q_2/r_{a2})]$$
$$= (9 \times 10^9\ \mathrm{C^2/N \cdot m^2})(0.20 \times 10^{-6}\ \mathrm{C})\{[(3.0 \times 10^{-6}\ \mathrm{C})/(0.541\ \mathrm{m})] + [(-3.0 \times 10^{-6}\ \mathrm{C})/(0.626\ \mathrm{m})]\}$$
$$= \boxed{1.36 \times 10^{-3}\ \mathrm{J.}}$$
When the charge is placed at (30 cm, 0 cm, 0 cm), the distances become
$$r_{b1} = r_{b2} = [(30\ \mathrm{cm})^2 + 0 + (5\ \mathrm{cm})^2]^{1/2} = 30.4\ \mathrm{cm}.$$
The potential energy is
$$U_b = (1/4\pi\varepsilon_0)[(q_1 q_3 / r_{b1}) + (q_2 q_3 / r_{b2})] = (1/4\pi\varepsilon_0)q_3[(q_1/r_{b1}) + (q_2/r_{b2})]$$
$$= (1/4\pi\varepsilon_0)(q_3/r_{b1})(q_1 + q_2) = \boxed{0,} \quad \text{because } q_1 = -q_2.$$

9. The distances between the two charges are
$$r_i = [(12\ \mathrm{cm} - 12\ \mathrm{cm})^2 + (60\ \mathrm{cm} - 25\ \mathrm{cm})^2 + (-50\ \mathrm{cm} - 0)^2]^{1/2} = 61.0\ \mathrm{cm};$$
$$r_f = [(12\ \mathrm{cm} - 12\ \mathrm{cm})^2 + (50\ \mathrm{cm} - 25\ \mathrm{cm})^2 + (25\ \mathrm{cm} - 0)^2]^{1/2} = 35.4\ \mathrm{cm}.$$
Because there is no change in kinetic energy, the work done by an external agent to move the second charge is
$$W = \Delta U = (1/4\pi\varepsilon_0)q_1 q_2[(1/r_f) - (1/r_i)]$$
$$= (9 \times 10^9\ \mathrm{C^2/N \cdot m^2})(1.5 \times 10^{-6}\ \mathrm{C})(-3 \times 10^{-6}\ \mathrm{C})\{[1/(0.351\ \mathrm{m})] - [1/(0.610\ \mathrm{m})]\} = \boxed{-4.8 \times 10^{-2}\ \mathrm{J.}}$$

17. We find the potential energy of the system of charges by adding the work required to bring the three charges in from infinity successively:
$$W_1 = q_1 V_0 = 0;$$
$$W_2 = q_2 V_1 = q_2(1/4\pi\varepsilon_0)q_1/r_{12} = (1/4\pi\varepsilon_0)q_1 q_2/r_{12};$$
$$W_3 = q_3 V_2 = q_3(1/4\pi\varepsilon_0)[(q_1/r_{13}) + (q_2/r_{23})] = (1/4\pi\varepsilon_0)[(q_1 q_3/r_{13}) + (q_2 q_3/r_{23})].$$
The total potential energy is
$$U = W_1 + W_2 + W_3 = (1/4\pi\varepsilon_0)[(q_1 q_2/r_{12}) + (q_1 q_3/r_{13}) + (q_2 q_3/r_{23})]$$
$$= (9 \times 10^9\ \mathrm{C^2/N \cdot m^2})\{[(2\ \mathrm{mC})(0.5\ \mathrm{mC})/(1\ \mathrm{m})] + [(2\ \mathrm{mC})(-1.5\ \mathrm{mC})/(0.5\ \mathrm{m})] +$$
$$[(0.5\ \mathrm{C})(-1.5\ \mathrm{C})/(1.5\ \mathrm{m})]\}(10^{-3}\ \mathrm{C/mC})^2 = \boxed{-5.0 \times 10^4\ \mathrm{J.}}$$

19. (a) The diameter of a circle subtends an angle of $90°$ at any point on the
circumference. Thus the distance from the negative charge to the point is
$r_2 = [(2R)^2 - r_1^2]^{1/2} = [(50\ cm)^2 - (30\ cm)^2]^{1/2} = 40\ cm.$
The potential at the point is
$V = (1/4\pi\varepsilon_0)[(q_1/r_1) + (q_2/r_2)]$
$\quad = (9 \times 10^9\ C^2/N \cdot m^2)\{[(24 \times 10^{-8}\ C)/(0.30\ m)]\ +$
$\quad\quad\quad\quad\quad [(-10 \times 10^{-8}\ C)/(0.40\ m)]\} = \boxed{+5.0 \times 10^3\ V.}$

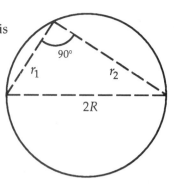

(b) The work required is
$W = q\,\Delta V = (-0.2 \times 10^{-6}\ C)(5.0 \times 10^3\ V - 0) = \boxed{-1.0 \times 10^{-3}\ J.}$
The negative value indicates that the negative charge wants
to "fall" to the higher potential.

21. If we let $q = 10^{-6}$ C, the charges are $q_1 = 2q$, $q_2 = -3q$, $q_3 = 5q$, $q_4 = 3q$.
The distances from each charge to the point are
$r_1 = [(30\ cm)^2 + (30\ cm)^2]^{1/2} = 42.4\ cm;$
$r_2 = r_4 = [(30\ cm)^2 + (16\ cm)^2]^{1/2} = 34.0\ cm;$
$r_3 = [(16\ cm)^2 + (16\ cm)^2]^{1/2} = 22.6\ cm.$
The potential at the point is
$V = (1/4\pi\varepsilon_0)[(q_1/r_1) + (q_2/r_2) + (q_3/r_3) + (q_4/r_4)]$
$\quad = (1/4\pi\varepsilon_0)q[(2/r_1) + (-3/r_2) + (5/r_3) + (3/r_4)]$
$\quad = (1/4\pi\varepsilon_0)q[(2/r_1) + (5/r_3)]$
$\quad = (9 \times 10^9\ C^2/N \cdot m^2)(10^{-6}\ C)\{[2/(0.424\ m)] + [5/(0.226\ m)]\}$
$\quad = \boxed{+2.4 \times 10^5\ V.}$

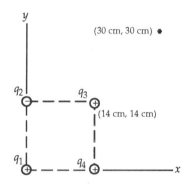

25. (a) From the symmetry of the charge distribution, we see that the electric field
is perpendicular to the slab and away from the centerline. This means that
$\boxed{E_A = 0.}$
To find the field at B, we construct a cylinder of height x with its axis
perpendicular to the slab as a Gaussian surface. One end of area A is placed
on the centerline, where the field is 0. Because the field is parallel to the
sides of the cylinder, there is flux only through the outer end, so we have
$\oint \mathbf{E} \cdot d\mathbf{A} = EA = Q_{enclosed}/\varepsilon_0 .$
If the end is at point B, the enclosed charge is ρAx and we have
$E_B = \rho x/\varepsilon_0 = (10^{-5}\ C/m^3)x/(8.85 \times 10^{-12}\ C^2/N \cdot m^2) = \boxed{(1.13 \times 10^6\ N/C \cdot m)x,\ x < 1\ cm.}$
If the end is at point C, the enclosed charge is $\rho Ad/2$ and we have
$E_C = \rho d/2\varepsilon_0 = (10^{-5}\ C/m^3)(2 \times 10^{-2}\ m)/2(8.85 \times 10^{-12}\ C^2/N \cdot m^2) = \boxed{1.13 \times 10^4\ N/C,\ x > 1\ cm.}$
As expected, the field outside the slab is uniform.

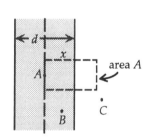

(b) We find the potential from the field by integrating over a path perpendicular to the slab:

$$V_B = V_A - \int_0^x \mathbf{E}_{inside} \cdot d\mathbf{s} = 0 - (1.13 \times 10^6\ N/C \cdot m)\int_0^x x'\,dx'$$
$$= \boxed{-(5.65 \times 10^5\ V/m^2)x^2,\ x < 1\ cm.}$$

(c)

The potential at the edge of the slab is
$V_{edge} = -(5.65 \times 10^5\ V/m^2)(0.01\ m)^2 = -56.5\ V.$
For the potential at point C we have

$$V_C = V_{edge} - \int_{edge}^x \mathbf{E}_{outside} \cdot d\mathbf{s} = -(56.5\ V) - (1.13 \times 10^4\ N/C)\int_{0.01\ m}^x dx'$$
$$= -56.5\ V - (1.13 \times 10^4\ V/m)(x - 0.01\ m),\ x > 1\ cm$$
$$= \boxed{+56.5\ V - (1.13 \times 10^4\ V/m)x,\ x > 1\ cm.}$$

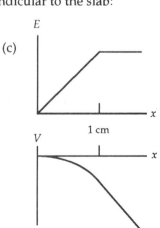

27. We consider the sphere to consist of an infinite number of spherical shells with thickness dr and charge $dq = \rho 4\pi r^2\, dr$, where the density of charge is

$$\rho = 3Q/4\pi R^3.$$

We choose the potential reference level at infinity.

At a point outside the sphere, all of the shells, and thus the sphere, are equivalent to point charges:

$$\boxed{V_{\text{outside}} = Q/4\pi\varepsilon_0 r, \text{ when } r > R.}$$

At a point inside the sphere, $r < R$, there are two contributions to the potential.

All of the shells with radius less than r are equivalent to point charges:

$$V_1 = q/4\pi\varepsilon_0 r = (\rho 4\pi r^3/3)/4\pi\varepsilon_0 r = \rho r^2/3\varepsilon_0.$$

For a shell of thickness dr with radius greater than r, the potential anywhere inside is constant and equal to the potential on the shell:

$$dV = dq/4\pi\varepsilon_0 r = \rho 4\pi r^2\, dr/4\pi\varepsilon_0 r = \rho r\, dr/\varepsilon_0.$$

We find the potential contribution from all of the shells with $r < r' < R$ by integrating:

$$V_2 = \frac{\rho}{\varepsilon_0}\int_r^R r'\, dr' = \frac{\rho}{\varepsilon_0}\left(\frac{r'^2}{2}\right)\Bigg|_r^R = \frac{\rho}{2\varepsilon_0}\left(R^2 - r^2\right).$$

The total potential is

$$\begin{aligned}
V_{\text{inside}} &= V_1 + V_2 = (\rho r^2/3\varepsilon_0) + [\rho(R^2 - r^2)/2\varepsilon_0] \\
&= (\rho/\varepsilon_0)[(R^2/2) - (r^2/6)] = (3Q/4\pi\varepsilon_0 R^3)[(R^2/2) - (r^2/6)] \\
&= \boxed{(Q/8\pi\varepsilon_0 R)[3 - (r/R)^2], \text{ when } r < R.}
\end{aligned}$$

If we compare the values at $r = R$, we get

$$V_{\text{outside}} = Q/4\pi\varepsilon_0 R \quad\text{and}\quad V_{\text{inside}} = (Q/8\pi\varepsilon_0 R)(3 - 1) = Q/4\pi\varepsilon_0 R = V_{\text{outside}}.$$

41. With the dipole pointing in the x-direction, the potential is

$$V = (p\cos\theta)/4\pi\varepsilon_0 r^2 = px/4\pi\varepsilon_0 r^3.$$

We find the components of the electric field from the partial derivatives of V.

For the x-component, we have

$$E_x = -\partial V/\partial x = -(p/4\pi\varepsilon_0 r^3) - [-(3px/4\pi\varepsilon_0 r^4)(\partial r/\partial x)].$$

From $r^2 = x^2 + y^2 + z^2$, we have

$$2r(\partial r/\partial x) = 2x, \quad\text{or}\quad \partial r/\partial x = x/r, \text{ so we get}$$

$$E_x = -(p/4\pi\varepsilon_0 r^3) + (3px^2/4\pi\varepsilon_0 r^5).$$

Similarly, we have

$$\begin{aligned}
E_y &= -\partial V/\partial y = -[-(3px/4\pi\varepsilon_0 r^4)(\partial r/\partial y)] \\
&= +3pxy/4\pi\varepsilon_0 r^5; \\
E_z &= -\partial V/\partial z = -[-(3px/4\pi\varepsilon_0 r^4)(\partial r/\partial z)] \\
&= +3pxz/4\pi\varepsilon_0 r^5.
\end{aligned}$$

Along the bisector (the y-axis), $x = 0$, so we have

$$\boxed{\mathbf{E} = -(p/4\pi\varepsilon_0 r^3)\mathbf{i}.}$$

Note that the symmetry along the y-axis shows us that the field there has only an x-component.

43. From the symmetry of the charge distribution, we know that the electric field is radial, so we find the electric field from

$$E_r = -\partial V/\partial r.$$

For $r < R$, we have

$$\begin{aligned}
V_{r < R} &= (Q/4\pi\varepsilon_0 R)[-2 + 3(r/R)^2]; \\
E_{r < R} &= -(Q/4\pi\varepsilon_0 R)(+6r/R^2) = -(Q/4\pi\varepsilon_0)(6r/R^3), \quad\text{or}\quad \boxed{E_{r < R} = (Q/4\pi\varepsilon_0)(6r/R^3)\hat{\mathbf{r}}.}
\end{aligned}$$

For $r > R$, we have

$$\begin{aligned}
V_{r > R} &= Q/4\pi\varepsilon_0 r; \\
E_{r > R} &= -(-Q/4\pi\varepsilon_0 r^2), \quad\text{or}\quad \boxed{E_{r > R} = (Q/4\pi\varepsilon_0 r^2)\hat{\mathbf{r}}.}
\end{aligned}$$

51. We choose the point P as the origin. The potential at P from a
differential element of the rod, which has a charge $dq = (q/L)\,dx$, is
$$dV = (1/4\pi\varepsilon_0)(dq/x).$$
To find the potential at P, we add (integrate) the
contributions from all elements:

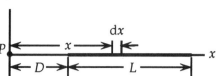

$$V = \frac{1}{4\pi\varepsilon_0}\int_D^{D+L}\left(\frac{q}{L}\right)\frac{dx}{x} = \frac{q}{4\pi\varepsilon_0 L}\ln\left(\frac{D+L}{D}\right)$$

$$= (9\times 10^9\ \text{N·m}^2/\text{C}^2)\frac{(2\times 10^{-6}\ \text{C})}{0.2\ \text{m}}\ln\left(\frac{0.1\ \text{m}+0.2\ \text{m}}{0.1\ \text{m}}\right) = \boxed{9.9\times 10^4\ \text{V.}}$$

53. (a) The potential for the two charges is the sum:

$$\boxed{V = \frac{1}{4\pi\varepsilon_0}\left(\frac{3q_0}{x-x_0}\right) + \frac{1}{4\pi\varepsilon_0}\left(\frac{-q_0}{x+\frac{x_0}{2}}\right) = \frac{q_0}{4\pi\varepsilon_0}\left[\left(\frac{3}{x-x_0}\right)-\left(\frac{1}{x+\frac{x_0}{2}}\right)\right].}$$

(b) When $x \gg x_0$, we use the approximation $1/(1\pm u) \simeq 1\mp u + u^2 \mp u^3 + \ldots$, when $u \ll 1$:

$$V = \frac{q_0}{4\pi\varepsilon_0 x}\left[\left(\frac{3}{1-\frac{x_0}{x}}\right)-\left(\frac{1}{1+\frac{x_0}{2x}}\right)\right]$$

$$= \frac{q_0}{4\pi\varepsilon_0}\left\{\frac{3}{x}\left[1+\frac{x_0}{x}+\left(\frac{x_0}{x}\right)^2+\left(\frac{x_0}{x}\right)^3+\ldots\right]-\frac{1}{x}\left[1-\frac{x_0}{2x}+\left(\frac{x_0}{2x}\right)^2-\left(\frac{x_0}{2x}\right)^3+\ldots\right]\right\}$$

$$= \frac{q_0}{4\pi\varepsilon_0}\left[\frac{2}{x}+\frac{7}{2}\frac{x_0}{x^2}+\frac{11}{4}\frac{x_0^2}{x^3}+\frac{25}{8}\frac{x_0^3}{x^4}+\ldots\right].$$

(c) At large values of x, the contribution of each term to the total decreases.
The first term from part (b) is the potential of a point charge, with $\boxed{q_{net} = 2q_0.}$
The second term is the potential of a dipole, with $\boxed{p = 7q_0x_0/2.}$

(d) For the point charge plus the dipole to be within 1% of the exact answer, we have

$$\frac{2}{x}+\frac{7}{2}\frac{x_0}{x^2} = 0.99\left[\frac{3}{x-x_0}-\frac{1}{x+(x_0/2)}\right] = \frac{0.99}{x}\left[\frac{3}{1-(x_0/x)}-\frac{2}{2+(x_0/x)}\right].$$

If we let $x_0/x = y$, we have

$$2+\frac{7}{2}y = 0.99\left[\frac{3(2+y)-2(1-y)}{(1-y)(2+y)}\right] = 0.99\left(\frac{4+5y}{2+y-y^2}\right).$$

A numerical solution gives $y = -0.0825, +0.0875$. The values of x are $-12.1x_0, +11.4x_0$.
Thus if $\boxed{|x| > 12.1x_0,}$ the point charge plus the dipole will be within 1% of the exact answer.

55. There is no change in the kinetic energy. We use the expression for the potential on the axis of a ring:
$$W_{a\to b} = q(V_b - V_a) = q(Q/4\pi\varepsilon_0)[(R^2+x_b^2)^{-1/2}-(R^2+x_a^2)^{-1/2}]$$
$$= (-8.5\times 10^{-8}\ \text{C})(3.5\times 10^{-7}\ \text{C})(9\times 10^9\ \text{C}^2/\text{N·m}^2)\times$$
$$\{[(0.24\ \text{m})^2+(0.85\ \text{m})^2]^{-1/2}-[(0.24\ \text{m})^2+(0.28\ \text{m})^2]^{-1/2}\} = \boxed{+4.2\times 10^{-4}\ \text{J.}}$$

57. After the connection, the two spheres must have the same potential:
$V = (1/4\pi\varepsilon_0)(q_1'/r_1) = (1/4\pi\varepsilon_0)(q_2'/r_2)$, or $q_1' = (r_1/r_2)q_2'$.
Because charge is conserved we have
$q_1 + q_2 = q_1' + q_2'$.
When we combine these two equations, we get
$q_2' = (q_1 + q_2)[r_2/(r_1 + r_2)]$.
The amount of charge that moves between the two spheres is the change in charge on either sphere,
so we have
$\Delta q_2 = q_2' - q_2 = \boxed{(q_1 r_2 - q_2 r_1)/(r_1 + r_2).}$
If we were to solve for q_1', we would get $\Delta q_1 = q_1' - q_1 = -(q_1 r_2 - q_2 r_1)/(r_1 + r_2)$; the negative of Δq_2.

61. (a) The wire connecting the spheres means that the potentials of the spheres are the same:
$q_1/4\pi\varepsilon_0 R_1 = q_2/4\pi\varepsilon_0 R_2$;
$q_1 = (R_1/R_2)q_2 = [(20 \text{ mm})/(100 \text{ mm})]q_2$, which gives $q_2 = 5q_1$.
The repulsive Coulomb force is
$F = (1/4\pi\varepsilon_0)(q_1 q_2/r^2)$;
$3.5 \text{ N} = (9 \times 10^9 \text{ C}^2/\text{N} \cdot \text{m}^2)(5q_1^2)/(0.25 \text{ m})^2$, which gives
$q_1 = 2.2 \times 10^{-6} \text{ C} = \boxed{2.2 \ \mu\text{C}}$ and $q_2 = 5q_1 = \boxed{11 \ \mu\text{C.}}$
    (b) The electric fields at the surfaces of the spheres are
$E_1 = (1/4\pi\varepsilon_0)(q_1/R_1^2) = (9 \times 10^9 \text{ C}^2/\text{N} \cdot \text{m}^2)(2.2 \times 10^{-6} \text{ C})/(0.020 \text{ m})^2 = \boxed{4.95 \times 10^7 \text{ V/m, radial.}}$
$E_2 = (1/4\pi\varepsilon_0)(q_2/R_2^2) = (9 \times 10^9 \text{ C}^2/\text{N} \cdot \text{m}^2)(11 \times 10^{-6} \text{ C})/(0.100 \text{ m})^2 = \boxed{9.90 \times 10^6 \text{ V/m, radial.}}$

65. (a) The maximum potential is reached when the charge on the sphere creates an electric field large
enough to break down the air. At the surface of a sphere, we have
$E = (1/4\pi\varepsilon_0)(Q/R^2)$ and $V = (1/4\pi\varepsilon_0)(Q/R)$, or
$V = ER = (3 \times 10^6 \text{ V/m})(1.3 \text{ m}) = \boxed{3.9 \times 10^6 \text{ V.}}$
    (b) The increase in kinetic energy comes from the decrease in potential energy, which means the proton
must go from high to low potential:
$\Delta K = K - 0 = -\Delta U = -q \Delta V$;
$K = -(+1 \text{ e})(-3.9 \times 10^6 \text{ V}) = +3.9 \times 10^6 \text{ eV} = \boxed{3.9 \text{ MeV} \ (6.2 \times 10^{-13} \text{ J}).}$
    (c) We find the charge on the sphere from
$V = (1/4\pi\varepsilon_0)(Q/R)$;
$3.9 \times 10^6 \text{ V} = (9 \times 10^9 \text{ C}^2/\text{N} \cdot \text{m}^2)Q/(1.3 \text{ m})$, which gives $\boxed{Q = 5.6 \times 10^{-4} \text{ C.}}$

71. From Table 25–1, with $V = 0$ at $r = \infty$, we have the potential inside a nonconducting sphere:
$V = (Q/8\pi\varepsilon_0 R)[3 - (r^2/R^2)]$.
The potential energy of a charge $-q$ is
$U = -qV = \boxed{-(qQ/8\pi\varepsilon_0 R)[3 - (r^2/R^2)].}$
The variable part of the potential energy has the form of the elastic potential energy of a spring:
$U = \tfrac{1}{2}kr^2$,
so the motion can be an oscillation, like the mass on a spring.
Comparing the coefficients, we have $\boxed{k = qQ/4\pi\varepsilon_0 R^3.}$

73. If we consider the sodium ion in the upper left corner, we see from symmetry that the net force must be along the diagonal:

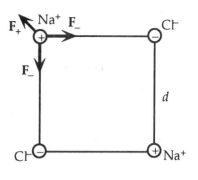

$$F_{net} = 2F_- \cos 45° - F_+ = (e^2/4\pi\varepsilon_0)\{(2\cos 45°/d^2) - [1/(d\sqrt{2})^2]\}$$
$$= (e^2/4\pi\varepsilon_0 d^2)(\sqrt{2} - \tfrac{1}{2})$$
$$= [(1.6 \times 10^{-19}\text{ C})^2(9 \times 10^9\text{ C}^2/\text{N}\cdot\text{m}^2)/(2.5 \times 10^{-19}\text{ m})^2](\sqrt{2} - \tfrac{1}{2})$$
$$= \boxed{3.4 \times 10^{-9}\text{ N toward the other Na}^+.}$$

We find the work required from the potential energy change:

$$W = \Delta U = e(V_\infty - V_{corner}) = (e/4\pi\varepsilon_0)\{0 - [-(2e/d) + (e/d\sqrt{2})]\}$$
$$= (e^2/4\pi\varepsilon_0 d)[2 - (1/\sqrt{2})]$$
$$= [(1.6 \times 10^{-19}\text{ C})^2(9 \times 10^9\text{ C}^2/\text{N}\cdot\text{m}^2)/(2.5 \times 10^{-19}\text{ m})][2 - (1/\sqrt{2})]$$
$$= \boxed{1.2 \times 10^{-18}\text{ J} \ (7.4\text{ eV}).}$$

79. (a) From the force diagram, we apply $\Sigma\mathbf{F} = 0$:

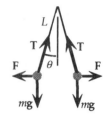

horizontal: $T\sin\theta = F = kqq/r^2$;
vertical: $T\cos\theta = mg$.

If we divide the two equations, we get

$$\tan\theta = F/mg = kq^2/r^2 mg = kq^2/(2L\sin\theta)^2 mg$$
$$\tan 30° = (9 \times 10^9\text{ N}\cdot\text{m}^2/\text{C}^2)(2.0 \times 10^{-6}\text{ C})^2/[2(0.80\text{ m})\sin 30°]^2 m(9.8\text{ m/s}^2),$$

which gives $\boxed{m = 9.9 \times 10^{-3}\text{ kg.}}$

(b) With the electric potential reference level at infinity and the gravitational potential reference level at $\theta = 0°$, we have

$$U = qV + mgy = (1/4\pi\varepsilon_0)(q^2/2L\sin\theta) + mg(L - L\cos\theta)$$
$$= [(9 \times 10^9\text{ N}\cdot\text{m}^2/\text{C}^2)(2 \times 10^{-6}\text{ C})^2/2(0.80\text{ m})(\sin\theta)] + (9.9 \times 10^{-3}\text{ kg})(9.8\text{ m/s}^2)(0.80\text{ m})(1 - \cos\theta)$$
$$= \boxed{(0.023/\sin\theta) + 0.078(1 - \cos\theta).}$$

81. We use the analogy to the charged spherical shell. When we are outside a charged cylindrical shell of radius r', the potential is that of a line charge: $V = -(\lambda/2\pi\varepsilon_0)\ln(r/a)$ with $V = 0$ at $r = a$. When we are inside a charged cylindrical shell of radius r', the potential is the potential on the surface: $V = -(\lambda/2\pi\varepsilon_0)\ln(r'/a)$. For a point inside the cylinder, $r < R$, the potential has two contributions: the sum (integral) of the shells inside r and the sum (integral) of the shells outside r. For a shell of radius r', the linear charge density is $d\lambda = \rho 2\pi r'\,dr'$. We find the potential from

$$V = \int_0^r -\frac{\rho 2\pi r'\,dr'}{2\pi\varepsilon_0}\ln\left(\frac{r}{a}\right) + \int_r^R -\frac{\rho 2\pi r'\,dr'}{2\pi\varepsilon_0}\ln\left(\frac{r'}{a}\right)$$

$$= -\frac{\rho}{\varepsilon_0}\ln\left(\frac{r}{a}\right)\int_0^r r'\,dr' - \frac{\rho}{\varepsilon_0}\int_r^R r'\ln\left(\frac{r'}{a}\right)dr'$$

$$= -\frac{\rho}{\varepsilon_0}\ln\left(\frac{r}{a}\right)\frac{r^2}{2} - \frac{\rho}{\varepsilon_0}\left\{\frac{r'^2}{2}\left[\ln\left(\frac{r'}{a}\right) - \frac{1}{2}\right]\right\}\Bigg|_r^R$$

$$= -\frac{\rho r^2}{2\varepsilon_0}\ln\left(\frac{r}{a}\right) - \frac{\rho R^2}{2\varepsilon_0}\left[\ln\left(\frac{R}{a}\right) - \frac{1}{2}\right] + \frac{\rho r^2}{2\varepsilon_0}\left[\ln\left(\frac{r}{a}\right) - \frac{1}{2}\right]$$

$$= \boxed{-\frac{\rho r^2}{4\varepsilon_0} - \frac{\rho R^2}{2\varepsilon_0}\left[\ln\left(\frac{R}{a}\right) - \frac{1}{2}\right], \ r < R.}$$

For a point outside the cylinder, $r > R$, all of the cylindrical shells appear to be line charges:

$$V = \int_0^R -\frac{\rho 2\pi r'\,dr'}{2\pi\varepsilon_0}\ln\left(\frac{r}{a}\right) = -\frac{\rho}{\varepsilon_0}\ln\left(\frac{r}{a}\right)\int_0^R r'\,dr'$$

$$= \boxed{-\frac{\rho R^2}{2\varepsilon_0}\ln\left(\frac{r}{a}\right), \ r > R.}$$

CHAPTER 26

7. Because the potential from the outer conductor is constant inside, the potential difference between the two conductors is due only to the inner conductor, which is equivalent to a point charge:
   $V = (Q/4\pi\varepsilon_0)[(1/r) - (1/R)]$, so the capacitance is
   $C = Q/V = 4\pi\varepsilon_0 rR/(R - r)$.
   (a) When r is finite and $R \to \infty$, the denominator becomes R, so we have
   $$\boxed{C \to 4\pi\varepsilon_0 r,}$$ which is the capacitance of the inner sphere.
   (b) When $(R - r) \ll r$, we have $R \to r$, so
   $$\boxed{C \to 4\pi\varepsilon_0 r^2/(R - r) = \varepsilon_0 A/d,}$$ which is the capacitance of parallel plates with separation d.

11. When the capacitor is isolated, the charge must be constant, so we have
    $Q = C_{min}V_{max} = C_{max}V_{min};$
    $V_{max} = (C_{max}/C_{min})V_{min}$
    $\qquad = [(0.2\ \mu\text{F})/(0.01\ \mu\text{F})](300\ \text{V}) = 6 \times 10^3\ \text{V} = \boxed{6\ \text{kV.}}$

17. (a) For a coaxial cable, we have
    $C = L2\pi\varepsilon_0/\ln(b/a)$
    $\qquad = (10\ \text{m})2\pi(8.85 \times 10^{-12}\ \text{F/m})/\ln[(8\ \text{mm}/3\ \text{mm})] = \boxed{5.67 \times 10^{-10}\ \text{F.}}$
    (b) The energy stored in 10 m of cable is
    $U_1 = \frac{1}{2}CV^2$
    $\qquad = \frac{1}{2}(5.67 \times 10^{-10}\ \text{F})(10^3\ \text{V})^2 = \boxed{2.83 \times 10^{-4}\ \text{J.}}$
    Because the capacitance is directly proportional to the length, the energy stored in 1 km of cable is
    $U_2 = [(10^3\ \text{m})/(10\ \text{m})]U_1 = \boxed{2.83 \times 10^{-2}\ \text{J.}}$

19. (a) When the plates of the disconnected capacitor are pulled apart, the charge does not change; both the capacitance and the potential difference will change. The initial capacitance and charge are
    $C_0 = \varepsilon_0 A/d_0$; $Q_0 = C_0 V_0 = \varepsilon_0 A V_0/d_0 = Q$.
    If we express the energy stored in the capacitor as
    $U = \frac{1}{2}Q^2/C = \frac{1}{2}Q^2 d/\varepsilon_0 A$,
    the change in stored energy is
    $\Delta U = \frac{1}{2}(Q_0^2/\varepsilon_0 A)(d_1 - d_0) = \boxed{(\varepsilon_0 A V_0^2/2d_0^2)(d_1 - d_0).}$
    (b) Because the external force is the only interaction with the capacitor, we have
    $W_F = \Delta U = \boxed{(\varepsilon_0 A V_0^2/2d_0^2)(d_1 - d_0).}$
    (c) If the plates stay connected to the battery, the potential difference does not change, while both the charge and capacitance will change. The change in stored energy is
    $\Delta U = \frac{1}{2}V_0^2(C - C_0) = (\varepsilon_0 A V_0^2/2)[(1/d_1) - (1/d_0)] = \boxed{(\varepsilon_0 A V_0^2/2d_1 d_0)(d_0 - d_1).}$
    (d) Even though there must still be work done by the external force to separate the opposite charges on the plates, the energy stored in the capacitor decreases ($d_0 < d_1$). The charge on the plates has decreased and energy has been stored in the battery.

25. For a conducting sphere, we have
$V = (1/4\pi\varepsilon_0)(q/r)$:
8.3×10^3 V $= (9 \times 10^9$ N$\cdot$m^2/C$^2)q/(18 \times 10^{-2}$ m$)$, which gives $\boxed{q = 1.7 \times 10^{-7} \text{ C.}}$
The energy density outside the sphere is
$$u = \tfrac{1}{2}\varepsilon_0 E^2 = \tfrac{1}{2}\varepsilon_0[(1/4\pi\varepsilon_0)(q/r^2)]^2$$
$$= \tfrac{1}{2}(8.85 \times 10^{-12} \text{ C}^2/\text{N}\cdot\text{m}^2)[(9 \times 10^9 \text{ N}\cdot\text{m}^2/\text{C}^2)(1.7 \times 10^{-7} \text{ C})/r^2]^2$$
$$= \boxed{1.0 \times 10^{-5}/r^4 \text{ J/m}^3,} \text{ with } r \text{ in m.}$$
Because there is no field inside the sphere, we find the total energy in the electric field by adding the energies of external spherical shells with radius r and thickness dr:

$$U = \int_R^\infty \frac{1.0 \times 10^{-5} \text{ J}\cdot\text{m}}{r^4} 4\pi r^2 \, dr = \left(4.0\pi \times 10^{-5} \text{ J}\cdot\text{m}\right)\int_R^\infty \frac{dr}{r^2} = \left(4.0\pi \times 10^{-5} \text{ J}\cdot\text{m}\right)\left(-\frac{1}{r}\right)\Big|_R^\infty$$

$$= \left(4.0\pi \times 10^{-5} \text{ J}\cdot\text{m}\right)\left(-\frac{1}{\infty} + \frac{1}{R}\right) = \left(4.0\pi \times 10^{-5} \text{ J}\cdot\text{m}\right)\left(-0 + \frac{1}{R}\right) = \frac{4.0\pi \times 10^{-5} \text{ J}\cdot\text{m}}{0.18 \text{ m}} = \boxed{7.0 \times 10^{-4} \text{ J.}}$$

Note that this is $\tfrac{1}{2}qV$, the energy stored in the spherical capacitor.

27. (a) For a coaxial cable, we have
$C = L2\pi\varepsilon_0/\ln(b/a)$
$= (0.15 \text{ m})2\pi(8.85 \times 10^{-12} \text{ F/m})/\ln[(2 \text{ cm}/0.02 \text{ cm})] = 1.8 \times 10^{-12}$ F $= \boxed{1.8 \text{ pF.}}$
(b) The energy that recharges the tube is stored in the capacitor:
$U = \tfrac{1}{2}CV^2 = \tfrac{1}{2}(1.8 \times 10^{-12} \text{ F})(5 \times 10^2 \text{ V})^2 = \boxed{2.3 \times 10^{-7} \text{ J.}}$

29. (a) We find the electric field between two parallel plates from the potential gradient:
$E = V/d$
$= (1500 \text{ V})/(0.5 \times 10^{-2} \text{ m}) = \boxed{3.00 \times 10^5 \text{ V/m.}}$
(b) In terms of the charge density on the plates, the field is $E = \sigma/\varepsilon_0$, which gives
$Q = \sigma A = \varepsilon_0 EA = (8.85 \times 10^{-12} \text{ F/m})(3.00 \times 10^5 \text{ V/m})(400 \times 10^{-4} \text{ m}^2) = \boxed{1.06 \times 10^{-7} \text{ C.}}$
(c) We have to remember that the field between the plates is produced by both plates, but the field that one plate produces at the other is one-half of this. The force exerted on one plate is
$F = Q\tfrac{1}{2}E = \tfrac{1}{2}(1.06 \times 10^{-7} \text{ C})(3.00 \times 10^5 \text{ V/m}) = \boxed{1.59 \times 10^{-2} \text{ N.}}$
(d) When the plates are pulled apart, the charge does not change, while both the capacitance and the potential difference will change. If we express the energy stored in the capacitor as
$U = \tfrac{1}{2}Q^2/C = \tfrac{1}{2}Q^2 d/\varepsilon_0 A$,
the change in stored energy is
$\Delta U = \tfrac{1}{2}(Q^2/\varepsilon_0 A)(d_2 - d_1)$
$= \tfrac{1}{2}[(1.06 \times 10^{-7} \text{ C})^2/(8.85 \times 10^{-12} \text{ C}^2/\text{N}\cdot\text{m}^2)(400 \times 10^{-4} \text{ m}^2)](0.20)(0.5 \times 10^{-2} \text{ m}) = \boxed{1.59 \times 10^{-5} \text{ J.}}$
To pull the plates apart requires a force to balance the attractive force from part (c). The work done by this force is
$W = F \, \Delta d = (1.59 \times 10^{-2} \text{ N})(0.2)(0.5 \times 10^{-2} \text{ m}) = 1.59 \times 10^{-5}$ J,
so the result is consistent with part (c).

33. From the redrawn circuit, we see that C_3 and C_4 are in series.

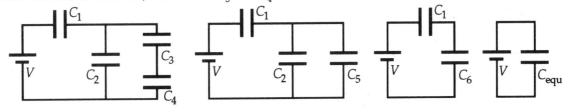

We find their equivalent capacitance from
$$1/C_5 = (1/C_3) + (1/C_4)$$
$$= [1/(2\ \mu F)] + [1/(5\ \mu F)],\ \text{which gives}\ C_5 = 1.43\ \mu F.$$
We redraw the circuit and find the equivalent capacitance of C_2 and C_5, which are in parallel:
$$C_6 = C_2 + C_5 = 4\ \mu F + 1.43\ \mu F = 5.43\ \mu F.$$
We redraw the circuit again and find the equivalent capacitance of C_1 and C_6, which are in series:
$$1/C_{equ} = (1/C_1) + (1/C_6) = [1/(3\ \mu F)] + [1/(5.43\ \mu F)],\ \text{which gives}\ \boxed{C_{equ} = 1.93\ \mu F.}$$

35. Because the potential from the outer conductor is constant inside, the potential difference between the two conductors is due to the inner conductor, which is equivalent to a point charge:

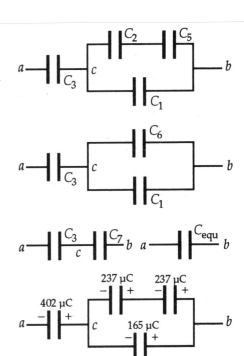

(a)

(b)

$V = (Q/4\pi\varepsilon_0)[(1/r) - (1/R)]$, so the capacitance is
$$C = Q/V = 4\pi\varepsilon_0 rR/(R - r)$$
$$= 4\pi(8.85 \times 10^{-12}\ \text{F/m})(3 \times 10^{-3}\ \text{m})(12 \times 10^{-3}\ \text{m})/[(12 - 3) \times 10^{-3}\ \text{m}]$$
$$= 4.5 \times 10^{-13}\ \text{F} = \boxed{0.45\ \text{pF.}}$$
When a wire connects the two spheres, they must be at the same potential and all the charge will be on the outer sphere. The potential of the charged outer sphere is
$V = Q/4\pi\varepsilon_0 R$, so the capacitance is
$$C = Q/V = 4\pi\varepsilon_0 R$$
$$= 4\pi(8.85 \times 10^{-12}\ \text{F/m})(12 \times 10^{-3}\ \text{m})$$
$$= 1.33 \times 10^{-12}\ \text{F} = \boxed{1.33\ \text{pF.}}$$

37. (a) From the circuit, we see that C_2 and C_5 are in series and find their equivalent capacitance from
$$1/C_6 = (1/C_2) + (1/C_5)$$
$$= [1/(2\ \mu F)] + [1/(5\ \mu F)],\ \text{which gives}\ C_6 = 1.43\ \mu F.$$
From the new circuit, we see that C_1 and C_6 are in parallel, with an equivalent capacitance
$$C_7 = C_1 + C_6 = 1\ \mu F + 1.43\ \mu F = 2.43\ \mu F.$$
From the new circuit, we see that C_3 and C_7 are in series and find their equivalent capacitance from
$$1/C_{equ} = (1/C_3) + (1/C_7)$$
$$= [1/(3\ \mu F)] + [1/(2.43\ \mu F)],\ \text{which gives}$$
$$\boxed{C_{equ} = 1.34\ \mu F.}$$

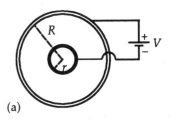

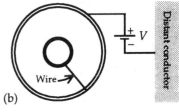

(b) The charge on the equivalent capacitor is also the charge on C_3 and C_7:
$$Q_{equ} = Q_3 = Q_7 = C_{equ}V_{ab} = (1.34\ \mu F)(300\ V) = 402\ \mu C.$$
We find the potential difference between c and b from
$$V_{cb} = Q_7/C_7 = (402\ \mu C)/(2.43\ \mu F) = 165\ V.$$
The charge on C_6 is also the charge on C_2 and C_5:
$$Q_6 = Q_2 = Q_5 = C_6 V_{cb} = (1.43\ \mu F)(165\ V) = 237\ \mu C.$$
The charge on C_1 is
$$Q_1 = C_1 V_{cb} = (1\ \mu F)(165\ V) = 165\ \mu C.$$
Because point b is at the higher potential, the charges are as shown in the diagram.

39. To find the equivalent capacitance between a and b,
   we redraw the circuit, and see that there are two
   capacitors in series in the top branch:
   $1/C_1 = (1/C) + (1/C)$, which gives $C_1 = \frac{1}{2}C$.
   For the two capacitors in parallel between d and b we have
   $C_2 = C_1 + C = \frac{1}{2}C + C = \frac{3}{2}C$.
   For the two capacitors in series between a and b we have
   $1/C_3 = (1/C) + (1/C_2)$, which gives $C_3 = 0.6C$.
   For the equivalent capacitance, we have
   $C_{equ,ab} = C + C_3 = C + 0.6C = \boxed{1.6C.}$
   To find the equivalent capacitance between a and c, we
   redraw the circuit, and use symmetry to simplify the circuit.
   The top and bottom paths from a to c are equivalent, so we have
   $V_b = V_d$,
   which means there is no potential difference across and no
   charge on the middle capacitor. The circuit will not change
   if we remove it.
   The top and bottom branches have two capacitors in series:
   $1/C_4 = (1/C) + (1/C)$, which gives $C_4 = \frac{1}{2}C$.
   We combine these two capacitors in parallel to find the
   equivalent capacitance of the circuit:
   $C_{equ,ac} = C_4 + C_4 = \frac{1}{2}C + \frac{1}{2}C = \boxed{C.}$
   To find the equivalent capacitance between b and d, we
   redraw the circuit, and see that the top and bottom paths
   have two capacitors in series:
   $1/C_5 = (1/C) + (1/C)$, which gives $C_5 = \frac{1}{2}C$.
   We now have three capacitors in parallel,
   with equivalent capacitance
   $C_{equ,bd} = C_5 + C_5 + C = \frac{1}{2}C + \frac{1}{2}C + C = \boxed{2C.}$

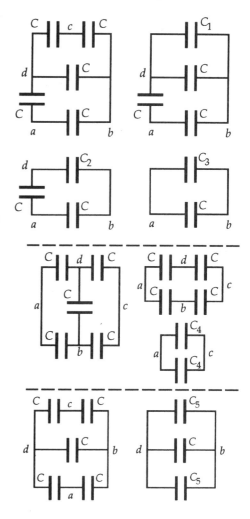

47. The capacitance of an isolated sphere in air is $C_0 = 4\pi\varepsilon_0 R$. When it is embedded in a dielectric, we have
   $C = \kappa C_0$, so the change is
   $\boxed{C - C_0 = (\kappa - 1)C_0 = (\kappa - 1)4\pi\varepsilon_0 R.}$
   The sign of the induced charge on the surface of the dielectric will be opposite to that of the original
   charge. If $E_0 = \sigma/\varepsilon_0$ is the original electric field just outside the sphere, we have
   $E = E_0/\kappa = E_0 + E_{ind}$;
   $\sigma/\kappa\varepsilon_0 = (\sigma/\varepsilon_0) + (\sigma_{ind}/\varepsilon_0)$, which gives $\boxed{\sigma_{ind}/\sigma = (\kappa - 1)/\kappa.}$

55. Because the potential from the outer shell is constant inside, in vacuum the potential difference between
   the two shells is due only to the inner shell, which is equivalent to a point charge:
   $V = (Q/4\pi\varepsilon_0)[(1/r_1) - (1/r_2)]$, so the capacitance in vacuum is $C = Q/V = 4\pi\varepsilon_0 r_1 r_2/(r_2 - r_1)$.
   When the space is filled with a dielectric, we have
   $C = \kappa C_0 = \kappa 4\pi\varepsilon_0 r_1 r_2/(r_2 - r_1)$.
   With air between the shells, the energy is
   $U_0 = \frac{1}{2}Q^2/C_0$.
   When the dielectric is added, the charge does not change, so the energy is
   $U = \frac{1}{2}Q^2/C = \frac{1}{2}Q^2/\kappa C_0$.
   The change in energy is
   $U - U_0 = \frac{1}{2}(Q^2/C_0)[(1/\kappa) - 1] = \frac{1}{2}(Q^2/\kappa C_0)(1 - \kappa) = \boxed{(1 - \kappa)Q^2/2C \ \text{(a decrease)}.}$

57. From the diagram, we see that the arrangement is equivalent to 9 capacitors in parallel:

$$C = 9C_1 = 9(\varepsilon_0 A/d)$$
$$= 9(8.85 \times 10^{-12}\ \text{F/m})(6.0 \times 10^{-2}\ \text{m})(8.0 \times 10^{-2}\ \text{m})/(1.2 \times 10^{-3}\ \text{m})$$
$$= 3.2 \times 10^{-10}\ \text{F} = \boxed{0.32\ \text{nF.}}$$

If the region is filled with a dielectric, we have
$$C' = \kappa C = 2.8(0.32\ \text{nF}) = \boxed{0.90\ \text{nF.}}$$

61. For a polar dielectric we have
$$\kappa = 1 + (a/T),\ \text{so}$$
$$C = \kappa\varepsilon_0 A/d = \kappa C_0 = [1 + (a/T)]C_0.$$
With the given data we have
$$3.2\ \mu\text{F} = [1 + (a/296\ \text{K})]C_0\ \text{and}\ 2.65\ \mu\text{F} = [1 + (a/360\ \text{K})]C_0.$$
When we solve these two equations for the two unknowns, we get
$$a = 8640\ \text{K}\ \text{and}\ C_0 = 0.106\ \mu\text{F}.$$
At a temperature of 48°C, we have
$$C = [1 + (8640\ \text{K}/321\ \text{K})](0.106\ \mu\text{F}) = \boxed{2.96\ \mu\text{F.}}$$

67. We find the capacitance of the strip of the dielectric at x, with width dx, from
$$dC = \kappa\varepsilon_0\,dA/D = \kappa\varepsilon_0 L\,dx/D.$$
The strips that make up the capacitor are in parallel, so the equivalent capacitance is

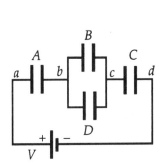

$$C = \int dC = \int_0^L \frac{\kappa\varepsilon_0 L}{D}\,dx = \varepsilon_0 \int_0^L \left[\kappa_0 + \frac{(\kappa_1 - \kappa_0)x}{L}\right]\frac{L}{D}\,dx$$

$$= \frac{\varepsilon_0 L}{D}\left[\kappa_0 x + \frac{(\kappa_1 - \kappa_0)x^2}{2L}\right]\Bigg|_0^L = \frac{\varepsilon_0 L}{D}\left[\kappa_0 L + \frac{(\kappa_1 - \kappa_0)L^2}{2L}\right],\ \text{which reduces to}$$

$$\boxed{C = \tfrac{1}{2}(\kappa_0 + \kappa_1)(\varepsilon_0 L^2/d).}$$

69. We find the equivalent capacitance of the circuit.
    B and D are in parallel:
$$C_1 = C_B + C_D = (4.3\ \mu\text{F}) + (2.1\ \mu\text{F}) = 6.4\ \mu\text{F}.$$
We now have three capacitors in series:
$$1/C_{equ} = (1/C_A) + (1/C_1) + (1/C_C)$$
$$= (1/5.4\ \mu\text{F}) + (1/6.4\ \mu\text{F}) + (1/3.2\ \mu\text{F}),$$
which gives $C_{equ} = 1.53\ \mu\text{F}.$

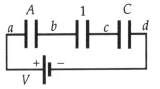

We find the charge on the equivalent capacitor, which is also the charge on each capacitor in series, from
$$Q_{equ} = Q_A = Q_1 = Q_C = C_{equ}V_{ab}$$
$$= (1.53\ \mu\text{F})(3000\ \text{V}) = 4.6 \times 10^3\ \mu\text{C}.$$
We find the potential differences from
$$V_A = V_{ac} = Q_A/C_A = (4.6 \times 10^3\ \mu\text{C})/(5.4\ \mu\text{F}) = \boxed{8.5 \times 10^2\ \text{V;}}$$
$$V_B = V_D = V_{cd} = Q_1/C_1 = (4.6 \times 10^3\ \mu\text{C})/(6.4\ \mu\text{F}) = \boxed{7.2 \times 10^2\ \text{V;}}$$
$$V_C = V_{db} = Q_C/C_C = (4.6 \times 10^3\ \mu\text{C})/(3.2\ \mu\text{F}) = \boxed{1.43 \times 10^3\ \text{V.}}$$

71. The initial energy stored in the capacitor is

$$U_0 = \tfrac{1}{2}C_0V_0^2 = \tfrac{1}{2}(3.0 \times 10^{-6}\text{ F})(1500\text{ V})^2 = \boxed{3.4\text{ J.}}$$

Because the source is disconnected, the charge on the capacitor does not change, and we have

$$C = \kappa C_0; \quad V = Q/C = Q/\kappa C_0 = V_0/\kappa.$$

The energy stored after the dielectric is inserted is

$$U = \tfrac{1}{2}CV^2 = \tfrac{1}{2}\kappa C_0(V_0/\kappa)^2 = (1/\kappa)(\tfrac{1}{2}C_0V_0^2) = (1/\kappa)U_0.$$

We find the work required to insert the dielectric from

$$\begin{aligned} W \; &= \Delta U = [(1/\kappa) - 1]U_0 \\ &= [(1/2.8) - 1](3.4\text{ J}) = \boxed{-2.2\text{ J.}} \end{aligned}$$

The negative value means that the dielectric is drawn into the region between the plates.

75. We find the equivalent capacitance for a series arrangement from

$$\frac{1}{C_{equ}} = \sum_i \left(\frac{1}{C_i}\right).$$

If we multiply by the value of the jth capacitance, we get

$$\frac{C_j}{C_{equ}} = \sum_i \left(\frac{C_j}{C_i}\right) = 1 + \sum_{i \neq j}\left(\frac{C_j}{C_i}\right).$$

Because the summation is positive, we have

$$C_j/C_{equ} > 1, \text{ for any value of } j.$$

Thus the equivalent capacitance is less than any of the individual capacitances.

CHAPTER 27

7. From the relation between the current density and the drift speed, we have
$J = I/A = n_q q v_d;$
$(1.2 \text{ A})/\pi(1.8 \times 10^{-3} \text{ m})^2 = (8.5 \times 10^{28} \text{ electrons}/\text{m}^3)(1.6 \times 10^{-19} \text{ C}/\text{electron})v_d$, which gives
$\boxed{v_d = 8.7 \times 10^{-6} \text{ m/s}.}$
For the second wire, the only change is the area, so we have
$v_{d2} = v_d R_1^2/R_2^2 = (8.7 \times 10^{-6} \text{ m/s})(1.8 \text{ mm})^2/(1.2 \text{ mm})^2 = \boxed{2.0 \times 10^{-5} \text{ m/s}.}$

11. Because the current density is a function of the distance from the axis, we choose a circular ring for the differential area and integrate to find the current:

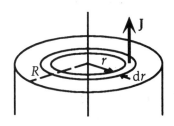

$$I = \iint_{\text{area}} \mathbf{J} \cdot d\mathbf{A} = \iint_{\text{area}} J \, dA = \int_0^R J_0\left(1 - \frac{r^2}{R^2}\right) 2\pi r \, dr = 2\pi J_0 \int_0^R \left(r - \frac{r^3}{R^2}\right) dr$$

$$= 2\pi J_0 \left(\frac{r^2}{2} - \frac{r^4}{4R^2}\right)\Bigg|_0^R = \boxed{\frac{\pi}{2} J_0 R^2.}$$

13. Because a mol of NaCl contributes an Avogadro's number of positive ions and an equal number of negative ions, we find the density for each carrier from
$n_+ = n_- = n = (0.1 \text{ mol/L})(6.02 \times 10^{23} \text{ ions/mol})(10^3 \text{ L/m}^3) = 6.02 \times 10^{25} \text{ ions/m}^3.$
Because both types of carriers are present, we have
$J = n_+ q_+ v_+ + n_- q_- v_- = nev_+ + n(-e)v_- = ne[v_+ - (-1.5v_+)] = 2.5nev_+ \, ;$
$40 \text{ A/m}^2 = (6.02 \times 10^{25} \text{ ions/m}^3)(1.6 \times 10^{-19} \text{ C/ion})(2.5v_+)$, which gives
$\boxed{v_+ = 1.7 \times 10^{-6} \text{ m/s}, \ v_- = -2.5 \times 10^{-6} \text{ m/s}.}$

19. The total current must be the same on each side of the junction:
$I_{\text{total}} = 2I_1 = 3I_2 \, ;$
$2(3 \text{ A}) = 3I_2$, which gives $I_2 = 2$ A in each of the smaller wires.
We find the drift speed in the larger wires from
$v_{\text{in}} = J/n_e e = I_1/A_1 n_e e$
$\quad = (3 \text{ A})/\pi(0.1 \text{ cm})^2(7 \times 10^{22} \text{ electrons/cm}^3)(1.60 \times 10^{-19} \text{ C/electron})$
$\quad = 8.5 \times 10^{-3} \text{ cm/s} = \boxed{8.5 \times 10^{-5} \text{ m/s}.}$
We find the drift speed in the smaller wires from
$v_{\text{out}} = J/n_e e = I_2/A_2 n_e e$
$\quad = (2 \text{ A})/\pi(0.05 \text{ cm})^2(7 \times 10^{22} \text{ electrons/cm}^3)(1.60 \times 10^{-19} \text{ C/electron})$
$\quad = 2.3 \times 10^{-2} \text{ cm/s} = \boxed{2.3 \times 10^{-4} \text{ m/s}.}$
The combined area of the smaller wires is less than the combined area of the larger wires. Charge conservation is equivalent to mass conservation in water flow, so the smaller area requires a greater speed.

21. With $d \ll R$, from symmetry, the current in the tube is uniformly distributed over an area $2\pi R d$. The current density in the tube is
$\boxed{J_{\text{tube}} = I/A = I/2\pi R d}$ along the tube.
The current density in the top plate is radial. At a distance r from the center, the current passes through an area $2\pi r d$. Thus the current density is
$\boxed{J_{\text{plate}} = I/A = I/2\pi r d}$ radial.

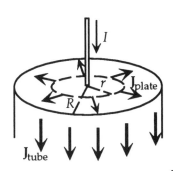

29. We find the resistance from
$$R_{20} = \rho_{20} L / A$$
$$= (1.72 \times 10^{-8} \, \Omega \cdot m)(2 \, m)/(36 \times 10^{-6} \, m) = \boxed{9.6 \times 10^{-4} \, \Omega.}$$
The increase in resistance is
$$\Delta R = R_{20}\alpha(T - 20°C) = (9.6 \times 10^{-4} \, \Omega)(0.0039/C°)(80°C) = \boxed{3.0 \times 10^{-4} \, \Omega.}$$

31. The power consumed in a resistor is
$$P = IV = I^2 R = V^2/R.$$
With a fixed potential difference, we have
$$(P_2 - P_1)/P_1 = [(1/R_2) - (1/R_1)]/(1/R_1) = (R_1 - R_2)/R_2.$$
If we assume that the temperature coefficient does not change with temperature, we get
$$(P_2 - P_1)/P_1 = \{R_{20}[1 + \alpha(T_1 - 20°C)] - R_{20}[1 + \alpha(T_2 - 20°C)]\}/R_{20}[1 + \alpha(T_2 - 20°C)]$$
$$= \alpha(T_1 - T_2)/[1 + \alpha(T_2 - 20°C)] = (0.0045/°C)(-400°C)/[1 + (0.0045/°C)(1180°C)]$$
$$= \boxed{-0.27.}$$
Because the resistance has increased, the power consumption has decreased.

33. We find the current from
$$R = V/I = \rho L / A = \rho_0(1 + \alpha \, \Delta T)L/A.$$
Table 27–2 gives the resistivity at 20°C. At 25°C, we have
$$(50 \, V)/I_{25} = (100 \times 10^{-8} \, \Omega \cdot m)[1 + (4 \times 10^{-4} \, /C°)(5°C)](0.50 \, m)/[\tfrac{1}{4}\pi(0.5 \times 10^{-3} \, m)^2],$$
which gives $\boxed{I_{25} = 19.6 \, A.}$
At 400°C, we have
$$(50 \, V)/I_{400} = (100 \times 10^{-8} \, \Omega \cdot m)[1 + (4 \times 10^{-4} \, /C°)(380°C)](0.50 \, m)/[\tfrac{1}{4}\pi(0.5 \times 10^{-3} \, m)^2],$$
which gives $\boxed{I_{400} = 17.1 \, A.}$

39. We find the resistance from
$$R_{Al} = \rho_{Al} L / A_{Al}$$
$$= (2.8 \times 10^{-8} \, \Omega \cdot m)(80 \, m)/\tfrac{1}{4}\pi(0.12 \times 10^{-2} \, m)^2 = \boxed{2.0 \, \Omega.}$$
The mass of the wire is
$$m_{Al} = \rho_{m,Al} A_{Al} L$$
$$= (2.7 \times 10^3 \, kg/m^3)[\tfrac{1}{4}\pi(0.12 \times 10^{-2} \, m)^2](80 \, m) = \boxed{0.24 \, kg.}$$
We find the area of the copper wire with the same resistance from
$$R_{Cu} = \rho_{Cu} L / A_{Cu};$$
$$2.0 \, \Omega = (1.7 \times 10^{-8} \, \Omega \cdot m)(80 \, m)/A_{Cu}, \text{ which gives } A_{Cu} = 6.8 \times 10^{-7} \, m^2.$$
The mass of the copper wire is
$$m_{Cu} = \rho_{m,Cu} A_{Cu} L$$
$$= (8.9 \times 10^3 \, kg/m^3)(6.8 \times 10^{-7} \, m^2)(80 \, m) = \boxed{0.48 \, kg.}$$

43. For the resistance of the cylindrical shell, we have
$$R_1 = \rho L / A_1 = \rho L / \pi(r_{outside}^2 - r_{inside}^2)$$
$$= (1.72 \times 10^{-8} \, \Omega \cdot m)(1 \, m)/\pi[(0.2 \times 10^{-2} \, m)^2 - (0.1 \times 10^{-2} \, m)^2] = \boxed{1.82 \times 10^{-3} \, \Omega.}$$
For the solid wire, we have
$$R_2 = \rho L / A_2 = \rho L / \pi r^2.$$
If we divide the two equations, we get
$$R_2/R_1 = 1 = r^2/(r_{outside}^2 - r_{inside}^2), \text{ which becomes}$$
$$r^2 = r_{outside}^2 - r_{inside}^2 = (0.2 \, cm)^2 - (0.1 \, cm)^2, \text{ which gives } \boxed{r = 0.17 \, cm.}$$
The ratio of masses is
$$m_2/m_1 = \rho_m L A_2 / \rho_m L A_1 = A_2/A_1 = r^2/(r_{outside}^2 - r_{inside}^2) = 1.$$
$$\boxed{\text{The masses are the same.}}$$

49. From kinetic theory, we have the average kinetic energy of the electron:
$K = \frac{3}{2}kT$,
which we use as the energy necessary to cross the energy gap. For the given elements, we have
Si: $(1.1 \text{ eV})(1.6 \times 10^{-19} \text{ J/eV}) = \frac{3}{2}(1.38 \times 10^{-23} \text{ J/K})T_{Si}$, which gives $\boxed{T_{Si} = 8.5 \times 10^3 \text{ K.}}$
Ge: $(0.7 \text{ eV})(1.6 \times 10^{-19} \text{ J/eV}) = \frac{3}{2}(1.38 \times 10^{-23} \text{ J/K})T_{Ge}$, which gives $\boxed{T_{Ge} = 5.4 \times 10^3 \text{ K.}}$
C: $(6 \text{ eV})(1.6 \times 10^{-19} \text{ J/eV}) = \frac{3}{2}(1.38 \times 10^{-23} \text{ J/K})T_C$, which gives $\boxed{T_C = 4.6 \times 10^4 \text{ K.}}$

59. For an ohmic resistor, we have
$P = IV = V^2/R$, or $V^2 = PR = P\rho L/A$;
$(110 \text{ V})^2 = (1250 \text{ W})(10^{-6} \; \Omega \cdot \text{m})L/(0.2 \times 10^{-6} \text{ m}^2)$, which gives $\boxed{L = 1.9 \text{ m.}}$

61. (a) For an ohmic resistor, we have
$P = IV$, so the maximum power is
$P_{max} = I_{max}V = (15 \text{ A})(110 \text{ V}) = 1.65 \times 10^3 \text{ W} = \boxed{1.65 \text{ kW.}}$
(b) We find the maximum number of light bulbs from
$N = P_{max}/P_{bulb} = (1.65 \times 10^3 \text{ W})/(75 \text{ W}) = 22 \rightarrow \boxed{22 \text{ bulbs.}}$

63. Assuming a constant resistance, for an ohmic resistor we have
$P = IV = V^2/R$, or
$P_2/P_1 = (V_2/V_1)^2(R_1/R_2) = (V_2/V_1)^2$;
$P_2/(500 \text{ W}) = [(105 \text{ V})/(115 \text{ V})]^2$, which gives $\boxed{P_2 = 417 \text{ W.}}$

65. We find the total energy used in the month from
$U = \$25.33/(\$0.08/\text{kWh}) = 317 \text{ kWh.}$
The average current during the month was
$I = P/V = U/Vt$, so the charge that passed through the meter was
$Q = It = U/V$, and the number of electrons was
$N = Q/e = U/Ve$
$= (317 \text{ kWh})(10^3 \text{ W/kW})(3600 \text{ s/h})/(120 \text{ V})(1.6 \times 10^{-19} \text{ C}) = \boxed{5.94 \times 10^{25} \text{ electrons.}}$

67. (a) We find the number of protons from
$N = Q/e = It/e$
$= (5 \times 10^{-6} \text{ A})(1 \text{ h})(3600 \text{ s/h})/(1.6 \times 10^{-19} \text{ C}) = \boxed{1.1 \times 10^{17} \text{ protons.}}$
(b) Because each proton has an energy of 4 MeV, the total energy is
$U_{total} = NU$
$= (1.1 \times 10^{17} \text{ protons})(4 \times 10^6 \text{ eV/proton})(1.6 \times 10^{-19} \text{ J/eV}) = \boxed{7.2 \times 10^4 \text{ J.}}$
(c) We find the power of the beam from
$P = U_{total}/t$
$= (7.2 \times 10^4 \text{ J})/(1 \text{ h})(3600 \text{ s/h}) = \boxed{20 \text{ W.}}$

71. We find the power dissipated from $P = IV$, estimating values from the plot.
Note the change in scales on the two sides of the plot.

In the ideal diode, either the potential or the current is zero, so the power is zero.

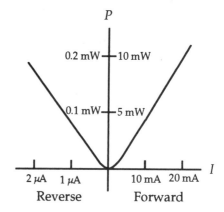

73. The resistance of the wire is $R = \rho L/A$, and the current is $I = JA = nev_d A$. The power dissipated, which becomes thermal energy, is

$$P = I^2 R = (nev_d A)^2 \rho L/A = (nev_d)^2 A \rho L$$
$$= [(8.5 \times 10^{28} \text{ electrons/m}^3)(1.6 \times 10^{-19} \text{ C})(1.2 \times 10^{-5} \text{m/s})]^2 [\pi(0.5 \times 10^{-3} \text{ m})^2](1.72 \times 10^{-8} \Omega \cdot \text{m})(3 \text{ m})$$
$$= 1.1 \times 10^{-3} \text{ W}.$$

For the wire to maintain its temperature, thermal energy must be removed at this rate, $\boxed{1.1 \times 10^{-3} \text{ W}.}$

75. (a) From the conservation of charge, we know that the current must be constant along the wire. Because the area is also constant, we have
$$E = \rho J = \rho_0 J e^{-x/L} = (\rho_0 I/A) e^{-x/L} = \boxed{E_0 e^{-x/L}.}$$

(b) We take the reference level for V to be $V = 0$ at $x = L$, so $V = V_0$ at $x = 0$. We integrate the relation between the field and the potential, $E = -dV/dx$:

$$\int_{V_0}^{V} dV = -\int_0^x E \, dx' = -E_0 \int_0^x e^{-x'/L} \, dx';$$

$$V - V_0 = -E_0(-L)e^{-x'/L}\Big|_0^x = +E_0 L\left(e^{-x/L} - 1\right) = -E_0 L\left(1 - e^{-x/L}\right).$$

We can determine E_0 from our reference level at $x = L$:
$-V_0 = -E_0 L(1 - e^{-1})$, which gives $E_0 L = V_0/(1 - e^{-1})$.
The potential is
$$V = V_0 - V_0(1 - e^{-x/L})/(1 - e^{-1}) = \boxed{V_0(e^{-x/L} - e^{-1})/(1 - e^{-1}).}$$

(c) We choose a differential segment of the wire at x with length dx. Because the resistivity is a function of x, while the area is constant, we find the total resistance by integration:

$$R = \int \frac{\rho \, dL}{A} = \int_0^L \frac{\rho_0 e^{-x/L} \, dx}{A} = -\frac{\rho_0 L}{A} e^{-x/L}\Big|_0^L$$

$$= -\frac{\rho_0 L}{A}(e^{-1} - 1) = \boxed{\frac{\rho_0 L}{A}(1 - e^{-1}).}$$

CHAPTER 28

It is important to draw a well-labeled circuit diagram. Applications of the loop rule and the junction rule are made by referring to the diagram for each term, with great care being taken with signs.

9. The terminal voltage of the battery is
$$V_{\text{term}} = \mathcal{E} - Ir = \mathcal{E} - I(\alpha + \beta I).$$
The power dissipated within the battery is
$$P = I^2 r = I^2(\alpha + \beta I).$$
For $I = 1.0$ A, we have
$$V_1 = (12.0 \text{ V}) - (1.0 \text{ A})[(0.15 \ \Omega) + (0.018 \ \Omega/\text{A})(1.0 \text{ A})] = \boxed{11.8 \text{ V.}}$$
$$P_1 = (1.0 \text{ A})^2[(0.15 \ \Omega) + (0.018 \ \Omega/\text{A})(1.0 \text{ A})] = \boxed{0.17 \text{ W.}}$$
For $I = 10.0$ A, we have
$$V_{10} = (12.0 \text{ V}) - (10.0 \text{ A})[(0.15 \ \Omega) + (0.018 \ \Omega/\text{A})(10.0 \text{ A})] = \boxed{8.7 \text{ V.}}$$
$$P_{10} = (10.0 \text{ A})^2[(0.15 \ \Omega) + (0.018 \ \Omega/\text{A})(10.0 \text{ A})] = \boxed{33 \text{ W.}}$$

11. If we go around the single loop in the direction shown, starting at point a, we have

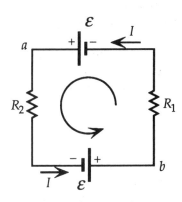

$$\Sigma \Delta V = -IR_2 + \mathcal{E} - IR_1 + \mathcal{E} = 0, \text{ which gives}$$

$$I = 2\mathcal{E}/(R_1 + R_2).$$
If we add the potentials from a to b, we have

$$V_a - IR_2 + \mathcal{E} = V_b = V_a, \text{ or } \mathcal{E} = IR_2 .$$
When we combine this with the expression for the current, we get

$$\mathcal{E} = 2\mathcal{E} R_2/(R_1 + R_2), \text{ which gives } \boxed{R_2 = R_1.}$$
From the expression for the current, we see that
$$I \rightarrow 0 \text{ when } \boxed{R_2 \rightarrow \infty.}$$

17. (a) If we assume initially there is no internal resistance, we have
$$I_0 = 2\mathcal{E}/R = 2(1.5 \text{ V})/(10 \ \Omega) = 0.30 \text{ A}.$$
The power delivered to the bulb is the power dissipated in the bulb:
$$P_0 = I_0^2 R = (0.30 \text{ A})^2(10 \ \Omega) = \boxed{0.90 \text{ W.}}$$
   (b) If the power delivered to the bulb, which is also the power dissipated in the bulb, decreases by one-third, we have
$$P = I^2 R;$$
$$\tfrac{2}{3}(0.90 \text{ W}) = I^2(10 \ \Omega), \text{ which gives } I = 0.25 \text{ A}.$$
For the single-loop circuit, we have
$$I = 2\mathcal{E}/(R + 2r);$$
$$0.25 \text{ A} = 2(1.5 \text{ V})/[(10 \ \Omega) + 2r], \text{ which gives } \boxed{r = 1.0 \ \Omega/\text{battery.}}$$

23. We assume the current directions shown in the diagram.
We use conservation of current at point a:
$$\Sigma I_{\text{in}} = 0;$$
$$I_1 - I_2 + I_3 = 0.$$
We apply the loop rule for the two loops indicated in the diagram:

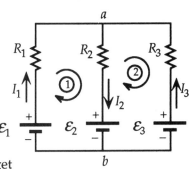

loop 1: $-I_2R_2 - \mathcal{E}_2 + \mathcal{E}_1 - I_1R_1 = 0;$
$\qquad -I_2(10 \ \Omega) - 6 \text{ V} + 12 \text{ V} - I_1(5 \ \Omega) = 0;$
loop 2: $+I_3R_3 - \mathcal{E}_3 + I_2R_2 + \mathcal{E}_2 = 0;$
$\qquad +I_3(12 \ \Omega) - 9 \text{ V} + I_2(10 \ \Omega) + 6 \text{ V} = 0.$
When we solve for the three unknown currents in these equations, we get
$$\boxed{I_1 = +0.45 \text{ A}, I_2 = +0.38 \text{ A}, I_3 = -0.068 \text{ A}.}$$

27. With identical batteries, the terminal voltage and the current through each battery are the same. When the batteries are connected in parallel, the terminal voltage is the voltage across the resistance, so we have

$\Sigma I_i = N I_i = I_a;$

$V_{ab} = \mathcal{E} - I_i r = I_a R.$

If we eliminate I_i from these equations, we get

$\boxed{I_a = \mathcal{E}/[R + (r/N)].}$

When the batteries are connected in series, the current through each battery is the current through the resistance, so we have

$\Sigma V_i = N V_i = V_{cd};$

$N(\mathcal{E} - I_b r) = I_b R,$ which gives

$\boxed{I_b = \mathcal{E}/[r + (R/N)].}$

In general, R will be much greater than r, so I_b will be greater than I_a.

29. To simplify the circuit, we combine R_2 and R_L, which are in parallel:

$1/R = (1/R_2) + (1/R_L)$, which gives $R = R_2 R_L/(R_2 + R_L)$.

We now have a single-loop circuit, so the current is

$I = \mathcal{E}/(R_1 + R).$

The voltage across the load is the voltage across R:

$V_L = IR = \mathcal{E}R/(R_1 + R).$

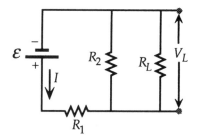

When we use the expression for R, we get

$V_L = \mathcal{E}R_2 R_L/[(R_1 R_2 + R_1 R_L + R_2 R_L)]$
$= (10\text{ V})(3.3\text{ k}\Omega)R_L/[(3.3\text{ k}\Omega)(3.3\text{ k}\Omega) + (3.3\text{ k}\Omega)R_L + (3.3\text{ k}\Omega)R_L]$
$= 33R_L/(10.9 + 6.6R_L)\text{ V, with }R_L\text{ in k}\Omega.$

For the given loads, we have

$V_{20\text{ k}\Omega} = (33)(20\text{ k}\Omega)/[10.9 + 6.6(20\text{ k}\Omega)] = 4.62\text{ V},\ \boxed{\Delta V = 0.38\text{ V};}$

$V_{200\text{ k}\Omega} = (33)(200\text{ k}\Omega)/[10.9 + 6.6(200\text{ k}\Omega)] = 4.96\text{ V},\ \boxed{\Delta V = 0.04\text{ V};}$

$V_{2\text{ M}\Omega} = (33)(2\times10^3\text{ k}\Omega)/[10.9 + 6.6(2\times10^3\text{ k}\Omega)] = 4.996\text{ V},\ \boxed{\Delta V = 0.004\text{ V}.}$

When $R_L \gg R_2$, the difference is small.

31. We can consider point a to be along the top and point b to be along the bottom, so the conservation of current gives

$\Sigma I_{in} = 0;$

junction a: $I_1 - I_2 - I_3 - I_4 = 0;$

junction b: $I_2 + I_3 + I_4 - I_1 = 0,$ which is the negative of the equation for junction a.

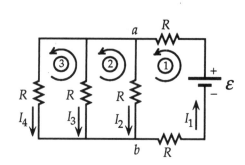

$\boxed{\text{There is only one independent junction.}}$

For the three loops indicated on the diagram, we have

loop 1: $-I_1 R - I_2 R - I_1 R + \mathcal{E} = 0;$
loop 2: $+I_2 R - I_3 R = 0;$
loop 3: $+I_3 R - I_4 R = 0.$

The solution of these four equations gives

$\boxed{I_1 = 3\mathcal{E}/7R,\ I_2 = I_3 = I_4 = \mathcal{E}/7R.}$

35. (a) We can reduce the circuit to a single loop by successively combining parallel and series combinations.
We combine R_3 and R_4, which are in parallel:
$$1/R_5 = (1/R_3) + (1/R_4)$$
$$= [1/(100\ \Omega)] + [1/(50\ \Omega)],$$
which gives $R_5 = 33.3\ \Omega$.
We combine R_1 and $R_2 + R_5$, which are in parallel:
$$1/R_6 = (1/R_1) + [1/(R_2 + R_5)]$$
$$= [1/(100\ \Omega)] + \{1/[(20\ \Omega) + (33.3\ \Omega)]\},$$
which gives $R_6 = 34.8\ \Omega$.
Because $V_{ab} = 6$ V, we find I_2 from
$$I_2 = V_{ab}/(R_5 + R_2) = (6\ \text{V})/[(33.3\ \Omega) + (20\ \Omega)] = 0.113\ \text{A}.$$
From the circuit we see that we can now find V_{cb} from
$$V_{cb} = I_2 R_5 = (0.113\ \text{A})(33.3\ \Omega) = 3.75\ \text{V}.$$
The current through the 50-Ω resistor is
$$I_4 = V_{cb}/R_4 = (3.75\ \text{V})/(50\ \Omega) = \boxed{0.075\ \text{A}.}$$

(b) For the conservation of current, we have
junction a: $\quad I - I_1 - I_2 = 0$;
junction c: $\quad I_2 - I_3 - I_4 = 0$.
For the three loops indicated on the diagram, we have

loop 1: $\quad \mathcal{E} - I_1 R_1 = 0$;
$$6\ \text{V} - I_1(100\ \Omega) = 0;$$
loop 2: $\quad I_1 R_1 - I_2 R_2 - I_3 R_3 = 0$;
$$I_1(100\ \Omega) - I_2(20\ \Omega) - I_3(100\ \Omega) = 0;$$
loop 3: $\quad -I_3 R_3 + I_4 R_4 = 0$;
$$-I_3(100\ \Omega) + I_4(50\ \Omega) = 0.$$
When we solve these equations, we get
$$I_1 = 0.060\ \text{A},\ I_2 = 0.113\ \text{A},\ I_3 = 0.038\ \text{A},\ I_4 = 0.075\ \text{A}.$$
Thus, the current through the 50-Ω resistor is $\boxed{0.075\ \text{A}.}$

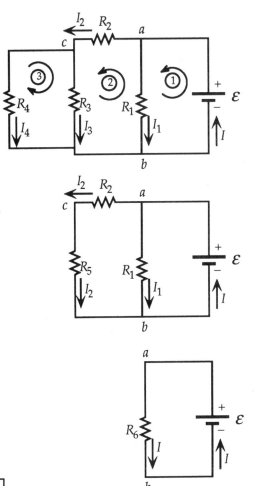

39. In the original configuration, there are no series or parallel combinations; however, from the symmetry of the resistors, we know that the current that goes into point A must split into thirds to go through the cube. The three points on the other side of the three resistors must be at the same potential, so we can connect them with a wire without changing the currents. In the same way, the other three corners of the cube must be at equal potentials, so we can connect them with a wire. From the redrawn circuit, we see that we have three parallel combinations, two of which are the same:
$$1/R_1 = 1/R_3 = (1/R) + (1/R) + (1/R),$$
which gives $R_1 = R_3 = R/3$;
$$1/R_2 = (1/R) + (1/R) + (1/R) + (1/R) + (1/R) + (1/R),$$
which gives $R_2 = R/6$;
We combine these three in series to get
$$R_{eq} = (R/3) + (R/6) + (R/3) = \boxed{5R/6.}$$
The current in the equivalent resistor is
$$I = V/(5R/6) = 6V/5R.$$
The current in a resistor connected to point A or point B is
$$\boxed{I_1 = I_3 = I/3 = 2V/5R.}$$
The current in each of the other resistors is
$$\boxed{I_2 = I/6 = V/5R.}$$

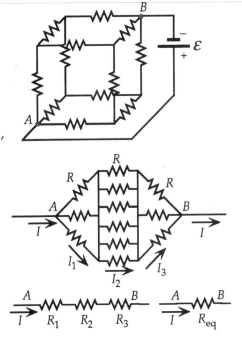

41. (a) For $n = 1$, we have two resistors in series:
$R_1 = R + R = \boxed{2R.}$
(b) For $n = 2$, we have a resistor in series with a parallel combination of a resistor and resistance R_1:
$R_2 = R + [R_1 R/(R_1 + R)]$
$= R + [2RR/(2R + R)] = \boxed{5R/3.}$
(c) For $n = 3$, we have a resistor in series with a parallel combination of a resistor and resistance R_2:
$R_3 = R + [R_2 R/(R_2 + R)]$
$= R + \{(5RR/3)/[(5R/3) + R]\} = \boxed{13R/8.}$
(d) For n rungs, we have a resistor in series with a parallel combination of a resistor and resistance R_{n-1}:
$R_n = R + [R_{n-1} R/(R_{n-1} + R)].$
In the limit of $n \to \infty$, we have
$R_n = R_{n-1} = R_{eq}$, which gives
$R_{eq} = R + [R_{eq} R/(R_{eq} + R)]$, which reduces to
$R_{eq}^2 - RR_{eq} - R^2 = 0.$
The solutions to this quadratic equation are
$R_{eq} = \frac{1}{2}(1 \pm \sqrt{5})R.$
Because the resistance cannot be negative, we have $\boxed{R_{eq} = \frac{1}{2}(1 + \sqrt{5})R = 1.618R.}$

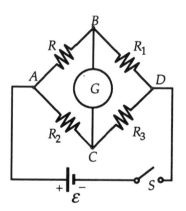

45. We find the equivalent resistance for R and R_V, which are in parallel, from
$1/R_{eq} = (1/R) + (1/R_V)$, which gives $R_{eq} = RR_V/(R + R_V)$.
(a) $R_{eq} = (10\,\Omega)(10^5\,\Omega)/[(10\,\Omega) + (10^5\,\Omega)] \approx \boxed{10\,\Omega.}$
(b) $R_{eq} = (10^5\,\Omega)(10^5\,\Omega)/[(10^5\,\Omega) + (10^5\,\Omega)] = \boxed{5 \times 10^4\,\Omega.}$
(c) $R_{eq} = (100 \times 10^6\,\Omega)(10^5\,\Omega)/[(100 \times 10^6\,\Omega) + (10^5\,\Omega)] \approx \boxed{10^5\,\Omega.}$
The equivalent resistance has the value of the resistor when $R_V \gg R$.

49. When there is no current through the galvanometer, we have $V_{BC} = 0$, a current I_1 through R and R_1, and a current I_2 through R_2 and R_3. Thus we have
$V_{AD} = I_1(R + R_1) = I_2(R_2 + R_3)$,   and
$V_{AB} = V_{AC}$ or $I_1 R = I_2 R_2$.
When we divide these two equations, we get
$(R + R_1)/R = (R_2 + R_3)/R_2$,   or
$1 + (R_1/R) = 1 + (R_3/R_2)$, which gives $R_1/R = R_3/R_2$.
The unknown resistance is
$\boxed{R = R_1 R_2/R_3.}$

55. With the time constant as the flash time, we have
time constant $= RC$;
$(1/500)\,\text{s} = R(600 \times 10^{-6}\,\text{F})$, which gives $\boxed{R = 3.3\,\Omega.}$

59. Because there is no internal resistance in the battery, the potential difference across R_2 and across the capacitor branch is $\mathcal{E}$. The current in R_2 is constant:

$$I_2 = \mathcal{E}/R_2.$$

The charging current in the capacitor branch is

$$I_1 = (\mathcal{E}/R_1)e^{-t/R_1C}.$$

Using the conservation of charge at the junction, we find the current in the battery:

$$\boxed{I_{\text{battery}} = I_1 + I_2 = (\mathcal{E}/R_1)e^{-t/R_1C} + (\mathcal{E}/R_2).}$$

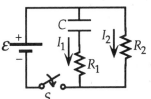

61. The possible capacitance values that we have are
$$C_1 = C_2 = 5\ \mu F;$$
$$C_{\text{parallel}} = C_1 + C_2 = 5\ \mu F + 5\ \mu F = 10\ \mu F;$$
$$C_{\text{series}} = C_1C_2/(C_1 + C_2) = (5\ \mu F)(5\ \mu F)/[(5\ \mu F) + (5\ \mu F)] = 2.5\ \mu F.$$
We need to combine the resistors to produce one of the following resistance values:
$$R_a = RC/C_1 = (1 \times 10^{-3}\ s)/(5 \times 10^{-6}\ F) = 200\ \Omega;$$
$$R_b = RC/C_{\text{parallel}} = (1 \times 10^{-3}\ s)/(10 \times 10^{-6}\ F) = 100\ \Omega;$$
$$R_c = RC/C_{\text{series}} = (1 \times 10^{-3}\ s)/(2.5 \times 10^{-6}\ F) = 400\ \Omega;$$
If we connect the 300-Ω resistors in parallel, we get
$$R_3 = R_2R_2/(R_2 + R_2) = (300\ \Omega)(300\ \Omega)/[(300\ \Omega) + (300\ \Omega)] = 150\ \Omega.$$
We see that we can produce R_c by putting this combination in series with the 250-Ω resistor:
$$R_c = R_1 + R_3 = 250\ \Omega + 150\ \Omega = 400\ \Omega.$$
Thus we connect the 300-Ω resistors in parallel with each other and in series with the 250-Ω resistor and the two capacitors.

69. For this single-loop circuit, we have
$$I = \mathcal{E}/(R + r) = \mathcal{E}/[R + (\alpha + \beta I)],\ \text{which is a quadratic equation for } I:\ \ \beta I^2 + (R + \alpha)I - \mathcal{E} = 0.$$
The positive solution is
$$I = \{-(R + \alpha) + [(R + \alpha)^2 + 4\beta\mathcal{E}]^{1/2}\}/2\beta.$$
The ratio of power delivered to the load to the power dissipated in the battery is
$$P_R/P_r = I^2R/I^2(\alpha + \beta I) = R/(\alpha + \beta I).$$

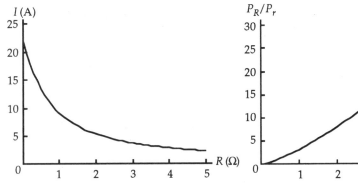 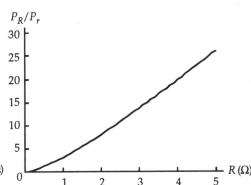

71. The two heating elements in the furnace are initially connected in parallel. The power dissipated in a resistor is $P = I^2R = V^2/R$. We find each resistance from
$$R_1 = V^2/P_1 = (110\ V)^2/(1000\ W) = 12.1\ \Omega.$$
$$R_2 = V^2/P_2 = (110\ V)^2/(2000\ W) = 6.05\ \Omega.$$
The power can be reduced by increasing the resistance, which means connecting them in series:
$$P = V^2/(R_1 + R_2) = (110\ V)^2/(12.1\ \Omega + 6.05\ \Omega) = \boxed{667\ W.}$$

75. Normally, this circuit would have six currents, one for each
    branch. We have used the symmetry of the circuit to
    reduce the number of currents to four, as shown in the diagram.
    For the junction equations, we have

    junction A (or D): $I - I_1 - I_2 = 0$;   (1)
    junction B (or C): $I_1 + I_3 - I_2 = 0$.   (2)

    For the loop equations, we have

    loop $ACDA$: $-I_2 R - I_1 R + \mathcal{E} = 0$;   (3)
    loop $BDCB$: $-I_2 R + I_1 R - I_3 R = 0$;   (4)

    When we combine Eq. (2) and Eq. (4), we find
    $\boxed{I_3 = 0}$ (as suggested by symmetry)   and $I_1 = I_2$.
    From Eq. (3), we get
    $\boxed{I_1 = I_2 = \mathcal{E}/2R.}$
    From Eq. (1), we get
    $\boxed{I = 2I_1 = \mathcal{E}/R.}$

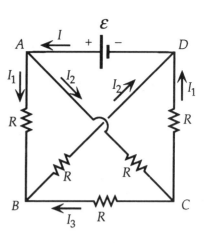

79. (a) With the switch open, we have a series circuit of the three
       resistors and the capacitor. For the time constant we have
       time constant $= R_{eq}C = (r + R_1 + R_2)C$
       $= [(0.04\ \Omega) + (0.1\ \Omega) + (2\ \Omega)](10\ \mu F) = \boxed{21.4\ \mu s.}$
    (b) After a long time, there will be no current in the circuit.
        The battery emf will be across the capacitor:
        $Q = C\mathcal{E} = (10\ \mu F)(240\ V) = 2400\ \mu C = \boxed{2.4\ mC.}$

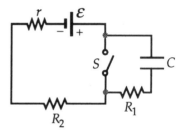

81. Because the emf has negligible resistance, the terminal voltage, which is the voltage across the capacitor,
    is the emf V_0. The resistance of the material in the capacitor is
    $R = \rho L/A = L/\sigma A = d/\sigma \pi r^2$.
    (a) We find the electric field between the capacitor plates from the potential gradient:
        $\boxed{E = V_0/d.}$
    (b) The current density depends on the electric field:
        $J = \sigma E = \sigma V_0/d.$
        The current is
        $I = JA = (\sigma V_0/d)\pi r^2 = \boxed{\sigma \pi r^2 V_0/d} = V_0/R.$

CHAPTER 29

3. Because the charge is negative, with our fingers pointing in the $-y$-direction, we want our thumb to point in the $+x$-direction. Thus we have to curl our fingers toward $-z$.

   | The magnetic field is in the $-z$-direction. |

   Formally, we have

   $\mathbf{F} = m\mathbf{a} = q(\mathbf{v} \times \mathbf{B})$;

   $-F\mathbf{i} = (-e)(-v\mathbf{j}) \times \mathbf{B}$, which gives $\mathbf{B} = -B\mathbf{k}$.

9. (a) The magnetic force produces an acceleration perpendicular to the original motion:

   $\mathbf{F} = q\mathbf{v} \times \mathbf{B} = m\mathbf{a}$,   or $a_\perp = evB/m$.

   For a small change in direction, we can take the force to be constant, so the perpendicular component of the velocity is

   $v_\perp = a_\perp \Delta t = (evB\,\Delta t)/m$.

   If D is the diameter of the region of the magnetic field, the time the particle spends in the field is

   $t = D/v$.

   The direction of motion is given by the direction of the velocity:

   $\tan \theta = v_\perp/v = (eB\,\Delta t)/m$.

   If the angle of deflection is small, we have

   $\tan \theta \approx$ | $\theta = (eB\,\Delta t)/m$. |

   (b) If D is the diameter of the region of the magnetic field, the time the particle spends in the field is

   $\Delta t = D/v$.

   The angle of deflection is

   $\theta = (eB\,\Delta t)/m = eBD/mv$;

   $0.1 \text{ rad} = (1.60 \times 10^{-19} \text{ C})B(0.1 \text{ m})/(1.7 \times 10^{-27} \text{ kg})(1.4 \times 10^7 \text{ m/s})$, which gives | $B = 0.15$ T. |

11. We choose vertically up to be the $+z$-axis. In the northern hemisphere, the magnetic field has a downward component, so

    $B = B_x\mathbf{i} + B_z\mathbf{k} = (24 \ \mu\text{T})\mathbf{i} - (18 \ \mu\text{T})\mathbf{k}$.

    We find the acceleration of the electron from $\mathbf{F} = q\mathbf{v} \times \mathbf{B} = m\mathbf{a}$:

    $\mathbf{F} = (-e)[(v_0\mathbf{i}) \times (B_x\mathbf{i} + B_z\mathbf{k})] = m\mathbf{a}$, which gives $\mathbf{a} = (ev_0B_z/m)\mathbf{j}$.

    The time the electron takes to reach the screen is

    $\Delta t = L/v_0 = (0.4 \text{ m})/(6 \times 10^7 \text{ m/s}) = 6.7 \times 10^{-9} \text{ s}$.

    If we assume a constant acceleration, the deflection of the electron is

    $\mathbf{d} \approx \frac{1}{2}\mathbf{a}t^2 = \frac{1}{2}(ev_0B_z/m)(\Delta t)^2 \ \mathbf{j}$

    $= \frac{1}{2}[(1.60 \times 10^{-19} \text{ C})(6 \times 10^7 \text{ m/s})(-18 \times 10^{-6} \text{ T})/(9.1 \times 10^{-31} \text{ kg})](6.7 \times 10^{-9} \text{ s})^2 \ \mathbf{j}$

    $= -(4.2 \times 10^{-3} \text{ m})\mathbf{j} =$ | $-(4.2 \text{ mm})\mathbf{j}$ (horizontal). |

    This small distance justifies taking the acceleration to be constant.

15. The magnetic force provides the centripetal acceleration:

    $qvB = mv^2/R$, so the radius of the path is

    $R = mv/qB$

        $= (9.1 \times 10^{-31} \text{ kg})(0.001)(3 \times 10^8 \text{ m/s})/(1.60 \times 10^{-19} \text{ C})(3 \times 10^7 \text{ T}) =$ | 5.7×10^{-14} m. |

    Note that this is much smaller than electron orbits in an atom.

    The magnitude of the magnetic force is

    $F = qvB$

        $= (1.60 \times 10^{-19} \text{ C})(0.001)(3 \times 10^8 \text{ m/s})(3 \times 10^7 \text{ T}) =$ | 1.4×10^{-6} N. |

21. (a) The speed of the electron is given by
$$v = (2K/m)^{1/2}$$
$$= [2(10 \times 10^3 \text{ eV})(1.60 \times 10^{-19} \text{ J/eV})/(9.1 \times 10^{-31} \text{ kg})]^{1/2} = 5.9 \times 10^7 \text{ m/s}.$$
If we assume that the deflection is small, the time the electron takes to reach the screen is
$$\Delta t = L/v = (0.40 \text{ m})/(5.9 \times 10^7 \text{ m/s}) = 6.8 \times 10^{-9} \text{ s}.$$
The magnetic force produces an acceleration perpendicular to the original motion:
$$a_\perp = evB/m.$$
For a small deflection, we can take the force to be constant, so the perpendicular component of the velocity is
$$v_\perp = a_\perp \Delta t = (evB\,\Delta t)/m$$
$$= (1.60 \times 10^{-19} \text{ C})(5.9 \times 10^7 \text{ m/s})(5 \times 10^{-5} \text{ T})(6.8 \times 10^{-9} \text{ s})/(9.1 \times 10^{-31} \text{ kg}) = 0.35 \times 10^7 \text{ m/s}.$$
Because this is small compared to the original speed, the speed is essentially constant. We find the angle of the velocity from the original direction from
$$\sin \theta = v_\perp/v = (0.35 \times 10^7 \text{ m/s})/(5.9 \times 10^7 \text{ m/s}) = 0.059, \text{ which gives } \theta = 3.4°.$$
The final velocity is 5.9×10^7 m/s, 3.4° from the original direction.

(b) Because we have assumed constant acceleration, we find the deflection of the electron from
$$d = v_{av}\Delta t = \tfrac{1}{2}(0 + v_\perp)\,\Delta t$$
$$= \tfrac{1}{2}(0.35 \times 10^7 \text{ m/s})(6.8 \times 10^{-9} \text{ s}) = 1.2 \times 10^{-2} \text{ m} = \boxed{1.2 \text{ cm.}}$$
This justifies our assumption of small deflection.

25. The component of the velocity parallel to the field does not change. The component perpendicular to the field produces a force which causes the circular motion. From the radius of the motion, $R = mv_\perp/qB$, we find the time for one revolution:
$$T = 2\pi R/v_\perp = 2\pi m/qB.$$
In this time, the distance the electron travels along the field is
$$d = v_{parallel}T = v_{parallel}2\pi m/qB$$
$$= (3 \times 10^5 \text{ m/s})(\cos 40°)2\pi(9.1 \times 10^{-31} \text{ kg})/(1.60 \times 10^{-19} \text{ C})(0.12 \text{ T}) = \boxed{6.8 \times 10^{-5} \text{ m.}}$$

27. The force produced by the magnetic field will always be perpendicular to the velocity and thus in the plane perpendicular to the axis of the tube. This will change the direction of the velocity but not its magnitude. When an electron reaches the outer cylinder, its kinetic energy must be equal to the decrease in potential energy:
$$K_f = -\Delta U = U_i = \boxed{500 \text{ eV.}}$$

31. (a) The radius of the path is $R = mv/qB$, so the cyclotron frequency is
$$f = 1/T = v/2\pi R = qB/2\pi m$$
$$= (1.60 \times 10^{-19} \text{ C})(1 \text{ T})/2\pi(1.7 \times 10^{-27} \text{ kg}) = \boxed{1.5 \times 10^7 \text{ Hz.}}$$
(b) We find the maximum velocity from
$$R_{max} = mv_{max}/qB = v_{max}/2\pi f;$$
$$0.50 \text{ m} = v_{max}/2\pi(1.5 \times 10^7 \text{ Hz}), \text{ which gives } \boxed{v_{max} = 4.8 \times 10^7 \text{ m/s tangential.}}$$
(c) The maximum kinetic energy is
$$K_{max} = \tfrac{1}{2}mv_{max}^2 = \tfrac{1}{2}(1.7 \times 10^{-27} \text{ kg})(4.8 \times 10^7 \text{ m/s})^2 = \boxed{1.9 \times 10^{-12} \text{ J} \ (1.2 \times 10^7 \text{ eV}).}$$
(d) In a full circle, the proton crosses the gap twice, so the energy gain in one cycle is
$$\Delta E = 2e\,\Delta V.$$
The number of circles is
$$n = K_{max}/\Delta E = (1.2 \times 10^7 \text{ eV})/2(1 \text{ e})(50 \times 10^3 \text{ V}) = \boxed{120.}$$
(e) The time the proton spends in the accelerator is
$$t = nT = n/f = (120)/(1.5 \times 10^7 \text{ Hz}) = \boxed{8.0 \times 10^{-6} \text{ s.}}$$

35. We find the speed of the particle from the kinetic energy:
$$v = (2K/m)^{1/2}.$$
The required magnetic field for the orbit of the particle with the maximum energy is
$$B_{max} = m_\alpha v_{max}/q_\alpha R = (2K_{max}m_\alpha)^{1/2}/q_\alpha R$$
$$= [2(6 \text{ MeV})(1.60 \times 10^{-13} \text{ J/MeV})4(1.67 \times 10^{-27} \text{ kg})]^{1/2}/2(1.60 \times 10^{-19} \text{ C})(0.10 \text{ m}) = \boxed{3.54 \text{ T.}}$$
Smaller kinetic energies require a smaller field.

37. The force of attraction between the two charges provides the centripetal acceleration:
$$F_0 = mR\omega_0^2.$$
A small uniform magnetic field perpendicular to the plane of the orbit, and thus the velocity, will add an additional radial force, with a magnitude
$$F_M = qvB = qR\omega B, \text{ where } \omega = \omega_0 + d\omega.$$
If the direction is such that this force is toward the center, we have
$$F_0 + F_M = mR\omega^2;$$
$$mR\omega_0^2 + qR(\omega_0 + d\omega)B = mR(\omega_0 + d\omega)^2.$$
If we can neglect the $B\,d\omega$ term, we get
$$mR\omega_0^2 + qRB\omega_0 = mR(\omega_0 + d\omega)^2 = mR\omega_0^2 + 2mR\omega_0\,d\omega + mR\,(d\omega)^2 \approx mR\omega_0^2 + 2mR\omega_0\,d\omega,$$
which gives
$$\boxed{d\omega = qB/2m.}$$

45. From $\mathbf{F} = I\mathbf{L} \times \mathbf{B}$, we see that the force on the wire produced by the magnetic field will be down, so the wire will move down. At the new equilibrium position, the magnetic force will be balanced by the increased elastic forces of the springs:
$$F_B = F_{elastic};$$
$$ILB = 2k\,\Delta y, \text{ which gives } \boxed{\Delta y = ILB/2k.}$$

55. (a) We find the current from the maximum torque:
$$\tau_{max} = \mu B = INAB;$$
$$3 \times 10^{-5} \text{ N} \cdot \text{m} = I(50)(6 \times 10^{-4} \text{ m}^2)(0.2 \text{ T}), \text{ which gives } \boxed{I = 5.0 \times 10^{-3} \text{ A} = 5.0 \text{ mA.}}$$
(b) We find the work required to rotate the coil from
$$W = \Delta U = (-\boldsymbol{\mu} \cdot \mathbf{B})_f - (-\boldsymbol{\mu} \cdot \mathbf{B})_i = \mu B(-\cos\theta_f + \cos\theta_i).$$
For a rotation of 180°, we have
$$W = \mu B[-\cos(180 + \theta_i) + \cos\theta_i]$$
$$= 2\mu B \cos\theta_i = 2\tau_{max}\cos\theta_i = \boxed{6 \times 10^{-5}\cos\theta_i \text{ J.}}$$
The work does depend on the initial angle.

59. (a) For the segment α, we have
$$\mathbf{F}_\alpha = I\mathbf{L} \times \mathbf{B} = I(2R\mathbf{i}) \times (-B\mathbf{k}) = \boxed{2IRB\,\mathbf{j}.}$$

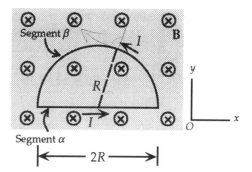

Segment β

Segment α

$\longleftarrow\!\!-\!\!2R\!-\!\!\longrightarrow$

(b) For the segment β, we choose a differential element ds at an angle θ from the x-axis. The force on every element will be directed toward the center of the arc, along the radius. By pairing elements symmetrically placed from the y-axis, which will have equal opposite horizontal force components, we see that the resultant force will be along the $-y$-axis. The force on the element is
$$d\mathbf{F}_\beta = I\,d\mathbf{s} \times \mathbf{B} = I(-\sin\theta\,ds\,\mathbf{i} + \cos\theta\,ds\,\mathbf{j}) \times (-B\mathbf{k})$$
$$= IB\,ds\,(-\sin\theta\,\mathbf{j} - \cos\theta\,\mathbf{i}).$$
The resultant force is
$$\mathbf{F}_\beta = IB\int(-\sin\theta\,ds)\mathbf{j}.$$
We could use $ds = R\,d\theta$ and perform the integration over θ. If we recognize that $\sin\theta\,ds = dx$, we can simplify the integral:
$$\mathbf{F}_\beta = -IB\int dx\,\mathbf{j} = -IB\,\Delta x\,\mathbf{j} = \boxed{-2IRB\,\mathbf{j}.}$$

(c) For the net force, we have
$$\mathbf{F}_{net} = \mathbf{F}_\alpha + \mathbf{F}_\beta = 2IRB\,\mathbf{j} - 2IRB\,\mathbf{j} = \boxed{0.}$$

(d) From the analysis of part (b), which did not use the shape of the wire, we see that the net force in a uniform field will be zero for a loop of any shape.

61. We consider 5° to be a small angle. The torque produced by the magnetic field will provide an angular acceleration to align the dipole with the magnetic field:
$$\tau = -\mu B \sin\theta \approx -\mu B\theta = I_M(d^2\theta/dt^2).$$
This is the equation of motion for angular harmonic motion, with angular frequency
$$\omega = (\mu B/I_M)^{1/2} = (IAB/I_M)^{1/2}$$
$$= [(50\times10^{-3}\,\text{A})(6.0\times10^{-4}\,\text{m}^2)(0.60\,\text{T})/(7.5\times10^{-7}\,\text{kg}\cdot\text{m}^2)]^{1/2} = 4.9\,\text{rad/s}.$$
If we choose $t = 0$ at the moment of release, we have
$$\boxed{\begin{aligned}&\theta = \theta_0\cos(\omega t),\text{ from which we get}\\&d\theta/dt = -\theta_0\omega\sin(\omega t),\quad\text{and}\quad(d^2\theta/dt^2) = -\theta_0\omega^2\cos(\omega t) = -\omega^2\theta.\end{aligned}}$$
The maximum angular speed of the coil is
$$(d\theta/dt)_{max} = \theta_0\omega = (5°)[(\pi\,\text{rad})/(180°)](4.9\,\text{rad/s}) = \boxed{0.43\,\text{rad/s}.}$$

67. We find the magnitude of the electric field of the point charge from
$$E = (1/4\pi\varepsilon_0)e/r^2 = (9\times10^9\,\text{N}\cdot\text{m}^2/\text{C}^2)(1.6\times10^{-19}\,\text{C})/(0.5\times10^{-10}\,\text{m})^2 = \boxed{5.8\times10^{11}\,\text{N/C}.}$$
The force produced by the electric field provides the centripetal acceleration:
$$eE = mv^2/r,\text{ which gives } v = (reE/m)^{1/2}.$$
For the magnetic field to provide the same centripetal acceleration, we have
$$evB = mv^2/r = eE,\text{ from which we get}$$
$$B = E/v = (mE/er)^{1/2}$$
$$= [(9.1\times10^{-31}\,\text{kg})(5.8\times10^{11}\,\text{N/C})/(1.6\times10^{-19}\,\text{C})(0.5\times10^{-10}\,\text{m})]^{1/2} = \boxed{2.6\times10^5\,\text{T}.}$$

69. To produce a force opposite to the electric field, the magnetic field must be directed in the x-direction. The speed of the deuteron is
$$v = (2K/m)^{1/2}.$$
The electric field of the capacitor is
$$E = \sigma/\varepsilon_0.$$
For the velocity selector, we have
$$v = E/B;$$
$$(2K/m)^{1/2} = \sigma/\varepsilon_0 B;$$
$$[2(60\,\text{keV})(1.6\times10^{-16}\,\text{J/keV})/(3.2\times10^{-27}\,\text{kg})]^{1/2} = (8.0\times10^{-7}\,\text{C/m}^2)/(8.85\times10^{-12}\,\text{C}^2/\text{N}\cdot\text{m}^2)B,$$
which gives $\boxed{B = 0.037\,\text{T}.}$

71. The radius of the path in the magnetic field is
$r = mv/eB$, or $mv = eBr$.
The kinetic energy of the electron is
$K = \frac{1}{2}mv^2 = \frac{1}{2}(eBr)^2/m$.
If the energy changes to $K - \Delta K$, the radius will change to $r - \Delta r$. If we form the ratio for the two energies, we have
$(K - \Delta K)/K = [(r - \Delta r)/r]^2$;
$1 - (\Delta K/K) = [1 - (\Delta r/r)]^2$;
$1 - 0.10 = [1 - (\Delta r/r)]^2$, which gives $\boxed{\Delta r/r = 0.05 \text{ (decrease)}.}$

75. (a) The angular momentum of the N electrons is
$\boxed{L = NmvR \text{ perpendicular to the orbit.}}$
(b) The period of the orbit is
$T = 2\pi R/v$.
The magnetic dipole moment is
$\mu = IA = (Ne/T)\pi R^2 = [Ne/(2\pi R/v)]\pi R^2 = \boxed{\frac{1}{2}NevR.}$
(c) The ratio is
$L/\mu = NmvR/\frac{1}{2}NevR = \boxed{2m/e.}$

77. From the first condition, we have
$\mathbf{F}_1 = q[\mathbf{E} + (\mathbf{v}_1 \times \mathbf{B})] = 0$, or $\mathbf{E} = -\mathbf{v}_1 \times \mathbf{B}$,
which means that $\mathbf{E}$ and $\mathbf{v}_1$ are perpendicular.
Because $\mathbf{v}_1 = v_1\mathbf{k}$, we can write the fields as
$\mathbf{E} = E_x\mathbf{i} + E_y\mathbf{j}$ and $\mathbf{B} = B_x\mathbf{i} + B_y\mathbf{j} + B_z\mathbf{k}$.
From the second condition, we have
$\mathbf{F}_2 = q[\mathbf{E} + (\mathbf{v}_2 \times \mathbf{B})] = m\mathbf{a}$;
$q\{E_x\mathbf{i} + E_y\mathbf{j} + [(v_2 \sin 42° \mathbf{j} + v_2 \cos 42° \mathbf{k}) \times (B_x\mathbf{i} + B_y\mathbf{j} + B_z\mathbf{k})]\} = ma(-\mathbf{i})$;
$q[E_x\mathbf{i} + E_y\mathbf{j} - (v_2B_x \sin 42°)\mathbf{k} + (v_2B_z \sin 42°)\mathbf{i} + (v_2B_x \cos 42°)\mathbf{j} - (v_2B_y \cos 42°)\mathbf{i}] = -ma\mathbf{i}$.
From the $\mathbf{k}$ term, we have
$B_x = 0$, so $\mathbf{B} = B_y\mathbf{j} + B_z\mathbf{k}$.
From the $\mathbf{j}$ terms, we have
$E_y = -v_2B_x \cos 42° = 0$, so $\mathbf{E} = E\mathbf{i}$.
From the $\mathbf{i}$ terms, we have
$q(E - v_2B_z \sin 42° - v_2B_y \cos 42°) = -ma$.
For the third condition of circular motion in the xy-plane, we can write the velocity as
$\mathbf{v}_3 = v_3 \cos(\omega t)\mathbf{i} + v_3 \sin(\omega t)\mathbf{j}$.
The force must lie in the xy-plane:
$\mathbf{F}_3 = q[\mathbf{E} + (\mathbf{v}_3 \times \mathbf{B})]$;
$F_{3x}\mathbf{i} + F_{3y}\mathbf{j} = q(E\mathbf{i} + \{[v_3 \cos(\omega t)\mathbf{i} + v_3 \sin(\omega t)\mathbf{j}] \times (B_y\mathbf{j} + B_z\mathbf{k})\})$;
$F_{3x}\mathbf{i} + F_{3y}\mathbf{j} = q\{E\mathbf{i} + [v_3B_y \cos(\omega t)]\mathbf{k} - [v_3B_z \cos(\omega t)]\mathbf{j} + [v_3B_z \sin(\omega t)]\mathbf{i}\}$;
From the $\mathbf{k}$ term, we have
$v_3B_y \cos(\omega t) = 0$, which gives $B_y = 0$, so $\mathbf{B} = B\mathbf{k}$.
If we use the result of the first condition, we get
$\mathbf{E} = -\mathbf{v}_1 \times \mathbf{B} = -(v_1\mathbf{k}) \times (B\mathbf{k}) = 0$.
Using the result from the $\mathbf{i}$ term of the second condition, we have
$q(E - v_2B_z \sin 42° - v_2B_y \cos 42°) = q(0 - v_2B \sin 42° - 0) = -ma$;
$-(1.60 \times 10^{-19} \text{ C})(8.0 \times 10^2 \text{ m/s})B \sin 42° = -(1.67 \times 10^{-27} \text{ kg})(3.5 \times 10^8 \text{ m/s}^2)$,
which gives $B = 6.8 \times 10^{-3}$ T.
Thus the values of the fields are $\boxed{\mathbf{E} = 0, \mathbf{B} = (6.8 \times 10^{-3} \text{ T})\mathbf{k}.}$

CHAPTER 30

The integration for Ampere's law is performed by choosing an appropriate path based on the symmetry of the current distribution, which allows some knowledge of the directions of the magnetic field. For parts of the path where **B** is zero or perpendicular to d**s**, the contribution to the integral is zero. For parts of the path where **B** is parallel to d**s** and constant, the contribution to the integral is Bs.

5. If the two currents are in the same direction, they will be attracted with a force per unit length of
$$F/L = \mu_0 I_1 I_2 / 2\pi d$$
$$= (4\pi \times 10^{-7}\ \text{T} \cdot \text{m/A})(5 \times 10^4\ \text{A})(100\ \mu\text{A})/2\pi(4\ \text{m}) = \boxed{0.25\ \mu\text{N/m attraction.}}$$

9. (a) The magnetic field of a wire with radius R is
$$B_{\text{inside}} = (\mu_0 I / 2\pi)(r/R^2),\quad B_{\text{outside}} = \mu_0 I / 2\pi r.$$
We see that the magnetic field inside increases as r increases and is maximal at $r = R$. The magnetic field outside decreases as r increases and is maximal at $r = R$. Thus the magnetic field is greatest at $\boxed{r = R.}$ (d)

    (b) The maximum magnetic field is
    $$\boxed{B_{\text{max}} = \mu_0 I / 2\pi R \text{ at } r = R.}$$

    (c) The minimum magnetic field is
    $$\boxed{B_{\text{min}} = 0 \text{ at } r = 0 \text{ and } r = \infty.}$$

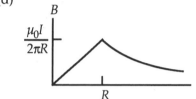

15. (a) The magnetic field of the wire is tangent to the circle of radius $r = (x^2 + y^2)^{1/2}$:
$$\mathbf{B} = \frac{\mu_0 I}{2\pi r}\left(-\sin\theta\,\mathbf{i} + \cos\theta\,\mathbf{j}\right) = \boxed{\frac{\mu_0 I}{2\pi(x^2 + y^2)}\left(-y\mathbf{i} + x\mathbf{j}\right).}$$

    (b) The total field from the two wires will be the sum of two expressions like the one in part (a), with x replaced by $x - a$ or $x + a$:
$$\boxed{\mathbf{B} = \frac{\mu_0 I}{2\pi}\left\{-\left[\frac{y}{(x-a)^2 + y^2} + \frac{y}{(x+a)^2 + y^2}\right]\mathbf{i} + \left[\frac{x-a}{(x-a)^2 + y^2} + \frac{x+a}{(x+a)^2 + y^2}\right]\mathbf{j}\right\}.}$$

    (c) If one of the currents is reversed, both components of the field from that wire will reverse:
$$\boxed{\mathbf{B} = \frac{\mu_0 I}{2\pi}\left\{\left[\frac{-y}{(x-a)^2 + y^2} + \frac{y}{(x+a)^2 + y^2}\right]\mathbf{i} + \left[\frac{x-a}{(x-a)^2 + y^2} - \frac{x+a}{(x+a)^2 + y^2}\right]\mathbf{j}\right\}.}$$

19. (a) We find the direction of the force due to the current in the wire from
F = q**v** × **B**. From the diagram, we see that the force, and thus the
deflection, of a negative charge will be | away from the wire. |

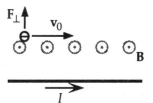

(b) We assume that the time of passage and the magnetic force
are small enough that the deflection is small. We can take
the magnetic field at the electron beam to be constant, which
gives a constant force. Because the force is away from the wire,
the speed parallel to the wire is constant, and the time of passage is
$$\Delta t = L/v_0 = (1.0 \text{ m})/(0.020)(3 \times 10^8 \text{ m/s}) = 1.67 \times 10^{-7} \text{ s}.$$
The magnetic field a distance d from the wire, $B = \mu_0 I/2\pi d$, produces a force directed away from the wire:
$$F_\perp = qv_0 B = ev_0 \mu_0 I/2\pi d.$$
The change in momentum created by the impulse of this force is
$$m \, \Delta v = F_\perp \, \Delta t = (ev_0 \mu_0 I/2\pi d) \, \Delta t.$$
The average speed away from the wire is
$$v_\perp = (0 + \Delta v)/2 = (ev_0 \mu_0 I/4\pi md) \, \Delta t, \text{ which produces a deflection}$$
$$s_\perp = v_\perp \, \Delta t = (ev_0 \mu_0 I/4\pi md)(\Delta t)^2 = (\mu_0/4\pi)(ev_0 I/md)(\Delta t)^2$$
$$= (10^{-7} \text{ T} \cdot \text{m/A})[(1.6 \times 10^{-19} \text{ C})(6.0 \times 10^6 \text{ m/s})(0.20 \text{ A})/$$
$$(9.1 \times 10^{-31} \text{ kg})(10.0 \times 10^{-2} \text{ m})](1.67 \times 10^{-7} \text{ s})^2$$
$$= \boxed{5.9 \times 10^{-3} \text{ m.}}$$
We see that our assumption of a small deflection is justified.

(c) Because the magnetic force is always perpendicular to the velocity, the speed does not change, and
the beam will have the same energy.

21. At a distance x from the wire, the magnetic field is directed into
the paper with magnitude
$$B = \mu_0 I/2\pi x.$$
Because the field is not constant over the square, we find the
magnetic flux by integration. We choose a differential element
parallel to the wire at position x with area $a \, dx$:

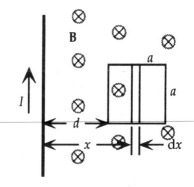

$$\Phi_B = \iint \mathbf{B} \cdot d\mathbf{A} = \iint B \, dA = \int_d^{a+d} \frac{\mu_0 I}{2\pi x} a \, dx$$
$$= \boxed{\frac{\mu_0 I a}{2\pi} \ln\left(\frac{a+d}{d}\right).}$$

29. The area of the disk is greater than the area of the solenoid. The magnetic field **B** is only inside the
solenoid along the axis of the solenoid and thus is parallel to the area **A**. The flux is
$$\Phi_B = \iint \mathbf{B} \cdot d\mathbf{A} = BA_{\text{solenoid}}$$
$$= (0.15 \text{ T})(4 \times 10^{-4} \text{ m}^2) = \boxed{6.0 \times 10^{-5} \text{ Wb.}}$$

31. The magnetic field inside the toroidal solenoid is circular and varies with the distance from the center of
the torus r:
$$B = \mu_0 NI/2\pi r.$$
We find the flux by integration. We choose a differential element at radius r with area $L \, dr$:

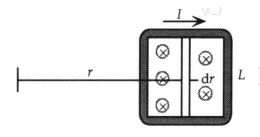

$$\Phi_B = \iint \mathbf{B} \cdot d\mathbf{A} = \int_R^{R+L} \frac{\mu_0 NI}{2\pi r} L \, dr$$
$$= \frac{\mu_0 NIL}{2\pi} \int_R^{R+L} \frac{dr}{r}, \text{ which gives}$$
$$\boxed{\Phi_B = (\mu_0 NIL/2\pi) \ln[(R+L)/R].}$$

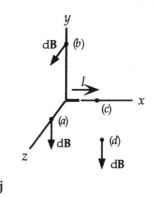

37. The differential element to be used in the Biot–Savart law is $d\boldsymbol{\ell} = d\ell\ \mathbf{i}$.

(a) $d\mathbf{B} = (\mu_0/4\pi)I[d\boldsymbol{\ell} \times (\mathbf{r}_a/r_a^3)] = (\mu_0/4\pi)I[(d\ell\ \mathbf{i}) \times (r_a\mathbf{k}/r_a^3)]$
$= [(\mu_0/4\pi)I\ dL/r_a^2](-\mathbf{j}) = -(10^{-7}\ \text{T}\cdot\text{m/A})(2\ \text{A})\ dL/(3 \times 10^{-2}\ \text{m})^2\ \mathbf{j}$
$= \boxed{-(2.2 \times 10^{-4}\ \text{T/m})\ dL\ \mathbf{j}.}$

(b) $d\mathbf{B} = (\mu_0/4\pi)I[d\boldsymbol{\ell} \times (\mathbf{r}_b/r_b^3)] = (\mu_0/4\pi)I[(d\ell\ \mathbf{i}) \times (r_b\mathbf{j}/r_b^3)]$
$= [(\mu_0/4\pi)I\ dL/r_b^2]\mathbf{k} = -(10^{-7}\ \text{T}\cdot\text{m/A})(2\ \text{A})\ dL/(6 \times 10^{-2}\ \text{m})^2\ \mathbf{k}$
$= \boxed{(0.56 \times 10^{-4}\ \text{T/m})\ dL\ \mathbf{k}.}$

(c) $d\mathbf{B} = (\mu_0/4\pi)I[d\boldsymbol{\ell} \times (\mathbf{r}_c/r_c^3)] = (\mu_0/4\pi)I[(d\ell\ \mathbf{i}) \times (r_c\mathbf{i}/r_c^3)]$
$= \boxed{0.}$

(d) $d\mathbf{B} = (\mu_0/4\pi)I[d\boldsymbol{\ell} \times (\mathbf{r}_d/r_d^3)] = (\mu_0/4\pi)I\{(d\ell\ \mathbf{i}) \times [(r_d/\sqrt{2})(\mathbf{i} + \mathbf{k})/r_d^3]\}$
$= [(\mu_0/4\pi)I\ dL/r_d^2\sqrt{2}](-\mathbf{j}) = -(10^{-7}\ \text{T}\cdot\text{m/A})(2\ \text{A})\ dL/(6\sqrt{2} \times 10^{-2}\ \text{m})^2\sqrt{2}\ \mathbf{j}$
$= \boxed{-(0.20 \times 10^{-4}\ \text{T/m})\ dL\ \mathbf{j}.}$

43. Because the point P is along the line of the two straight segments of the wire, there is no magnetic field from these segments.
The magnetic field at the point P is the sum of two fields:
$\mathbf{B} = \mathbf{B}_{\text{inner semicircle}} + \mathbf{B}_{\text{outer semicircle}}.$
Each field is half that of a circular loop, with the field of the inner semicircle into the page and that of the outer semicircle out of the page, so we subtract the two magnitudes:
$B = \tfrac{1}{2}(\mu_0 I/2R_{\text{inner}}) - \tfrac{1}{2}(\mu_0 I/2R_{\text{outer}}) = (\mu_0 I/4)[(1/R_{\text{inner}}) - (1/R_{\text{outer}})]$
$= (\pi \times 10^{-7}\ \text{T}\cdot\text{m/A})(12\ \text{A})[(1/0.05\ \text{m}) - (1/0.08\ \text{m})]$, which gives
$\boxed{B = 2.8 \times 10^{-5}\ \text{T}\quad\text{into the page.}}$

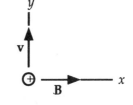

45. We choose the coordinate system shown in the figure.
The magnetic field of the dipole will be directed along the x-axis:
$\mathbf{B} = (\mu_0/2\pi)(\mu/d^3)\mathbf{i}.$
The force on the charge produces the acceleration of the charge:
$\mathbf{F} = q\mathbf{v} \times \mathbf{B} = m\mathbf{a};$
$q(v\mathbf{j}) \times [(\mu_0/2\pi)(\mu/d^3)\mathbf{i}] = qv(\mu_0/2\pi)(\mu/d^3)(-\mathbf{k}) = m\mathbf{a},$
so the acceleration of the charge is
$\mathbf{a} = -(\mu_0/2\pi)(qv\mu/md^3)\mathbf{k}.$
Because the acceleration is perpendicular to the velocity, the charge will start to move in a circle with radius
$R = mv/qB = mv/q(\mu_0/2\pi)(\mu/d^3) = \boxed{2\pi mvd^3/q\mu\mu_0.}$

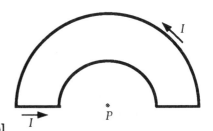

55. If we assume a parallel-plate capacitor, the electric field in the capacitor is
$E = V/d = (V_0/d)\cos(\omega t)$, so the electric flux is
$\Phi_E = EA = (AV_0/d)\cos(\omega t).$
The displacement current is
$I_d = \varepsilon_0\ d\Phi_E/dt = \varepsilon_0 A\ (dE/dt) = (\varepsilon_0 AV_0/d)[-\sin(\omega t)]\ \omega = \boxed{-CV_0\omega\sin(\omega t).}$
Note that this is the charging current.

57. From the charge density $\sigma = \sigma_0(1 - t/t_0)$, we find the rate at which the charge is changing:
$d\sigma/dt = -\sigma_0/t_0.$
The electric field at the surface is that of a point charge, $Q = \sigma 4\pi R^2$. We find the electric flux through a spherical surface just outside the conducting sphere:
$\Phi_E = EA = (1/4\pi\varepsilon_0)(\sigma 4\pi R^2/R^2)(4\pi R^2) = \sigma 4\pi R^2/\varepsilon_0.$
The displacement current is
$I_d = \varepsilon_0\ d\Phi_E/dt = \varepsilon_0(4\pi R^2/\varepsilon_0)\ d\sigma/dt = \boxed{-4\pi R^2\sigma_0/t_0.}$
The current in the wire is
$I = dq/dt = A\ d\sigma/dt = \boxed{-4\pi R^2\sigma_0/t_0 = I_d.}$

59. We call the radius of the inner wire R_1, the inside radius of the outer wire R_2, and the outside radius of the outer wire R_3. The current densities in the wires are

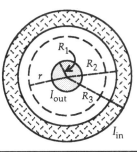

$$J_{inner} = I/\pi R_1^2 \quad \text{and} \quad J_{outer} = I/\pi(R_3^2 - R_2^2).$$

Because of the cylindrical symmetry, we know that the magnetic fields will be circular. In each case we apply Ampere's law to a circular path of radius r.

(a) Inside the inner wire, $r < R_1$:

$$\oint \mathbf{B} \cdot d\mathbf{s} = \mu_0 I_{enclosed};$$
$$B2\pi r = \mu_0 J_{inner}\pi r^2 = \mu_0 I\pi r^2/\pi R_1^2, \text{ which gives } \boxed{B = (\mu_0 I/2\pi R_1^2)r \text{ circular CCW}, r < R_1.}$$

(b) Between the wires, $R_1 < r < R_2$:

$$\oint \mathbf{B} \cdot d\mathbf{s} = \mu_0 I_{enclosed};$$
$$B2\pi r = \mu_0 I, \text{ which gives } \boxed{B = \mu_0 I/2\pi r \text{ circular CCW}, R_1 < r < R_2.}$$

(c) Inside the outer wire, $R_2 < r < R_3$:

$$\oint \mathbf{B} \cdot d\mathbf{s} = \mu_0 I_{enclosed};$$
$$B2\pi r = \mu_0[I - J_{outer}(\pi r^2 - \pi R_2^2)]$$
$$= \mu_0[I - I(\pi r^2 - \pi R_2^2)/\pi(R_3^2 - R_2^2)] = \mu_0 I[1 - (r^2 - R_2^2)/(R_3^2 - R_2^2)],$$

which gives $\boxed{B = (\mu_0 I/2\pi r)(R_3^2 - r^2)/(R_3^2 - R_2^2) \text{ circular CCW}, R_2 < r < R_3.}$

(d) Outside the outer wire, $R_3 < r$:

$$\oint \mathbf{B} \cdot d\mathbf{s} = \mu_0 I_{enclosed};$$
$$B2\pi r = \mu_0(I - I), \text{ which gives } \boxed{B = 0, R_3 < r.}$$

61. Because the magnetic field produced by the long wire depends only on the distance from the wire y, $B = \mu_0 I_1/2\pi y$, we find the force on the rectangular loop. The net force is

$$\mathbf{F} = \mathbf{F}_a + \mathbf{F}_b + \mathbf{F}_c + \mathbf{F}_d.$$

For segments c and d of the loop, the symmetry of the field and the opposite directions of the current I_2 give

$$\mathbf{F}_c + \mathbf{F}_d = 0.$$

Because the wires and the magnetic field are perpendicular,

we have

$$\mathbf{F} = \mathbf{F}_a + \mathbf{F}_b = I_2 L(\mu_0 I_1/2\pi a)(-\mathbf{j}) + I_2 L(\mu_0 I_1/2\pi b)(+\mathbf{j})$$
$$= (\mu_0/4\pi)(2I_1 I_2 L)[(1/b) - (1/a)]\mathbf{j}$$
$$= (10^{-7} \text{ T} \cdot \text{m/A})2(10 \text{ A})(5 \text{ A})(0.20 \text{ m})\{[1/(0.05 \text{ m})] - [1/(0.02 \text{ m})]\}\mathbf{j}$$
$$= \boxed{-6.0 \times 10^{-5}\mathbf{j} \text{ N (attraction).}}$$

63. (a) For the attractive magnetic force per unit length, we have

$$F_B/L = I(\mu_0 I/2\pi d) = (\mu_0/2\pi)(I^2/d)$$
$$= (2 \times 10^{-7} \text{ A} \cdot \text{m}^2)(1 \text{ A})^2/(1 \times 10^{-2} \text{ m}) = 2.0 \times 10^{-5} \text{ N/m}.$$

The linear charge density is

$$\lambda = (10^{21} \text{ carriers/cm})(10^2 \text{ cm/m})(1.6 \times 10^{-19} \text{ C/carrier}) = 1.6 \times 10^4 \text{ C/m}.$$

The electric field a distance d from an infinite line of charge is $E = (1/2\pi\varepsilon_0)(\lambda/d)$.

For the repulsive electric force per unit length, we have

$$F_E/L = \lambda E = \lambda(1/2\pi\varepsilon_0)(\lambda/d) = (1/2\pi\varepsilon_0)(\lambda^2/d)$$
$$= 2(9 \times 10^9 \text{ N} \cdot \text{m}^2/\text{C}^2)(1.6 \times 10^4 \text{ C/m})^2/(1 \times 10^{-2} \text{ m}) = 4.6 \times 10^{20} \text{ N/m}.$$

The ratio of the forces is

$$\boxed{F_B/F_E = 4.3 \times 10^{-26}.}$$

(b) For the forces to be equal, we have

$$\lambda^2/\varepsilon_0 = \mu_0 I^2, \quad \text{or} \quad \lambda^2 = \varepsilon_0\mu_0 I^2.$$

If we use $\varepsilon_0\mu_0 = 1/c^2$ and $\lambda = ne$, we get

$$n = I/ce = (1 \text{ A})/(3 \times 10^8 \text{ m/s})(1.6 \times 10^{-19} \text{ C/electron}) = 2.1 \times 10^{10} \text{ electrons/m}$$
$$= \boxed{2.1 \times 10^8 \text{ electrons/cm.}}$$

(c) For the fraction that is the excess, we have $f = (2.1 \times 10^8/\text{cm})/(10^{21}/\text{cm}) = \boxed{2.1 \times 10^{-13}.}$

65. (a) We choose $x = 0$ at the left coil. The magnetic fields on the axis from the coils are in the same direction, so we find the magnitude of the total field from

$$B(x) = \frac{\mu_0 I}{2}\left\{\frac{R^2}{\left(R^2 + x^2\right)^{3/2}} + \frac{R^2}{\left[R^2 + (R-x)^2\right]^{3/2}}\right\}$$

$$= \boxed{\frac{\mu_0 I}{2R}\left\{\frac{1}{\left[1 + (x/R)^2\right]^{3/2}} + \frac{1}{\left[2 - 2(x/R) + (x/R)^2\right]^{3/2}}\right\}.}$$

At $x = 0$, we have

$B(0) = (\mu_0 I/2R)[1 + (1/2^{3/2})] = \boxed{0.677\,\mu_0 I/R.}$

At $x = R/4$, we have

$B(R/4) = (\mu_0 I/2R)(\{1/[1 + (1/4)^2]^{3/2}\} + \{1/[2 - 2(1/4) + (1/4)^2]^{3/2}\}) = \boxed{0.713\,\mu_0 I/R.}$

At $x = R/2$, we have

$B(R/2) = (\mu_0 I/2R)(\{1/[1 + (1/2)^2]^{3/2}\} + \{1/[2 - 2(1/2) + (1/2)^2]^{3/2}\}) = \boxed{0.716\,\mu_0 I/R.}$

(b) When we differentiate the expression for B, we get

$$\frac{dB}{dx} = \frac{\mu_0 I}{2R}\left\{\frac{(-3/2)2(1/R)(x/R)}{\left[1 + (x/R)^2\right]^{5/2}} + \frac{(-3/2)2[(-1/R) + (1/R)(x/R)]}{\left[2 - 2(x/R) + (x/R)^2\right]^{5/2}}\right\}$$

$$= -\frac{3\mu_0 I}{2R^2}\left\{\frac{(x/R)}{\left[1 + (x/R)^2\right]^{5/2}} + \frac{(x/R) - 1}{\left[2 - 2(x/R) + (x/R)^2\right]^{5/2}}\right\};$$

$$\frac{d^2 B}{dx^2} = -\frac{3\mu_0 I}{2R^2}\left\{\frac{1/R}{\left[1 + (x/R)^2\right]^{5/2}} + \frac{(-5/2)2(1/R)(x/R)^2}{\left[1 + (x/R)^2\right]^{7/2}} \right.$$
$$\left. + \frac{1/R}{\left[2 - 2(x/R) + (x/R)^2\right]^{5/2}} + \frac{[(x/R) - 1](-5/2)2[(-1/R) + (1/R)(x/R)]}{\left[2 - 2(x/R) + (x/R)^2\right]^{7/2}}\right\}$$

$$= -\frac{3\mu_0 I}{2R^3}\left\{\frac{1}{\left[1 + (x/R)^2\right]^{5/2}} - \frac{5(x/R)^2}{\left[1 + (x/R)^2\right]^{7/2}} + \frac{1}{\left[2 - 2(x/R) + (x/R)^2\right]^{5/2}} - \frac{5[(x/R) - 1]^2}{\left[2 - 2(x/R) + (x/R)^2\right]^{7/2}}\right\}.$$

At $x = R/2$, we have

$$\frac{dB}{dx} = -\frac{3\mu_0 I}{2R^2}\left\{\frac{1/2}{\left[1 + (1/2)^2\right]^{5/2}} + \frac{(1/2) - 1}{\left[2 - 2(1/2) + (1/2)^2\right]^{5/2}}\right\} = -\frac{3\mu_0 I}{2R^2}\left\{\frac{1/2}{(5/4)^{5/2}} + \frac{-(1/2)}{(5/4)^{5/2}}\right\} = 0;$$

$$\frac{d^2 B}{dx^2} = -\frac{3\mu_0 I}{2R^3}\left\{\frac{1}{\left[1 + (1/2)^2\right]^{5/2}} - \frac{5(1/2)^2}{\left[1 + (1/2)^2\right]^{7/2}} + \frac{1}{\left[2 - 2(1/2) + (1/2)^2\right]^{5/2}} - \frac{5[(1/2) - 1]^2}{\left[2 - 2(1/2) + (1/2)^2\right]^{7/2}}\right\}.$$

$$= -\frac{3\mu_0 I}{2R^3}\left[\frac{1}{(5/4)^{5/2}} - \frac{5/4}{(5/4)^{7/2}} + \frac{1}{(5/4)^{5/2}} - \frac{5/4}{(5/4)^{7/2}}\right] = 0.$$

67. We choose a radius of $r = 10$ cm for the trajectory. This requires a tube with a diameter of ≈ 30 cm. If the accelerating potential of the electron gun is V, we find the speed of the electron from

$\frac{1}{2}mv^2 = eV$, or $v = (2eV/m)^{1/2}$.

From Example 29–2, we have

$e/m = v/Br = (2eV/m)^{1/2}/Br$, which becomes

$V = (e/2m)r^2B^2 = [(1.6 \times 10^{-19}\ C)/2(9.1 \times 10^{-31}\ kg)](0.10\ m)^2B^2 = (8.8 \times 10^8\ V/T^2)B^2$.

Some possible combinations are

B, T	V, V
5×10^{-4}	220
1×10^{-3}	880
1.5×10^{-3}	2000
2×10^{-3}	3500
5×10^{-3}	22000

We choose a convenient potential of $V = 2000$ V, so $B = 1.5 \times 10^{-3}$ T, which is $\approx 30\ B_{Earth}$.

We create the field with Helmholtz coils with a radius of 20 cm to accommodate the tube.

From the solution for Problem 65 we have

$B \approx 0.71\ \mu_0 NI/R$;

1.5×10^{-3} T $= 0.71(4\pi \times 10^{-7}\ T \cdot m/A)NI/(0.20\ m)$, which gives $NI = 336\ A \cdot$ turns.

If we wind 400 turns on the coils, the current is

$I = (336\ A \cdot turns)/(400\ turns) = 0.84$ A.

The concern is that the current not be so high that the heat generation is a problem.

CHAPTER 31

The direction of the induced emf is such as to oppose the cause producing it. Thus the induced emf will produce a flux that is opposite to the change in flux that is causing it, not necessarily opposite to the original flux. For a motional emf, there will be an induced current that produces a force to oppose the motion.

5. The magnetic flux through the loop is
$$\Phi_B = \iint \mathbf{B} \cdot d\mathbf{A} = \iint (B\mathbf{i}) \cdot (dA\,\mathbf{i}) = BA.$$
The magnitude of the current in the loop is
$$I = \mathcal{E}/R = (d\Phi_B/d\tau)/R = (A/R)\,dB/dt = (\pi r^2/R)\,dB/dt;$$
$$2\text{ A} = [\pi(3.5 \times 10^{-2}\text{ m})^2/(1.5 \times 10^{-3}\ \Omega)]\,dB/dt, \text{ which gives } \boxed{dB/dt = (0.78\text{ T/s})\mathbf{i}.}$$
Note that the direction of current is not specified, so the magnetic field could be increasing or decreasing.

7. We take $t = 0$ when the leading edge of the loop enters the magnetic field. The position of this edge is given by $x = -L + vt$. The leading edge will reach the z-axis at $t_1 = L/v$ and the other side at $t_2 = 2L/v$. The trailing edge will leave the magnetic field at $t_3 = 3L/v$.
    We have three regions to consider:
    Region I, $0 < t < (L/v)$.
        The magnetic flux through the loop is
        $$\Phi_B = BL(x + L). \quad \text{(Note that } x < 0.)$$
        The magnetic flux is up and increasing.
        The induced emf is
        $$\mathcal{E}_1 = -d\Phi_B/dt = -BL(dx/dt) = \boxed{-BLv \text{ (clockwise as viewed from above).}}$$
    Region II, $(L/v) < t < (2L/v)$.
        The magnetic flux through the loop is
        $$\Phi_B = [+B(L-x) - B(x)]L = BL(L - 2x). \quad \text{(Note that } x > 0.)$$
        Initially the magnetic flux is up but decreases and becomes down.
        The induced emf is
        $$\mathcal{E}_2 = -d\Phi_B/dt = -BL(-2\,dx/dt) = \boxed{+2BLv \text{ (counterclockwise).}}$$
    Region III, $(2L/v) < t < (3L/v)$.
        The magnetic flux through the loop is
        $$\Phi_B = -B(2L - x)L = B(x - 2L)L. \quad \text{(Note that } x > L.)$$
        The magnetic flux is down and decreasing.
        The induced emf is
        $$\mathcal{E}_3 = -d\Phi_B/dt = -BL(dx/dt) = \boxed{-BLv \text{ (clockwise).}}$$

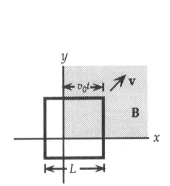

11. We take $t = 0$ when the leading corner of the loop passes the origin and enters the magnetic field. Before $t = 0$, there is no flux through the coil, and thus there is no emf. The trailing corner will enter the magnetic field at $t = L/v_0$. After this time the flux through the coil will be constant, and thus there is no emf. Between these two times, each side of the coil will have moved a distance $v_0 t$ into the magnetic field.
    The area of the coil through which there is a magnetic field is
    $$\mathbf{A} = (v_0 t)^2 \mathbf{k}.$$
    The magnetic flux through the coil is
    $$\Phi_B = \mathbf{B} \cdot \mathbf{A} = [B_0(\mathbf{i} + \mathbf{j} + \mathbf{k})] \cdot [(v_0 t)^2 \mathbf{k}] = B_0 v_0^2 t^2.$$
    The induced emf is
    $$\mathcal{E} = -d\Phi_B/dt = \boxed{-2B_0 v_0^2 t, \ 0 < t < L/v_0 \text{ (counterclockwise).}}$$

13. Because the magnetic field varies in the x-direction but not the y-direction, we find the flux through the loop at time t by choosing a strip with length D and thickness dx:

$$\Phi_B = \iint \mathbf{B} \cdot d\mathbf{A} = \int_0^{vt} (Cx\mathbf{j}) \cdot (D\,dx\mathbf{j}) = \tfrac{1}{2}CDv^2t^2.$$

The induced emf is

$\mathcal{E} = -d\Phi_B/dt = -CDv^2t$ clockwise.

The resistance of the loop is $R = \alpha L = \alpha(2D + 2vt)$.
The current in the loop is

$I = \mathcal{E}/R = -CDv^2t\,/2\alpha(D + vt)$ clockwise.

This differs from the result of Example 31–5 by the presence of a vt term in the numerator. Here the cause of the emf is both the increasing area and the increasing magnetic field.

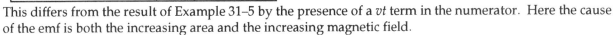

25. When the bar has moved a distance x, the area swept out by the bar is
    $A = xL \sin 60°$,
so the magnetic flux through the area is
    $\Phi_B = BxL \sin 60°$.
The potential difference is due to the emf induced by the changing flux:

$\mathcal{E} = -d\Phi_B/dt = -(dx/dt)BL \sin 60° = -(5.0 \text{ m/s})(5 \times 10^{-3} \text{ T})(0.7 \text{ m})(0.866) = -1.5 \times 10^{-2} \text{ V}.$

From Lenz's law, the emf would generate a current to oppose the increase in flux, clockwise around the area or down in the bar. The potential difference between the two ends of the bar is

1.5×10^{-2} V, with the bottom end at the higher potential.

27. We let $t = 0$ when the rod enters the magnetic field. When the rod has traveled a distance $x = vt$, the length of the rod in the magnetic field is $2R \sin \theta$, where

$\cos \theta = (R - vt)/R = 1 - (vt/R)$ and
$\sin \theta = [1 - (1 - vt/R)^2]^{1/2} = [(2vt/R) - (v^2t^2/R^2)]^{1/2}.$

Because $\mathbf{v}$, $\mathbf{B}$, and $\mathbf{L}$ are perpendicular, the motional emf is

$\mathcal{E} = -BvL = -Bv2R \sin \theta$

$\quad = -Bv2R[(2vt/R) - (v^2t^2/R^2)]^{1/2} = -2Bv(2Rvt - v^2t^2)^{1/2}.$

From Lenz's law, we see that the emf will be directed down. After the rod leaves the field at $t = 2R/v$, the emf will be zero:

$\mathcal{E} = -2Bv(2Rvt - v^2t^2)^{1/2}$ for $0 < t < 2R/v.$

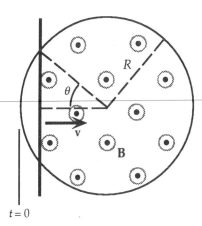

$t = 0$

33. (a) If L is the separation of the rails and D is the distance of the bar from the end of the rails, the magnetic flux through the loop is
    $\Phi_B = BA = BLD.$
    The induced emf is

$\mathcal{E} = -d\Phi_B/dt = -BL\,(dD/dt) = -BLv,$

    which produces a current

$I = \mathcal{E}/R = BLv/R.$

    The Lorentz force on this current is
    $F = ILB = B^2L^2v/R$, opposite to the direction of the motion.
    To maintain the motion, an equal and opposite external force is required:
    $F = (0.28 \text{ T})^2(0.30 \text{ m})^2(0.60 \text{ m/s})/(0.05 \text{ }\Omega) = \boxed{0.085 \text{ N.}}$

    (b) The rate of Joule heating in the resistor is
    $P = I^2R = (BLv/R)^2R = B^2L^2v^2/R = Fv$
    $\quad = (0.085 \text{ N})(0.60 \text{ m/s}) = \boxed{0.051 \text{ W.}}$ This is the rate at which the external force does work.

41. From the cylindrical symmetry, we know that the electric field depends only on r and must be circular. For a circular path just outside the solenoid, the magnetic field through the path is the magnetic field inside the solenoid, $B = \mu_0 nI$, so the flux is

$$\Phi_B = BA = B\pi r^2 = \mu_0 nI\pi r^2.$$

If we apply Faraday's law, we have

$$\oint \mathbf{E} \cdot d\mathbf{s} = -d\Phi_B/dt;$$
$$E \oint ds = -\mu_0 n\pi r^2 (dI/dt);$$
$$E2\pi r = -\mu_0 n\pi r^2 I_0 \omega [-\sin(\omega t)], \text{ which gives } \boxed{E = +(\mu_0 nI_0\omega r/2)\sin(\omega t), \text{ circular.}}$$

43. We take the clockwise path as the positive direction. The magnetic flux through the path is

$$\Phi_B = \mathbf{B} \cdot \mathbf{A} = -BA$$
$$= -B(1/4)[\pi(R/2)^2 - \pi(R/4)^2] = -3B\pi R^2/64.$$

The induced emf is

$$\mathcal{E} = -d\Phi_B/dt = +(3\pi R^2/64)\, dB/dt.$$

Because the magnetic field decreases linearly, we have

$$\mathcal{E} = +(3\pi R^2/64)\,\Delta B/\Delta t$$
$$= [3\pi(0.08 \text{ m})^2/64][(0.7 \text{ T} - 1.5 \text{ T})/(25\times10^{-3} \text{ s})]$$
$$= \boxed{-0.030 \text{ V (counterclockwise).}}$$

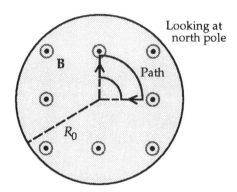

Looking at north pole

B

Path

R_0

Note that the flux created by an induced counterclockwise current is out of the page, to oppose the decrease in the flux of the electromagnet.

47. If we assume rolling without slipping, the tangential speed of the friction wheel is the tangential speed of the bicycle wheel, which is the linear speed of the bicycle. Thus the angular speed of the friction wheel is $\omega = v/r$, where r is the radius of the friction wheel. We can write the magnetic flux through each coil as

$$\Phi_B = NB_0 A \cos(\omega t),$$

For the two coils, the induced emf is

$$\mathcal{E} = 2(-d\Phi_B/dt) = -2NB_0 A\omega \sin(\omega t).$$

We find the speed of the bicycle from the maximum emf:

$$\mathcal{E}_{max} = 2NB_0 A\omega;$$
$$6.4 \text{ V} = 2(70)(0.1 \text{ T})(8\times10^{-4} \text{ m}^2)[v/(0.01 \text{ m})], \text{ which gives } \boxed{v = 5.7 \text{ m/s.}}$$

51. (a) The magnetic field of the wire depends on the distance from
the wire, $B = (\mu_0/2\pi)I/y$. When the moving rod is a distance
x from the resistor, we select a differential slice a distance y
from the wire and find the magnetic flux through the area
enclosed by the rod and resistor by integration:

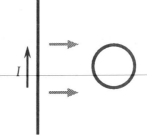

$$\Phi_B = \iint \mathbf{B} \cdot d\mathbf{A} = \int_a^b \left(\frac{\mu_0 I_0}{2\pi y}\right) x \, dy = \left(\frac{\mu_0 I_0 x}{2\pi}\right)\ln\left(\frac{b}{a}\right).$$

The induced emf in the rod is

$$\mathcal{E} = -d\Phi_B/dt = -(\mu_0 I/2\pi)(dx/dt)\ln(b/a)$$
$$= -(2 \times 10^{-7}\,\text{T} \cdot \text{m/A})(150\,\text{A})(45 \times 10^{-2}\,\text{m/s})\ln[(16\,\text{cm})/(8\,\text{cm})] = \boxed{-9.4 \times 10^{-6}\,\text{V (up)}.}$$

(b) To oppose the increase in flux into the page, the induced current in the loop will be counterclockwise,
with magnitude

$$I_{\text{ind}} = \mathcal{E}/R$$
$$= (9.4 \times 10^{-6}\,\text{V})/(0.20\,\Omega) = 4.7 \times 10^{-5}\,\text{A} = \boxed{47\,\mu\text{A}.}$$

(c) We find the time for the rod to move 100 cm from $\Delta t = \Delta x/v$. The rate at which work is done on the
rod must equal the Joule heating. The work done in Δt is

$$W = P\,\Delta t = I_{\text{ind}}^2 R\,\Delta t = I_{\text{ind}}^2 R\,\Delta x/v$$
$$= (4.7 \times 10^{-5}\,\text{A})^2(0.20\,\Omega)(1.00\,\text{m})/(45 \times 10^{-2}\,\text{m/s}) = \boxed{9.7 \times 10^{-10}\,\text{J}.}$$

The work must be done by an external force applied to the rod. The small value means that the
presence of any friction would require much more work by the external force.

55. The magnetic field from the wire is directed into the page with a
magnitude that depends on the distance D from the wire:

$$B = \mu_0 I/2\pi D.$$

As the wire is moved toward the loop, D decreases, so B at the loop
will increase. The magnetic flux through the loop, which is directed
into the page, will increase. The induced current will produce a magnetic
field out of the page, to oppose the increased flux.
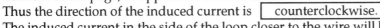
Thus the direction of the induced current is $\boxed{\text{counterclockwise.}}$
The induced current in the side of the loop closer to the wire will be
in a stronger magnetic field and will experience a greater Lorentz force.
Because the force on the side closer to the wire is directed away from the wire,
while the force on the side farther from the wire is directed toward the wire,
the net force will be $\boxed{\text{away from the wire.}}$

57. Because the change in flux through the coil is due to the change in the magnetic field, we have

$$\mathcal{E} = -d\Phi_B/dt = -NA\,dB/dt.$$

We find the charge that passes through the coil from

$$q = \int I\,dt = \int(\mathcal{E}/R)\,dt = \int(-NA/R)(dB/dt)\,dt$$
$$= -(NA/R)\int dB = -(NA/R)\,\Delta B = -(NA/R)(-2B) = 2NAB/R$$
$$= 2(200\,\text{turns})\pi(4.0 \times 10^{-2}\,\text{m})^2(1.4\,\text{T})/(5.6\,\Omega) = \boxed{0.50\,\text{C}.}$$

59. Because the flux through the loop is increasing, the induced emf and current will produce a flux into the paper and thus will be clockwise; positive charge will accumulate on the lower plate of the capacitor. The magnitude of the induced emf is

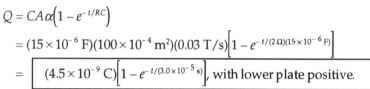

$$\mathcal{E} = d\Phi_B/dt = A\, dB/dt = A\, d(\alpha t)/dt = A\alpha.$$

We write a clockwise loop equation for the circuit, starting at point a:

$$A\alpha - IR - Q/C = 0, \text{ with } I = dQ/dt.$$

This is the equation for a series RC circuit, which has the solution

$$Q = CA\alpha\left(1 - e^{-t/RC}\right)$$

$$= (15 \times 10^{-6}\text{ F})(100 \times 10^{-4}\text{ m}^2)(0.03\text{ T/s})\left[1 - e^{-t/(2\,\Omega)(15 \times 10^{-6}\text{ F})}\right]$$

$$= \boxed{(4.5 \times 10^{-9}\text{ C})\left[1 - e^{-t/(3.0 \times 10^{-5}\text{ s})}\right]}, \text{ with lower plate positive.}$$

We find the current in the circuit from

$$I = \frac{dQ}{dt} = \frac{A\alpha}{R}e^{-t/RC}$$

$$= \frac{(100 \times 10^{-4}\text{ m}^2)(0.03\text{ T/s})}{(2\,\Omega)}e^{-t/(2\,\Omega)(15 \times 10^{-6}\text{ F})} = \boxed{(1.5 \times 10^{-4}\text{ A})\,e^{-t/(3.0 \times 10^{-5}\text{ s})}}, \text{ clockwise.}$$

61. If we choose $t = 0$ at the position shown, the magnetic flux through the circuit is

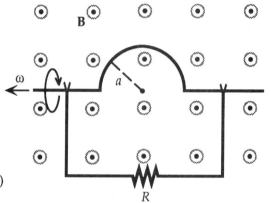

$$\Phi_B = \Phi_{B,\text{rectangle}} + \Phi_{B,\text{semicircle}}$$
$$= \Phi_{B,\text{rectangle}} + B(\pi a^2/2)\cos(\omega t).$$

The induced emf is

$$\mathcal{E} = -d\Phi_B/dt$$
$$= -(B\pi a^2/2)[-\sin(\omega t)]\omega = \boxed{(B\pi a^2\omega/2)\sin(\omega t).}$$

The current in the circuit is $I = \mathcal{E}/R$.

We find the average power dissipated from

$$P_{\text{av}} = I^2_{\text{av}}R = \mathcal{E}^2_{\text{av}}/R$$
$$= [(B\pi a^2\omega/2)^2/R][\sin^2(\omega t)]_{\text{av}} = [(B\pi a^2\omega/2)^2/R](1/2)$$
$$= \boxed{B^2\pi^2a^4\omega^2/8R.}$$

63. (a) The magnetic force provides the centripetal acceleration:

$$evB = mv^2/R, \text{ which gives}$$
$$v = eBR/m = (1.6 \times 10^{-19}\text{ C})(10^{-6}\text{ T})(1\text{ m})/(9.1 \times 10^{-31}\text{ kg}) = \boxed{1.8 \times 10^5\text{ m/s.}}$$

(b) The kinetic energy of the electron is

$$E = \tfrac{1}{2}mv^2 = \tfrac{1}{2}(9.1 \times 10^{-31}\text{ kg})(1.8 \times 10^5\text{ m/s})^2 = \boxed{1.4 \times 10^{-20}\text{ J.}}$$

(c) We find the circular electric field along the electron's path by applying Faraday's law:

$$\int \mathbf{E}_{\text{field}} \cdot d\mathbf{s} = -d\Phi_B/dt;$$

$$E_{\text{field}} \int ds = E_{\text{field}}\,2\pi R = -A\,\Delta B/\Delta t = -\pi R^2\,\Delta B/\Delta t, \text{ which gives } E_{\text{field}} = -\tfrac{1}{2}R\,\Delta B/\Delta t.$$

In a time Δt, the electron moves a circular distance $\Delta s = v\,\Delta t$, so the work done by the field, which changes the energy of the electron, is

$$W = -eE_{\text{field}}\,\Delta s = -e(-\tfrac{1}{2}R\,\Delta B/\Delta t)v\,\Delta t = \tfrac{1}{2}eRv\,\Delta B = \Delta E.$$

The fractional energy change is

$$\Delta E/E = (\tfrac{1}{2}eRv\,\Delta B)/(\tfrac{1}{2}mv^2) = (eR/mv)\,\Delta B.$$

If we use the result from part (a) for the speed, we have

$$\Delta E/E = [eR/m(eBR/m)]\,\Delta B = \Delta B/B, \text{ which is independent of } R \text{ and } v.$$

(d) From part (c), we have

$$\Delta E/E = \Delta B/B = \boxed{-10\%.}$$

CHAPTER 32

7. (a) We find the magnetic field from
    $$\mathbf{B} = \mu_0(1 + \chi_m)\mathbf{H} = (1 + \chi_m)\mathbf{B}_{ext} = (1 + \chi_m)\mathbf{B}_0 ,$$
    so the change in the magnetic field for silver is
    $$\Delta \mathbf{B} = \mathbf{B} - \mathbf{B}_0 = \chi_m\mathbf{B}_0 = \boxed{-(2.4 \times 10^{-5})\mathbf{B}_0.}$$
    (b) For cupric oxide, we have
    $$\Delta \mathbf{B} = \mathbf{B} - \mathbf{B}_0 = \chi_m\mathbf{B}_0 = \boxed{+(2.6 \times 10^{-4})\mathbf{B}_0.}$$

9. We find the magnetic field from
    $$\mathbf{B} = \mu_0(1 + \chi_m)\mathbf{H} = \mu\mathbf{H};$$
    $$B = \mu nI;$$
    $$1.6 \text{ T} = (1320)(4\pi \times 10^{-7}\text{ T}\cdot\text{m/A})(15 \times 10^2\text{ turns/m})I, \text{ which gives } \boxed{I = 0.64 \text{ A.}}$$

13. The magnetic moment of a circulating charge is
    $$m = \tfrac{1}{2}evr_0.$$
    For this to be equal to the Bohr magneton, we have
    $$m_B = \tfrac{1}{2}evr_0;$$
    $$9.27 \times 10^{-24}\text{ A}\cdot\text{m}^2 = \tfrac{1}{2}(1.6 \times 10^{-19}\text{ C})v(2.8 \times 10^{-15}\text{ m}), \text{ which gives } v = 4.1 \times 10^{10}\text{ m/s},$$
    which is greater than the speed of light by a factor of 100.

19. The magnetic intensity inside the solenoid is
    $$H = nI = (400\text{ turns/m})(0.5\text{ A}) = 200\text{ A/m}.$$
    (a) The magnetic field inside the iron bar is
    $$B = \mu H = (640)\mu_0 H$$
    $$= (640)(4\pi \times 10^{-7}\text{ T}\cdot\text{m/A})(200\text{ A/m}) = \boxed{0.16\text{ T.}}$$
    (b) The magnetic field inside the solenoid but outside the iron bar is
    $$B_0 = \mu_0 H = (4\pi \times 10^{-7}\text{ T}\cdot\text{m/A})(1.0 \times 10^3\text{ A/m}) = \boxed{2.5 \times 10^{-4}\text{ T.}}$$

21. If we treat the coil as a tightly wound coil, we have
    $$H = NI/2R$$
    $$= (50\text{ turns})(0.16\text{ A})/2(2.75 \times 10^{-2}\text{ m}) = \boxed{1.45 \times 10^2\text{ A/m.}}$$
    The magnetic field in the iron is
    $$B = \mu_0(1 + \chi_m)H$$
    $$= (4\pi \times 10^{-7}\text{ T}\cdot\text{m/A})[1 + (4.8 \times 10^3)](1.45 \times 10^2\text{ A/m}) = \boxed{0.88\text{ T.}}$$

23. The maximum magnetization occurs when all the dipoles are aligned:
    $$M_{sat} = (N/V)(2.2\mu_B)$$
    $$= [(7.87\text{ g/cm}^3)/(55.8\text{ g/mol})](6.02 \times 10^{23}\text{ atoms/mol})(2.2)(9.27 \times 10^{-24}\text{ A}\cdot\text{m}^2)(10^6\text{ cm}^3/\text{m}^3)$$
    $$= 1.73 \times 10^6\text{ A/m}.$$
    We estimate the maximum value of H by assuming that the susceptibility is constant up to this maximum magnetization:
    $$M_{sat} = \chi_m H_{max};$$
    $$1.73 \times 10^6\text{ A/m} = (6000)H_{max}, \text{ which gives } \boxed{H_{max} = 289\text{ A/m.}}$$

27. We take $+ z$-axis up and counterclockwise positive, so the electron's velocity is negative.
    From symmetry, we know that the electric field is circular.
    With the positive direction counterclockwise, we apply Faraday's law to the electron path:
    $\int \mathbf{E} \cdot d\mathbf{s} = E \int ds = - d\Phi_B/dt;$
    $E2\pi R = - \pi R^2\, dB/dt$, which gives $E = -\tfrac{1}{2}R\, dB/dt$.
    We have $E < 0$ (clockwise).
    From Newton's second law, we have
    $F_{\text{tangential}} = m_e a;$
    $- eE = - e(-\tfrac{1}{2}R\, dB/dt) = m_e\, dv/dt$, which gives $dv/dt = (eR/2m_e)\, dB/dt.$
    Because $v_i < 0$, with $dv/dt > 0$; the electron slows down.
    We integrate to find the final speed, with B_f the magnitude of the final magnetic field:

    $$\int_{v_i}^{v_f} dv = \int_0^{B_f} \frac{eR}{2m_e}\, dB;$$

    $$v_f - v_i = \frac{eR}{2m_e}\left(B_f - 0\right), \text{ which gives}$$

    $$v_f = v_i + (eR/2m_e)B_f.$$
    Because $v_i < 0$, the electron's speed decreases.
    The final magnetic field is $\mathbf{B}_f = + B_f\mathbf{k}$, and the orbital angular momentum of the electron is $\mathbf{L} = m_e vR\mathbf{k}$, so the change is
    $\Delta \mathbf{L} = \Delta(m_e vR\mathbf{k})$
    $\qquad = m_e(v_f - v_i)R\mathbf{k} = m_e(+ eR/2m_e B_f)R\mathbf{k} = + \tfrac{1}{2}eR^2\mathbf{B}_f.$
    The change in the orbital magnetic moment is
    $\Delta \mathbf{m} = - (e/2m_e)\,\Delta\mathbf{L} = - (e/2m_e)(\tfrac{1}{2}eR^2\mathbf{B}_f) = - e^2R^2\mathbf{B}_f/4m_e.$

31. The magnetic field produced by the free current is
    $B_{\text{ext}} = \mu_0 I/2\pi r$ circular.
    The magnetic intensity is
    $H = B_{\text{ext}}/\mu_0 = I/2\pi r$ circular
    $\qquad = (10 \times 10^{-3}\ \text{A})/2\pi r = \boxed{(1.6 \times 10^{-3}\ \text{A})/r \text{ circular.}}$
    The magnetic field in the material is
    $B = \mu H = \mu_0(1 + \chi_m)H$
    $\qquad = (4\pi \times 10^{-7}\ \text{T} \cdot \text{m/A})[1 + (2.6 \times 10^{-4})][(1.6 \times 10^{-3}\ \text{A})/r] = \boxed{(2.0 \times 10^{-9}\ \text{T} \cdot \text{m})/r \text{ circular.}}$
    The change in temperature will change the susceptibility, which we find from Curie's law:
    $\chi_m T = \mu_0 C = \chi_m' T';$
    $(2.6 \times 10^{-4})(300\ \text{K}) = \chi_m'(86\ \text{K})$, which gives $\chi_m' = 9.1 \times 10^{-4}.$
    The change in the magnetic field is
    $\Delta B = \mu_0 \Delta\chi_m H$
    $\qquad = (4\pi \times 10^{-7}\ \text{T} \cdot \text{m/A})[(9.1 \times 10^{-4}) - (2.6 \times 10^{-4})][(1.6 \times 10^{-3}\ \text{A})/r\ \text{A/m}] = \boxed{(1.3 \times 10^{-12}\ \text{T} \cdot \text{m})/r.}$

33. We use the ideal gas law to find the density of protons:
    $n = N_A/V = N_A p/RT$
    $\qquad = (6.02 \times 10^{23}\ \text{protons/mol})(1.01 \times 10^5\ \text{Pa})/(8.314\ \text{J/mol} \cdot \text{K})(273\ \text{K}) = 2.68 \times 10^{25}\ \text{protons/m}^3.$
    The magnetic moment of the proton is
    $m_p = g_S S = 5.58(e/2M_p)(\tfrac{1}{2}\hbar)$
    $\qquad = 5.58[1.60 \times 10^{-19}\ \text{C}/2(1.67 \times 10^{-27}\ \text{kg})]\tfrac{1}{2}(1.05 \times 10^{-34}\ \text{J} \cdot \text{s}) = 1.40 \times 10^{-26}\ \text{A} \cdot \text{m}^2.$
    When all protons are aligned, we have
    $M = n m_p = (2.68 \times 10^{25}\ \text{protons/m}^3)(1.40 \times 10^{-26}\ \text{A} \cdot \text{m}^2) = \boxed{0.38\ \text{A/m.}}$
    We find the magnetic field inside the gas from
    $B = \mu_0 M = (4\pi \times 10^{-7}\ \text{T} \cdot \text{m/A})(0.38\ \text{A/m}) = \boxed{4.7 \times 10^{-7}\ \text{T.}}$

35. When we use $\mathbf{B} = B\mathbf{k}$, we have
$$d\mathbf{m}/dt = g_S\mathbf{m} \times \mathbf{B} = g_S(m_x\mathbf{i} + m_y\mathbf{j} + m_z\mathbf{k}) \times (B\mathbf{k});$$
$$(dm_x/dt)\mathbf{i} + (dm_y/dt)\mathbf{j} + (dm_z/dt)\mathbf{k} = g_S(m_yB\mathbf{i} - m_xB\mathbf{j}).$$

   (a) From the **k**-terms, we have
$$dm_z/dt = 0, \text{ so } m_z \text{ is a constant.}$$

   (b) From the **i**- and **j**-terms, we have
$$dm_x/dt = g_Sm_yB \quad \text{and} \quad dm_y/dt = -g_Sm_xB.$$
If we differentiate $m^2 = \mathbf{m} \cdot \mathbf{m}$, we get
$$d(m_x^2 + m_y^2 + m_z^2)/dt = 2[(m_x\,dm_x/dt) + (m_y\,dm_y/dt) + (m_z\,dm_z/dt)].$$
$$= 2g_S[m_x(g_Sm_yB) + m_y(-g_Sm_xB) + 0] = 0,$$
so $m_x^2 + m_y^2 + m_z^2$ is a constant.
This is consistent with the original equation, where the change in **m** is perpendicular to **m**, so the magnitude of **m** is not changed.

   (c) We differentiate the proposed solution, $m_x = m_1\cos(\omega t)$ and $m_y = -m_1\sin(\omega t)$:
$$dm_x/dt = -m_1\omega\sin(\omega t) = +\omega m_y;$$
$$dm_y/dt = -m_1\omega\cos(\omega t) = -\omega m_x.$$
We see that these satisfy the **i**- and **j**-equations if $\omega = g_SB$.

39. Because the gap of width d is perpendicular to the magnetic field, we know from the result of Problem 38 that the magnetic field is constant around the torus. For the magnetic intensities, we have
$$H_{core} = B/\mu, \quad \text{and} \quad H_{gap} = B/\mu_0.$$
We apply Ampere's law to a path which is a circle with radius R, the mean radius of the torus. Every turn of the coil passes through the area enclosed by this path, so we have
$$\oint \mathbf{H} \cdot d\mathbf{s} = I_{enclosed};$$
$$\int_{core} \mathbf{H} \cdot d\mathbf{s} + \int_{gap} \mathbf{H} \cdot d\mathbf{s} = (B/\mu)(2\pi R - d) + (B/\mu_0)d = NI,$$
which becomes
$$\boxed{B = NI\mu_0/\{[(2\pi R - d)/(\mu/\mu_0)] + d\}.}$$
For a 1 mm gap, we have
$$B = (500)(8\text{ A})(4\pi \times 10^{-7}\text{ T} \cdot \text{m/A})/(\{[2\pi(0.30\text{ m}) - (0.001\text{ m})]/(1200)\} + (0.001\text{ m})) = \boxed{1.96\text{ T.}}$$
For a 3 cm gap, we have
$$B = (500)(8\text{ A})(4\pi \times 10^{-7}\text{ T} \cdot \text{m/A})/(\{[2\pi(0.30\text{ m}) - (0.03\text{ m})]/(1200)\} + (0.03\text{ m})) = \boxed{0.16\text{ T.}}$$
The magnetic field with no gap is 3.2 T, so we see that a small gap has a significant effect.

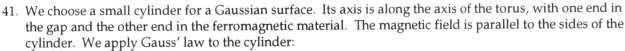

41. We choose a small cylinder for a Gaussian surface. Its axis is along the axis of the torus, with one end in the gap and the other end in the ferromagnetic material. The magnetic field is parallel to the sides of the cylinder. We apply Gauss' law to the cylinder:
$$\oint d\Phi_B = \oint \mathbf{B} \cdot d\mathbf{A} = 0;$$
$$\iint_{gap} \mathbf{B} \cdot d\mathbf{A} + \iint_{ferromagnetic} \mathbf{B} \cdot d\mathbf{A} = 0;$$
$$+ B_{gap}A - B_{ferromagnetic}A = 0, \text{ which gives } B_{gap} = B_{ferromagnetic}.$$

43. If we approximate the magnetic field as that along the axis of the dipole, we have
$$B = (\mu_0/4\pi)(2m/R^3);$$
$$0.6 \times 10^{-4}\text{ T} = (10^{-7}\text{ T} \cdot \text{m/A})2m/(6.4 \times 10^6\text{ m})^3, \text{ which gives } \boxed{m = 8 \times 10^{22}\text{ A} \cdot \text{m}^2.}$$
The magnetization of the core is
$$M = m/V$$
$$= (8 \times 10^{22}\text{ A} \cdot \text{m}^2)/(4\pi/3)(3.2 \times 10^6\text{ m})^3 = \boxed{6 \times 10^2\text{ A/m.}}$$
If the magnetic moment is due to a circulating current at the radius of the core, we have
$$m = IA;$$
$$8 \times 10^{22}\text{ A} \cdot \text{m}^2 = I\pi(3.2 \times 10^6\text{ m})^2, \text{ which gives } \boxed{I = 2 \times 10^9\text{ A.}}$$

45. For the charge moving in a ring of radius r_0, the effective current is
$$I = e/T = e/(2\pi/\omega) = e\omega/2\pi.$$
For the magnetic moment, we have
$$m_{\text{ring}} = IA = (e\omega/2\pi)\pi r_0^2 = \tfrac{1}{2}e\omega r_0^2.$$
The surface charge density on the disk is $\sigma = e/\pi r_0^2$. We consider the disk to be an infinite number of rings. For a differential element, we choose a ring of radius r and thickness dr, which has charge
$$dq = \sigma 2\pi r\, dr = (e/\pi r_0^2)2\pi r\, dr = (2er/r_0^2)\, dr.$$
The effective current of the element is
$$dI = dq/T = [(2er/r_0^2)\, dr]/(2\pi/\omega) = (e\omega/\pi r_0^2)r\, dr.$$
We find the magnetic moment by integration:
$$m_{\text{disk}} = \int dm = \int_0^{r_0} \left[\left(\frac{e\omega}{\pi r_0^2}\right)r\, dr\right]\pi r^2$$
$$= \left(\frac{e\omega}{r_0^2}\right)\int_0^{r_0}\left(r^3\, dr\right) = \left(\frac{e\omega}{r_0^2}\right)\left(\frac{r_0^4}{4}\right) = \frac{e\omega r_0^2}{4}.$$
For the ratio of the magnetic moments of the two models, we have
$$\boxed{m_{\text{ring}}/m_{\text{disk}} = (\tfrac{1}{2}e\omega r_0^2)/(e\omega r_0^2/4) = 2.}$$

47. (a) We find C from the normalization condition:
$$N = \int dN = C\int_0^{\pi} 2\pi \sin\theta\, d\theta\, e^{(mB\cos\theta)/kT}.$$
We change variable to $z = (mB\cos\theta)/kT$, with $dz = -(mB/kT)\sin\theta\, d\theta$:
$$N = C\int_{mB/kT}^{-mB/kT} -\frac{2\pi kT}{mB}e^z\, dz = -C\frac{2\pi kT}{mB}(e^z)\Big|_{mB/kT}^{-mB/kT}$$
$$= -C\frac{2\pi kT}{mB}\left(e^{-mB/kT} - e^{mB/kT}\right), \text{ which gives}$$
$$\boxed{C = \left(\frac{NmB}{2\pi kT}\right)\Big/\left(e^{mB/kT} - e^{-mB/kT}\right).}$$

(b) The integrand used in part (a) represents dN, the number of systems at an angle θ. We find $\langle\cos\theta\rangle$ by putting $\cos\theta$ into the integral, with the same change of variable:
$$\langle\cos\theta\rangle = \frac{\int \cos\theta\, dN}{\int dN} = \frac{C}{N}\int_0^{\pi}(\cos\theta)2\pi \sin\theta\, d\theta\, e^{(mB\cos\theta)/kT}$$
$$= \frac{C}{N}\int_{mB/kT}^{-mB/kT} -2\pi\left(\frac{kT}{mB}\right)^2 z e^z\, dz = -\frac{C2\pi}{N}\left(\frac{kT}{mB}\right)^2\left[e^z(z-1)\right]\Big|_{mB/kT}^{-mB/kT}$$
$$= -\frac{C2\pi}{N}\left(\frac{kT}{mB}\right)^2\left[e^{-mB/kT}\left(-\frac{mB}{kT}-1\right) - e^{mB/kT}\left(\frac{mB}{kT}-1\right)\right].$$
When we use the result for C/N from part (a), we have
$$\boxed{\langle\cos\theta\rangle = \frac{e^{mB/kT} + e^{-mB/kT}}{e^{mB/kT} - e^{-mB/kT}} - \frac{kT}{mB}.}$$

(c)

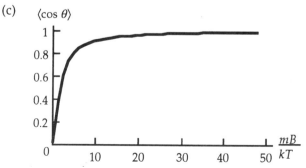

This curve shows the saturation that occurs at large magnetic field. When the value of $\langle\cos\theta\rangle \to 1$, $\theta \to 0$ for all dipoles; the dipoles are aligned.

CHAPTER 33

3. We consider the flux created in the loop from a current in the solenoid. Because there is a magnetic field only inside the solenoid, the magnetic flux through the single loop is
    $$\Phi_{B,\text{loop}} = B_{\text{solenoid}}A_{\text{solenoid}} = \mu_0 n I_{\text{solenoid}}A_{\text{solenoid}}.$$
    We find the mutual inductance from
    $$\Phi_{B,\text{loop}} = M I_{\text{solenoid}}, \text{ which gives}$$
    $$M = \mu_0 n A_{\text{solenoid}}$$
    $$= (4\pi \times 10^{-7} \text{ T} \cdot \text{m/A})[(600 \text{ turns})/(0.25 \text{ m})]\pi(1.8 \times 10^{-2} \text{ cm})^2 = 3.1 \times 10^{-6} \text{ H} = \boxed{3.1\ \mu\text{H.}}$$

7. (a) We find the mutual inductance from
    $$\Phi_B(1) = \Phi_{B1} + M_{12}I_2;$$
    $$0.012 \text{ T} \cdot \text{m}^2 = 0.010 \text{ T} \cdot \text{m}^2 + M_{12}(2 \text{ A}), \text{ which gives} \quad \boxed{M_{12} = 1.0 \text{ mH.}}$$
    (b) Because $M_{21} = M_{12}$, the flux through the second circuit is
    $$\Phi_B(2) = L_2 I_2 + M_{21}I_1$$
    $$= (1 \times 10^{-3} \text{ H})(2 \text{ A}) + (1.0 \text{ mH})(1 \text{ A}) = \boxed{3 \times 10^{-3} \text{ Wb.}}$$

9. (a) When only one coil is used, we use the expression for the inductance of a solenoid,
    $$L_1 = \mu_0 A \ell n_1^2 = \boxed{\mu_0 A N_1^2 / \ell;} \quad \text{or}$$
    $$L_2 = \mu_0 A \ell n_2^2 = \boxed{\mu_0 A N_2^2 / \ell,} \quad \text{depending on which one has a closed circuit.}$$
    (b) If the series windings are in the same direction, the magnetic fields of the two coils are in the same direction, so the total magnetic field is
    $$B_{\text{total}b} = \mu_0(n_1 + n_2)I = \mu_0(N_1 + N_2)I/\ell.$$
    Because this field passes through all the turns of both coils, we have
    $$\Phi_{Bb} = B_{\text{total}b}A(N_1 + N_2) = \mu_0 A(N_1 + N_2)^2 I/\ell, \text{ so the self inductance is}$$
    $$L_b = \Phi_{Bb}/I = \boxed{\mu_0 A(N_1 + N_2)^2/\ell.}$$
    (c) If the series windings are in opposite directions, the magnetic fields of the two coils are in opposite directions, so the total magnetic field is
    $$B_{\text{total}c} = \mu_0(n_1 - n_2)I = \mu_0(N_1 - N_2)I/\ell.$$
    Because this field passes through all the turns of both coils and the induced emfs will be in opposite directions, we have
    $$\Phi_{Bc} = B_{\text{total}c}A(N_1 - N_2) = \mu_0 A(N_1 - N_2)^2 I/\ell, \text{ so the self inductance is}$$
    $$L_c = \Phi_{Bc}/I = \boxed{\mu_0 A(N_1 - N_2)^2/\ell.}$$
    (d) The magnetic field from one coil is
    $$B_1 = \mu_0 n_1 I = \mu_0 N_1 I/\ell.$$
    Because this field passes through all the turns of the other coil, we have
    $$\Phi_{B21} = B_1 A N_2 = \mu_0 A N_1 N_2 I/\ell, \text{ so the mutual inductance is}$$
    $$M = \Phi_{B21}/I = \boxed{\mu_0 A N_1 N_2 / \ell.}$$

11. The flux produced by the varying currents through the coils will be through each coil, so we have
    $$d\Phi_{B1}/dt = d\Phi_{B2}/dt = d\Phi_B/dt.$$
    The induced emfs in the coils will be
    $$\mathcal{E}_1 = -N_1 d\Phi_B/dt, \quad \text{and} \quad \mathcal{E}_2 = -N_2 d\Phi_B/dt.$$
    Unless $N_1 = N_2$, we have $\mathcal{E}_1 \neq \mathcal{E}_2$. This difference in emf will create an internal current, limited only by the resistance of the coils, so the coils should not be connected in parallel. This is similar to the situation of two different batteries connected in parallel; one will charge the other.

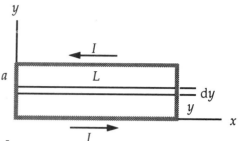

21. Because $a \ll L$, we can ignore the magnetic field from the short sides. The fields from the two long sides are in the same direction, so we can double the field from one when we calculate the flux through the circuit. For the field produced by the bottom wire of radius r, we have

$B = \mu_0 I y / 2\pi r^2, \quad y < r; \quad$ and $\quad B = \mu_0 I / 2\pi y, \quad y \geq r$.

To find the flux through the loop, we choose a strip dy a distance y from the wire:

$$\Phi_B = 2 \iint \mathbf{B} \cdot d\mathbf{A} = 2 \int_0^r \frac{\mu_0 I \ell}{2\pi r^2} y \, dy + 2 \int_r^a \frac{\mu_0 I \ell}{2\pi y} \, dy = \frac{\mu_0 I \ell}{\pi} \left[\tfrac{1}{2} + \ln\left(\tfrac{a}{r}\right) \right].$$

The self inductance is

$$\boxed{L = \frac{\mu_0 \ell}{\pi} \left[\tfrac{1}{2} + \ln\left(\tfrac{a}{r}\right) \right].}$$

If $r \to 0$, $\ln(a/r) \to \infty$. The radius of wire cannot be neglected.

23. We find the mutual inductance by finding the flux through the upper circuit (2) produced by the current in the lower circuit (1). Because $a \ll L$, and $b \ll L$, we can ignore the magnetic field from the short sides of circuit (1).
The magnetic field from a long wire is

$B = \mu_0 I / 2\pi y$.

The fields from the two wires are in opposite directions. To find the flux through circuit (2), we choose a strip dy a distance y from the bottom wire:

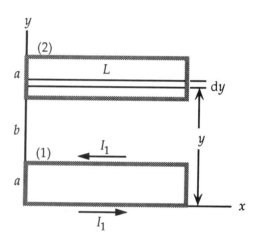

$$\Phi_{B21} = \iint \mathbf{B} \cdot d\mathbf{A} = \int_{a+b}^{2a+b} \left[\frac{\mu_0 I_1 L}{2\pi y} \, dy - \frac{\mu_0 I_1 L}{2\pi(y-a)} \, dy \right]$$

$$= \frac{\mu_0 I_1 L}{2\pi} \left[\ln\left(\frac{2a+b}{a+b}\right) - \ln\left(\frac{a+b}{b}\right) \right] = \frac{\mu_0 I_1 L}{2\pi} \ln\left[\frac{(2a+b)b}{(a+b)^2} \right].$$

Because $(a+b)^2 > (2a+b)b$, this flux is negative, that is, into the paper. For the mutual inductance, we have

$$M = \frac{|\Phi_{B21}|}{I_1} = \boxed{\frac{\mu_0 L}{2\pi} \ln\left[\frac{(a+b)^2}{(2a+b)b} \right].}$$

29. The energy in the inductor is

$U = \tfrac{1}{2} L I^2$.

Because of the decreasing current, the power expended is

$P = -dU/dt = -LI \, dI/dt = -L(I_0 e^{-\alpha t}) I_0(-\alpha) e^{-\alpha t} = L I_0^2 \alpha e^{-2\alpha t}$

$= (2 \text{ mH})(4.0 \text{ A})^2(0.02 \text{ s}^{-1}) e^{-2(0.02 \text{ /s})t} = \boxed{0.64 \, e^{-(0.04/\text{s})t} \text{ mW}.}$

33. (a) The energy stored in the solenoid is

$U = \tfrac{1}{2} L I^2$

$= \tfrac{1}{2}(8 \text{ H})(40 \text{ A})^2 = \boxed{6.4 \times 10^3 \text{ J}.}$

(b) The stored energy becomes thermal energy. We find the volume of helium evaporated from

$Q_v = L_v V;$

$6.4 \times 10^3 \text{ J} = (2.7 \times 10^3 \text{ J/L})V$, which gives $\boxed{V = 2.4 \text{ L}.}$

37. The energy density of the magnetic field is

$$u_B = \tfrac{1}{2}B^2/\mu_0 = \tfrac{1}{2}(\mu_0 I/2\pi r)^2/\mu_0 = \mu_0 I^2/8\pi^2 r^2$$
$$= (4\pi \times 10^{-7}\ \text{T}\cdot\text{m/A})(20\ \text{A})^2/8\pi^2 r^2 = \boxed{(6.4 \times 10^{-6}\ \text{J/m})/r^2}.$$

The energy density of the electric field of the capacitor is

$$u_C = \tfrac{1}{2}\varepsilon_0 E^2 = \tfrac{1}{2}\varepsilon_0(V/d)^2 = \tfrac{1}{2}\varepsilon_0(Q/Cd)^2$$
$$= \tfrac{1}{2}(8.85 \times 10^{-12}\ \text{F/m})[(10^{-7}\ \text{C})/(6.3 \times 10^{-9}\ \text{F})(1.5 \times 10^{-3}\ \text{m})]^2 = 5.0 \times 10^{-4}\ \text{J/m}^3.$$

For the energies to be equal, we have

$$(6.4 \times 10^{-6}\ \text{J/m})/r^2 = 5.0 \times 10^{-4}\ \text{J/m}^3,\ \text{which gives } r = 0.11\ \text{m} = \boxed{11\ \text{cm.}}$$

51. The scaling of any length is $a_2 = 10a_1$, and the scaling of an area is $A_2 = 100A_1$.

For the resistance we have

$$R_2 = \rho \ell_2/A_2 = \rho(10\ell_1)/100A_1 = R_1/10.$$

For the inductance we have

$$L_2 = \mu A_2 N^2/\ell_2 = \mu(100A_1)N^2/10\ell_1 = 10L_1.$$

For the capacitance we have

$$C_2 = \varepsilon A_2/d_2 = \varepsilon(100A_1)/10d_1 = 10C_1.$$

(a) For the undamped frequency we have

$$\omega_2 = 1/(L_2 C_2)^{1/2} = 1/(10L_1\,10C_1)^{1/2} = \boxed{\omega_1/10.}$$

(b) For the damping factor we have

$$\alpha_2 = R_2/2L_2 = (R_1/10)/2(10L_1) = \boxed{\alpha_1/100.}$$

(c) For the damped frequency we have

$$\boxed{\omega_2'^2 = \omega_2^2 - \alpha_2^2 = (\omega_1^2/100) - (\alpha_1/100)^2.}$$

There is no direct proportionality. Damping becomes less important for the larger circuit.

55. For critical damping, we have

$$R_c = 2(L/C)^{1/2},\ \text{and}\ \ \alpha = R_c/2L = 1/(LC)^{1/2}.$$

The charge on the capacitor is

$$Q = Q_0 e^{-\alpha t}\cos(\omega' t + \phi) = Q_0 e^{-\alpha t}\cos(\omega' t),\ \text{so } Q = Q_0\ \text{when } t = 0.$$

The current is

$$I = dQ/dt = Q_0 e^{-\alpha t}[-\omega'\sin(\omega' t)] + Q_0(-\alpha)e^{-\alpha t}\cos(\omega' t),\ \text{so } I_0 = -Q_0\alpha\ \text{when } t = 0.$$

For the instantaneous power consumption in the resistor, we have

$$P = I^2 R_c = (-Q_0\alpha e^{-\alpha t})^2 R_c = Q_0^2(1/LC)e^{-2\alpha t}[2(L/C)^{1/2}] = \boxed{2Q_0^2/(LC^3)^{1/2}e^{-2\alpha t}.}$$

57. From the expression for the charge,

$$Q = Q_0 e^{-\alpha t}\cos(\omega' t + \phi),\ \text{we get}$$
$$I = dQ/dt = Q_0 e^{-\alpha t}[-\omega'\sin(\omega' t + \phi)] + Q_0(-\alpha)e^{-\alpha t}\cos(\omega' t + \phi)$$
$$= Q_0 e^{-\alpha t}[-\omega'\sin(\omega' t + \phi) - \alpha\cos(\omega' t + \phi)];$$
$$dI/dt = Q_0 e^{-\alpha t}[-\omega'^2\cos(\omega' t + \phi) - \alpha(-\omega')\sin(\omega' t + \phi)] +$$
$$Q_0(-\alpha)e^{-\alpha t}[-\omega'\sin(\omega' t + \phi) - \alpha\cos(\omega' t + \phi)]$$
$$= Q_0 e^{-\alpha t}[(\alpha^2 - \omega'^2)\cos(\omega' t + \phi) + 2\alpha\omega'\sin(\omega' t + \phi)].$$

For the three terms on the left-hand side of Eq. (33–20), we have

$$-L\,dI/dt = +Q_0 L e^{-\alpha t}[(\omega'^2 - \alpha^2)\cos(\omega' t + \phi) + 2\alpha\omega'\sin(\omega' t + \phi)];$$
$$-IR = Q_0 R e^{-\alpha t}[+\omega'\sin(\omega' t + \phi) + \alpha\cos(\omega' t + \phi)];$$
$$-Q/C = -(Q_0/C)e^{-\alpha t}\cos(\omega' t + \phi).$$

When we add the three terms, we get

$$(-L\,dI/dt) - IR - (Q/C) =$$
$$Q_0 e^{-\alpha t}\{[(\omega'^2 - \alpha^2)L + R\alpha - (1/C)]\cos(\omega' t + \phi) + [(-2\alpha\omega' L + R\omega')\sin(\omega' t + \phi)]\}.$$

We use Eqs. (33–27) and (33–28) in the coefficients of the trigonometric functions:

cosine term: $(\omega'^2 - \alpha^2)L + R\alpha - (1/C) = [(1/LC) - \alpha^2 - \alpha^2]L + R\alpha - (1/C)$
$$= \alpha(-2\alpha L + R) = \alpha(-R + R) = 0;$$

sine term: $-2\alpha\omega' L + R\omega' = \omega'(-2\alpha L + R) = 0.$

Thus the three terms add to zero, which is Eq. (33–20).

59. The angular frequency of the damped circuit is
$$\omega' = (1/LC - R^2/4L^2)^{1/2} = (\omega^2 - \alpha^2)^{1/2} = \omega[1 - (\alpha^2/\omega^2)]^{1/2}.$$
If $\alpha \ll \omega$, we use the approximation $(1 - x)^{1/2} \approx 1 - \tfrac{1}{2}x$:
$$\omega' \approx \omega[1 - \tfrac{1}{2}(\alpha^2/\omega^2)] \approx \omega[1 - \tfrac{1}{2}(R^2/4L^2)(LC)] \approx \omega - R^2(C/L)^{1/2}/8L.$$
The period of the undamped circuit is
$$T = 2\pi/\omega.$$
For the slightly damped case, we have
$$T' = 2\pi/\omega' = 2\pi/(\omega^2 - \alpha^2)^{1/2} = (2\pi/\omega)[1 - (\alpha^2/\omega^2)]^{-1/2}$$
$$\approx T[1 + \tfrac{1}{2}(\alpha^2/\omega^2)] = T[1 + \tfrac{1}{2}(R^2/4L^2)(LC)] = \boxed{T + \tfrac{1}{4}\pi R^2(C^3/L)^{1/2}.}$$

61. (a) For the oscillating charge on the capacitor, we can write
$$Q = Q_0 \cos(\omega t), \text{ where } Q_0 \text{ is the initial charge at } t = 0.$$
The current in the circuit is
$$I = dQ/dt = -Q_0\omega \sin(\omega t).$$
The maximum current is
$$I_{max} = Q_0\omega = Q_0/(LC)^{1/2} = (30 \times 10^{-9}\,\text{C})/[(2 \times 10^{-5}\,\text{H})(20 \times 10^{-9}\,\text{F})]^{1/2} = 4.7 \times 10^{-2}\,\text{A} = \boxed{47\,\text{mA.}}$$
(b) The maximum energy stored in the inductor is
$$U_{L\,max} = \tfrac{1}{2}LI_{max}^2 = \tfrac{1}{2}(2 \times 10^{-5}\,\text{H})(4.7 \times 10^{-2}\,\text{A})^2 = \boxed{2.2 \times 10^{-8}\,\text{J.}}$$
(c) The ratio of the two maximum energies is
$$U_{L\,max}/U_{C\,max} = \tfrac{1}{2}LI_{max}^2/(\tfrac{1}{2}Q_0^2/C) = L(Q_0\omega)^2/(Q_0^2/C) = LC\omega^2 = \boxed{1.}$$
Note that this must be so from the conservation of energy, which oscillates between the capacitor and the inductor.

63. (a) For the angular frequency of an RLC circuit, we have
$$\omega'^2 = \omega^2 - \alpha^2, \text{ where } \alpha = R/2L.$$
If the resistance is very small, we have $\omega' \approx \omega$.
The charge on the capacitor is
$$Q = Q_0 e^{-\alpha t}\cos(\omega' t) \approx Q_0 e^{-\alpha t}\cos(\omega t).$$
At $t = 0$, $Q = Q_0$, and at the end of each oscillation, $\cos(\omega t) = 1$. So after 100 oscillations, we have
$$5\,\mu\text{C} = (30\,\mu\text{C})e^{-\alpha t}, \text{ with } t = 100(2\pi/\omega), \text{ which gives}$$
$$\alpha t = 1.79 = \alpha(100)2\pi/\omega = \alpha 200\pi(LC)^{1/2};$$
$$\alpha 200\pi[(1.5 \times 10^{-3}\,\text{H})(3 \times 10^{-3}\,\text{F})]^{1/2} = 1.79, \text{ which gives } \alpha = 1.34\,\text{s}^{-1}.$$
We find the resistance from
$$\alpha = R/2L;$$
$$1.34\,\text{s}^{-1} = R/2(1.5 \times 10^{-3}\,\text{H}), \text{ which gives } \boxed{R = 4.0 \times 10^{-3}\,\Omega.}$$
(b) We find the energies of the circuit from the maximum energies of the capacitor:
$$U_0 = \tfrac{1}{2}Q_0^2/C = \tfrac{1}{2}(30 \times 10^{-6}\,\text{C})^2/(3 \times 10^{-3}\,\text{F}) = \boxed{1.5 \times 10^{-7}\,\text{J at } t = 0;}$$
$$U = \tfrac{1}{2}Q^2/C = \tfrac{1}{2}(5 \times 10^{-6}\,\text{C})^2/(3 \times 10^{-3}\,\text{F}) = \boxed{4.2 \times 10^{-9}\,\text{J at } t = 100 \text{ oscillations.}}$$
(c) The two energies are not the same because energy has been lost from Joule heating as the current flows through the resistance.

67. (a) When the switch has been closed for a long time, the currents will be constant. Because there is no change in current through the inductor, there will be no emf in the inductor and thus no potential difference across it. Because the resistor R_2 is in parallel with the inductor, the potential difference across it must also be zero; the current through the 24-Ω resistor is

$$\boxed{I_2 = 0.}$$

For the outside loop, we have

$$I_1 = I_\varepsilon = \varepsilon/R_1 = (12\ \text{V})/(5 \times 10^3\ \Omega) = 2.4 \times 10^{-3}\ \text{A, so}$$
$$\boxed{I_1 = I_\varepsilon = I_L = 2.4\ \text{mA.}}$$

(b) When the switch is opened, there is no current through the battery and R_1, so we have

$$I_L = I_0 e^{-R_2 t/L};$$
$$\tfrac{1}{2}I_0 = I_0 e^{-(24\ \Omega)(8 \times 10^{-6}\ \text{s})/L},\ \text{which gives}$$
$$L = 2.8 \times 10^{-4}\ \text{H} = \boxed{0.28\ \text{mH.}}$$

(c) Because the switch is open, there will be no current through R_1 and the battery:

$$\boxed{I_1 = I_\varepsilon = 0.}$$

For the inductor, we have

$$I_L = I_0 e^{-R_2 t/L} = \left(2.4 \times 10^{-3}\ \text{A}\right) e^{-(24\ \Omega)(12 \times 10^{-6}\ \text{s})/(2.8 \times 10^{-4}\ \text{H})},\ \text{which gives}$$
$$\boxed{I_L = I_2 = 8.6 \times 10^{-4}\ \text{A} = 0.86\ \text{mA.}}$$

73.

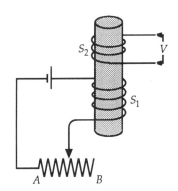
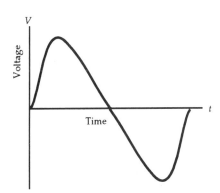

While the slider moves from A to B, the current in solenoid S_1 decreases, causing a decrease in the magnetic flux in the core. This decrease in flux generates an induced emf in solenoid S_2, which opposes the decrease and depends on the mutual inductance: $V = -M\,dI/dt$.

The time dependence of dI/dt is determined by the specific motion of the slider and the self-inductance of solenoid S_1. We assume a smooth motion, with starting and stopping regions, that will give a maximum rate of change of the current a short time after starting. Note that, as the resistance increases, the rate of the fractional change in the resistance will decrease. When the motion of the slider turns around to return to A, the current and the flux will go through a minimum; the induced emf will be zero. Then the flux will start to increase, and the sign of the voltage will change. Assuming the same type of motion, the voltage will be the reverse of the first stage.

75. We assume that the magnetic field due to the primary winding of the torus is constant:

$$B_1 = \mu n_1 I_1,$$

so the flux from the primary winding through the N_2 turns of the secondary winding is

$$\Phi_{B21} = N_2 B_1 A = N_2 \mu n_1 I_1 A.$$

The mutual inductance is

$$M = \Phi_{B21}/I_1 = N_2 \mu n_1 I_1 A/I_1 = \mu N_2 N_1 A/\ell_1$$
$$= 2500(4\pi \times 10^{-7}\ \text{T} \cdot \text{m/A})(40\ \text{turns})(220\ \text{turns})(4 \times 10^{-4}\ \text{m}^2)/(0.35\ \text{m}) = \boxed{3.2 \times 10^{-2}\ \text{H.}}$$

The iron core increases B and thus also the linked flux and M, and concentrates flux in the core so that a difference in cross-sectional area is not important.

CHAPTER 34

5. When all turns of the secondary coil are used, the voltages of both coils are the same, so there must be 1200 turns on the primary coil. We find the number of turns of the secondary to be used for 45 V from

$\mathcal{E}_2/\mathcal{E}_1 = N_2/N_1$;

$(45\text{V})/(120 \text{ V}) = N_2/(1200 \text{ turns})$, which gives $\boxed{N_2 = 450 \text{ turns.}}$

We find the current in the secondary coil from

$\mathcal{E}_2/\mathcal{E}_1 = I_1/I_2$;

$(45\text{V})/(120 \text{ V}) = (10 \text{ A})/I_2$, which gives $\boxed{I_2 = 27 \text{ A.}}$

7. We find the voltage of the secondary coil from

$\mathcal{E}_2/\mathcal{E}_1 = N_2/N_1$;

$\mathcal{E}_2/(220 \text{ V}) = (40 \text{ turns})/(1200 \text{ turns})$, which gives $\mathcal{E}_2 = 7.33 \text{ V}$.

We find the current in the secondary coil from

$P_2 = I_2^2 R = I_2 \mathcal{E}_2$;

$88 \text{ W} = I_2(7.33 \text{ V})$, which gives $I_2 = 12 \text{ A}$.

We find the current in the primary coil from

$N_2/N_1 = I_1/I_2$;

$(40 \text{ turns})/(1200 \text{ turns}) = I_1/(12 \text{ A})$, which gives $\boxed{I_1 = 0.40 \text{ A.}}$

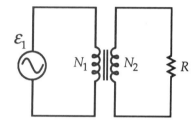

9. If we call the secondary windings N_2 and N_3, we have

$\mathcal{E}_2/\mathcal{E}_1 = N_2/N_1$ and $\mathcal{E}_3/\mathcal{E}_1 = N_3/N_1$, which combines to

$\mathcal{E}_3/\mathcal{E}_2 = N_3/N_2$;

$(11 \text{ V})/(220 \text{ V}) = N_3/(1000 \text{ turns})$, which gives $\boxed{N_3 = 50 \text{ turns.}}$

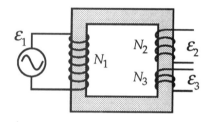

13. If we write the current as $I = I_0 \sin(\omega t)$, the emf in the solenoid is

$\mathcal{E} = -L \, dI/dt = -LI_0\omega \cos(\omega t) = -\mathcal{E}_0 \cos(\omega t)$.

We find the angular frequency from

$\mathcal{E}_0 = LI_0\omega = I_0 X_L$;

$330 \text{ V} = (2 \text{ A})(15 \times 10^{-3} \text{ H})\omega$, which gives $\boxed{\omega = 1.1 \times 10^4 \text{ rad/s.}}$

21. (a) For phasor C we have $C \sin(\omega t + \phi)$.
    We can write
    $D \cos(\omega t + \phi) = D \sin[\omega t + \phi + (\pi/2)]$,
    so $\boxed{\text{D is more advanced in phase.}}$
    (b) The phase difference is $\pi/2$, or
    $\boxed{90°. \text{ D leads.}}$
    (c) We can write
    $f(t) = A \cos(\omega t) + B \sin(\omega t)$
    $= f \sin \delta \cos(\omega t) + f \cos \delta \sin(\omega t)$
    $= f \sin(\omega t + \delta)$,
    where $f = (A^2 + B^2)^{1/2}$, and $\tan \delta = A/B$.
    Whether **f** leads or lags **C** depends on whether $\delta > \phi$, or $\delta < \phi$.

(a)

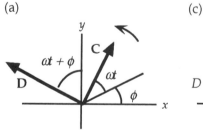

(c)

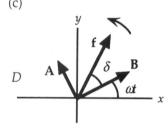

29. $X_C = 1/\omega C = 1/2\pi(1200.0 \text{ Hz})(2 \times 10^{-6} \text{ F}) = \boxed{66.3 \ \Omega.}$

$X_L = \omega L = 2\pi(1200.0 \text{Hz})(92 \times 10^{-3} \text{ H}) = \boxed{694 \ \Omega.}$

$Z = [(X_L - X_C)^2 + R^2]^{1/2} = \{[(694 \ \Omega) - (66.3 \ \Omega)]^2 + (500 \ \Omega)^2\}^{1/2} = \boxed{802 \ \Omega.}$

$Q_{max} = V_0/\omega Z = (80 \text{ V})/2\pi(1200.0\text{Hz})(802 \ \Omega) = 1.3 \times 10^{-5} \text{ C} = \boxed{13 \ \mu\text{C}.}$

$\tan \phi = (X_L - X_C)/R = [(694 \ \Omega) - (66.3 \ \Omega)]/(500 \ \Omega) = +1.26$, which gives $\boxed{\phi = +51.5°.}$

$I_{max} = V_0/Z = (80 \text{ V})/(802 \ \Omega) = \boxed{0.10 \text{ A}.}$

31. Because the emf is turned on at $t = 0$, we have

$V = V_0 \sin(\omega t)$.

All elements have zero voltage at $t = 0$ s. For $f = 1200.0$ Hz, 0.10000 seconds is 120 cycles, so any transient currents will have subsided, and the steady-state current will have been reached.

The voltage on the capacitor is

$V_C = Q/C = -(Q_{max}/C) \cos(\omega t - \phi)$

$= -[(13 \times 10^{-6} \text{ C})/(2 \times 10^{-6} \text{ F})] \cos[2400\pi(0.10000 \text{ s})(180°/\pi) - (+51.5°)] = \boxed{-4.05 \text{ V}.}$

The current in the circuit is

$I = (V_0/Z) \sin(\omega t - \phi)$,

so the voltage across the inductor is

$V_L = L \, dI/dt = +(LV_0/Z)\omega \cos(\omega t - \phi) = +(X_L V_0/Z) \cos(\omega t - \phi)$

$= [(694 \ \Omega)(80 \text{ V})/(802 \ \Omega)] \cos[2400\pi(0.10000 \text{ s})(180°/\pi) - (+51.5°)] = \boxed{+43.1 \text{ V}.}$

Note that the voltage across the resistor is

$V_R = IR = (V_0 R/Z) \sin(\omega t - \phi)$

$= [(80 \text{ V})(500 \ \Omega)/(802 \ \Omega)] \sin[2400\pi(0.10000 \text{ s})(180°/\pi) - (+51.5°)] = -39.0 \text{ V}$,

so the sum of the voltages is 0, which is V at $t = 0.10000$ s.

33. We find the reactances and the impedance:

$X_C = 1/\omega C = 1/2\pi(60 \text{ Hz})(2 \times 10^{-6} \text{ F}) = 1.33 \times 10^3 \ \Omega$;

$X_L = \omega L = 2\pi(60 \text{ Hz})(0.8 \text{ H}) = 302 \ \Omega$;

$Z = [(X_L - X_C)^2 + R^2]^{1/2}$

$= \{[(302 \ \Omega) - (1.33 \times 10^3 \ \Omega)]^2 + (600 \ \Omega)^2\}^{1/2} = 1.19 \times 10^3 \ \Omega.$

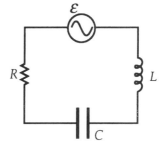

We find the maximum current from

$I_{max} = V_0/Z = (110 \text{ V})/(1.19 \times 10^3 \ \Omega) = 0.092 \text{ A} = \boxed{92 \text{ mA}.}$

The maximum potential drops are

$V_{R\max} = I_{max}R = (0.092 \text{ A})(600 \ \Omega) = \boxed{55.5 \text{ V}.}$

$V_{C\max} = I_{max}X_C = (0.092 \text{ A})(1.33 \times 10^3 \ \Omega) = \boxed{123 \text{ V}.}$

$V_{L\max} = I_{max}X_L = (0.092 \text{ A})(302 \ \Omega) = \boxed{27.9 \text{ V}.}$

Note that these do not add to 110 V. The maxima occur at different times.

39. We find the inductance from the resonant frequency:

$\omega_0^2 = 1/LC$;

$[2\pi(60 \text{ Hz})]^2 = 1/L(16 \times 10^{-6} \text{ F})$, which gives $L = 0.44$ H.

At resonance the reactances are equal:

$X_C = X_L = \omega L = 2\pi(60 \text{ Hz})(0.44 \text{ H}) = 166 \ \Omega$, and $Z = R = 30 \ \Omega.$

The voltage across the capacitor is

$V_{C\max} = I_{max}X_C = V_0X_C/Z = (12 \text{ V})(166 \ \Omega)/(30 \ \Omega) = \boxed{66 \text{ V},}$

which is also the voltage across the inductance. Because the inductance voltage lags the resistance voltage, $V_{R\max} = I_{max}R$, by $\pi/2$, we find the voltage across the inductor-resistor combination from

$V_{RL\max}^2 = V_{R\max}^2 + V_{L\max}^2$

$= (12 \text{ V})^2 + (66 \text{ V})^2$, which gives $\boxed{V_{RL\max} = 67 \text{ V}.}$

43. With $V(t) = V_0 \sin(\omega t)$ and $I(t) = I_0 \sin(\omega t - \phi)$, we find the phase difference between the two from
$\tan \phi = (X_L - X_C)/R$.

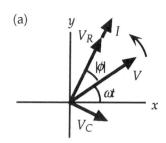

(a)

(a) For a circuit with a resistor and capacitor, we have
$\tan \phi = (X_L - X_C)/R = -X_C/R$.
Thus $\phi < 0$, and the current leads the voltage.
The potential drop across the resistor is
$V_R = RI = RI_0 \sin(\omega t - \phi)$.
The potential drop across the capacitor is
$V_C = Q/C = -(Q_0/C)\cos(\omega t - \phi)$.

(b) For a circuit with a resistor and inductor, we have
$\tan \phi = (X_L - X_C)/R = +X_L/R$.
Thus $\phi > 0$, and the current lags the voltage.
The potential drop across the resistor is
$V_R = RI = RI_0 \sin(\omega t - \phi)$.
The potential drop across the inductor is
$V_L = L\,dI/dt = LI_0 \cos(\omega t - \phi)$.

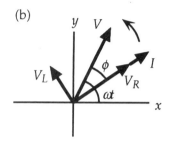

(b)

47. The average power in an RLC circuit is
$\langle P \rangle = \tfrac{1}{2}V_0^2 R/Z = \tfrac{1}{2}V_0^2 R/[(X_L - X_C)^2 + R^2]^{1/2} = \tfrac{1}{2}V_0^2 R/\{[(\omega L) - (1/\omega C)]^2 + R^2\}^{1/2}$.

(a) When $\omega \to \infty$, $Z \to \omega L$, and we have

$$\boxed{\langle P \rangle \to \tfrac{1}{2}V_0^2 R/\omega L \to 0.}$$

The current is a very small because there is a large induced emf in the inductor.

(b) When $\omega \to 0$, $Z \to 1/\omega C$, and we have

$$\boxed{\langle P \rangle \to \tfrac{1}{2}V_0^2 R\omega C \to 0.}$$

The current is a very small because the system approaches a DC circuit, which would have no current through the capacitor.

53. We find the impedance of the coil from
$\langle P \rangle = (V_{rms1}^2/Z_1)\cos\phi$;
$200\,\text{W} = [(110\,\text{V})^2/Z_1](0.6)$, which gives $Z_1 = 36.3\,\Omega$.
The resistance of the coil is
$R = Z_1 \cos\phi = (36.3\,\Omega)(0.6) = 21.8\,\Omega$.
We find the inductive reactance from
$Z_1^2 = X_L^2 + R^2$;
$(36.3\,\Omega)^2 = X_L^2 + (21.8\,\Omega)^2$, which gives $X_L = 29.0\,\Omega$.
When the capacitor is added to the circuit, we find the impedance from
$\langle P \rangle = (V_{rms2}^2/Z_2)\cos\phi_2 = V_{rms2}^2 R/Z_2^2$;
$200\,\text{W} = (220\,\text{V})^2(21.8\,\Omega)/Z_2^2$, which gives $Z_2 = 72.6\,\Omega$.
We find the capacitive reactance from
$Z_2^2 = (X_L - X_C)^2 + R^2$;
$(72.6\,\Omega)^2 = [(29.0\,\Omega) - X_C]^2 + (21.8\,\Omega)^2$, which gives $X_C = 98.2\,\Omega$.
The capacitance is
$C = 1/\omega X_C = 1/2\pi(60\,\text{Hz})(98.2\,\Omega) = 2.7\times10^{-5}\,\text{F} = \boxed{27\,\mu\text{F.}}$
To maintain the same power factor with the same resistance by adding a capacitor, we have
$R/Z_1 = R/Z_3$,   or $Z_1 = Z_3$;
$X_L^2 = (X_L - X_{C3})^2$, which has two solutions:
$X_{C3} = 0$, which is the original circuit,   and $X_{C3} = 2X_L = 2(29.0\,\Omega) = 58.0\,\Omega$.
The capacitance is
$C_3 = 1/\omega X_{C3} = 1/2\pi(60\,\text{Hz})(58.0\,\Omega) = 4.6\times10^{-5}\,\text{F} = \boxed{46\,\mu\text{F.}}$

55. We find the impedance and resistance of the motor from
$\langle P \rangle = (V_0^2/2Z_1) \cos \phi$;
$5 \times 10^3 \text{ W} = [(220 \text{ V})^2/2Z_1](0.8)$, which gives $Z_1 = 3.87 \, \Omega$;
$R_1 = Z_1 \cos \phi = (3.87 \, \Omega)(0.8) = 3.10 \, \Omega$.
The current in the circuit is
$I_0 = V_0/Z_1 = (220 \text{ V})/(3.87 \, \Omega) = 56.8 \text{ A}$.
If we include the transmission line, we find the impedance from
$Z_2^2 = (X_L - X_C)^2 + (R_1 + R_2)^2 = Z_1^2 - R_1^2 + (R_1 + R_2)^2$;
$\quad = (3.87 \, \Omega)^2 - (3.10 \, \Omega)^2 + [(3.10 \, \Omega) + (2.5 \, \Omega)]^2$, which gives $Z_2 = 6.07 \, \Omega$.
The voltage that must be supplied at the input is
$V_{02} = I_0 Z_2 = (56.8 \text{ A})(6.07 \, \Omega) = \boxed{345 \text{ V.}}$
The power that must be supplied at the input is
$\langle P_2 \rangle = V_{02}^2(R_1 + R_2)/2Z_2^2 = (345 \text{ V})^2[(3.10 \, \Omega) + (2.5 \, \Omega)]/2(6.07 \, \Omega)^2 = 9.05 \times 10^3 \text{ W} = \boxed{9.05 \text{ kW.}}$

59. (a) We find the reactances and the impedance:
$X_C = 1/\omega C = 1/2\pi(60 \text{ Hz})(20 \times 10^{-6} \text{ F}) = 133 \, \Omega$;
$X_L = \omega L = 2\pi(60 \text{ Hz})(10 \times 10^{-3} \text{ H}) = 3.77 \, \Omega$;
$Z = [(X_L - X_C)^2 + R^2]^{1/2} = \{[(3.77 \, \Omega) - (133 \, \Omega)]^2 + (50 \, \Omega)^2\}^{1/2} = 139 \, \Omega$.
The power factor is
$\cos \phi = R/Z = (50 \, \Omega)/(139 \, \Omega) = 0.36$.
We find the power absorbed from
$\langle P \rangle = I_{\text{rms}}^2 Z \cos \phi = (V_{\text{rms}}^2/Z) \cos \phi \quad = [(110 \text{ V})^2/(139 \, \Omega)](0.36) = \boxed{31 \text{ W.}}$
(b) If the resistance is halved, we find the new impedance:
$Z = [(X_L - X_C)^2 + R^2]^{1/2} = \{[(3.77 \, \Omega) - (133 \, \Omega)]^2 + (25 \, \Omega)^2\}^{1/2} = 132 \, \Omega$.
The power factor is
$\cos \phi = R/Z = (25 \, \Omega)/(132 \, \Omega) = 0.19$.
We find the power absorbed from
$\langle P \rangle = I_{\text{rms}}^2 Z \cos \phi = (V_{\text{rms}}^2/Z) \cos \phi \quad = [(110 \text{ V})^2/(132 \, \Omega)](0.19) = \boxed{17 \text{ W.}}$
(c) The maximum power drawn is
$P_{\text{max}} = 2\langle P \rangle = 2(17 \text{ W}) = \boxed{34 \text{ W};}$

65. For a high-pass filter, we put the inductor and resistor in series and
use the voltage across the inductor as the output. At low frequency, the
reactance of the inductor will be small and the voltage across the
inductor will be small. At high frequencies, the reactance of the inductor
will be large and the voltage across the resistor will be small.
If we set $X_L = R$, we have
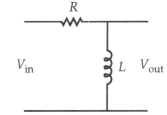
$Z = (X_L^2 + R^2)^{1/2} = X_L \sqrt{2}$, so $V_{\text{out}} = IX_L = V_{\text{in}} X_L /Z = V_{\text{in}} \sqrt{2}$,
which we will use as the condition at our cut-off frequency:
$R = 2\pi f L$, which gives $R/L = 2\pi f = 2\pi(8 \times 10^3 \text{ Hz}) = 5.0 \times 10^4 \text{ s}^{-1}$.
We can satisfy this requirement with many combinations, e. g., $\boxed{R = 2.5 \text{ k}\Omega \text{ and } L = 50 \text{ mH.}}$

69. The impedance of the circuit is
    $Z = (X_C^2 + R^2)^{1/2} = [(1/\omega C)^2 + R^2]^{1/2}$.
    The input voltage is $V_{in} = I_0 Z$, and the output voltage is
    $V_{out} = I_0 X_C$, which gives
    $V_{out}/V_{in} = X_C/Z = (1/\omega C)/[(1/\omega C)^2 + R^2]^{1/2}$.
    At low frequency, $X_C \to \infty$, $Z \to X_C$, so $V_{out} \to V_{in}$.
    At high frequency, $X_C \to 0$, $Z \to R$, so $V_{out} \to 0$.
    Thus at high frequencies, X_C is small, and V_{out} is small.

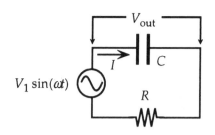

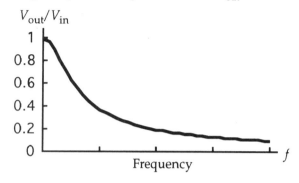

73. The maximum current found in Problem 72 is $I_0 = 1.55$ A.
    (a) For the capacitors, we have
    $V_{C1} = I_0 X_{C1} = I_0/\omega C_1 = (1.55 \text{ A})/2\pi(8000 \text{ Hz})(6 \times 10^{-6} \text{ F}) =$ $\boxed{5.14 \text{ V};}$
    $V_{C2} = I_0 X_{C2} = I_0/\omega C_2 = (1.55 \text{ A})/2\pi(8000 \text{ Hz})(14 \times 10^{-6} \text{ F}) =$ $\boxed{2.20 \text{ V}.}$
    (b) For the inductor, we have
    $V_L = I_0 X_L = (1.55 \text{ A})(1.51 \, \Omega) =$ $\boxed{2.34 \text{ V}.}$
    Note that, because there is a π phase difference between the capacitors and the inductor, we have
    $V_{C1} + V_{C2} - V_L = V_0$.

75. The reactance of the capacitor is
    $X_C = 1/\omega C = 1/2\pi(3000 \text{ Hz})(2 \times 10^{-6} \text{ F}) = 26.5 \, \Omega$.
    The total impedance of the circuit is
    $Z = [X_C^2 + (R_1 + R_2)^2]^{1/2}$
    $= [(26.5 \, \Omega)^2 + (15 \, \Omega + 8 \, \Omega)^2]^{1/2} = 35.1 \, \Omega$.
    The current in the circuit is
    $I_0 = V_0/Z = (3 \text{ V})/(35.3 \, \Omega) = 8.5 \times 10^{-2} \text{ A}$, and $I_{rms} = I_0/\sqrt{2}$.
    The power dissipated in Z_2 is
    $P_2 = I_{rms}^2 R_2 = \tfrac{1}{2}(8.5 \times 10^{-2} \text{ A})^2(8 \, \Omega) = 2.9 \times 10^{-2} \text{ W} =$ $\boxed{29 \text{ mW}.}$

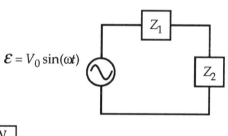

79. We find the inductance from the resonant frequency:
    $\omega_0 = 2\pi f_0 = (1/LC)^{1/2}$;
    $2\pi(18 \times 10^6 \text{ Hz}) = [1/L(33 \times 10^{-12} \text{ F})]^{1/2}$, which gives $\boxed{L = 2.4 \times 10^{-6} \text{ H} = 2.4 \, \mu\text{H}.}$
    We find the resistance from the width of the power vs. frequency curve:
    $\Delta\omega = 2\pi \, \Delta f = R/L$;
    $2\pi(40 \times 10^3 \text{ Hz}) = R/(2.4 \times 10^{-6} \text{ H})$, which gives $\boxed{R = 0.060 \, \Omega.}$

81. The reactance of the capacitor is
    $X_C = 1/\omega C = 1/2\pi(60\ \text{Hz})(15 \times 10^{-6}\ \text{F}) = 177\ \Omega.$
    The impedance of the circuit is
    $Z = (X_C^2 + R^2)^{1/2} = [(177\ \Omega)^2 + R^2]^{1/2},$
    so the rms current is
    $I_{rms} = V_{rms}/Z = (110\ \text{V})/[(177\ \Omega)^2 + R^2]^{1/2}.$
    The power delivered to the circuit is the power
    dissipated in the resistor:
    $P = I_{rms}^2 R = V_{rms}^2 R/Z^2 = V_{rms}^2 R/(X_C^2 + R^2).$
    To find the value of R for which the power is maximum,
    we set $dP/dR = 0$:
    $$dP/dR = [V_{rms}^2/(X_C^2 + R^2)] + [V_{rms}^2 R(-2R)/(X_C^2 + R^2)^2]$$
    $$= V_{rms}^2(X_C^2 + R^2 - 2R^2)/(X_C^2 + R^2)^2 = V_{rms}^2(X_C^2 - R^2)/(X_C^2 + R^2)^2 = 0,$$
    which gives $\boxed{R = X_C = 177\ \Omega.}$

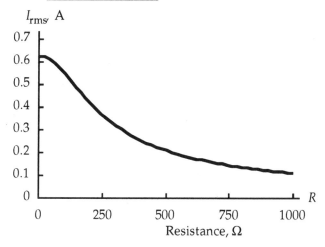

83. (a) For the complex charge, we have
    $$\boxed{Q_c(t) = CV_c(t) = CV_0 e^{i\omega t}.}$$
    For the complex current, we have
    $$\boxed{I_c(t) = dQ_c(t)/dt = i\omega C V_0 e^{i\omega t}.}$$
    (b) For the current, we have
    $$I = \text{Im}[I_c(t)] = \text{Im}\{i\omega C V_0\,[\cos(\omega t) + i\sin(\omega t)]\} = \text{Im}[i\omega C V_0 \cos(\omega t) + i^2\omega C V_0 \sin(\omega t)]$$
    $$= \text{Im}[i\omega C V_0 \cos(\omega t) - \omega C V_0 \sin(\omega t)] = \omega C V_0 \cos(\omega t).$$

CHAPTER 35

5. From the argument of the cosine function, we see that the wave is traveling in $-z$-direction. Because $\mathbf{E}$ and $\mathbf{B}$ are perpendicular to each other and to the direction of propagation, $\mathbf{B}$ can have only an x-component, with magnitude $B_0 = E_0/c$:

$\boxed{\mathbf{B} = (E_0/c)\cos(kz + \omega t)\mathbf{i}, \text{ traveling in the } -z\text{-direction.}}$

The positive $\mathbf{i}$-direction is chosen so the wave travels in the $-z$-direction.

11. We assume that $\mathbf{E}$ is the same along a wavefront perpendicular to the direction of propagation. We choose a pillbox for a Gaussian surface with the plane surfaces perpendicular to the direction of propagation and one of the plane surfaces at a wavefront where $\mathbf{E} = \mathbf{B} = 0$. With no free charge within the pillbox, we have

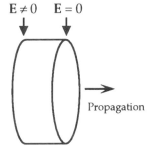

$\oiint \mathbf{E} \cdot d\mathbf{A} = \iint_{planes} \mathbf{E} \cdot d\mathbf{A} + \iint_{side} \mathbf{E} \cdot d\mathbf{A} = 0.$

If $\mathbf{E}$ at some point on the side is not perpendicular to $d\mathbf{A}$, there will be a point on the opposite side where $\mathbf{E}$ is in the same direction but $d\mathbf{A}$ is in the opposite direction. Thus the surface integral over the side is zero, so we have

$\iint \mathbf{E} \cdot d\mathbf{A} = 0$ for the plane surface where $\mathbf{E} \neq 0$.

This requires that $\mathbf{E}$ is perpendicular to the plane surface and thus is perpendicular to the direction of propagation.

15. If we choose the direction of propagation as the x'-axis, the electric field is

$\mathbf{E} = \mathbf{E}_0 \cos(kx' - \omega t + \phi).$

From the figure, we see that

$x' = x \cos\theta + y \sin\theta$, so we have

$\mathbf{E} = \mathbf{E}_0 \cos[k(x\cos\theta + y\sin\theta) - \omega t + \phi]$

$= \mathbf{E}_0 \cos(kx\cos\theta + ky\sin\theta - \omega t + \phi).$

$\mathbf{E}_0$ can have any direction in the plane that is perpendicular to the

direction of propagation, which is $\boxed{\text{the plane formed by the } z\text{-axis and the line } y = -(\tan\theta)x.}$

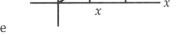

17. For a magnetic field in the y-direction that is maximum at $z = 0$ and $t = 0$ and traveling in the $+z$-direction, we write

$\mathbf{B} = B_0 \cos(kz - \omega t)\mathbf{j}.$

The electric field will be perpendicular to $\mathbf{B}$: $\mathbf{E} = E_x\mathbf{i} = \pm E_0 \cos(kz - \omega t)\mathbf{i}.$

We use Eq. (35–5) to determine if E_x is positive when B_y is positive:

$-\partial B_y/\partial z = \mu_0\varepsilon_0 \partial E_x/\partial t;$

$+ kB_0 \sin(kz - \omega t) = -\mu_0\varepsilon_0(\pm E_0)(-\omega)\sin(kz - \omega t)$, which requires the positive sign.

For the given wavelength we have

$k = 2\pi/\lambda = 2\pi/(600 \times 10^{-9}\,\text{m}) = 1.05 \times 10^7\,\text{m}^{-1}$ and

$\omega = ck = (3 \times 10^8\,\text{m/s})(1.05 \times 10^7\,\text{m}^{-1}) = 3.15 \times 10^{15}\,\text{rad/s}.$

Thus we have

$\mathbf{E} = E_0 \cos(kz - \omega t)\mathbf{i} = cB_0 \cos(kz - \omega t)\mathbf{i}.$

$= (3 \times 10^8\,\text{m/s})(10^{-8}\,\text{T})\cos[(1.05 \times 10^7\,\text{m}^{-1})z - (3.15 \times 10^{15}\,\text{rad/s})t]\mathbf{i}$

$= \boxed{(3\,\text{V/m})\cos[(1.05 \times 10^7\,\text{m}^{-1})z - (3.15 \times 10^{15}\,\text{rad/s})t]\mathbf{i}.}$

25. We assume the beam is perpendicular to the surface. Because the light totally reflects from the surface, the momentum of the beam reverses direction. We find the radiation pressure from

$F/A = 2I/c = 2\langle u \rangle = 2(\tfrac{1}{2}\varepsilon_0 E_0^2) = \varepsilon_0 E_0^2$

$= (8.85 \times 10^{-12}\,\text{C}^2/\text{N}\cdot\text{m}^2)(3\,\text{V/m})^2 = \boxed{8.0 \times 10^{-11}\,\text{N/m}^2.}$

29. We find the intensity from
$$I = \langle S \rangle = (c/\mu_0)\langle B^2 \rangle = cB_{rms}^2/\mu_0$$
$$= (3 \times 10^8 \text{ m/s})(7 \times 10^{-9} \text{ T})^2/(4\pi \times 10^{-7} \text{ T} \cdot \text{m/A}) = \boxed{1.2 \times 10^{-2} \text{ W/m}^2.}$$
The energy transported in one minute is
$$U = IAt = (1.2 \times 10^{-2} \text{ W/m}^2)(0.1 \text{ m}^2)(1 \text{ min})(60 \text{ s/min}) = \boxed{0.070 \text{ J.}}$$

33. We assume the light is perpendicular to the surface. Because the light totally reflects from the surface, the momentum of the beam reverses direction. We find the radiation pressure from
$$F/A = 2\langle u \rangle = 2I/c = 2P/4\pi R^2 c$$
$$= 2(3.8 \times 10^{26} \text{ W})/4\pi(10^{10} \text{ m})^2(3 \times 10^8 \text{ m/s}) = \boxed{2.0 \times 10^{-3} \text{ N/m}^2.}$$

37. The force exerted by the radiation is
$$F_{radiation} = (\text{pressure})A = uA = (I/c)A = P/c.$$
To suspend the paper, the force from the radiation must balance the force of gravity:
$$P/c = mg;$$
$$P/(3 \times 10^8 \text{ m/s}) = (0.2 \times 10^{-3} \text{ kg})(9.8 \text{ m/s}^2), \text{ which gives } \boxed{P = 5.9 \times 10^5 \text{ W.}}$$
At this rate of energy absorption, the paper would burn or vaporize.

39. We consider the beam that falls on an area A of the surface. The cross-sectional area of the beam is $A \cos \theta$. In a time Δt, the volume of the beam that reflects from the surface is $(A \cos \theta) c \Delta t$.
The momentum in this volume is
$$p = (S/c^2)(A \cos \theta) c \Delta t.$$
Upon reflection, only the component of momentum perpendicular to the surface reverses, so the momentum change is
$$\Delta p = 2p \cos \theta.$$
The momentum transferred to the surface per unit area is $\boxed{\Delta p/A = 2(S/c) \cos^2 \theta \; \Delta t.}$

41. We first consider the situation where the rectangle is perpendicular to the beam. Because the pressure is uniform, the force exerted by the beam on the bright side, where the radiation reflects, is
$$F_{bright} = 2uA_{bright} = 2(S/c)H(w/2).$$
The force is uniform over the surface, so the resultant force acts at the center of the bright side, which means a moment arm of $w/4$ about the axis and a torque of
$$\tau_{bright} = + (w/4)(S/c)Hw = + SHw^2/4c.$$
The force exerted by the beam on the dark side, where the radiation is absorbed, is
$$F_{dark} = uA_{dark} = (S/c)H(w/2).$$
The resultant force acts at the center of the dark side, which means a moment arm of $w/4$ and a torque opposite to that on the bright side of
$$\tau_{dark} = - (w/4)(S/c)H(w/2) = - SHw^2/8c.$$
Thus $\boxed{\text{there is a net torque,}}$ which is maximum when the surface is perpendicular to the beam:
$$\tau_{max} = + (SHw^2/4c) - (SHw^2/8c) = SHw^2/8c.$$
When the surface has rotated an angle θ, the torque will decrease for two reasons: The area presented by the rectangle to the beam decreases by a factor of $\cos \theta$, and the moment arm decreases by a factor of $\cos \theta$. The torque becomes
$$\tau = \tau_{max} \cos^2 \theta.$$
Because opposite sides are reversed, as the other side of the rectangle becomes illuminated, the torque will be in the same direction. The average torque over a full rotation is
$$\tau_{average} = \tau_{max} \langle \cos^2 \theta \rangle = \tfrac{1}{2}\tau_{max} = SHw^2/16c.$$
$$= (0.5 \text{ kg/s}^3)(3.0 \times 10^{-2} \text{ m})(1.0 \times 10^{-2} \text{ m})^2/16(3 \times 10^8 \text{ m/s}) = \boxed{3.1 \times 10^{-16} \text{ N} \cdot \text{m.}}$$

45. We place the dipole at the origin of the coordinate system.
    (a) The electric field is parallel to the antenna,

    $+ y$-direction.

    (b) $\mathbf{E} \times \mathbf{B}$ is in the z-direction. The magnetic field is perpendicular to the electric field and the direction of propagation:

    $- x$-direction.

    (c) The Poynting vector is in the direction of propagation:

    $+ z$-direction.

    (d) A half-cycle later, the electric and magnetic fields will have reversed, but the direction of propagation will be the same:

    electric field in $- y$-direction,
    magnetic field in $+ x$-direction,
    Poynting vector in $+ z$-direction.

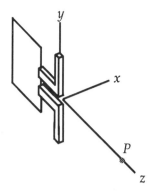

55. (a) The first Polaroid sheet will reduce the intensity of the original beam to

    $I_1 = (1/2)I_0.$

    (b) Through the second sheet, we have
    $I_2 = I_1 \cos^2 \theta.$
    For the intensity to be zero, we have
    $\cos^2 \theta = 0$, which gives $\theta = 90°.$

    (c) When we insert the third sheet, we have
    $I_3 = I_2 \cos^2 (90° - \theta)$, which gives
    $I_3 = [\cos^2 (90° - \theta)](\cos^2 \theta)(1/2)I_0 = (1/2)(\sin^2 \theta \cos^2 \theta)I_0 =$ $(1/8)I_0 \sin^2 (2\theta).$

    (d) The intensity will be zero when
    $\sin^2(2\theta) = 0$, which gives $2\theta = 0°$ or $180°$, and thus $\theta = 0°$ or $90°.$

57. The frequency and energy of a photon at this wavelength are
    $f = c/\lambda = (3 \times 10^8 \text{ m/s})/(630 \times 10^{-9} \text{ m}) = 4.8 \times 10^{14} \text{ Hz};$
    $hf = (6.63 \times 10^{-34} \text{ J} \cdot \text{s})(4.8 \times 10^{14} \text{ Hz}) = 3.2 \times 10^{-19} \text{ J}.$
    The time for a photon to travel through the length of the apparatus is $\Delta t = L/c$. If we want one photon to be present in the apparatus, the energy of one photon will pass through the apparatus in this time. The power of the beam is
    $P = hf/\Delta t = hfc/L$,   and the intensity is
    $I = \langle S \rangle = P/A = hfc/LA$
    $= (3.2 \times 10^{-19} \text{ J})(3 \times 10^8 \text{ m/s})/(2 \text{ m})(1 \times 10^{-6} \text{ m}^2) =$ $4.7 \times 10^{-5} \text{ W/m}^2.$

59. We choose the axis of the solenoid as the z-axis.
    (a) The magnitude of the magnetic field is
    $B = \mu_0 n I = \mu_0 n I_0 \cos(\omega t)$, so we have
    $$\boxed{\mathbf{B} = \mu_0 n I_0 \cos(\omega t)\mathbf{k} \text{ (along the axis).}}$$
    (b) From the cylindrical symmetry, the electric field will be circular. We apply Faraday's law to a circular path with radius $r < R$:
    $$\oint \mathbf{E} \cdot \mathbf{ds} = -(d/dt) \int \mathbf{B} \cdot \mathbf{dA} = -(d/dt) \int B \, dA = -(d/dt)BA = -A(dB/dt);$$
    $E 2\pi r = -\pi r^2 \mu_0 n I_0 (d/dt) \cos(\omega t) = \pi r^2 \mu_0 n I_0 \omega \sin(\omega t)$, which gives
    $$\boxed{\mathbf{E} = \tfrac{1}{2}\mu_0 n I_0 \omega r \sin(\omega t) \text{ (circular).}}$$
    (c) We find the Poynting vector from
    $$\mathbf{S} = (1/\mu_0)\mathbf{E} \times \mathbf{B} = (1/\mu_0)EB\,\hat{\mathbf{r}} = \tfrac{1}{2}\mu_0 n^2 I_0^2 \omega r \sin(\omega t) \cos(\omega t)\hat{\mathbf{r}};$$
    $$\boxed{\mathbf{S} = (1/4)\mu_0 n^2 I_0^2 \omega r \sin(2\omega t)\hat{\mathbf{r}}.}$$
    The period of oscillation of the fields is $T = 2\pi/\omega$. For each quarter-cycle, we have
    $$\boxed{\begin{aligned} &0 < t < \tfrac{1}{4}T: \quad \mathbf{S} \text{ is } +\hat{\mathbf{r}}; \\ &\tfrac{1}{4}T < t < \tfrac{1}{2}T: \quad \mathbf{S} \text{ is } -\hat{\mathbf{r}}; \\ &\tfrac{1}{2}T < t < \tfrac{3}{4}T: \quad \mathbf{S} \text{ is } +\hat{\mathbf{r}}; \\ &\tfrac{3}{4}T < t < T: \quad \mathbf{S} \text{ is } -\hat{\mathbf{r}}. \end{aligned}}$$

61. We assume that the radiation totally reflects from the sail. The radiation force will be away from the sun, with magnitude
    $$F_{\text{radiation}} = 2uA = 2IA/c.$$
    Because the energy through a sphere of radius r must be independent of r, we have
    $I 4\pi r^2 = k'$, or $I = k'/4\pi r^2$.
    The gravitational force is toward the sun, so the net force is
    $$\begin{aligned} F_{\text{net}} &= F_{\text{radiation}} - F_{\text{grav}} \\ &= (2k'A/4\pi c r^2) - (GMm/r^2) = k/r^2, \text{ where } k = (k'A/2\pi c) - GMm. \end{aligned}$$
    The value of k will not change, so the force will always have the same sign.

63. (a) We find the angular frequency of the wave:
    $$\omega = 2\pi f = 2\pi \times 10^{14} \text{ rad/s}.$$
    (b) We find the speed of the wave:
    $$v = c/n = (3 \times 10^8 \text{ m/s})/1.4 = 2.14 \times 10^8 \text{ m/s}.$$
    The wave number is
    $$k = \omega/v = (2\pi \times 10^{14} \text{ rad/s})/(2.14 \times 10^8 \text{ m/s}) = 2.93 \times 10^6 \text{ m}^{-1}.$$
    (c) If x' is the direction of propagation, we have
    $$x' = x \cos 30° + y \sin 30°.$$
    (d) For a wave polarized along the z-axis, we have
    $$\mathbf{E} = E\mathbf{k}.$$
    (e) For a dielectric medium the wave travels at a speed $v = c/n$, so $B = E/v = nE/c$. We find the amplitude of the electric field from the magnitude of the Poynting vector:
    $$S = (1/\mu_0)EB = (n/\mu_0 c)E^2 = \tfrac{1}{2}nc\varepsilon_0 E_0^2;$$
    $$500 \text{ W/m}^2 = \tfrac{1}{2}(1.4)(3 \times 10^8 \text{ m/s})(8.85 \times 10^{-12} \text{ C}^2/\text{N} \cdot \text{m}^2)E_0^2, \text{ which gives } E_0 = 5.19 \times 10^2 \text{ V/m}.$$
    The electric field is
    $$\begin{aligned} \mathbf{E} &= E_0 \cos(kx' - \omega t)\mathbf{k} = E_0 \cos[k(x \cos 30° + y \sin 30°) - \omega t]\mathbf{k} \\ &= (5.19 \times 10^2 \text{ V/m}) \cos[(2.93 \times 10^6 \text{ m}^{-1}) \cos 30° \, x + (2.93 \times 10^6 \text{ m}^{-1}) \sin 30° \, y - (2\pi \times 10^{14} \text{ s}^{-1})t]\mathbf{k} \\ &= \boxed{(5.19 \times 10^2 \text{ V/m}) \cos[(2.54 \times 10^6 \text{ m}^{-1})x + (1.47 \times 10^6 \text{ m}^{-1})y - (2\pi \times 10^{14} \text{ s}^{-1})t]\mathbf{k}.} \end{aligned}$$

69. We let M be the mass of the mirror, v' be the final speed of the mirror, and f' be the frequency of the reflected photons. For energy conservation, we have

$Nhf + \frac{1}{2}Mv^2 = Nhf' + \frac{1}{2}Mv'^2$, which becomes

$v' = v\{1 + [2Nh(f - f')/Mv^2]\}^{1/2}$.

Because the photon energies are much smaller than the mirror kinetic energy, we get

$v' \approx v\{1 + \frac{1}{2}[2Nh(f - f')/Mv^2]\} = v + [Nh(f - f')/Mv]$.

For momentum conservation, we have

$(Nhf/c) + Mv = -(Nhf'/c) + Mv'$.

If we substitute the result from energy conservation, we get

$(Nhf/c) + Mv = -(Nhf'/c) + Mv + [Nh(f - f')/Mv]$, which becomes

$f/c = -(f'/c) + (f/v) - (f'/v)$, which gives

$$\boxed{f' = [(c - v)/(c + v)]f.}$$

71. We choose the axis of the circular plates as the z-axis.
We find the electric field from the charge density:

$\mathbf{E} = (\sigma/\varepsilon_0)\mathbf{k} = (Q/\pi R^2 \varepsilon_0)\mathbf{k}$.

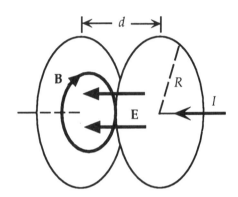

(a) From the cylindrical symmetry, we know that the magnetic field will be circular, centered on the z-axis. We choose a circular path with radius $r < R$ to apply Ampere's law:

$\oint \mathbf{B} \cdot d\mathbf{s} = \mu_0 I_{\text{enclosed}} + \mu_0\varepsilon_0 (d/dt)\oint \mathbf{E} \cdot d\mathbf{A}$;

$B2\pi r = 0 + \mu_0\varepsilon_0 (d/dt)[(Q/\pi R^2 \varepsilon_0)\pi r^2]$, which gives

$$\boxed{\mathbf{B} = (\mu_0 r/2\pi R^2)\, dQ/dt \text{ circular, for } r < R.}$$

(b) The Poynting vector,

$\mathbf{S} = (1/\mu_0)\mathbf{E} \times \mathbf{B}$,

will be directed toward the axis of the plates:

$\mathbf{S} = -(1/\mu_0)(Q/\pi R^2 \varepsilon_0)(\mu_0 r/2\pi R^2)(dQ/dt\,)\hat{\mathbf{r}} =$ 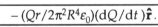 $\boxed{-(Qr/2\pi^2 R^4 \varepsilon_0)(dQ/dt\,)\hat{\mathbf{r}}.}$

(c) The energy flow into the capacitor through the cylinder at $r = R$ is

$P = S(\text{surface area}) = (Qr/2\pi^2\varepsilon_0 R^4)(dQ/dt\,)(2\pi R d)$

$= [Q/(\varepsilon_0 \pi R^2/d)]\, dQ/dt = (d/dt)(Q^2/2C)$,

which is the rate of change of the capacitor energy.

CHAPTER 36

5. The transit time of the light beam must equal the time for the circumference of the wheel to travel the distance between openings:

$\Delta t = 2D/c = L/R\omega;$

$(1000 \text{ m})/(3 \times 10^8 \text{ m/s}) = (1.5 \times 10^{-3} \text{ m})/(15.0 \times 10^{-2} \text{ m})\omega$, which gives

$\omega = 3.0 \times 10^3 \text{ rad/s}$, or $\boxed{f = 2.9 \times 10^4 \text{ rev/min.}}$

This is an extremely high rotation rate. The wheel must have the material strength to provide the necessary internal centripetal force to prevent the wheel from flying apart.

23. When total internal reflection occurs, for the glass–air interface we have

$n \sin \theta_i = n_{air} \sin \theta_{air} = \sin \theta_{air} > 1.$

When we add the stack of layers, at the first layer we have

$n \sin \theta_i = n_1 \sin \theta_1.$

If the value of n_1 is such that $\sin \theta_1 > 1$, total reflection will occur at this surface. If not, light will go to the next surface. At each surface, if total internal reflection does not occur, we continue until we get to air:

$n_1 \sin \theta_1 = n_2 \sin \theta_2;$
$n_2 \sin \theta_2 = n_3 \sin \theta_3;$

$\quad\cdot$
$\quad\cdot$
$\quad\cdot$

$n_k \sin \theta_k = n_{air} \sin \theta_{air} = \sin \theta_{air};$ so, when we add all of the equations, we get

$n \sin \theta_i = \sin \theta_{air};$ which is the original condition, so total internal reflection must occur.

If total internal reflection does not occur at the glass–air interface, we have

$n \sin \theta_i = n_{air} \sin \theta_{air} = \sin \theta_{air} < 1.$

When we add the layers, at the first layer we have

$n \sin \theta_i = n_1 \sin \theta_1 < 1.$

Because $n_1 > 1$, $\sin \theta_1 < 1$, so total reflection will not occur at this surface. The light will go to the next surface. The same reasoning shows that at each surface total internal reflection does not occur. As before, we continue until we get to air and add the equations to get

$n \sin \theta_i = \sin \theta_{air};$ which is the original condition, so total internal reflection does not occur.

$\boxed{\text{The property of the interface cannot be changed}}$ by adding a stack of layers.

25. At the first surface, we have

$n_{air} \sin \theta_1 = n \sin \theta_2;$

$(1.00) \sin 40° = (1.6) \sin \theta_2$, which gives $\theta_2 = 23.7° = \theta_3.$

From the diagram we see that the ray strikes the back of the sphere at a latitude of

$\alpha = 180° - [40° + (180° - 2(23.7°)] = \boxed{7.4°.}$

From the isosceles triangle, we see that the angle of incidence at the back of the sphere is $\theta_3 = 23.7°$. The critical angle is

$n \sin \theta_c = n_{air} \sin 90°;$

$(1.6) \sin \theta_c = (1.00)(1)$, which gives $\theta_c = 38.7°.$

$\boxed{\text{Because } \theta_3 < \theta_c, \text{ there will not be total internal reflection.}}$

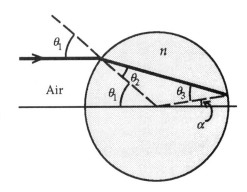

27. Light that strikes the top half of the prism will be refracted downward. From the symmetry of the prism and its placement, light that strikes the bottom half will be refracted upward at the same angle. Thus we find what happens to the light passing through the top half and invert the result for the bottom half. At the first surface of the prism, we have

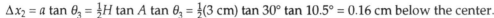
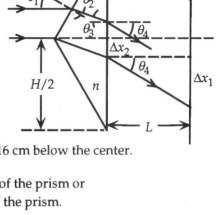

$n_{air} \sin \theta_1 = n \sin \theta_2$;
$(1.00) \sin 30° = (1.5) \sin \theta_2$, which gives $\theta_2 = 19.5°$.

We find the angle of incidence at the second surface from
$(90° - \theta_2) + (90° - \theta_3) + A = 180°$, which gives
$\theta_3 = A - \theta_2 = 30° - 19.5° = 10.5°$.

For the refraction at the second surface, we have
$n \sin \theta_3 = n_{air} \sin \theta_4$;
$(1.5) \sin 10.5° = (1.00) \sin \theta_4$, which gives $\theta_4 = 15.9°$.

The ray that passes through the top edge of the prism will strike the screen a distance
$\Delta x_1 = L \tan \theta_4 = (2 \text{ cm}) \tan 15.9° = 0.57 \text{ cm}$ below the top.

The ray that passes just above the center of the prism will strike the second surface a distance
$\Delta x_2 = a \tan \theta_3 = \frac{1}{2}H \tan A \tan \theta_3 = \frac{1}{2}(3 \text{ cm}) \tan 30° \tan 10.5° = 0.16 \text{ cm}$ below the center.

This ray will strike the screen 0.57 cm below this point, which is
$\Delta x = \Delta x_1 + \Delta x_2 = 0.16 \text{ cm} + 0.57 \text{ cm} = 0.73 \text{ cm}$ below the center of the prism or
$\frac{1}{2}H - \Delta x = 1.5 \text{ cm} - 0.73 \text{ cm} = 0.77 \text{ cm}$ above the bottom edge of the prism.

From the symmetry of the pattern, we have the following:

> Outside the "shadow" of the prism, there is direct illumination.
> For a distance of 0.57 cm from each edge, there will be no illumination.
> For the next distance of 0.77 cm – 0.57 cm = 0.20 cm, there will be illumination from one half.
> For the next distance of 0.73 cm to the center of the pattern, there will be illumination from both halves.

29. For the refraction from water into air, we have

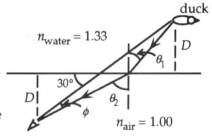

$n_{air} \sin \theta_1 = n_{water} \sin \theta_2$;
$(1.00) \sin \theta_1 = (1.33) \sin \theta_2$.

Both angles are unknown, so we obtain another relation from the distance along the glass:
$D \tan \theta_1 + D \tan \theta_2 = 2D/\tan 30° = 3.464 \text{ m}$.
$\tan \theta_1 + \tan \theta_2 = 2/\tan 30° = 3.464$.

Because of the presence of the trigonometric functions, we solve the two equations numerically to get
$\theta_1 = 44.3°$ and $\theta_2 = 68.3°$.

The angle between the lines of sight is
$\phi = \theta_2 - \theta_1 = 68.3° - 44.3° = \boxed{8.3°.}$

31. For the refraction at the first surface, we have

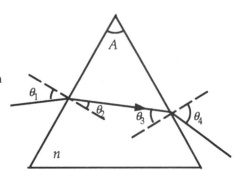

$n_{air} \sin \theta_1 = n \sin \theta_2$;
$(1.00) \sin 35° = (1.55) \sin \theta_2$, which gives $\theta_2 = 21.7°$.

We use the fact that the sum of the angles of a triangle must be 180° to find the angle of incidence at the second surface from
$(90° - \theta_2) + (90° - \theta_3) + A = 180°$, which gives
$\theta_3 = A - \theta_2 = 60° - 21.7° = 38.3°$.

For the refraction at the second surface, we have
$n \sin \theta_3 = n_{air} \sin \theta_4$;
$(1.55) \sin 38.3° = (1.00) \sin \theta_4$, which gives

$\boxed{\theta_4 = 74° \text{ from the normal.}}$

33. We find the angle inside the glass from
$$n_{\text{air}} \sin \theta_1 = n_{\text{glass}} \sin \theta_2;$$
$(1.00) \sin 60° = (1.54) \sin \theta_2$, which gives $\theta_2 = 34.2°$.
For the reflection, the angle of incidence is equal to the angle
of reflection. The refraction when the ray leaves the glass
is the reverse of the initial refraction. Relative to the
entrance position, we find the distance along the glass where
the exit beam leaves from
$$L_2 = 2h \tan \theta_2 = 2(5 \text{ mm}) \tan 34.2° = 6.8 \text{ mm}.$$
We can use the same analysis to find the distance along the
same line where the incident beam would have been
without the glass,
by setting $\theta_2 = \theta_1$:
$$L_1 = 2h \tan \theta_1 = 2(5 \text{ mm}) \tan 60° = 17.3 \text{ mm}.$$
The difference in the distance is
$$D = L_1 - L_2 = 17.3 \text{ mm} - 6.8 \text{ mm} = 10.5 \text{ mm along the glass surface}.$$
The displacement perpendicular to the direction of the reflected beam is
$$d = D \cos \theta_1 = (10.5 \text{ mm}) \cos 60° = \boxed{5.3 \text{ mm.}}$$

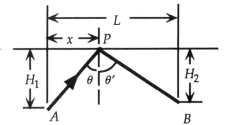

41. The ray travels from point A, a distance H_1 from the surface,
to point B, a distance H_2 from the surface, after reflecting
from the surface. The component of the distance between A and
B parallel to the surface is L. If x is the component of the
distance between A and the point P, where the ray meets the
surface, parallel to the surface, the time of travel is
$$t_{AB} = t_{AP} + t_{PB}$$
$$= [(x^2 + H_1^2)^{1/2}/c] + \{[(L-x)^2 + H_2^2]^{1/2}/c\}.$$
We find the value of x for the minimum time from
$$dt_{AB}/dx = [(1/c)\tfrac{1}{2}(2x)/(x^2 + H_1^2)^{1/2}] + \{(1/c)\tfrac{1}{2}(-2)(L-x)/[(L-x)^2 + H_2^2]^{1/2}\} = 0,$$
which reduces to
$$x/(x^2 + H_1^2)^{1/2} = (L-x)/[(L-x)^2 + H_2^2]^{1/2}.$$
From the diagram, we see that
$$x/(x^2 + H_1^2)^{1/2} = \sin \theta \text{ and } (L-x)/[(L-x)^2 + H_2^2]^{1/2} = \sin \theta',$$
so we have $\sin \theta = \sin \theta'$, or $\theta = \theta'$.

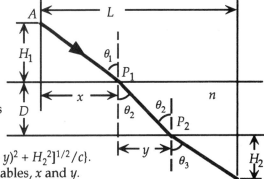

43. The ray travels from point A in air, a distance H_1
from the top surface, to point B in air, a distance H_2
from the bottom surface. The component of the distance
between A and B parallel to the surface is L. If x is the
component of the distance between A and the point P_1
parallel to the surface and y is the component of the
distance between P_1 and the point P_2, where the ray leaves
the glass, parallel to the surface, the time of travel is
$$t_{AB} = t_{AP1} + t_{P1P2} + t_{P2B}$$
$$= [(x^2 + H_1^2)^{1/2}/c] + [(y^2 + D^2)^{1/2}/(c/n)] + \{[(L-x-y)^2 + H_2^2]^{1/2}/c\}.$$
We see that the time depends on the two independent variables, x and y.
For the minimum time due to variation in x, we have
$$\partial t_{AB}/\partial x = [(1/c)\tfrac{1}{2}(2x)/(x^2 + H_1^2)^{1/2}] + \{(1/c)\tfrac{1}{2}(-2)(L-x-y)/[(L-x-y)^2 + H_2^2]^{1/2}\} = 0,$$
which reduces to
$$x/(x^2 + H_1^2)^{1/2} = (L-x-y)/[(L-x-y)^2 + H_2^2]^{1/2}.$$
From the diagram, we see that
$$x/(x^2 + H_1^2)^{1/2} = \sin \theta_1 \text{ and } (L-x-y)/[(L-x-y)^2 + H_2^2]^{1/2} = \sin \theta_3,$$
so we have $\sin \theta_1 = \sin \theta_3$, or $\theta_1 = \theta_3$. The beam emerges parallel to the original direction.

47. (a) For the refraction from glass to air, we have
$$n \sin \theta = \sin \theta_{air},$$
which gives
$$(\sin \theta)_{max} = 1/n \quad \text{and}$$
$$(\cos^2 \theta)_{min} = 1 - (\sin^2 \theta)_{max} = 1 - \frac{1}{n^2} = \frac{n^2 - 1}{n^2} = \frac{C/(\omega_0^2 - \omega^2 - C)}{(\omega_0^2 - \omega^2)/(\omega_0^2 - \omega^2 - C)} = \frac{C}{\omega_0^2 - \omega^2}.$$
Because $\cos \theta \leq 1$, we have
$$\omega_0^2 - \omega^2 \geq C, \quad \text{or} \quad \omega^2 \leq \omega_0^2 - C;$$
$$\omega^2 \leq (685 \times 10^{30} \text{ rad}^2/\text{s}^2) - (529 \times 10^{30} \text{ rad}^2/\text{s}^2), \text{ which gives } \boxed{\omega \leq 12.5 \times 10^{15} \text{ rad/s.}}$$
   (b) For an angle of 90° in air at this frequency, we have
$$(\cos^2 \theta)_{min} = C/(\omega_0^2 - \omega_{max}^2);$$
$$= (529 \times 10^{30} \text{ rad}^2/\text{s}^2)/[(685 \times 10^{30} \text{ rad}^2/\text{s}^2) - (3.2 \times 10^{15} \text{ rad/s})^2],$$
which gives $\boxed{\theta = 27.7°.}$

49. For the refraction at the first surface, we have
$$n_{air} \sin \theta_1 = n \sin \theta_2;$$
$$(1.00) \sin (\tfrac{1}{2}A) = n \sin \theta_2.$$
We find the angle of incidence at the second surface from
$$\theta_3 = A - \theta_2.$$
For the refraction at the second surface, we have
$$n \sin \theta_3 = n_{air} \sin \theta_4 = (1.00) \sin \theta_4.$$
If we assume an average index of 1.5 and $A = 60°$, we find
$$\theta_2 = 19.5°, \quad \theta_3 = 40.5°, \quad \text{and} \quad \theta_4 = 77.0°.$$
Because the changes in angles and indices are small, we approximate them by differentials.
From the three equations, we have
$$0 = (dn \sin \theta_2) + [n(\cos \theta_2) \, d\theta_2];$$
$$d\theta_3 = -d\theta_2;$$
$$(dn \sin \theta_3) + [n(\cos \theta_3) \, d\theta_3] = (\cos \theta_4) \, d\theta_4.$$
When we combine these equations, we get
$$(\cos \theta_4) \, d\theta_4 = (dn \sin \theta_3) - [n(\cos \theta_3) \, d\theta_2]$$
$$= (dn \sin \theta_3) + [n(\cos \theta_3)(dn \sin \theta_2)/n(\cos \theta_2)]$$
$$= dn[(\sin \theta_3)(\cos \theta_2) + (\cos \theta_3)(\sin \theta_2)]/(\cos \theta_2).$$
From the sine of the sum of two angles, we have
$$(\sin \theta_3)(\cos \theta_2) + (\cos \theta_3)(\sin \theta_2) = \sin(\theta_3 + \theta_2) = \sin A,$$
so our result is
$$dn = [(\cos \theta_4)(\cos \theta_2)/(\sin A)] \, d\theta_4.$$
For a separation of 2°, we have
$$dn = [(\cos 77.0°)(\cos 19.5°)/(\sin 60°)](2°)(\pi \text{ rad}/180°) = \boxed{0.009.}$$

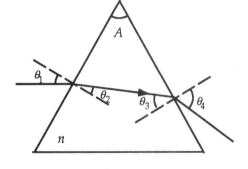

51. We find the critical angle for light leaving the water:
$$n \sin \theta_1 = \sin \theta_2;$$
$$(1.33) \sin \theta_c = \sin 90°, \text{ which gives } \theta_c = 48.8°.$$
We see from the diagram that the highest point on the pin from which light will emerge from the water is determined by a ray that just misses the edge of the cork and reaches the water surface at the critical angle:
$$R/H = \tan \theta_c;$$
$$(1.5 \text{ cm})/H = \tan 48.8°, \text{ which gives } \boxed{H = 1.31 \text{ cm.}}$$

53. For the refraction at the first surface, the angle of incidence is ϕ, so we have

$n_{air} \sin \theta_1 = n \sin \theta_2$;

$(1.00) \sin \phi = n \sin \theta_2$.

We find the angle of incidence at the second surface from

$\theta_3 = 2\phi - \theta_2$.

For the refraction at the second surface, we have

$n \sin \theta_3 = n_{air} \sin \theta_4 = (1.00) \sin \theta_4$.

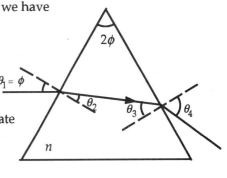

Because the changes in angles and indices are small, we approximate them by differentials. From the three equations, we have

$0 = (dn \sin \theta_2) + [n(\cos \theta_2)\, d\theta_2]$;

$d\theta_3 = -d\theta_2$;

$(dn \sin \theta_3) + [n(\cos \theta_3)\, d\theta_3] = (\cos \theta_4)\, d\theta_4$.

When we combine these equations, we get

$(\cos \theta_4)\, d\theta_4 = (dn \sin \theta_3) - [n(\cos \theta_3)\, d\theta_2]$

$= (dn \sin \theta_3) + [n(\cos \theta_3)(dn \sin \theta_2)/n(\cos \theta_2)]$

$= dn[(\sin \theta_3)(\cos \theta_2) + (\cos \theta_3)(\sin \theta_2)]/(\cos \theta_2)$.

From the sine of the sum of two angles, we have

$(\sin \theta_3)(\cos \theta_2) + (\cos \theta_3)(\sin \theta_2) = \sin(\theta_3 + \theta_2) = \sin(2\phi)$,

so our result is

$\boxed{\begin{array}{l} \Delta n = [(\cos \theta_4)(\cos \theta_2)/\sin(2\phi)]\, \Delta\theta, \text{ where} \\ \cos \theta_2 = \cos\{\sin^{-1}[(\sin \phi)/n]\} \text{ and} \\ \cos \theta_4 = \cos\{\sin^{-1}[n \sin(2\phi - \{\sin^{-1}[(\sin \phi)/n]\})]\}. \end{array}}$

55. For the refraction at the first surface, we have

$\sin \theta_i = n \sin \theta_2$.

We find the angle of incidence at the second surface from

$\theta_3 = 2\phi - \theta_2$.

For the refraction at the second surface, we have

$n \sin \theta_3 = \sin \theta_f$.

The total deflection angle is the sum of the deflections that take place at each surface:

$\Theta = (\theta_f - \theta_3) + (\theta_i - \theta_2) = \theta_i + \theta_f - 2\phi$

$= \sin^{-1}(n \sin \theta_2) + \sin^{-1}[n \sin(2\phi - \theta_2)] - 2\phi$.

This gives the deflection as a function of θ_2. To find the angle θ_2 that minimizes the deflection, we set $d\Theta/d\theta_2 = 0$:

$\dfrac{d\Theta}{d\theta_2} = \dfrac{n \cos \theta_2}{\sqrt{1 - (n \sin \theta_2)^2}} - \dfrac{n \cos(2\phi - \theta_2)}{\sqrt{1 - [n \sin(2\phi - \theta_2)]^2}} + 0 = 0, \quad \text{or}$

$\dfrac{n \cos \theta_2}{\sqrt{1 - (n \sin \theta_2)^2}} = \dfrac{n \cos(2\phi - \theta_2)}{\sqrt{1 - [n \sin(2\phi - \theta_2)]^2}}$.

We can see by inspection that we have $\theta_2 = 2\phi - \theta_2$, or $\theta_2 = \phi$.

From the first refraction equation, we have

$\sin \theta_i = n \sin \theta_2 = n \sin \phi$, or $\boxed{\theta_i = \sin^{-1}(n \sin \phi).}$

Note that, for minimum deflection, the ray goes through the prism symmetrically, with the internal ray perpendicular to the bisector of the apex angle.

57. (a) We find the distance Earth moves during a period of Io from

$d = v_{Earth}T$

$= (30 \text{ km/s})(42.5 \text{ h})(3600 \text{ s/h}) = \boxed{4.6 \times 10^6 \text{ km.}}$

(b) When Earth is moving toward Jupiter, the light that indicates the end of the period travels a shorter distance than the light that indicated the start of the period, by the distance that Earth has moved. This means that the measured period is smaller. We find the speed of light from

$c = d/\Delta t = (4.6 \times 10^9 \text{ m})/(15 \text{ s}) = \boxed{3.1 \times 10^8 \text{ m/s.}}$

CHAPTER 37

The sign convention for distances and radii must be followed. Object distances are positive when the object is on the same side of the surface or lens as the incident or incoming light. Image distances, radii of curvature, and focal points are positive if they are on the same side of the surface or lens as the outgoing light, i. e., reflected light for mirrors and transmitted light for refraction. A negative object distance denotes a virtual object. A negative image distance denotes a virtual image.

5. We label the three images formed by one reflection I_1. With double reflections, each of these images forms two images from reflections off the other mirrors, which we label I_2. The distance from the axis for the I_1 images is

> $L_1 = 2(\tfrac{1}{2}a) \tan 30° = 0.577a.$

The distance from the axis of the I_2 images is

> $L_2 = 2d \cos 30° = 2(\tfrac{1}{2}a/\cos 30°) \cos 30° = a,$

which can be seen from the figure.

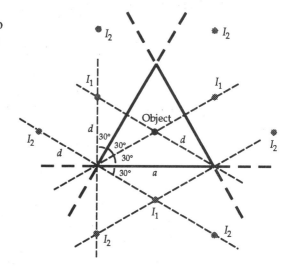

17. Because the glass has parallel sides and is thin, it will have no effect on the refraction. We can consider refraction directly from water to air. The plane surface is treated as a sphere with an infinite radius:
    $(n_1/s) + (n_2/i) = (n_2 - n_1)/R;$
    $[(1.33)/(40 \text{ cm})] + [(1.0)/i] = (1.0 - 1.33)/(\infty)$, which gives $\boxed{i = -30 \text{ cm (behind the glass).}}$

19. We find the location of the image of the fault from
    $(n_1/s) + (n_2/i) = (n_2 - n_1)/R;$
    $[(1.6)/(2.3 \text{ cm})] + [(1.0)/i] = (1.0 - 1.6)/(-0.8 \text{ cm})$, which gives $\boxed{i = +18.4 \text{ cm.}}$
    This image is 18.4 cm in front of the glass rod. If we assume that a person can see clearly objects that are at least 25 cm away, you should be at least $\boxed{43 \text{ cm from the surface.}}$

23. (a) For the refraction at the convex surface, for which $R > 0$, we have
$(n_1/s) + (n_2/i) = (n_2 - n_1)/R.$
For an object very far from the surface, $s \to \infty$, which gives
$0 + (n_2/i) = (n_2 - n_1)/R,$ or $i = n_2R/(n_2 - n_1).$
Because $n_2 > n_1$, $i > 0$, the image is real. The magnification is
$M = -i/s = -n_2R/(n_2 - n_1)s.$
Because $n_2 > n_1$, $M < 0$; the image is inverted.
Because s is very large, the image is reduced.

　(b) For the refraction at the convex surface, for which $R > 0$, we have
$(n_1/s) + (n_2/i) = (n_2 - n_1)/R.$
For an image very far from the surface, $i \to \infty$, which gives
$(n_1/s) + 0 = (n_2 - n_1)/R,$ or $\boxed{s = n_1R/(n_2 - n_1).}$

　(c) If the object position is δ less than the critical value from part (b), we have
$s = [n_1R/(n_2 - n_1)] - \delta = (n_1R - x)/(n_2 - n_1),$ where $x = (n_2 - n_1)\delta.$
For the refraction at the convex surface, we have
$(n_1/s) + (n_2/i) = (n_2 - n_1)/R;$
$[n_1(n_2 - n_1)/(n_1R - x)] + (n_2/i) = (n_2 - n_1)/R,$ which reduces to
$i = n_2R(n_1R - x)/(n_2 - n_1)(-x).$
When x is small, we have
$i \approx -n_2n_1R^2/(n_2 - n_1)x = -n_2n_1R^2/(n_2 - n_1)^2\delta.$
When $\delta \to 0$, $i \to -\infty$. $\boxed{\text{The image is very far in front of the boundary.}}$
Because the image is on the side opposite to the outgoing light, $\boxed{\text{the image is virtual.}}$

　(d) As s decreases to 0, $\delta \to n_1R/(n_2 - n_1)$, and $x \to n_1R$. Thus, $i \to 0$.
$\boxed{\text{The position of the image approaches the boundary.}}$

25. From the diagram, we have
$\theta_2 = \beta + \alpha$, $\theta_1 = \beta + \gamma$, $AB = BC \tan \beta = SB \tan \gamma = IB \tan \alpha$, where
$BC = R$, $SB = s$, and $IB = -i$ (from the sign convention).
When the angles are small, $\sin \phi \approx \tan \phi \approx \phi$, so we have
$AB = R\beta = s\gamma = -i\alpha.$
For the refraction at the surface, we have
$n_1 \sin \theta_1 = n_2 \sin \theta_2$. For small angles, this becomes
$n_1\theta_1 = n_2\theta_2;$
$n_1(\beta + \gamma) = n_2(\beta + \alpha);$
$n_1[(AB/R) + (AB/s)] = n_2[(AB/R) - (AB/i)],$
which gives
$(n_1/s) + (n_2/i) = (n_2 - n_1)/R.$

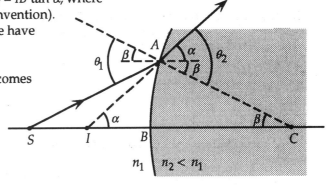

31. (a) For a real image, $s > 0$, and $i > 0$. The magnification is
$M = -i/s = -2$, which gives $i = 2s$.
For the refraction of the thin lens, we have
$$\frac{1}{s} + \frac{1}{i} = \frac{1}{f};$$
$$\frac{1}{s} + \frac{1}{2s} = \frac{1}{+25 \text{ cm}},$$ which gives $\boxed{s = +38 \text{ cm.}}$

　(b) For a virtual image, $s > 0$, and $i < 0$. The magnification is
$M = -i/s = +2$, which gives $i = -2s$.
For the refraction of the thin lens, we have
$$\frac{1}{s} + \frac{1}{i} = \frac{1}{f};$$
$$\frac{1}{s} + \frac{1}{-2s} = \frac{1}{+25 \text{ cm}},$$ which gives $\boxed{s = +13 \text{ cm.}}$

33. For a thin lens, the centers of the two spherical surfaces are at the same location. The surfaces for the two refractions of a ray are parallel, so the ray is displaced but undeviated. Two parallel rays will still be parallel after passing through the lens. Thus the focal length is infinite. From the refraction of the thin lens, we have

$$i = -s,$$

which means that a real object produces a

| virtual image. |

Note that this is as if the lens were not there.

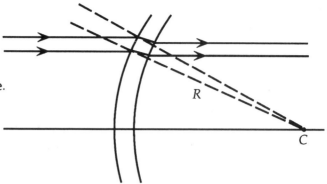

35. The image of a distant source will be at the focal point, so the image distance will be the focal length, which we find from

$$1/f = (n - 1)[(1/R_1) - (1/R_2)].$$

Lens a:

(a) $1/f_a = (1.55 - 1)\{[1/(+ 25 \text{ cm})] - [1/(- 60 \text{ cm})]\}$,

which gives | $f_a = 32.1$ cm. |

The image will be 32.1 cm beyond the lens.

(b) The positive image distance means that the image is

| inverted and real. |

(c) | $M = -i/s = -(+)/(+) = -,$ | which is consistent with part (b).

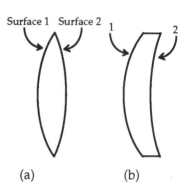

(a) (b)

Lens b:

(a) $1/f_b = (1.55 - 1)\{[1/(+ 25 \text{ cm})] - [1/(+ 60 \text{ cm})]\}$,

which gives | $f_b = + 77.9$ cm. |

The image will be 77.9 cm beyond the lens.

(b) The positive image distance means that the image is

| inverted and real. |

(c) | $M = -i/s = -(+)/(+) = -,$ | which is consistent with part (b).

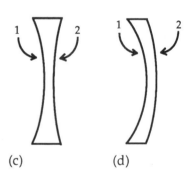

(c) (d)

Lens c:

(a) $1/f_c = (1.55 - 1)\{[1/(- 25 \text{ cm})] - [1/(+ 60 \text{ cm})]\}$,

which gives | $f_c = - 32.1$ cm. |

The image will be 32.1 cm in front of the lens.

(b) The negative image distance means that the image is

| upright and virtual. |

(c) | $M = -i/s = -(-)/(+) = +,$ | which is consistent with part (b).

Lens d:

(a) $1/f_d = (1.55 - 1)\{[1/(- 25 \text{ cm})] - [1/(- 60 \text{ cm})]\}$,

which gives | $f_d = - 77.9$ cm. |

The image will be 77.9 cm in front of the lens.

(b) The negative image distance means that the image is

| upright and virtual. |

(c) | $M = -i/s = -(-)/(+) = +,$ | which is consistent with part (b).

37. Lens *a*:

We locate the image from

$(1/s_a) + (1/i_a) = 1/f_a;$ $[1/(65 \text{ cm})] + (1/i_a) = 1/(+32.1 \text{ cm})$, which gives $i_a = +63.4 \text{ cm}.$

The magnification is

$M_a = -i_a/s_a = -(+63.4)/(65 \text{ cm}) = -0.98.$

> The image is 63.4 cm to the left of the lens, inverted, real, with $M = -0.98$.

Lens *b*:

We locate the image from

$(1/s_b) + (1/i_b) = 1/f_b;$ $[1/(65 \text{ cm})] + (1/i_b) = 1/(+77.9 \text{ cm})$, which gives $i_b = -393 \text{ cm}.$

The magnification is

$M_b = -i_b/s_b = -(-393 \text{ cm})/(65 \text{ cm}) = +6.0.$

> The image is 393 cm to the right of the lens, upright, virtual, with $M = +6.0$.

Lens *c*:

We locate the image from

$(1/s_c) + (1/i_c) = 1/f_c;$ $[1/(65 \text{ cm})] + (1/i_c) = 1/(-32.1 \text{ cm})$, which gives $i_c = -21.5 \text{ cm}.$

The magnification is

$M_c = -i_c/s_c = -(-21.5 \text{ cm})/(65 \text{ cm}) = +0.33.$

> The image is 21.5 cm to the right of the lens, upright, virtual, with $M = +0.33$.

Lens *d*:

We locate the image from

$(1/s_d) + (1/i_d) = 1/f_d;$ $[1/(65 \text{ cm})] + (1/i_d) = 1/(-77.9 \text{ cm})$, which gives $i_d = -35.4 \text{ cm}.$

The magnification is

$M_d = -i_d/s_d = -(-35.4 \text{ cm})/(65 \text{ cm}) = +0.55.$

> The image is 35.4 cm to the right of the lens, upright, virtual, with $M = +0.55$.

47. From the diagram, we have

$\angle CAF = \theta$ and $\angle ACF = \theta.$

This means that the triangle *ACF* is isosceles, with a base of length *R*. By dropping a perpendicular from *F* to the base *AC*, we have

$CF = (\tfrac{1}{2}R)/(\cos\theta).$

The distance from the mirror to the point *F* is

$f = R - CF = R - [(\tfrac{1}{2}R)/(\cos\theta)]$, which gives

$f = R\{1 - [1/(2\cos\theta)]\}.$

For small θ, $\cos\theta \approx 1$, so we have

$f = R[1 - (\tfrac{1}{2})] = \tfrac{1}{2}R.$

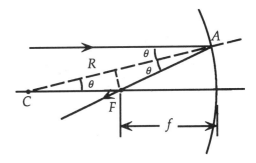

49. We could calculate the focal lengths directly; however, the changes in index are small, so we can use differentials. We differentiate the direct expression for the focal length:

$$f = \frac{1}{(n-1)}\left(\frac{1}{R_1} - \frac{1}{R_2}\right)^{-1};$$

$$df = -\frac{dn}{(n-1)^2}\left(\frac{1}{R_1} - \frac{1}{R_2}\right)^{-1}.$$

If we use the average index, we get

$$df = -\frac{(1.48135 - 1.48523)}{(1.48329 - 1)^2}\left(\frac{1}{20.00 \text{ cm}} - \frac{1}{28.75 \text{ cm}}\right)^{-1} = \boxed{1.09 \text{ cm.}}$$

51. We find the focal lengths of the lenses from

$$\frac{1}{f} = (n-1)\left(\frac{1}{R_1} - \frac{1}{R_2}\right);$$

$$\frac{1}{f_1} = (1.50-1)\left(\frac{1}{+35 \text{ cm}} - \frac{1}{-35 \text{ cm}}\right), \text{ which gives} \quad f_1 = +35 \text{ cm};$$

$$\frac{1}{f_2} = (1.50-1)\left(\frac{1}{-35 \text{ cm}} - \frac{1}{+35 \text{ cm}}\right), \text{ which gives} \quad f_2 = -35 \text{ cm}.$$

If the object is near the <u>positive</u> lens, for the two refractions, we have

$$\frac{1}{s_1} + \frac{1}{i_1} = \frac{1}{f_1};$$

$$\frac{1}{+10 \text{ cm}} + \frac{1}{i_1} = \frac{1}{+35 \text{ cm}}, \text{ which gives} \quad i_1 = -14 \text{ cm};$$

$$\frac{1}{s_2} + \frac{1}{i_2} = \frac{1}{f_2};$$

$$\frac{1}{+29 \text{ cm}} + \frac{1}{i_2} = \frac{1}{-35 \text{ cm}}, \text{ which gives} \quad i_2 = -15.8 \text{ cm}.$$

> The image is 15.8 cm in front of the negative lens, or 0.8 cm from the positive lens on the object side.

If the object is near the <u>negative</u> lens, for the two refractions, we have

$$\frac{1}{s_3} + \frac{1}{i_3} = \frac{1}{f_3};$$

$$\frac{1}{+10 \text{ cm}} + \frac{1}{i_3} = \frac{1}{-35 \text{ cm}}, \text{ which gives} \quad i_3 = -7.78 \text{ cm};$$

$$\frac{1}{s_4} + \frac{1}{i_4} = \frac{1}{f_4};$$

$$\frac{1}{+22.8 \text{ cm}} + \frac{1}{i_4} = \frac{1}{+35 \text{ cm}}, \text{ which gives} \quad i_4 = -65.4 \text{ cm}.$$

> The image is 65.4 cm in front of the positive lens, or 50.4 cm from the negative lens on the object side.

The order of the lenses is important.

53. The focal lengths of the two mirrors are
$f_1 = R_1/2 = 16 \text{ cm}$;
$f_2 = R_2/2 = 7.0 \text{ cm}$.
(a) For the first reflection, we have

$$\frac{1}{s_1} + \frac{1}{i_1} = \frac{1}{f_1};$$

$$\frac{1}{+7 \text{ cm}} + \frac{1}{i_1} = \frac{1}{+16 \text{ cm}}, \text{ which gives} \quad i_1 = -12.4 \text{ cm}.$$

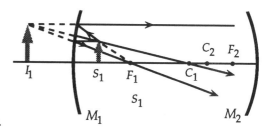

> The first image is 12.4 cm behind M_1.

(b) The object distance for the second reflection is 62.4 cm, so we have

$$\frac{1}{s_2} + \frac{1}{i_2} = \frac{1}{f_2};$$

$$\frac{1}{+62.4 \text{ cm}} + \frac{1}{i_2} = \frac{1}{+7.0 \text{ cm}}, \text{ which gives} \quad i_2 = +7.9 \text{ cm}.$$

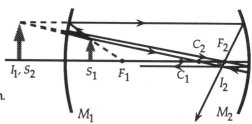

> The second image is 7.9 cm in front of M_2.

55. We find the focal length by finding the image distance for an object very far away.
    For the refraction by the lens, we have
    $(1/s_1) + (1/i_1) = 1/f_1$;
    $0 + (1/i_1) = 1/f_1$, or, as expected, $i_1 = f_1$.
    The first image is the object for reflection from the mirror, with an object distance of $d - f_1$.
    For the mirror, we have
    $(1/s_2) + (1/i_2) = 1/f_2$;
    $[1/(d - f_1)] + (1/i_2) = 1/f_2$, which gives $i_2 = f_2(d - f_1)/[d - (f_1 + f_2)]$.
    Because the reflected light must pass through the lens, we have a second refraction.
    The second image is the object for the lens, with an object distance of $d - i_2$.
    For the lens, we have
    $(1/s_3) + (1/i_3) = 1/f_1$;
    $[1/(d - i_2)] + (1/i_3) = 1/f_1$.
    Because this image must be at the focal length of the combination, we have
    $1/f = (1/f_1) - [1/(d - i_2)] = (1/f_1) - \{1/[d - f_2(d - f_1)/(d - f_1 - f_2)]\}$.
    After some algebra, we have

    $$f = [f_1(d^2 - f_1 d - 2f_2 d + f_1 f_2)]/(d^2 - 2f_1 d - 2f_2 d + 2f_1 f_2 + f_1^2).$$

59. We find the focal length of the lens from

    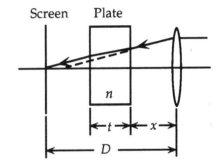

    Screen   Plate

    $1/f = (n - 1)[(1/R_1) - (1/R_2)]$
         $= (1.4 - 1)\{[1/(25 \text{ cm})] - [1/(-25 \text{ cm})]\}$, which gives
    $f = 31.25$ cm.
    For a distant object, the image produced by the lens will be
    at the focal point: $i_1 = f = 31.25$ cm.
    The light coming through the lens will refract at the first
    surface of the plate, so this image is a virtual object for
    refraction at the front surface of the plate. We let x be the
    distance from the lens to this surface.
    The object distance is $-(f - x) = x - f$.
    For the radius of the flat surface, we use $R = \infty$:
    $(1/s_2) + (n/i_2) = (n - 1)/\infty$;
    $[1/(x - f)] + (n/i_2) = 0$, which gives $i_2 = n(f - x)$.
    This image is a virtual object for refraction at the back surface of the plate.
    The object distance is $-(i_2 - t) = t - i_2$. For the radius of the flat surface, we use $R = \infty$:
    $(n/s_3) + (1/i_3) = (1 - n)/\infty$;
    $\{n/(t - [n(f - x)]\} + (1/i_3) = 0$, which gives $i_3 = (f - x) - (t/n)$.
    For this final image to be on the screen, we have
    $i_3 + t + x = D$;
    $(f - x) - (t/n) + t + x = D$;
    $f - (t/n) + t = D$;
    $(31.25 \text{ cm}) - (t/1.4) + t = 35$ cm, which gives $t = 13.1$ cm.
    Because the answer does not depend on x,

    the plate, with thickness 13.1 cm, may be placed anywhere between the lens and the screen.

CHAPTER 38

For thin-film interference, for each of the interfering waves we find the phase change relative to the wave incident on the first surface. Each phase change may have two contributions: one due to additional path and one of 0° or 180° due to reflection. The net phase difference between the interfering waves is the difference of their phase changes.

7. For constructive interference, the path-length difference is a multiple of the wavelength:
    $$\Delta L = d \sin \theta = n\lambda.$$
    We find the location of the bright spots on the screen from $y = R \tan \theta$.
    For small angles, we have
    $$\sin \theta \simeq \tan \theta, \text{ which gives } y = R(n\lambda/d) = nR\lambda/d.$$
    The separation of bright spots is
    $$\Delta y = \Delta n\, R\lambda/d.$$
    (a) When $\lambda_2 = 2\lambda_1$, we have $\Delta y_2 = 2\,\Delta y_1$; | the separation doubles. |
    (b) When $d_2 = 2d_1$, we have $\Delta y_2 = \frac{1}{2}\Delta y_1$; | the separation reduces by $\frac{1}{2}$. |
    (c) When $R_2 = 2R_1$, we have $\Delta y_2 = 2\,\Delta y_1$; | the separation doubles. |
    (d) The separation of maxima does not depend on the intensity; | there is no change. |

11. The overlapping slits form a system of two point sources. As the double slit is rotated, the distance between the sources changes. From the figure we have
    $$d' = d/\cos\phi.$$
    Because the system is equivalent to the double slit, we use the result from Problem 7:
    $$y_{\max} = nR\lambda/d' = nR\lambda\,(\cos\phi)/d,\quad n = 0, \pm 1, \pm 2, \dots .$$
    As ϕ increases, $\cos\phi$ and thus $y_{\max}$ decreases;
    the fringes move | inward. |

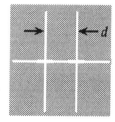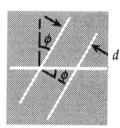

13. The phase difference at a point on the screen at an angle θ from the line perpendicular to the sources has two contributions. The additional distance gives a phase difference of $2\pi(d \sin\theta)/\lambda$. The difference in frequency will create a phase difference which is a function of time: $(\Delta\omega)t = 2\pi t\,\Delta f$.
    For maxima we have
    $$\phi = [2\pi(d \sin\theta)/\lambda] + 2\pi t\,\Delta f = n2\pi,\quad n = 0, \pm 1, \pm 2, \dots .$$
    The location of the first non-central maximum is
    $$y = R \tan\theta \approx R \sin\theta = (R\lambda/d)(1 - t\,\Delta f) = (Rv/fd)(1 - t\,\Delta f).$$
    We find the speed of the fringe movement from
    | $v_{\text{fringe}} = |dy/dt| = (Rv/d)\,\Delta f/f.$ |
    For the ripple tank we have
    $$v_{\text{fringe,ripple}} = [(1\text{ m})(0.15\text{ m/s})/(0.05\text{ m})](10^{-6}) = \boxed{3 \times 10^{-6}\text{ m/s.}}$$
    For the optical experiment we have
    $$v_{\text{fringe,optical}} = [(1\text{ m})(3 \times 10^8\text{ m/s})/(0.25\ 10^{-3}\text{ m})](10^{-6}) = \boxed{1.2 \times 10^6\text{ m/s.}}$$
    Because the speed for the ripple tank is unnoticeable, the frequency difference can be much larger, which means less coherence, before the effect is noticeable. In the optical experiment, the fringe pattern would be completely blurred. The coherence of the two sources must be many orders of magnitude better to have a noticeable interference pattern.

21. To distinguish the slit separation from differentials, we let the separation be D. When the screen is far from the slits, the angles are small, so the intensity a distance y from the central maximum becomes
$$I = 4I_0 \cos^2[(\pi D \sin\theta)/\lambda] \simeq 4I_0 \cos^2(\pi y D/\lambda R).$$
The average intensity over the screen is
$$I_{av} = \frac{\displaystyle\int_{screen} I\,dy}{\displaystyle\int_{screen} dy} = \frac{\displaystyle\int_{screen} 4I_0 \cos^2(\pi y D/\lambda R)\,dy}{\displaystyle\int_{screen} dy}.$$
If we change variable to $\beta = \pi y D/\lambda R$, we have $d\beta = (\pi D/\lambda R)\,dy$, and the average intensity becomes
$$I_{av} = \frac{\displaystyle\int_{screen} 4I_0(\lambda R/\pi D)\cos^2\beta\,d\beta}{\displaystyle\int_{screen} (\lambda R/\pi D)\,d\beta} = 4I_0\langle\cos^2\beta\rangle.$$
For many fringes on the screen, β varies over many cycles, and the average value of $\cos^2\beta$ is $\frac{1}{2}$. The average intensity is
$$\boxed{I_{av} = 4I_0\langle\cos^2\beta\rangle = 4I_0(\tfrac{1}{2}) = 2I_0.}$$

23. The intensity of the pattern is
$$I = 4I_0\cos^2[(\pi d\sin\theta)/\lambda].$$
At the central maximum, we have $\theta = 0$ and $I = 4I_0$.
We find the angle where the intensity is half its maximum value from
$$I = \tfrac{1}{2}(4I_0) = 4I_0\cos^2[(\pi d\sin\theta_{1/2})/\lambda], \quad \text{or} \quad \cos[(\pi d\sin\theta_{1/2})/\lambda] = 1/\sqrt{2};$$
$(\pi d\sin\theta_{1/2})/\lambda = \pi/4$, which gives $\theta_{1/2} = \sin^{-1}(\lambda/4d)$.
The angular width is twice this angle:
$$\boxed{\Delta\theta = 2\sin^{-1}(\lambda/4d).}$$
The other angles where the intensity is $2I_0$ are given by
$$(\pi d\sin\theta)/\lambda = 3\pi/4,\, 5\pi/4,\, \ldots\,.$$
In general, the widths will not be the same, except for maxima near the central maximum, where the angles are small, so $\sin\theta \simeq \theta$.

27. For distances which are much larger than $d = 2.5$ m, we assume that the amplitudes of the electric fields are the same, as in the double-slit analysis. The path-length difference from the two slits is
$$\Delta L = (x^2 + d^2)^{1/2} - x = x[1 + (d^2/x^2)]^{1/2} - x$$
$$\simeq x + (d^2/2x) - x = d^2/2x.$$
Because the sources are out of phase by 45° or $\pi/4$ rad, the total phase difference is
$$\phi = (2\pi\,\Delta L/\lambda) + (\pi/4) = (\pi d^2/x\lambda) + (\pi/4).$$
The intensity is
$$I = 4I_0\cos^2(\phi/2) = 4I_0\cos^2[(\pi d^2/2x\lambda) + (\pi/8)]$$
$$= 4I_0\cos^2\{[\pi(2.5\text{ m})^2/2x(0.020\text{ m})] + (\pi/8)\}$$
$$= 4I_0\cos^2[(156\pi/x) + (\pi/8)].$$

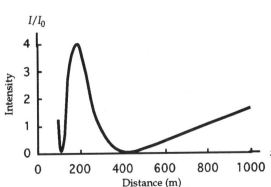

33. At a distance r from the center of the lens, the thickness of
the air space is h, and the phase difference for the reflected
waves from the path-length difference and the reflection at
the bottom surface is
$$\phi = (2h/\lambda)2\pi - \pi.$$
For the first dark ring, we have
$$\phi = (2h/\lambda)2\pi - \pi = \pi, \quad \text{or}$$
$$h = \tfrac{1}{2}\lambda.$$
From the triangle in the diagram, we have
$$r^2 + (R - h)^2 = R^2, \quad \text{or} \quad r^2 = 2hR - h^2 \simeq 2hR, \text{ when } h \ll R.$$
We find the radius of the first dark ring from
$$r^2 = 2(\tfrac{1}{2}\lambda)R = \lambda R, \text{ which gives } \boxed{r = (\lambda R)^{1/2}.}$$

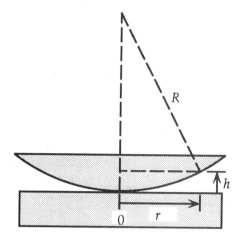

35. When the separation of the plate and the center of the lens is D, the phase difference from the
path-length difference and the reflection from the flat plate is
$$\phi = (2D/\lambda)2\pi + \pi.$$
For the maxima, this must equal $m2\pi$. Because the spot was dark initially, m counts the number of
maxima that pass:
$$(2D/\lambda)2\pi + \pi = m2\pi, \quad \text{or}$$
$$m = (2D/\lambda) + \tfrac{1}{2} = 2(0.25 \times 10^{-3}\text{ m})/(500 \times 10^{-6}\text{ m}) + \tfrac{1}{2} = (1.0 \times 10^3) + \tfrac{1}{2}.$$
Thus $\boxed{1000 \text{ maxima pass.}}$
A particular fringe corresponds to a particular phase difference. As the thickness of the air layer
increases away from the center, the phase difference increases. When the lens is pulled away, the
increase in thickness means a particular phase difference occurs closer to the center, so the
corresponding fringe moves closer to the center. $\boxed{\text{The rings move in to the center.}}$

39. At a distance x from the center of the lens, the thickness of
the air space is y, and the phase difference for the reflected
waves from the path-length difference and the reflection at the
bottom surface is
$$\phi = (2y/\lambda)2\pi + \pi.$$
For the dark rings, this phase difference must be an odd
multiple of π, so we have
$$\phi = (2y/\lambda)2\pi + \pi = (2n + 1)\pi, \quad n = 0, 1, 2, \ldots, \quad \text{or}$$
$$y = \tfrac{1}{2}n\lambda, \quad n = 0, 1, 2, \ldots .$$
Because $n = 0$ corresponds to the dark center, n represents the
number of the ring.
From the triangle in the diagram, we have
$$x^2 + (R - y)^2 = R^2, \quad \text{or} \quad x^2 = 2yR - y^2 \simeq 2yR, \text{ when } y \ll R.$$
The position of the nth dark ring is $\boxed{x = (n\lambda R)^{1/2}.}$

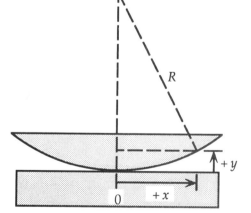

43. With respect to the incident wave, the wave that reflects from the oil at the top surface has a phase change of

$\phi_1 = \pi$.

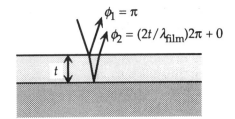

With respect to the incident wave, the wave that reflects from the water at the bottom surface of the oil has a phase change due to the additional path-length but, because the index of the oil is greater than the index of water, no phase change on reflection:

$\phi_2 = (2t/\lambda_{film})2\pi + 0$.

For constructive interference, the net phase change is

$\phi = (2t/\lambda_{film})2\pi - \pi = m2\pi$, $m = 0, 1, 2, \ldots$, or

$t = \frac{1}{2}\lambda_{film}(m + \frac{1}{2}) = \frac{1}{2}(\lambda/n_{film})(m + \frac{1}{2})$, $m = 0, 1, 2, \ldots$.

For the two wavelengths, we have

$t = \frac{1}{2}(\lambda_1/n_{film})(m_1 + \frac{1}{2}) = \frac{1}{2}(\lambda_2/n_{film})(m_2 + \frac{1}{2})$, which gives

$(2m_1 + 1)/(2m_2 + 1) = \lambda_2/\lambda_1 = (682 \text{ nm})/(434 \text{ nm}) = 1.571$.

By trying the various integers, with $m_1 > m_2$, we find that the set of smallest integers that satisfies the equation is $m_1 = 5$ and $m_2 = 3$.

We find the minimum thickness from

$t = \frac{1}{2}(\lambda_1/n_{film})(m_1 + \frac{1}{2}) = \frac{1}{2}[(434 \text{ nm})/1.51](5 + \frac{1}{2}) = \boxed{790 \text{ nm.}}$

45. With respect to the incident wave, the wave that reflects from the top surface of the soap bubble has a phase change of

$\phi_1 = \pi$.

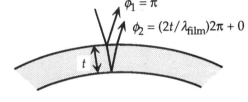

With respect to the incident wave, the wave that reflects from the air at the bottom surface of the bubble has a phase change due to the additional path-length but, because the index of the soap film is greater than the index of air, no phase change on reflection:

$\phi_2 = (2t/\lambda_{film})2\pi + 0$.

For constructive interference, the net phase change is

$\phi = (2t/\lambda_{film})2\pi - \pi = m_c 2\pi$, $m_c = 0, 1, 2, \ldots$, or

$t = \frac{1}{2}\lambda_{film}(m_c + \frac{1}{2}) = \frac{1}{2}(\lambda_1/n_{film})(m_c + \frac{1}{2})$, $m_c = 0, 1, 2, \ldots$.

For destructive interference, the net phase change is

$\phi = (2t/\lambda_{film})2\pi - \pi = (2m_d - 1)\pi$, $m_d = 1, 2, 3, \ldots$, or

$t = \frac{1}{2}\lambda_{film}m_d = \frac{1}{2}(\lambda_2/n_{film})m_d$, $m_d = 1, 2, 3, \ldots$.

Because the maximum and minimum are adjacent, $m_d = m_c + 1$, so we have

$t = \frac{1}{2}(\lambda_1/n_{film})(m_c + \frac{1}{2}) = \frac{1}{2}(\lambda_2/n_{film})(m_c + 1)$;

$(666 \text{ nm})(m_c + \frac{1}{2}) = (555 \text{ nm})(m_c + 1)$, which gives $m_c = 2$, and thus $m_d = 3$.

We find the thickness from

$t = \frac{1}{2}(\lambda_2/n_{film})m_d = \frac{1}{2}[(555 \text{ nm})/1.34](3) = \boxed{621 \text{ nm.}}$

47. (a) With respect to the incident wave, the wave that reflects
    from the top surface of the oil film has a phase change of
    $\phi_1 = \pi$.
    With respect to the incident wave, the wave that reflects
    from the air at the bottom surface of the bubble has a phase
    change due to the additional path-length but no phase
    change on reflection:
    $\phi_2 = (2t/\lambda_{film})2\pi + 0$.
    For constructive interference, the net phase change is
    $\phi = (2t/\lambda_{film})2\pi - \pi = m_c 2\pi$, $m_c = 0, 1, 2, \ldots$, or
    $t = \tfrac{1}{2}\lambda_{film}(m_c + \tfrac{1}{2}) = \tfrac{1}{2}(\lambda_1/n_{film})(m_c + \tfrac{1}{2})$, $m_c = 0, 1, 2, \ldots$.
    Because the minimum thickness occurs for $m = 0$, we have
    $t_{min} = \tfrac{1}{2}[(550 \text{ nm})/1.2](0 + \tfrac{1}{2}) = \boxed{115 \text{ nm.}}$

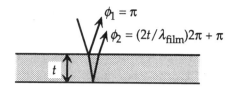

   (b) To maintain maximally reflected light, if n were increased, the wavelength would have to be
       $\boxed{\text{longer,}}$ so the ratio remained constant.
   (c) Because the index of the film is less than the index of water, there would be an additional phase
       change of π introduced from the reflection at the bottom surface. What was maximally reflected
       would now be minimally reflected.
       $\boxed{\text{There would be no reflection of green light, but some reflection of red and blue, giving a purple hue.}}$

51. We find the angle of refraction from
    $\sin \theta = n_{coat} \sin \theta'$;
    $\sin 30° = 1.25 \sin \theta'$, which gives $\theta' = 23.6°$.
    We must find the phase difference for the two rays along a
    common wavefront, indicated on the figure. We do this by
    referring each ray to the incident point, where the two rays
    separate. The wave that reflects from the coating has a
    phase change of
    $\phi_1 = (\ell/\lambda)2\pi + \pi$.
    The wave that reflects from the glass has a phase change of
    $\phi_2 = (2d/\lambda_{coat})2\pi + \pi = (2dn_{coat}/\lambda)2\pi + \pi$.
    For destructive interference, the net phase change is
    $\phi = (2dn_{coat}/\lambda)2\pi + \pi - [(\ell/\lambda)2\pi + \pi] = (2m-1)\pi$, $m = 1, 2, 3, \ldots$.
    The minimum thickness means minimum d and thus $m = 1$, so we have
    $(2dn_{coat}/\lambda) - (\ell/\lambda) = \tfrac{1}{2}$.
    From the figure, we see that
    $D = 2t \tan \theta'$; $\ell = D \sin \theta = 2t \tan \theta' \sin \theta$; $d = t/\cos \theta'$.
    When we substitute these in the above equation, we have
    $(2t/\lambda)\{[n/(\cos \theta')] - (\tan \theta' \sin \theta)\} = \tfrac{1}{2}$;
    $[2t/(550 \text{ nm})]\{[1.25/(\cos 23.6°)] - (\tan 23.6° \sin 30°)\} = \tfrac{1}{2}$, which gives $\boxed{t = 120 \text{ nm.}}$

55. The number of fringe shifts produced by a mirror movement of ΔL is
    $N = 2\Delta L/\lambda$.
    For the minimum mirror movement, the minimum number of fringes is
    $N_{min} = 2(0.03 \times 10^{-3} \text{ m})/(590 \times 10^{-9} \text{ m}) = 102$.
    The uncertainty in the number due to the uncertainty in the wavelength is
    $dN = -(2\Delta L/\lambda^2)\, d\lambda$, or $dN/N = -d\lambda/\lambda = 0.1\%$.
    $\boxed{\text{A minimum of approximately 100 fringes must be counted to 0.1 of a fringe.}}$

59. We neglect the variation in amplitude:

$E_1 \simeq E_2 = E$.

Because the sources radiate in phase, they will be in phase at the midpoint, $x' = 0$.

Between the sources the waves travel in opposite directions, so we have

$E_{net} = E \sin(kx' - \omega t) + E \sin(kx' + \omega t) = E[\sin(kx' - \omega t) + \sin(kx' + \omega t)]$

$= 2E \sin(kx') \cos(\omega t) = 2E \sin(2\pi f x'/c) \cos(2\pi f t)$

We find the energy density from

$u = \langle \frac{1}{2}\varepsilon_0 E_{net}^2 \rangle = 2\varepsilon_0 E^2 \sin^2(2\pi f x'/c) \langle \cos^2(2\pi f t) \rangle$.

The time average of $\langle \cos^2 \theta \rangle = \frac{1}{2}$. If we change to the distance from one source, $x = x' + \frac{1}{2}L$, we have

$\boxed{u \propto \sin^2[2\pi f(x - \frac{1}{2}L)/c].}$

61. Because the obstacles are much smaller than the wavelength, the reflections radiate uniformly in all directions. The maxima will be determined by the phase shift only. When the incoming wavefront reaches the top obstacle, it still has to travel a distance $L_1 = d \sin\theta$ to reach the bottom obstacle. When it reaches the bottom obstacle, the wave reflecting from the top obstacle will have traveled a distance $L_2 = d \sin\theta'$. For maxima, the net path-length difference must be a multiple of the wavelength:

$L_1 - L_2 = (d \sin\theta) - (d \sin\theta')$

$= m\lambda$, $m = 0, \pm 1, \pm 2, \ldots$, which gives

$\boxed{\sin\theta' = \sin\theta - (m\lambda/d), \quad m = 0, \pm 1, \pm 2, \ldots .}$

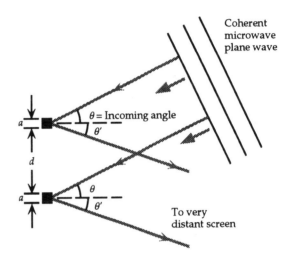

65. When the signals are 90° out of phase, the total phase difference is

$\phi = (\Delta L/\lambda)2\pi + (\pi/2)$

$= \{[(\lambda/4)\cos\theta]/\lambda\}2\pi + (\pi/2) = (\pi/2)(1 + \cos\theta)$.

The radiated intensity is

$I = 4I_0 \cos^2(\phi/2) = 4I_0 \cos^2[(\pi/4)(1 + \cos\theta)]$.

The intensity will be maximum when

$(\pi/4)(1 + \cos\theta) = 0$ or π, which gives

$1 + \cos\theta = 0$ or 4; $\cos\theta = -1$ or 3.

Because $\cos\theta = 3$ is impossible, there is only one maximum at $\theta = 180°$.

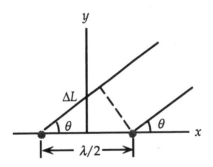

69. We find the angle of refraction from
$$\sin \theta = n \sin \theta'.$$
We must find the phase difference for the two rays along a common wavefront, indicated on the figure. We do this by referring each ray to the incident point, where the two rays separate. The wave that reflects from the coating has a phase change of

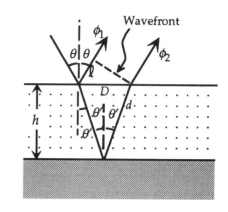
Wavefront

$$\phi_1 = (\ell/\lambda)2\pi + \pi.$$
The wave that reflects from the mirror has a phase change of
$$\phi_2 = (2d/\lambda_{glass})2\pi + 0 = (2nd/\lambda)2\pi.$$
For constructive interference, the net phase change is
$$\phi = (2nd/\lambda)2\pi - [(\ell/\lambda)2\pi + \pi] = m2\pi, \quad m = 1, 2, 3, \ldots;$$
$$(2nd/\lambda) - (\ell/\lambda) = m - \tfrac{1}{2}, \quad m = 1, 2, 3, \ldots.$$
From the figure, we see that
$$D = 2h \tan \theta'; \quad \ell = D \sin \theta = 2h \tan \theta' \sin \theta; \quad d = h/\cos \theta'.$$
When we substitute these in the above equation, we have
$$(2h/\lambda)\{[n/(\cos \theta')] - (\tan \theta' \sin \theta)\} = m - \tfrac{1}{2}, \quad m = 1, 2, 3, \ldots.$$
We reduce the term in { } by using the result from the refraction:
$$[n/(\cos \theta')] - (\tan \theta' \sin \theta) = (n - \sin \theta' \sin \theta)/\cos \theta'$$
$$= [n - (\sin^2 \theta)/n]/[1 - (\sin^2 \theta)/n^2]^{1/2} = (n^2 - \sin^2 \theta)^{1/2}.$$
The condition for maxima becomes
$$(2h/\lambda)(n^2 - \sin^2 \theta)^{1/2} = m - \tfrac{1}{2}, \quad m = 1, 2, 3, \ldots, \text{ which gives}$$
$$\sin^2 \theta = n^2 - [\lambda(m - \tfrac{1}{2})/2h]^2;$$

$$\boxed{\theta = \sin^{-1}\{n^2 - [\lambda(m - \tfrac{1}{2})/2h]^2\}^{1/2}, \quad m = 1, 2, 3, \ldots.}$$

CHAPTER 39

Note: For problems involving the resolution of the diffraction pattern of a circular aperture, we will use the approximation $\theta_{min} \simeq \lambda/D$.

9.   We find the number of lines from
$R = mN$;
$10^4 = (1)N$, which gives $N = 10^4$ lines, so $d = \ell/N = (2 \text{ cm})/(10^4) = 2 \times 10^{-4} \text{ cm} = 2 \times 10^{-6} \text{ m}$.
We find the angles for the two orders from
$\sin \theta = m\lambda/d$;
$\sin \theta_1 = (1)(580 \times 10^{-9} \text{ m})/(2 \times 10^{-6} \text{ m}) = 0.290$, which gives $\theta_1 = 16.9°$;
$\sin \theta_2 = (2)(580 \times 10^{-7} \text{ m})/(2 \times 10^{-6} \text{ m}) = 0.580$, which gives $\theta_2 = 35.5°$.
The angular separation of the two orders is $\boxed{\Delta\theta = 18.6°.}$

13.   The phase difference for rays through adjacent slits is due to the path-length difference. For a maximum, the path-length difference is a multiple of the wavelength. From the diagram, the additional distance to the grating is $d \sin \alpha$ for the ray on the right and $d \sin \beta$ for the ray on the left. For the maxima, we have
$d \sin \alpha - d \sin \beta = m\lambda, \quad m = 0, \pm 1, \pm 2, \ldots$, which gives
$\boxed{\sin \beta = \sin \alpha - (m\lambda/d), \quad m = 0, \pm 1, \pm 2, \ldots.}$

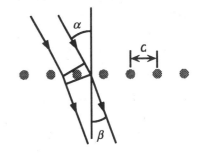

17.   The intensity pattern for the single slit is
$I/I_0 = (\sin^2 \alpha)/\alpha^2$,
where $\alpha = (\pi a \sin \theta)/\lambda$, and the minima occur when $\alpha_{min} = n\pi$, $n = \pm 1, \pm 2, \ldots$.
If we take the maxima to be halfway between the minima, we have
$\alpha_{max} \simeq (n + \tfrac{1}{2})\pi$, with $\alpha_1 = \tfrac{3}{2}\pi$ for the peak next to the central peak.
We find the ratio of intensities from
$I_1/I_0 = (\sin^2 \alpha_1)/\alpha_1^2 = (-1)^2/(\tfrac{3}{2}\pi)^2$, which gives $\boxed{I_0/I_1 = 22.2.}$
We find the angle where the intensity is half the maximum by finding α from
$I/I_{max} = \tfrac{1}{2} = (\sin^2 \alpha)/\alpha^2$, or $\alpha = \sqrt{2} \sin \alpha$.
We can find α by plotting α (in rad) and $\sqrt{2} \sin \alpha$ as a function of α and finding the intersection, or by trial and error. The result is $\alpha = 1.39 \text{ rad} = 79.7°$.
We find the angle from the horizontal from
$\alpha = (\pi a \sin \theta)/\lambda$;
$1.39 \text{ rad} = [\pi(10 \times 10^{-6} \text{ m}) \sin \theta]/(470 \times 10^{-9} \text{ m})$, which gives $\sin \theta = 0.0208$. $\boxed{\theta = 1.19°.}$
Because $\sin \theta$ is proportional to the wavelength and $\sin \theta$ increases as θ increases, an increase in the wavelength will $\boxed{\text{increase } \theta.}$

21.   The angular position of the minima for a single slit are given by
$a \sin \theta = m\lambda$, or, for small angles, $\theta = m\lambda/a$, $m = \pm 1, \pm 2, \ldots$.
The angular spread of the first minimum is
$\Delta\theta_1 = 2\theta_1 = 2\lambda/a$.
To find the angular width at half-maximum, we find the phase at half-maximum:
$I = I_{max}(\sin^2 \alpha_h)/\alpha_h^2 = \tfrac{1}{2}I_{max}$,   or $\alpha_h^2 = 2 \sin^2 \alpha_h$.
This equation can be solved graphically or numerically to get $\alpha_h = 1.392 \text{ rad}$.
We find the corresponding angle from
$\alpha_h = (\pi a/\lambda) \sin \theta_h \simeq \pi a \theta_h/\lambda$, or
$\Delta\theta_h = 2\theta_h = 2\lambda\alpha_h/\pi a = 0.886\lambda/a$.
Thus we have $\boxed{\Delta\theta_h/\Delta\theta_1 = 0.443.}$

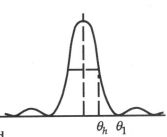

23. The path-length difference between the top and bottom of the slit for
the incident wave is
$$a \sin \theta_i.$$
The path-length difference between the top and bottom of
the slit for the diffracted wave is
$$a \sin \theta.$$

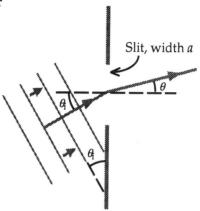

Slit, width a

When the net path-length difference is a multiple of a
wavelength, there will be an even number of segments of the
wave which will have a path-length difference of $\lambda/2$; there
will be minima given by
$$(a \sin \theta_i) - (a \sin \theta) = m\lambda, \quad m = \pm 1, \pm 2, \ldots, \quad \text{or}$$
$$\boxed{\sin \theta = \sin \theta_i - (m\lambda/a), \text{ where } m = \pm 1, \pm 2, \ldots .}$$
When $\theta = \theta_i$, the net path-length difference is zero, and there will be
constructive interference. $\boxed{\text{There is a "central maximum."}}$

31. We find the minimum size of the aperture from
$$\theta_{\min} \approx \lambda/D = S/L;$$
$$(525 \times 10^{-9} \text{ m})/D = (10 \text{ in})(2.54 \times 10^{-2} \text{ m/in})/(220 \text{ mi})(1.6 \times 10^3 \text{ m/mi}), \text{ which gives } \boxed{D = 0.73 \text{ m.}}$$
It would be better if the film were sensitive to $\boxed{\text{shorter}}$ wavelengths, allowing S to be smaller.

33. The image distance for a distant object will be the focal length.
We find the image distance for the object 5 m away from
$$(1/s) + (1/i) = 1/f:$$
$$(1/5 \text{ m}) + (1/i) = 1/0.050 \text{ m}, \text{ which gives } i = 0.0505 \text{ m} = 50.5 \text{ mm.}$$
If the aperture is d, from similar triangles we find the diameter of
the image on the film b from
$$d/i = b/(i-f), \text{ which gives } b = d(i-f)/i.$$
We estimate the diameter of the diffraction circle b' from
$$\theta_{\min} \approx \lambda/d = \tfrac{1}{2}b'/f, \text{ which gives } b' = 2\lambda f/d.$$
For the two contributions to be equal, we use an average
wavelength of 550 nm and have
$$b = b'; \quad d(i-f)/i = 2\lambda f/d, \quad \text{or}$$
$$d^2 = 2\lambda f i/(i-f) = 2(550 \times 10^{-9} \text{ m})(50 \times 10^{-3} \text{ m})(50.5 \text{ mm})/(50.5 \text{ mm} - 50 \text{ mm}), \text{ which gives}$$
$$\boxed{d = 2.4 \times 10^{-3} \text{ m} = 2.4 \text{ mm.}}$$

37. The maxima of the grating are given by
$$\beta = (\pi d \sin \theta)/\lambda = n\pi, \quad n = 0, \pm 1, \pm 2, \ldots .$$
The minima of the single-slit pattern are given by
$$\alpha = (\pi a \sin \theta)/\lambda = m\pi, \quad m = \pm 1, \pm 2, \ldots .$$
The angles where the maxima of the grating fall on the minima of the single slit, so that the order is
missing, are given by
$$\sin \theta = n\lambda/d = m\lambda/a, \text{ which becomes}$$
$$n/m = d/a = 3.$$
For the first minimum of the single slit, $m = 1$, we have $n = 3$, so the $\boxed{\text{third order}}$ of the double-slit
pattern is missing.

39. Because the angles are small, we have
$$\sin\theta \simeq \theta \simeq \tan\theta = y/R$$
$$= y/(3.0\times10^3\text{ mm}) = 3.33\times10^{-4}\,y,\text{ with }y\text{ in mm}.$$
The phase for the double-slit pattern is
$$\beta = (\pi d\sin\theta)/\lambda \simeq \pi d\,\theta/\lambda$$
$$= \pi(1.1\times10^{-3}\text{ m})(3.33\times10^{-4}\,y)/(690\times10^{-9}\text{ m}) = 1.67\,y,\text{ with }y\text{ in mm};$$
and the phase for the single-slit pattern is
$$\alpha = (\pi a\sin\theta)/\lambda \simeq \pi a\,\theta/\lambda$$
$$= \pi(0.20\times10^{-3}\text{ m})(3.33\times10^{-4}\,y)/(690\times10^{-9}\text{ m}) = 0.303\,y,\text{ with }y\text{ in mm}.$$
The intensity of the pattern is
$$I = I_0\{[\sin(2\beta)]/\sin\beta\}^2\,[(\sin\alpha)/\alpha]^2 = 4I_0(\cos^2\beta)[(\sin\alpha)/\alpha]^2.$$
For the four locations, we have

y, mm	θ, rad	α, rad	β, rad	I/I_0
0.050	1.67×10^{-5}	0.0152	0.0835	3.97
0.50	1.67×10^{-4}	0.152	0.835	1.79
1.5	5.00×10^{-4}	0.455	2.51	2.43
3.0	1.00×10^{-3}	0.909	5.01	0.26

41. The multiple-slit pattern has two small peaks between the large peaks. The smaller peaks occur when there is constructive interference between non-adjacent slits, but not between adjacent slits, so the intensity is much less. Because there are two peaks, there must be three spacings or   four slits.
This can be seen from the plot of $I = I_{max}[\sin(N\beta)/\sin\beta]^2$ for $N = 4$ shown in Figure 39–9.
The larger peaks occur when the rays from all slits are in phase, so the path difference between adjacent slits is $n\lambda$. This is the same as the condition for the double slit. For the small angles shown, the maxima of the grating are given by
$$\sin\theta \simeq \theta \simeq n\lambda/d,\ n = 0,\pm1,\pm2,\dots .$$
We use the data from the fifth (including the missing third) large peak:
$$0.01\text{ rad} = (5)(600\text{ nm})/d,\text{ which gives}\quad d \simeq 3\times10^5\text{ nm} = 0.3\text{ mm}.$$
We see that the third and sixth orders are missing. From Problem 37 we have
$$d/a = n/m = 3/1 = 6/2 = 3,\text{ so}\quad a = d/3 = 0.1\text{ mm}.$$

43. The maxima of the grating are given by
$$\sin\theta = n\lambda/d,\ n = 0,\pm1,\pm2,\dots .$$
The minima of the single-slit pattern are given by
$$\sin\theta = m\lambda/a,\ m = \pm1,\pm2,\dots .$$
Orders will be missing if
$$n\lambda/d = m\lambda/a,\text{ or }n = (d/a)m.$$
Because $d/a = (1.2\text{ mm})/(0.4\text{ mm}) = 3$ is an integer, the missing orders are
$$n = 3m = 3,6,9,\dots .$$
Because the angles are small, we find the angles for the missing orders from
$$\sin\theta \simeq \theta = m\lambda/a = m(589\times10^{-9}\text{ m})/(0.4\times10^{-3}\text{ m}) = (1.47\times10^{-3})m\text{ rad},\ m = 3,6,9,\dots .$$
The angles are $0.084°,\ 0.169°,\ 0.253°,\dots .$

45. The phase for the double-slit pattern is
$$\beta = (\pi d \sin \theta)/\lambda \simeq \pi d\, \theta/\lambda,$$
and the phase for the single-slit pattern is
$$\alpha = (\pi a \sin \theta)/\lambda \simeq \pi a\, \theta/\lambda,\ \text{which gives}\ \ \beta/\alpha = d/a = (0.30\ \text{mm})/(0.25\ \text{mm}) = 1.20.$$
The intensity of the pattern is
$$I = I_0\{[\sin(2\beta)]/(\sin \beta)\}^2\,[(\sin \alpha)/\alpha)]^2 = 4I_0(\cos^2 \beta)[(\sin \alpha)/\alpha]^2.$$
We find the value of α from
$$I = \tfrac{1}{2}I_{max} = 2I_0,\ \ \text{or}\ \ (\cos^2 \beta)[(\sin \alpha)/\alpha]^2 = 1/2.$$
If we take the square root and use $\beta = 1.20\alpha$, we get a transcendental equation for α:
$$\alpha/\sqrt{2} = \cos(1.20\alpha) \sin \alpha.$$
A numerical solution gives $\alpha = 0.600$ rad.
We find the angle from the central axis from
$$\sin \theta = \lambda\alpha/\pi a = (625 \times 10^{-9}\ \text{m})(0.600\ \text{rad})/\pi(0.25 \times 10^{-3}\ \text{m}) = 0.000478,$$
which gives $\boxed{\theta = 4.78 \times 10^{-4}\ \text{rad} = 0.0274°.}$

55. (a) This system is equivalent to a grating of N lines with a spacing of $d = \lambda/(N-1)$. Our expression for the phase uses the angle measured from the normal to the grating, so the phase for this array is
$$\beta = [\pi d \sin(90° - \theta)]/\lambda$$
$$= (\pi d \cos \theta)/\lambda = (\pi \cos \theta)/(N-1).$$
Because the antennas are perfect radiators, radiating uniformly in all directions, there is no single-slit effect. The intensity is
$$I = I_0[\sin(N\beta)/\sin \beta]^2$$
$$= \boxed{I_0\{\sin[(N\pi \cos \theta)/(N-1)]/\sin[(\pi \cos \theta)/(N-1)]\}^2.}$$

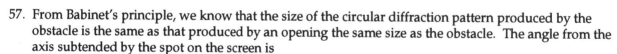

N antennas.

spacing $\dfrac{\lambda}{N-1}$

   (b) The maxima of the pattern occur at
$$\sin(90° - \theta) = \cos \theta = m\lambda/d = m(N-1),\ m = 0, \pm 1, \pm 2, \ldots .$$
Because $N > 1$, and $\cos \theta \leq 1$, the only possible value of m is 0.
The maxima occur at
$$\cos \theta = 0,\ \theta = \pi/2\ \text{and}\ \ 3\pi/2,\ \text{with a magnitude of}\ I = N^2 I_0.$$
At $\theta = 0$ and 180°, we have
$$I = I_0\{\sin[N\pi/(N-1)]/\sin[\pi/(N-1)]\}^2 \to I_0,\ \text{for}\ N > 2.$$
> A numerical calculation shows a slight decrease in I for θ close to 0° and 180° and then a buildup to the maxima at 90° and 270°.

57. From Babinet's principle, we know that the size of the circular diffraction pattern produced by the obstacle is the same as that produced by an opening the same size as the obstacle. The angle from the axis subtended by the spot on the screen is
$$\theta \simeq \tan \theta = \tfrac{1}{2}D/L.$$
We estimate the size of the obstacle d from
$$\theta \simeq \lambda/d = \tfrac{1}{2}D/L;$$
$$(633 \times 10^{-9}\ \text{nm})/d = \tfrac{1}{2}(0.70 \times 10^{-2}\ \text{m})/(2.5\ \text{m}),\ \text{which gives}\ \boxed{d = 4.5 \times 10^{-4}\ \text{m} = 0.45\ \text{mm}.}$$

59. The wavelength of the radiation is
$$\lambda = c/f = (3 \times 10^8\ \text{m/s})/(1.25 \times 10^{23}\ \text{Hz}) = 2.4 \times 10^{-15}\ \text{m}.$$
Because this is comparable to the size of the nucleus, we do not expect the first minimum to be at a small angle. From Babinet's principle, we estimate the angle from the first minimum of the diffraction pattern of a single slit the size of the nucleus:
$$\sin \theta \simeq \lambda/a$$
$$= (2.4 \times 10^{-15}\ \text{m})/2(3.2 \times 10^{-15}\ \text{m}) = 0.375,\ \text{which gives}\ \boxed{\theta = 22°.}$$
From the discussion of the Airy disk, the value obtained by using a factor of 1.22 in the expression for $\sin \theta$ is 27°.

CHAPTER 40

3. The speed of the light spot is the tangential speed of the rotation:
    $$v = r\omega = (70 \times 10^3 \text{ m})(\cos 25°)(900 \text{ rev}/\text{s})(2\pi \text{ rad}/\text{rev}) = \boxed{3.6 \times 10^8 \text{ m/s.}}$$
    Because this is not the speed of any energy or mass, there is no violation of the limitation of the speed of light. The photons are traveling radially away from the source. The spot is really the image of successive photons reflecting from the cloud.

7. To the spaceship's passenger, the diameter of the galaxy is contracted:
    $$L = L_0[1 - (u/c)^2]^{1/2} = (2.5 \times 10^{19} \text{ m})[1 - (0.999)^2]^{1/2} = 1.1 \times 10^{18} \text{ m.}$$
    The galaxy travels toward the spaceship with a speed $0.999c$, so the time for the galaxy to pass the spaceship is
    $$t = L/u = (1.1 \times 10^{18} \text{ m})/0.999c = \boxed{3.7 \times 10^9 \text{ s} \approx 120 \text{ yr.}}$$

13. (a) The observer on the satellite will measure the contracted length:
    $$L = L'[1 - (u/c)^2]^{1/2} = (30 \text{ m})[1 - (0.6)^2]^{1/2} = \boxed{24 \text{ m.}}$$
    (b) In the frame of the satellite, the spaceship travels at $0.6c$, so we find the time for point B to reach the satellite from
    $$t = L/0.6c = (24 \text{ m})/0.6(3 \times 10^8 \text{ m/s}) = \boxed{+ 13.3 \times 10^{-8} \text{ s.}}$$
    (c) In the frame of the spaceship, the light flash travels the rest length of the spaceship:
    $$t_1' = L'/c = (30 \text{ m})/(3 \times 10^8 \text{ m/s}) = \boxed{10 \times 10^{-8} \text{ s.}}$$
    (d) To the observer on the satellite, the point B was 24 m away and moving toward the satellite with a speed $0.6c$. The light flash travels at a speed c, so we have
    $$L = ut_1 + ct_1 = (u + c)t_1;$$
    $$24 \text{ m} = (0.6c + c)t_1 = 1.6(3 \times 10^8 \text{ m/s})t_1, \text{ which gives } \boxed{t_1 = 5.0 \times 10^{-8} \text{ s.}}$$

15. The last sight of the arch will be from the light that is coming vertically down when it reaches the eye. In the reference frame of the sprinter, the time from when this light left the arch is
    $$t = h/c.$$
    From the reference frame of the arch, the clocks in the sprinter's frame run slower, so the time for the light to travel in the reference frame of the arch is
    $$t' = t/[1 - (u/c)^2]^{1/2}.$$
    In this time the sprinter will move a horizontal distance
    $$x' = vt' = vh/c[1 - (u/c)^2]^{1/2} = \gamma hv/c.$$

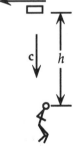

    Note that, even though the light is emitted directly above the sprinter, it does not travel vertically in the reference frame of the arch. The greater time is because the light must travel a distance greater than h. The speed of light is still c in the reference frame of the arch.

21. (a) We find the speed of the star from the Doppler redshift:
    $$\lambda/\lambda_0 = \{[1 + (u/c)]/[1 - (u/c)]\}^{1/2};$$
    $$(611.7 \text{ nm})/(587.6 \text{ nm}) = \{[1 + (u/c)]/[1 - (u/c)]\}^{1/2}, \text{ which gives}$$
    $$u = 0.0402c = \boxed{1.2 \times 10^7 \text{ m/s.}}$$
    (b) We estimate the distance from
    $$D = u/H = (1.2 \times 10^7 \text{ m/s})/(2.5 \times 10^{-18} \text{ s}^{-1}) = \boxed{4.8 \times 10^{24} \text{ m} \quad (5.1 \times 10^8 \text{ ly}).}$$

25. During the outward part of the journey, we find the frequency at which Jessica receives the birthday messages from

$$f_1/f_0 = [(1 - \beta)/(1 + \beta)]^{1/2} = \{[1 - (12/13)]/[1 + (12/13)]\}^{1/2} = (1/25)^{1/2} = 1/5.$$

Because $f_0 = 1 \text{ yr}^{-1}$, $f_1 = (1/5) \text{ yr}^{-1}$. The interval between receiving cards is $\Delta T_1 = 1/f_1 = \boxed{5 \text{ yr.}}$

During the return trip, we find the frequency at which Jessica receives the birthday messages from

$$f_2/f_0 = [(1 + \beta)/(1 - \beta)]^{1/2} = \{[1 + (12/13)]/[1 - (12/13)]\}^{1/2} = (25)^{1/2} = 5.$$

Because $f_0 = 1 \text{ yr}^{-1}$, $f_2 = 5 \text{ yr}^{-1}$. The interval between receiving cards is $\Delta T_2 = 1/f_2 = \boxed{(1/5) \text{ yr.}}$

Jessica will notice a change in frequency when she is halfway through the trip. Because she must receive 52 messages in all, we find the total time T in Jessica's reference frame from

$$N = [\tfrac{1}{2}T/(5 \text{ yr})] + \{\tfrac{1}{2}T/[(1/5) \text{ yr}]\} = 52, \text{ which gives } \boxed{T = 20 \text{ yr.}}$$

29. We choose the train as the S reference frame with the origin at the point A at the back of the train. The S' reference frame fixed to the ground is in standard orientation and moves with velocity $u = -0.7c$, so $\gamma = [1 - (u/c)^2]^{-1/2} = [1 - (0.7)^2]^{-1/2} = 1.40$.

(a) Because the bags are released simultaneously on the train, they are released at different times in S' and thus $\boxed{\text{the distance between the bags is not a measured length.}}$

(b) In the S frame the bags are dropped at

$$x_A = 0, t_A = 0; x_B = L_0 = 100 \text{ m}, \quad t_B = 0.$$

We use the Lorentz transformation to find the positions and times when the bags are dropped. In the S' frame the bags are dropped at

$$x_A' = 0, t_A' = 0;$$
$$x_B' = \gamma(x_B - ut_B) = (1.40)[(100 \text{ m}) - (-0.7c)(0)] = 140 \text{ m},$$
$$t_B' = \gamma[t_B - (ux_B/c^2)] = (1.40)\{0 - [(-0.7)(100 \text{ m})/(3 \times 10^8 \text{ m/s})]\} = 3.27 \times 10^{-7} \text{ s}.$$

The distance between bags on the ground is $\boxed{140 \text{ m.}}$ Note that this is longer than the rest length of the train. Because the bags are dropped at different times in the reference frame of the ground, we do not have a contracted length.

33. We find the speed of the *Enterprise* with respect to XG4T from the addition of the speed of the Klingon ship with respect to XG4T and the speed of the *Enterprise* with respect to the Klingon ship:

$$v = (v_1 + v_2)/(1 + v_1v_2/c^2) = (0.20c + 0.25c)/[1 + (0.20)(0.25)] = 0.43c.$$

The relative speed of the *Enterprise* with respect to the Klingon ship, as observed on the planet, is

$$v_{\text{rel}} = 0.43c - 0.20c = \boxed{0.23c.}$$

35. With $\Delta x = 0$, we use the invariant

$$(\Delta x')^2 - (c \Delta t')^2 = (\Delta x)^2 - (c \Delta t)^2 = -c \Delta t^2, \text{ which we rearrange:}$$
$$(c \Delta t')^2 = (\Delta x')^2 + (c \Delta t)^2.$$

Because all terms are positive, we have $|\Delta t'| > |\Delta t|$.

39. From the Lorentz transformation $x' = \gamma(x - ut)$, $t' = \gamma[t - (xu/c^2)]$, we find the transformations of the partial derivatives:

$$c\frac{\partial}{\partial x} = c\frac{\partial x'}{\partial x}\frac{\partial}{\partial x'} + c\frac{\partial t'}{\partial x}\frac{\partial}{\partial t'} = \gamma c\frac{\partial}{\partial x'} - \frac{\gamma u}{c}\frac{\partial}{\partial t'} = \gamma c\left(\frac{\partial}{\partial x'} - \frac{\gamma u}{c^2}\frac{\partial}{\partial t'}\right),$$

$$c^2\frac{\partial^2}{\partial x^2} = \gamma^2 c^2\left(\frac{\partial}{\partial x'} - \frac{\gamma u}{c^2}\frac{\partial}{\partial t'}\right)\left(\frac{\partial}{\partial x'} - \frac{\gamma u}{c^2}\frac{\partial}{\partial t'}\right) = \gamma^2\left(c^2\frac{\partial^2}{\partial x'^2} - 2u\frac{\partial^2}{\partial t'\partial x'} + \frac{u^2}{c^2}\frac{\partial^2}{\partial t'^2}\right);$$

$$\frac{\partial}{\partial t} = \frac{\partial x'}{\partial t}\frac{\partial}{\partial x'} + \frac{\partial t'}{\partial t}\frac{\partial}{\partial t'} = -\gamma u\frac{\partial}{\partial x'} + \gamma\frac{\partial}{\partial t'} = \gamma\left(-u\frac{\partial}{\partial x'} + \frac{\partial}{\partial t'}\right),$$

$$\frac{\partial^2}{\partial t^2} = \gamma^2\left(-u\frac{\partial}{\partial x'} + \frac{\partial}{\partial t'}\right)\left(-u\frac{\partial}{\partial x'} + \frac{\partial}{\partial t'}\right) = \gamma^2\left(u^2\frac{\partial^2}{\partial x'^2} - 2u\frac{\partial^2}{\partial t'\partial x'} + \frac{\partial^2}{\partial t'^2}\right).$$

When we use these in the wave equation, we get

$$\frac{\partial^2 E}{\partial t^2} - c^2\frac{\partial^2 E}{\partial x^2} = \gamma^2\left(1 - \frac{u^2}{c^2}\right)\frac{\partial^2 E}{\partial t'^2} - \gamma^2\left(1 - \frac{u^2}{c^2}\right)c^2\frac{\partial^2 E}{\partial x'^2} = \gamma^2\left(1 - \frac{u^2}{c^2}\right)\left(\frac{\partial^2 E}{\partial t'^2} - c^2\frac{\partial^2 E}{\partial x'^2}\right).$$

Because $\gamma^2 = [1 - (u^2/c^2)]^{-1}$, we have $\dfrac{\partial^2 E}{\partial t^2} - c^2\dfrac{\partial^2 E}{\partial x^2} = \dfrac{\partial^2 E}{\partial t'^2} - c^2\dfrac{\partial^2 E}{\partial x'^2}$.

The wave equation for the electric field has the $\boxed{\text{same form}}$ in both reference frames.

47. We find the work required to provide the kinetic energy change: $W = \Delta K = mc^2(\gamma_f - \gamma_i)$.

(a) $W = mc^2(\{1/[1 - (v_f/c)^2]^{1/2}\} - \{1/[1 - (v_i/c)^2]^{1/2}\})$
$= (938\text{ MeV})\{[1/(1 - 0.012)^{1/2}] - [1/(1 - 0)^{1/2}]\} = \boxed{0.047\text{ MeV.}}$

(b) $W = mc^2(\{1/[1 - (v_f/c)^2]^{1/2}\} - \{1/[1 - (v_i/c)^2]^{1/2}\})$
$= (938\text{ MeV})\{[1/(1 - 0.81^2)^{1/2}] - [1/(1 - 0.8^2)^{1/2}]\} = \boxed{36\text{ MeV.}}$

(c) $W = mc^2(\{1/[1 - (v_f/c)^2]^{1/2}\} - \{1/[1 - (v_i/c)^2]^{1/2}\})$
$= (938\text{ MeV})\{[1/(1 - 0.91^2)^{1/2}] - [1/(1 - 0.9^2)^{1/2}]\} = \boxed{110\text{ MeV.}}$

(d) To give the proton a speed of c would require increasing its kinetic energy to ∞. Because this requires an infinite energy source, $\boxed{\text{it is not possible to increase the speed to } c.}$

49. The rest energy of the proton is
$mc^2 = (1.67 \times 10^{-27}\text{ kg})(3 \times 10^8\text{ m/s})^2/(1.6 \times 10^{-13}\text{ J/MeV}) = 938\text{ MeV}$.

(a) We find the speed from
$p = m\gamma v = mc^2\gamma v/c^2$;
$(746 \times 10^3\text{ MeV}/c) = (938\text{ MeV}/c)\gamma v/c$, which gives $\gamma v/c = 794$.
From the definition of γ, we have
$\gamma^2[1 - (v/c)^2] = 1$, or $\gamma^2 - 1 = (\gamma v/c)^2 = (794)^2$, which gives $\gamma = 794$.
Because v is almost c, we write $v/c = 1 - x$, $x \ll 1$. From the definition of γ, we have
$\gamma^2[1 - (1 - x)^2] \approx \gamma^2 2x = 1$, or $x = 1/2\gamma^2 = 1/2(794)^2 = 7.9 \times 10^{-7}$.
The speed of the proton is $\boxed{[1 - (7.9 \times 10^{-7})]c.}$

(b) We find the kinetic energy of the proton from
$K = mc^2(\gamma - 1) = (938\text{ MeV})(794 - 1) = \boxed{745\text{ GeV.}}$

55. We carry more significant figures than justified by the data to show the differences.
    (a) If we ignore the recoil, all of the released energy will be in the kinetic energy of the He nucleus:
    $$K = \tfrac{1}{2}m_{He}v^2 = \tfrac{1}{2}p_{He}^2/m_{He}, \text{ or}$$
    $$p_{He}^2c^2 = 2m_{He}c^2K = 2[4(938 \text{ MeV})](6 \text{ MeV}), \text{ which gives } \boxed{p_{He}c = 212.19 \text{ MeV.}}$$
    (b) If the Np nucleus recoils, for momentum conservation we have
    $$P_{Np} - p_{He} = 0, \text{ or } P_{Np} = p_{He}.$$
    For energy conservation we have
    $$K = \tfrac{1}{2}m_{He}v^2 + \tfrac{1}{2}M_{Np}V^2 = (\tfrac{1}{2}p_{He}^2/m_{He}) + (\tfrac{1}{2}P_{Np}^2/M_{Np}) = \tfrac{1}{2}p_{He}^2c^2[(1/m_{He}c^2) + (1/M_{Np}c^2)];$$
    $$6 \text{ MeV} = \tfrac{1}{2}p_{He}^2c^2\{[1/4(938 \text{ MeV})] + [1/237(938 \text{ MeV})]\}, \text{ which gives } \boxed{p_{He}c = 210.42 \text{ MeV.}}$$
    (c) If the Np nucleus recoils, for momentum conservation we have
    $$P_{Np} - p_{He} = 0, \text{ or } P_{Np} = p_{He}.$$
    For energy conservation treated relativistically we have
    $$K + M_{Am}c^2 = (p_{He}^2c^2 + m_{He}^2c^4)^{1/2} + (P_{Np}^2c^2 + M_{Np}^2c^4)^{1/2}$$
    $$= (p_{He}^2c^2 + m_{He}^2c^4)^{1/2} + (p_{He}^2c^2 + M_{Np}^2c^4)^{1/2};$$
    $$6 \text{ MeV} + 241(938 \text{ MeV}) = \{p_{He}^2c^2 + [4(938 \text{ MeV})]^2\}^{1/2} + \{p_{He}^2c^2 + [237(938 \text{ MeV})]^2\}^{1/2}.$$
    When this equation is solved, either by successive squaring or numerically, we get
    $$\boxed{p_{He}c = 210.50 \text{ MeV.}}$$
    Because 6 MeV is much smaller than the rest masses, the nonrelativistic and relativistic results do not differ very much. It is more important to take into account the recoil.

57. The relation between the momentum and the kinetic energy of a nucleus is
    $$(p_Mc)^2 = E^2 - (Mc^2)^2 = (K_M + Mc^2)^2 - M^2c^4 = K_M^2 + 2K_MMc^2.$$
    For the conservation laws, we have
    energy: $M^*c^2 = E_{rad} + K_M + Mc^2;$
    momentum: $0 = p_M - (E_{rad}/c),$ which gives
    $$E_{rad} = p_Mc = (K_M^2 + 2K_MMc^2)^{1/2}.$$
    If we substitute this in the energy equation, we have
    $$M^*c^2 = (K_M^2 + 2K_MMc^2)^{1/2} + K_M + Mc^2, \text{ which we rearrange:}$$
    $$(M^* - M)c^2 - K_M = (K_M^2 + 2K_MMc^2)^{1/2}.$$
    After squaring and reducing, we get
    $$-2K_M(M^* - M)c^2 + (M^* - M)^2c^4 = 2K_MMc^2, \text{ which gives } \boxed{K_M = (M^* - M)^2c^2/2M^*.}$$

61. In the stationary frame, there is an acceleration toward the center of the disk, $a = R\omega^2$. To an observer on the disk, there is an outward gravitational force, $F = mR\omega^2$. We find the gravitational potential from
    $$F/m = -\partial\phi/\partial R = R\omega^2, \text{ which gives } \boxed{\phi = -\tfrac{1}{2}R^2\omega^2.}$$
    We find the frequency change from
    $$\Delta f/f = \phi/c^2 = -\tfrac{1}{2}R^2\omega^2/c^2.$$
    Because f decreases, the moving clock will be $\boxed{\text{slow}}$ compared to a clock at the center.

63. The rest energy of an electron or positron is 0.511 MeV. Because $E \gg mc^2$, the speed of each particle is $v \simeq c$, in opposite directions. With respect to one of the particles, the storage ring is moving at speed c. The other particle is moving at speed c in the same direction with respect to the storage ring. We find the relative speed from the addition of velocities:
    $$v' = (v_1 + v_2)/[1 + (v_1v_2/c^2)] = (c + c)/[1 + (c^2/c^2)] = \boxed{c.}$$

67. (a) We find the speed of Earth with respect to the cosmic ray, which is the speed with which the cosmic ray approaches Earth, from the contracted diameter:

$D = D_0/\gamma = (3/7)D_0$, which gives $\gamma = 7/3 = 2.33$;

$(v/c)^2 = 1 - (1/\gamma^2) = 1 - (3/7)^2$, which gives $\boxed{v = 0.90c.}$

(b) We find the energy of the approaching proton from

$E = mc^2\gamma = (1 \text{ GeV})(2.33) = \boxed{2.33 \text{ GeV.}}$

Note that the kinetic energy of the proton is 1.33 GeV.

73. (a) We find the factor γ from

$E = K + mc^2 = mc^2\gamma$;

$(250 \text{ MeV}) + (140 \text{ MeV}) = (140 \text{ MeV})\gamma$, which gives $\gamma = 2.79$.

The speed of the pi meson is

$v = c[1 - (1/\gamma^2)]^{1/2}$

$\quad = c[1 - (1/2.79)^2]^{1/2} = \boxed{0.933c.}$

(b) We find the momentum from

$p = m\gamma v = mc^2\gamma v/c^2$

$\quad = (140 \text{ MeV})(2.79)(0.933)/c = \boxed{3.6 \times 10^2 \text{ MeV}/c \quad (1.9 \times 10^{-19} \text{ kg} \cdot \text{m/s}).}$

(c) Because the clock (the decay) of the pi meson runs slower, the half-life is dilated by a factor of γ.

The distance before half of the pi mesons decay is

$L = v\tau = v\gamma\tau_0 = (0.933)(3 \times 10^8 \text{ m/s})(2.79)(1.5 \times 10^{-8}\text{ s}) = \boxed{12 \text{ m.}}$

75. Because the four products stick together and have the same mass, their momenta must be the same. From momentum conservation, we have

$\mathbf{p}_1 + \mathbf{p}_2 = 4\mathbf{p}'$, or $p + 0 = 4p'$.

From energy conservation, we have

$E_1 + mc^2 = 4E'$;

$(p^2c^2 + m^2c^4)^{1/2} + mc^2 = 4(p'^2c^2 + m^2c^4)^{1/2}$.

When we substitute the momentum result for p', we have

$(p^2c^2 + m^2c^4)^{1/2} + mc^2 = 4[(p^2c^2/16) + m^2c^4]^{1/2} = (p^2c^2 + 16m^2c^4)^{1/2}$.

After we square both sides and cancel common terms, we have

$(p^2c^2 + m^2c^4)^{1/2} = 7mc^2$;

$(pc)^2 + (mc^2)^2 = 49(mc^2)^2$, which gives

$p = (\sqrt{48})mc^2/c = (\sqrt{48})(938 \text{ MeV})/c = 6.5 \times 10^3 \text{ MeV}/c = \boxed{6.5 \text{ GeV}/c \quad (3.47 \times 10^{-18} \text{ kg} \cdot \text{m/s}).}$

CHAPTER 41 Regular Version

Note: At the atomic scale, it is most convenient to have energies in electron-volts and wavelengths in nanometers. A useful expression for the energy of a photon in terms of its wavelength is
$E = hf = hc/\lambda = (6.63 \times 10^{-34}\text{ J}\cdot\text{s})(3 \times 10^8\text{ m/s})(10^{-9}\text{ nm/m})/(1.60 \times 10^{-19}\text{ J/eV})\lambda;$
$E = (1.24 \times 10^3\text{ eV}\cdot\text{nm})/\lambda.$

9. (a) We estimate the smallest angle from
$$\theta_{min} \approx \lambda/D = h/Dp = h/D(2m_eK)^{1/2}$$
$$= (6.63 \times 10^{-34}\text{ J}\cdot\text{s})/(3.5 \times 10^{-4}\text{ m})[2(9.1 \times 10^{-31}\text{ kg})(2.5 \times 10^4\text{ eV})(1.6 \times 10^{-19}\text{ J/eV})]^{1/2}$$
$$= 2.2 \times 10^{-8}\text{ rad} = \boxed{(1.3 \times 10^{-6})^\circ\ (0.0046\text{''}).}$$

(b) We assume that the two objects can be resolved if the wavelength of the electrons is less than the separation of the objects. We find the energy from
$$K = p^2/2m_e = h^2/2m_e\lambda^2 = h^2/2m_ed^2$$
$$= (6.63 \times 10^{-34}\text{ J}\cdot\text{s})^2/2(9.1 \times 10^{-31}\text{ kg})(5.0 \times 10^{-9}\text{ m})^2 = 9.7 \times 10^{-21}\text{ J} = \boxed{0.06\text{ eV.}}$$

15. We find the momentum of the electron from
$K = p^2/2m_e;$
$(3 \times 10^3\text{ eV})(1.6 \times 10^{-19}\text{ J/eV}) = p^2/2(9.1 \times 10^{-31}\text{ kg})$, which gives $p = 3.0 \times 10^{-23}\text{ kg}\cdot\text{m/s}.$
We find the minimum uncertainty in its position from
$$\Delta x_{min} = \hbar/\Delta p_x$$
$$= (1.05 \times 10^{-34}\text{ J}\cdot\text{s})/(0.02)(3.0 \times 10^{-23}\text{ kg}\cdot\text{m/s}) = 1.8 \times 10^{-10}\text{ m} = \boxed{0.18\text{ nm.}}$$

19. The uncertainty in the electron's position is r, so the uncertainty in the momentum is
$\Delta p > \hbar/r.$
This puts a lower bound on the kinetic energy:
$K > (\Delta p)^2/2m = \hbar^2/2mr.$
The total energy of the electron is
$$E = U + K > -\frac{e^2}{4\pi\varepsilon_0 r} + \frac{\hbar^2}{2mr^2}.$$
To find the value of r that minimizes the energy, we set $dE/dr = 0$:
$$\frac{dE}{dr} = +\frac{e^2}{4\pi\varepsilon_0 r^2} - \frac{2\hbar^2}{2mr^3} = 0,\text{ which gives } r = \frac{4\pi\varepsilon_0\hbar^2}{me^2}.$$
When we substitute this value into the energy, we get
$$E_{min} = -\left(\frac{e^2}{4\pi\varepsilon_0}\right)\left(\frac{me^2}{4\pi\varepsilon_0\hbar^2}\right) + \left(\frac{\hbar^2}{2m}\right)\left(\frac{me^2}{4\pi\varepsilon_0\hbar^2}\right)^2 = -\left(\frac{m}{2}\right)\left(\frac{e^2}{4\pi\varepsilon_0\hbar}\right)^2.$$
The numerical values are
$$r = \frac{4\pi\varepsilon_0\hbar^2}{me^2} = \frac{(1.055 \times 10^{-34}\text{ J}\cdot\text{s})^2}{(9 \times 10^9\text{ N}\cdot\text{m}^2/\text{C}^2)(9.1 \times 10^{-31}\text{ kg})(1.6 \times 10^{-19}\text{ C})^2} = 0.53 \times 10^{-10}\text{ m};$$
$$E_{min} = -\frac{m}{2}\left(\frac{e^2}{4\pi\varepsilon_0\hbar}\right)^2 = -\frac{9.1 \times 10^{-31}\text{ kg}}{2}\left[\frac{(1.6 \times 10^{-19}\text{ C})^2(9 \times 10^9\text{ N}\cdot\text{m}^2/\text{C}^2)}{(1.055 \times 10^{-34}\text{ J}\cdot\text{s})}\right]^2 \frac{1}{(1.6 \times 10^{-19}\text{ J/eV})} = \boxed{-13.6\text{ eV.}}$$

21. If we write the potential energy as
    $U = U_0 |x|/a$,
    we see that the potential energy is symmetric about $x = 0$. If the position of the particle is undetermined
    to an accuracy b, so that $\Delta x = b$, the uncertainty in the momentum is
    $\Delta p > \hbar/b$,
    which we take to be the minimum momentum of the particle. The energy of the particle is
    $E = (p^2/2m) + U = [(\hbar/b)^2/2m] + (U_0 b/a) = (\hbar^2/2mb^2) + (U_0 b/a)$.
    To find the value of b that minimizes the energy, we set $dE/db = 0$:
    $dE/db = -(2\hbar^2/2mb^3) + (U_0/a) = 0$, which gives $b = (\hbar^2 a/mU_0)^{1/3}$.
    The estimate for the lowest energy of the particle is
    $E = [\hbar^2/2m(\hbar^2 a/mU_0)^{2/3}] + [U_0(\hbar^2 a/mU_0)^{1/3}/a] = \boxed{\frac{3}{2}(\hbar^2 U_0^2/ma^2)^{1/3}}$.

27. The photoelectric current is initiated when the photon energy becomes greater than the work function.
    The maximum wavelength corresponds to this minimum energy:
    $E = hf_{min} = hc/\lambda_{max} = W$;
    Aluminum: $(1.24 \times 10^3 \text{ eV} \cdot \text{nm})/\lambda_{max, Al} = 4.28 \text{ eV}$, which gives $\boxed{\lambda_{max, Al} = 290 \text{ nm}}$;
    Cesium: $(1.24 \times 10^3 \text{ eV} \cdot \text{nm})/\lambda_{max, Cs} = 2.14 \text{ eV}$, which gives $\boxed{\lambda_{max, Cs} = 579 \text{ nm}}$;
    Nickel: $(1.24 \times 10^3 \text{ eV} \cdot \text{nm})/\lambda_{max, Ni} = 5.15 \text{ eV}$, which gives $\boxed{\lambda_{max, Ni} = 241 \text{ nm}}$;
    Lead: $(1.24 \times 10^3 \text{ eV} \cdot \text{nm})/\lambda_{max, Pb} = 4.25 \text{ eV}$, which gives $\boxed{\lambda_{max, Pb} = 292 \text{ nm}}$.

35. At the threshold wavelength, the kinetic energy of the photoelectrons is zero, so we have
    $K = hf - W = 0$;
    $W = hf = hc/\lambda$
    $= (1.24 \times 10^3 \text{ eV} \cdot \text{nm})/\lambda = (1.24 \times 10^3 \text{ eV} \cdot \text{nm})/(270 \text{ nm}) = \boxed{4.60 \text{ eV} \ (7.37 \times 10^{-19} \text{ J})}$.
    We find the maximum kinetic energy for a wavelength of 120 nm from
    $K_{max} = hf - W = hc/\lambda - W$
    $= [(1.24 \times 10^3 \text{ eV} \cdot \text{nm})/(120 \text{ nm})] - (4.60 \text{ eV}) = \boxed{5.75 \text{ eV} \ (9.21 \times 10^{-19} \text{ J})}$.

37. (a) Because the photon gives up half its energy, we have
    $hf = \frac{1}{2}hf_0$, or $hc/\lambda = \frac{1}{2}hc/\lambda_0$, which gives $\lambda = 2\lambda_0$.
    From the Compton scattering analysis, we have
    $\lambda - \lambda_0 = (h/m_e c)(1 - \cos\theta)$;
    $2\lambda_0 - \lambda_0 = (2.43 \times 10^{-12} \text{ m})(1 - \cos 180°)$, which gives $\lambda_0 = 4.85 \times 10^{-12} \text{ m}$.
    The frequency of the incident photon is
    $f_0 = c/\lambda_0 = (3 \times 10^8 \text{ m/s})/(4.85 \times 10^{-12} \text{ m}) = \boxed{6.18 \times 10^{19} \text{ Hz}}$.
    The energy of the incident photon is
    $E_0 = hf_0 = (6.63 \times 10^{-34} \text{ J} \cdot \text{s})(6.18 \times 10^{19} \text{ Hz}) = 4.10 \times 10^{-14} \text{ J} = \boxed{0.256 \text{ MeV}}$.
    (b) The kinetic energy of the electron is
    $K = \frac{1}{2}hf_0 = \frac{1}{2}(0.256 \text{ MeV}) = 0.128 \text{ MeV}$.
    Because this is on the order of $m_e c^2$, we use relativistic expressions to find the velocity:
    $E = K + m_e c^2 = m_e c^2/[1 - (v/c)^2]^{1/2}$;
    $0.128 \text{ MeV} + 0.511 \text{ MeV} = (0.511 \text{ MeV})/[1 - (v/c)^2]^{1/2}$, which gives
    $\boxed{v = 0.60c \text{ in the direction of the original photon}}$.

43. We find the energy of the emitted photon from
    $hf = E_i - E_f = [(-13.6 \text{ eV})/n_i^2] - [(-13.6 \text{ eV})/n_f^2] = (13.6 \text{ eV})[(1/n_f^2) - (1/n_i^2)]$.
    For the jump from $n = 3$ to $n = 1$, we have
    $hf = (13.6 \text{ eV})[(1/1^2) - (1/3^2)] = \boxed{12.1 \text{ eV}}$;
    $\lambda = (1.24 \times 10^3 \text{ eV} \cdot \text{nm})/(12.1 \text{ eV}) = \boxed{103 \text{ nm}}$.
    For the jump from $n = 5$ to $n = 3$, we have
    $hf = (13.6 \text{ eV})[(1/3^2) - (1/5^2)] = \boxed{0.97 \text{ eV}}$;
    $\lambda = (1.24 \times 10^3 \text{ eV} \cdot \text{nm})/(0.97 \text{ eV}) = \boxed{1280 \text{ nm}}$.

47. (a) We find the dimensions of α from

$$[\alpha] = [e^2]/[\varepsilon_0][\hbar][c]$$
$$= [Q^2]/[Q^2 T^2 M^{-1} L^{-3}][ML^2 T^{-1}][LT^{-1}] = 1,$$

so $\boxed{\alpha \text{ is a dimensionless constant.}}$

The value of α is

$$\alpha = e^2/4\pi\varepsilon_0 \hbar c$$
$$= (1.60 \times 10^{-19}\text{ C})^2/4\pi(8.85 \times 10^{-12}\text{ C}^2/\text{N} \cdot \text{m}^2)(1.055 \times 10^{-34}\text{ J} \cdot \text{s})(3 \times 10^8\text{ m/s}) = \boxed{0.00731.}$$

The value of $1/\alpha = 1/0.00731 = \boxed{137.}$

(b) The energy of the nth hydrogen level is

$$E_n = -(m_e/2n^2)(e^2/4\pi\varepsilon_0 \hbar)^2 = \boxed{-m_e c^2 \alpha^2/2n^2.}$$

(c) The kinetic energy of the lowest level is

$$K_1 = -E_1;$$
$$\tfrac{1}{2}m_e v^2 = m_e c^2 \alpha^2/2(1)^2, \text{ which gives } \boxed{v_1 = \alpha c.}$$

49. The potential energy of the mass is

$$U(r) = mgh = mg\alpha r^2.$$

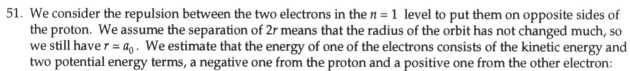

If we compare this to the potential energy of a harmonic oscillator,
$\tfrac{1}{2}kx^2$, we see that the motion in r will be simple harmonic, with a
force constant of $k = 2mg\alpha$. The angular frequency of the motion will be

$$\omega = (k/m)^{1/2} = (2g\alpha)^{1/2}.$$

From the result of Example 41–6, the allowed energies of this system are

$$E_n = n\hbar\omega,$$

so the separation of allowed energies is

$$\Delta E = (\Delta n)\,\hbar\omega = \hbar(2g\alpha)^{1/2} = (1.055 \times 10^{-34}\text{ J} \cdot \text{s})[2(9.8\text{ m/s}^2)(0.25 \times 10^2\text{ m}^{-1})]^{1/2}$$
$$= \boxed{2.3 \times 10^{-33}\text{ J } (1.4 \times 10^{-14}\text{ eV}).}$$

51. We consider the repulsion between the two electrons in the $n = 1$ level to put them on opposite sides of
the proton. We assume the separation of $2r$ means that the radius of the orbit has not changed much, so
we still have $r \simeq a_0$. We estimate that the energy of one of the electrons consists of the kinetic energy and
two potential energy terms, a negative one from the proton and a positive one from the other electron:

$$E \simeq (p^2/2m_e) - (e^2/4\pi\varepsilon_0 a_0) + [e^2/4\pi\varepsilon_0(2a_0)].$$

If we estimate the kinetic energy as that in the Bohr atom, we have

$$p^2/2m_e = e^2/8\pi\varepsilon_0 a_0, \text{ so we have}$$
$$E \simeq (e^2/8\pi\varepsilon_0 a_0) - (e^2/4\pi\varepsilon_0 a_0) + (e^2/8\pi\varepsilon_0 a_0) \simeq 0.$$

The values of the terms are $+13.6$ eV, -27.2 eV, and $+13.6$ eV, so we would expect that the energy
required to ionize one of the electrons would be $\boxed{< 1 \text{ eV.}}$

53. We consider each hydrogen atom to be in a cube with side L. We find L from

$$L^3 = m_p/\rho = (1.67 \times 10^{-27}\text{ kg})/(10^{-10}\text{ g/cm}^3)(10^{-3}\text{ kg/g})(10^6\text{ cm}^3/\text{m}^3), \text{ which gives } L = 2.6 \times 10^{-7}\text{ m.}$$

This is the interatomic spacing, which must be three times the diameter of the atom:

$$L \simeq 3D = 3(2n^2 a_0);$$
$$2.6 \times 10^{-7}\text{ m} \simeq 6n^2(0.53 \times 10^{-10}\text{ m}), \text{ which gives } \boxed{n \simeq 28.}$$

55. We consider the n and ℓ levels, with $2(2\ell + 1)$ electrons per level:

$n = 1, \ell = 0$:   2 electrons,   2 electrons total;
$n = 2, \ell = 0$:   2 electrons,   4 electrons total;
$n = 2, \ell = 1$:   6 electrons,   10 electrons total;
$n = 3, \ell = 0$:   2 electrons,   12 electrons total;
$n = 3, \ell = 1$:   6 electrons,   18 electrons total;
$n = 3, \ell = 2$:   10 electrons,   28 electrons total.
$n = 4, \ell = 0$:   2 electrons,   30 electrons total;
$n = 4, \ell = 1$:   6 electrons,   36 electrons total;
$n = 4, \ell = 2$:   10 electrons,   46 electrons total;
$n = 4, \ell = 3$:   14 electrons,   60 electrons total;
$n = 5, \ell = 0$:   2 electrons,   62 electrons total;
$n = 5, \ell = 1$:   6 electrons,   68 electrons total;
$n = 5, \ell = 2$:   10 electrons,   78 electrons total;
$n = 5, \ell = 3$:   1 electron,   79 electrons total.

The least strongly bound electron will have $\boxed{n = 5, \ell = 3.}$

57. From the results of Problem 55, the maximum number of electrons in each of the n shells is

$n = 1$:   2 electrons;
$n = 2$:   8 electrons;
$n = 3$:   18 electrons;
$n = 4$:   32 electrons.

The Z values for the elements with filled shells are

$n = 1$ filled: $\boxed{Z = 2;}$
$n = 2$ filled: $\boxed{Z = 10;}$
$n = 3$ filled: $\boxed{Z = 28;}$
$n = 4$ filled: $\boxed{Z = 60.}$

61. For particles with $E = pc$, we set the closest distance equal to one-half the wavelength:

$$d = n_e^{-1/3} = \tfrac{1}{2}\lambda_F = \tfrac{1}{2}h/p_F = \tfrac{1}{2}hc/E_F, \text{ which gives } E_F = \pi\hbar c n_e^{1/3}.$$

The energy of the system is

$$E = C E_F N_e = C\pi\hbar c n_e^{1/3} N_e = \boxed{C\pi\hbar c N_e^{4/3} V^{-1/3}.}$$

We find the pressure from

$$p = -\,dE/dV = -C\pi\hbar c N_e^{4/3}(-\tfrac{1}{3})V^{-4/3} = \boxed{\tfrac{1}{3}C\pi\hbar c n_e^{4/3}.}$$

63. The Fermi energy for sodium is

$$E_F = \tfrac{1}{2}\hbar^2\pi^2 n_e^{2/3}/m_e$$
$$= \tfrac{1}{2}[(1.055 \times 10^{-34}\ \text{J}\cdot\text{s})^2\pi^2(2.657 \times 10^{28}\ \text{m}^{-3})^{2/3}/(9.1 \times 10^{-31}\ \text{kg}) = 5.3 \times 10^{-19}\ \text{J}.$$

This energy is small enough that we use the nonrelativistic expression to find the speed:

$$E_F = \tfrac{1}{2}m_e v^2;$$
$$5.3 \times 10^{-19}\ \text{J} = \tfrac{1}{2}(9.1 \times 10^{-31}\ \text{kg})v^2, \text{ which gives } \boxed{v = 1.1 \times 10^6\ \text{m/s.}}$$

69. The lifetime of the radioactive potassium is

$$\tau = T_{1/2}/\ln(2) = (1.3 \times 10^9\ \text{yr})/\ln(2) = 1.88 \times 10^9\ \text{yr}.$$

If N_0 potassium nuclei were part of the eruption, the number now present is

$$N_K = N_0 e^{-t/\tau}.$$

Because every potassium nucleus that decayed produced one stable argon nucleus, the number of argon nuclei now present is the number of potassium nuclei that have decayed:

$$N_{Ar} = N_0 - N_K = N_0(1 - e^{-t/\tau}), \text{ which gives}$$
$$N_K/N_{Ar} = e^{-t/\tau}/(1 - e^{-t/\tau});$$
$$10 = e^{-t/\tau}/(1 - e^{-t/\tau}), \text{ which gives } t = 0.0953\tau = (0.0953)(1.88 \times 10^9\ \text{yr}) = \boxed{1.8 \times 10^8\ \text{yr.}}$$

CHAPTER 41 Extended Version

Note: At the atomic scale, it is most convenient to have energies in electron-volts and wavelengths in nanometers. A useful expression for the energy of a photon in terms of its wavelength is
$E = hf = hc/\lambda = (6.63 \times 10^{-34} \text{ J} \cdot \text{s})(3 \times 10^8 \text{ m/s})(10^{-9} \text{ nm/m})/(1.60 \times 10^{-19} \text{ J/eV})\lambda;$
$E = (1.24 \times 10^3 \text{ eV} \cdot \text{nm})/\lambda.$

5. (a) The maxima are given by the Bragg condition
$2d \sin \theta = n\lambda = nh/p = nh/(2mK)^{1/2}, \quad n = 1, 2, \ldots .$
Because the maximum possible angle is 90°, we have
$2(0.074 \times 10^{-9} \text{ m}) \sin 90° = (4)(6.63 \times 10^{-34} \text{ J} \cdot \text{s})/[2(9.1 \times 10^{-31} \text{ kg})K]^{1/2}$, which gives
$\boxed{K = 1.8 \times 10^{-16} \text{ J} \quad (1.1 \times 10^3 \text{ eV}).}$
(b) For neutrons, we have
$2(0.074 \times 10^{-9} \text{ m}) \sin 90° = (4)(6.63 \times 10^{-34} \text{ J} \cdot \text{s})/[2(1.67 \times 10^{-27} \text{ kg})K]^{1/2}$, which gives
$\boxed{K = 9.6 \times 10^{-20} \text{ J} \quad (0.60 \text{ eV}).}$

11. To have diffraction effects, the wavelength must be on the order of the spacing:
$\lambda = h/p = h/(2mK)^{1/2} \approx d.$
For an electron, we have
$(6.63 \times 10^{-34} \text{ J} \cdot \text{s})/[2(9.1 \times 10^{-31} \text{ kg})K]^{1/2} \approx 10^{-10} \text{ m}$, which gives $\boxed{K \approx 2.4 \times 10^{-17} \text{ J} \ (150 \text{ eV}).}$
For a neutron, we have
$(6.63 \times 10^{-34} \text{ J} \cdot \text{s})/[2(1.67 \times 10^{-27} \text{ kg})K]^{1/2} \approx 10^{-10} \text{ m}$, which gives $\boxed{K \approx 1.3 \times 10^{-20} \text{ J} \ (0.082 \text{ eV}).}$

13. We estimate the fraction of electrons that tunnel through the barrier from
$F \approx e^{-(2/\hbar)a\sqrt{2m((U)-E)}}$
$\approx e^{-[2/(1.05 \times 10^{-34} \text{ J} \cdot \text{s})](0.10 \times 10^{-9} \text{ m})\sqrt{2(9.1 \times 10^{-31} \text{ kg})(2.0 \text{ eV} - 1.0 \text{ eV})(1.60 \times 10^{-19} \text{ J/eV})}} = e^{-1.0} = \boxed{0.36.}$

15. The energy of the truck is
$E = \tfrac{1}{2}mv^2 = \tfrac{1}{2}(3000 \text{ kg})(2.0 \text{ m/s})^2 = 6.0 \times 10^3 \text{ J}.$
The potential energy associated with the bump is
$U = mgh = (3000 \text{ kg})(9.8 \text{ m/s}^2)(0.25 \text{ m}) = 7.35 \times 10^3 \text{ J}.$
We estimate the tunneling factor from
$F \approx e^{-(2/\hbar)a\sqrt{2m((U)-E)}}$
$\approx e^{-[2/(1.05 \times 10^{-34} \text{ J} \cdot \text{s})](0.10 \text{ m})\sqrt{2(3000 \text{ kg})(7.35 \times 10^3 \text{ J} - 6.0 \times 10^3 \text{ J})}} \approx \boxed{e^{-5.4 \times 10^{36}} \approx 0.}$
Trucks do not tunnel!

21. We find the uncertainty in the momentum from
$\Delta p \approx \hbar/\Delta x \approx (1.05 \times 10^{-34} \text{ J} \cdot \text{s})/(1.5 \times 10^{-14} \text{ m}) \approx 7.0 \times 10^{-21} \text{ kg} \cdot \text{m/s}.$
Assuming that $\langle p \rangle = \Delta p$, we estimate the lowest energy as the uncertainty in the kinetic energy:
$K = \Delta K \approx (\Delta p)^2/2m \approx (7.0 \times 10^{-21} \text{ kg} \cdot \text{m/s})^2/2(1.67 \times 10^{-27} \text{ kg}) \approx \boxed{1.5 \times 10^{-14} \text{ J} \ (0.092 \text{ MeV}).}$

25. From $E = hc/\lambda$, we relate the energy spread to the wavelength spread:
$\Delta E = (hc/\lambda^2) \Delta\lambda.$
From the Heisenberg uncertainty relation, we have
$\Delta E = (hc/\lambda^2) \Delta\lambda = \hbar/\Delta t = h/(2\pi \Delta t), \quad \text{or}$
$\Delta\lambda = \lambda^2/(2\pi c \Delta t) = (720 \times 10^{-9} \text{ m})^2/2\pi(3 \times 10^8 \text{ m/s})(10^{-9} \text{ s}) = 2.7 \times 10^{-13} \text{ m} = \boxed{2.7 \times 10^{-4} \text{ nm}.}$

27. For an electron confined within $2R$, we find the uncertainty in the momentum from
$$\Delta p \approx \hbar/\Delta x \approx \hbar/2R$$
$$\approx (1.05 \times 10^{-34}\ \text{J} \cdot \text{s})/(2 \times 10^{-14}\ \text{m}) \approx 5.3 \times 10^{-21}\ \text{kg} \cdot \text{m/s},$$
which we take to be the momentum of the electron. For an electron, we have
$$m_e c = (9.1 \times 10^{-31}\ \text{kg})(3 \times 10^8\ \text{m/s}) = 2.7 \times 10^{-22}\ \text{kg} \cdot \text{m/s}.$$
Because $p \gg m_e c$, we must use relativistic expressions, and thus $v \approx c$.
The distance from the origin that this speed will produce in 1 s is
$$R' \approx ct = (3 \times 10^8\ \text{m/s})(1\ \text{s}) = \boxed{3 \times 10^8\ \text{m.}}$$
If the electron is initially measured within a sphere of radius $R = 10^{-10}$ m, we have
$$\Delta p \approx \hbar/\Delta x \approx \hbar/2R$$
$$\approx (1.05 \times 10^{-34}\ \text{J} \cdot \text{s})/(2 \times 10^{-10}\ \text{m}) \approx 5.3 \times 10^{-25}\ \text{kg} \cdot \text{m/s},$$
which we take to be the momentum of the electron. Because $p \ll m_e c$, we can use nonrelativistic expressions, and thus $v = p/m_e$. The distance from the origin that this speed will produce in 1 s is
$$R' \approx (p/m_e)t = [(5.3 \times 10^{-25}\ \text{kg} \cdot \text{m/s})/(9.1 \times 10^{-31}\ \text{kg})](1\ \text{s}) = \boxed{6 \times 10^5\ \text{m.}}$$

29. We choose the y-axis parallel to the width of the slit. The uncertainty in the y-position of an electron passing through the slit is the width of the slit a. We find the uncertainty in the y-momentum from
$$\Delta p_y \approx \hbar/\Delta y \approx \hbar/a,$$
which we take to be the y-momentum of the electron. The time for the electron to reach the screen is
$$t = Dm_e/p.\ \text{In this time the displacement in the } y\text{-direction is}$$
$$y = (\Delta p_y/m_e)t = (\hbar/am_e)(Dm_e/p) = \hbar D/pa = hD/2\pi pa.$$
Because the y-displacement could be in either direction, the total spread of the beam on the screen is
$$\Delta y = a + 2y = \boxed{a + (hD/\pi pa).}$$
To find the value of a that minimizes the spread, we set $\mathrm{d}(\Delta y)/\mathrm{d}a = 0$:
$$\mathrm{d}(\Delta y)/\mathrm{d}a = 1 - (hD/\pi pa^2) = 0,\ \text{which gives} \boxed{a_{\min} = (hD/\pi p)^{1/2}.}$$

31. For a neutron confined within $2R$, we find the uncertainty in the momentum from
$$\Delta p \approx \hbar/\Delta x \approx \hbar/2R$$
$$\approx (1.05 \times 10^{-34}\ \text{J} \cdot \text{s})/2(6.0 \times 10^{-15}\ \text{m}) \approx 8.8 \times 10^{-21}\ \text{kg} \cdot \text{m/s},$$
which we take to be the minimum momentum of the particle. The energy of the neutron is
$$E = (p^2/2m) + U.$$
If we assume that the lowest energy ≈ 0, we have
$$U_{\min} \approx -p^2/2m$$
$$= -[(8.8 \times 10^{-21}\ \text{kg} \cdot \text{m/s})^2/2(1.67 \times 10^{-27}\ \text{kg})]/(1.6 \times 10^{-19}\ \text{J/eV}) = -1.4 \times 10^5\ \text{eV} = \boxed{-0.14\ \text{MeV.}}$$

37. The maximum kinetic energy of the photoelectron is the energy of the photon less the energy to free the electron, which is the work function. We find the work function from
$$K_{\max} = hf - W;$$
$$1.7\ \text{eV} = (6.63 \times 10^{-34}\ \text{J} \cdot \text{s})(0.85 \times 10^{15}\ \text{Hz})/(1.6 \times 10^{-19}\ \text{J/eV}) - W,\ \text{which gives} \boxed{W = 1.8\ \text{eV.}}$$

43. We find the rate at which photons are emitted from
$$P = nhf = nhc/\lambda;$$
$$4 \times 10^{26}\ \text{W} = n[(1.24 \times 10^3\ \text{eV} \cdot \text{nm})/(500\ \text{nm})](1.6 \times 10^{-19}\ \text{J/eV}),\ \text{which gives} \boxed{n = 1.0 \times 10^{45}\ \text{photons/s.}}$$

45. We estimate 3 mm for the radius of the pupil of the night-adapted eye. We find the intensity from
$$I = nhf/\pi r^2 = nhc/\lambda \pi r^2$$
$$= (5\ \text{photons/s})(6.63 \times 10^{-34}\ \text{J} \cdot \text{s})(3 \times 10^8\ \text{m/s})/(550 \times 10^{-9}\ \text{m})\pi(3 \times 10^{-3}\ \text{m})^2 = \boxed{6 \times 10^{-14}\ \text{W/m}^2.}$$
For the intensity ratio, we have
$$I/I_0 = (6 \times 10^{-14}\ \text{W/m}^2)/(1.4 \times 10^3\ \text{W/m}^2) = \boxed{4 \times 10^{-17}.}$$

49. At the threshold wavelength, the kinetic energy of the photoelectrons is zero, so we have
$K = hf - W = 0$;
$W = hf = hc/\lambda$
   $= (1.24 \times 10^3 \text{ eV} \cdot \text{nm})/\lambda = (1.24 \times 10^3 \text{ eV} \cdot \text{nm})/(270 \text{ nm}) = \boxed{4.60 \text{ eV} \ (7.37 \times 10^{-19} \text{ J}).}$
We find the maximum kinetic energy for a wavelength of 120 nm from
$K_{max} = hf - W = hc/\lambda - W$
   $= [(1.24 \times 10^3 \text{ eV} \cdot \text{nm})/(120 \text{ nm})] - (4.60 \text{ eV}) = \boxed{5.75 \text{ eV} \ (9.21 \times 10^{-19} \text{ J}).}$

53. For a small wavelength or frequency range, the energy density is
$U = u(f, T) \, df.$
To change to the wavelength, we use $f = c/\lambda$, which gives
$df = - (c/\lambda^2) \, d\lambda.$
Because we are interested in the magnitude of the range, we drop the negative sign.
The energy density becomes
$$U = \frac{8\pi h}{c^3} \frac{f^3}{e^{hf/kT} - 1} \, df = \frac{8\pi h}{\lambda^3} \frac{1}{e^{hc/\lambda kT} - 1} \frac{c}{\lambda^2} \, d\lambda = \frac{8\pi hc}{\lambda^5} \frac{1}{e^{hc/\lambda kT} - 1} \, d\lambda.$$
For the wavelength range 690 nm to 710 nm, we use the average wavelength and $d\lambda = 20$ nm:
$$U = \frac{8\pi (6.63 \times 10^{-34} \text{ J} \cdot \text{s})(3 \times 10^8 \text{ m/s})}{(700 \times 10^{-9} \text{ m})^5} \frac{1}{e^{(6.63 \times 10^{-34} \text{ J} \cdot \text{s})(3 \times 10^8 \text{ m/s})/(700 \times 10^{-9} \text{ m})(1.38 \times 10^{-23} \text{ J/K})(6000 \text{ K})} - 1} (20 \times 10^{-9} \text{ m})$$
$$= \boxed{2.0 \times 10^{-2} \text{ J/m}^3.}$$
For the wavelength range 440 nm to 460 nm, we use the average wavelength and $d\lambda = 20$ nm:
$$U = \frac{8\pi (6.63 \times 10^{-34} \text{ J} \cdot \text{s})(3 \times 10^8 \text{ m/s})}{(450 \times 10^{-9} \text{ m})^5} \frac{1}{e^{(6.63 \times 10^{-34} \text{ J} \cdot \text{s})(3 \times 10^8 \text{ m/s})/(450 \times 10^{-9} \text{ m})(1.38 \times 10^{-23} \text{ J/K})(6000 \text{ K})} - 1} (20 \times 10^{-9} \text{ m})$$
$$= \boxed{2.6 \times 10^{-2} \text{ J/m}^3.}$$
The ratio of the energy densities is
$U_{700}/U_{450} = (2.0 \times 10^{-2} \text{ J/m}^3)/(2.6 \times 10^{-2} \text{ J/m}^3) = \boxed{0.77.}$

55. If d is the atomic spacing in the cubic lattice, the volume occupied by an atom is d^3.
We find d from the density:
$d^3 = Am/\rho$
   $= (64)(1.67 \times 10^{-27} \text{ kg})/(8900 \text{ kg/m}^3)$, which gives $d = 2.29 \times 10^{-10}$ m.
In a layer of the lattice, each atom presents an area d^2. In 5 layers the number of atoms per unit area is
$n = 5(1/d^2) = 5[1/(2.29 \times 10^{-10} \text{ m})^2] = 9.53 \times 10^{19} \text{ atoms/m}^2.$
We find the number of atoms, on average, that furnish 1 photoelectron/s from the rate R at which photoelectrons are emitted:
$N = n/R$
   $= (9.53 \times 10^{19} \text{ atoms/m}^2)/(8.3 \times 10^{10} \text{ photoelectrons/m}^2 \cdot \text{s})$
   $= \boxed{1.15 \times 10^9 \text{ atoms/(photoelectron/s)}.}$

59. Because animal cells are constantly being replaced, the concentration of ^{14}C in the living animal will be the concentration in the atmosphere until the animal dies. After that time no carbon atoms will be replaced, so the concentration will decrease with the half-life of ^{14}C.
    We find the elapsed time from
    $N(t) = 0.15N_0 = N_0 e^{-t/\tau},$ or
    $\ln 0.15 = -t/\tau = -0.693t/T_{1/2}$, which gives
    $t = T_{1/2}(-\ln 0.15)/0.693 = (5730 \text{ yr})(1.90)/(0.693) =$ $\boxed{1.57 \times 10^4 \text{ yr ago.}}$
    We find the year the mammoth lived from
    date $\approx (1.57 \times 10^4 \text{ yr}) - (0.20 \times 10^4 \text{ yr}) \approx$ $\boxed{13{,}700 \text{ B. C.}}$

67. (a) We find the peak wavelength for the Sun from
    $\lambda_{max} = (2.9 \times 10^{-3} \text{ m} \cdot \text{K})/T = (2.9 \times 10^{-3} \text{ m} \cdot \text{K})/(6000 \text{ K}) =$ $\boxed{0.48 \times 10^{-6} \text{ m} = 480 \text{ nm,}}$
    which is in the visible range.
    (b) For "red–hot," we use a peak wavelength of 650 nm:
    $T = (2.9 \times 10^{-3} \text{ m} \cdot \text{K})/\lambda_{max} \approx (2.9 \times 10^{-3} \text{ m} \cdot \text{K})/(650 \times 10^{-9} \text{ m}) \approx$ $\boxed{4500 \text{ K.}}$
    This is too high an estimate. Even at much lower temperatures (e.g., 1500 K, where the peak is at 2000 nm), there is significant radiation in the visible range. From the shape of the energy density curve given in Problem 53, there will be greater radiation at the red end (long wavelength) of the visible spectrum, so the object will appear red.

69. The work function is the energy that must be provided so free electrons in the tungsten can escape. We can treat the crack as a potential barrier with a height of 4.6 eV. The probability of tunneling through the barrier is

    $P \approx e^{-(2/\hbar)a\sqrt{2m(U-E)}}$

    $\approx e^{-[2/(1.05 \times 10^{-34} \text{ J} \cdot \text{s})](0.2 \times 10^{-9} \text{ m})\sqrt{2(9.1 \times 10^{-31} \text{ kg})(4.6 \text{ eV})(1.6 \times 10^{-19} \text{ J/eV})}} = e^{-4.4} = 0.012 =$ $\boxed{1.2\%.}$

CHAPTER 42 Extended Version

Note: At the atomic scale, it is most convenient to have energies in electron-volts and wavelengths in nanometers. A useful expression for the energy of a photon in terms of its wavelength is
$E = hf = hc/\lambda = (6.63 \times 10^{-34}\,\text{J}\cdot\text{s})(3 \times 10^{8}\,\text{m/s})(10^{-9}\,\text{nm/m})/(1.60 \times 10^{-19}\,\text{J/eV})\lambda;$
$E = (1.24 \times 10^{3}\,\text{eV}\cdot\text{nm})/\lambda.$
A factor that appears in the analysis of electron energies is
$e^{2}/4\pi\varepsilon_{0} = (1.60 \times 10^{-19}\,\text{C})^{2}(9 \times 10^{9}\,\text{N}\cdot\text{m}^{2}/\text{C}^{2}) = 2.30 \times 10^{-28}\,\text{J}\cdot\text{m}.$

7.   (a) Because the longest wavelengths come from the smallest energy changes, the states are
$n = 2, 3, \text{and } 4.$

(b) We use the energies from the hydrogen energy-level diagram to find the wavelengths:
$\lambda_{n} = hc/\Delta E = (1.24 \times 10^{3}\,\text{eV}\cdot\text{nm})/(E_{n} - E_{1});$
$\lambda_{2} = (1.24 \times 10^{3}\,\text{eV}\cdot\text{nm})/[(-3.40\,\text{eV}) - (-13.6\,\text{eV})]$
$= \boxed{122\,\text{nm};}$
$\lambda_{3} = (1.24 \times 10^{3}\,\text{eV}\cdot\text{nm})/[(-1.51\,\text{eV}) - (-13.6\,\text{eV})]$
$= \boxed{103\,\text{nm};}$
$\lambda_{4} = (1.24 \times 10^{3}\,\text{eV}\cdot\text{nm})/[(-0.85\,\text{eV}) - (-13.6\,\text{eV})]$
$= \boxed{97\,\text{nm.}}$
These wavelengths are in the $\boxed{\text{ultraviolet region.}}$

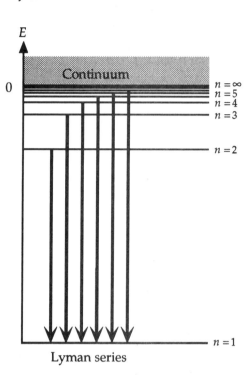

Lyman series

9.   The energy levels for triply ionized beryllium are given by
$E_{n} = -(13.6\,\text{eV})Z^{2}/n^{2} = -(13.6\,\text{eV})4^{2}/n^{2} = -(218\,\text{eV})/n^{2}.$
The longest wavelength comes from the smallest energy change, the $n = 2 \rightarrow n = 1$ transition:
$\lambda = hc/\Delta E = (1.24 \times 10^{3}\,\text{eV}\cdot\text{nm})/(E_{2} - E_{1})$
$= (1.24 \times 10^{3}\,\text{eV}\cdot\text{nm})/\{-(218\,\text{eV})[(1/2^{2}) - (1/1^{2})]\} = \boxed{7.58\,\text{nm.}}$

13. (a) We find the dimensions of α from
$[\alpha] = [e^{2}]/[\varepsilon_{0}][\hbar][c]$
$= [Q^{2}]/[Q^{2}T^{2}M^{-1}L^{-3}][ML^{2}T^{-1}][LT^{-1}] = 1,$
so $\boxed{\alpha \text{ is a dimensionless constant.}}$
The value of α is
$\alpha = e^{2}/4\pi\varepsilon_{0}\hbar c$
$= (2.30 \times 10^{-28}\,\text{J}\cdot\text{m})/(1.055 \times 10^{-34}\,\text{J}\cdot\text{s})(3 \times 10^{8}\,\text{m/s}) = \boxed{0.00731.}$
The value of $1/\alpha = 1/0.00731 = \boxed{137.}$

(b) The energy of the nth hydrogen level is
$E_{n} = -(m_{e}/2n^{2})(e^{2}/4\pi\varepsilon_{0}\hbar)^{2} = \boxed{-m_{e}c^{2}\alpha^{2}/2n^{2}.}$

(c) The kinetic energy of the lowest level is
$K_{1} = -E_{1};$
$\tfrac{1}{2}m_{e}v_{1}^{2} = m_{e}c^{2}\alpha^{2}/2(1)^{2},$ which gives $\boxed{v_{1} = \alpha c.}$

19. We find the frequency of the radiation for a jump from level $n + k$ to a level n from

$hf = 2\pi\hbar f = E_{n+k} - E_n;$

$f = (m_e/4\pi\hbar^3)(e^2/4\pi\varepsilon_0)^2\{(1/n^2) - [1/(n+k)^2]\}.$

We rewrite this as

$$\boxed{f = (m_e/4\pi\hbar^3)(e^2/4\pi\varepsilon_0)^2(1/n^2)(1 - \{1/[1 + (k/n)^2]\}).}$$

When $n \gg k$, we have

$f \simeq (m_e/4\pi\hbar^3)(e^2/4\pi\varepsilon_0)^2(1/n^2)\{1 - [1 - (2k/n)]\} \simeq (m_e/2\pi\hbar^3)(e^2/4\pi\varepsilon_0)^2(k/n^3), \quad n \gg k.$

If we use $L = n\hbar$, we have

$$\boxed{f \simeq (m_e/2\pi L^3)(e^2/4\pi\varepsilon_0)^2 k, \quad n \gg k.}$$

For the classical frequency, we have

$f_{cl} = 1/T = (m_e/2\pi L^3)(e^2/4\pi\varepsilon_0)^2,$

so $\boxed{f(n \gg k) = k(1/T).}$ We see that the two frequencies will be the same when $k = 1$.

21. For the deflection of a beam of low-energy electrons in a magnetic field gradient, we have

$\theta = (D\mu_B/mv^2)\, dB/dz = (D\mu_B/2K)\, dB/dz$

$= [(0.2\text{ m})(9.27 \times 10^{-24}\text{ J/T})/2(100\text{ eV})(1.6 \times 10^{-19}\text{ J/eV})](100\text{ T/m}) = 5.8 \times 10^{-6}\text{ rad} = \boxed{(3.3 \times 10^{-4})°.}$

This corresponds to a splitting of about 10 μm on a screen 1 m away.

23. The energies of the hydrogen atom are

$E_n = -\{1/[1 + (m_e/m_p)]\}(1/2n^2)(m_e e^4/(4\pi\varepsilon_0\hbar)^2 = -\{1/[1 + (m_e/m_p)]\}(13.6/n^2)\text{ eV}.$

The wavelength of the emitted photon for the $n = 2$ to $n = 1$ transition is

$\lambda = hc/\Delta E = (1.24 \times 10^3\text{ eV} \cdot \text{nm})/\Delta E$

$= [1 + (m_e/m_p)](1.24 \times 10^3\text{ eV} \cdot \text{nm})/(13.6\text{ eV})[(1/1^2) - (1/2^2)] = (122\text{ nm})[1 + (m_e/m_p)].$

If we apply the same analysis to deuterium, we have

$\lambda = (122\text{ nm})[1 + (m_e/m_d)] = (122\text{ nm})[1 + (m_e/2m_p)].$

The difference in the wavelengths is

$\Delta\lambda = (122\text{ nm})\{[1 + (m_e/m_p)] - [1 + (m_e/2m_p)]\}$

$= (122\text{ nm})(m_e/m_p)(1 - \tfrac{1}{2}) = (122\text{ nm})[(9.1 \times 10^{-31}\text{ kg})/(1.67 \times 10^{-27}\text{ kg})]\tfrac{1}{2} = \boxed{0.033\text{ nm.}}$

25. Because the masses of the two particles are the same, we use the reduced mass

$M = m_e m_e/(m_e + m_e) = \tfrac{1}{2}m_e,$

in the expression for the hydrogen energy levels:

$E_n = -(M/2n^2)(e^2/2\varepsilon_0 h)^2 = -(\tfrac{1}{2}m_e/2n^2)(e^2/2\varepsilon_0 h)^2 = -\tfrac{1}{2}(13.6\text{ eV})/n^2 = \boxed{-(6.8\text{ eV})/n^2.}$

The first few energy levels are $E_1 = -6.8\text{ eV}$, $E_2 = -1.7\text{ eV}$, $E_3 = -0.76\text{ eV}$.

For the visible wavelengths from 400 nm to 700 nm, we find the required energy range for the photon from

$E = hc/\lambda = (1.24 \times 10^3\text{ eV} \cdot \text{nm})/\lambda;$

$E_{400} = (1.24 \times 10^3\text{ eV} \cdot \text{nm})/(400\text{ nm}) = 3.10\text{ eV};$

$E_{700} = (1.24 \times 10^3\text{ eV} \cdot \text{nm})/(700\text{ nm}) = 1.77\text{ eV}$, so the visible energy range is 1.77 eV to 3.10 eV.

Photon energies greater than 3.10 eV will be in the ultraviolet (UV);

photon energies less than 1.77 eV will be in the infrared (IR).

The photon energy radiated in the transition will be the difference in the energies of the two levels.

For the $n = 1$ series, we have

$\Delta E_{min} = 6.8\text{ eV} - 1.7\text{ eV} = 5.1\text{ eV}$, so all $\Delta E > 3.10\text{ eV}$ and $\boxed{\text{all transitions are in the UV.}}$

For the $n = 2$ series, we have

$\Delta E_{max} = 1.7\text{ eV}$, so all $\Delta E < 1.77\text{ eV}$ and $\boxed{\text{all transitions are in the IR.}}$

For all other series, $\Delta E < 1.77\text{ eV}$ and $\boxed{\text{all transitions are in the IR.}}$

29. We fill the n and ℓ levels, with $2(2\ell + 1)$ electrons per level:
$n = 1, \ell = 0$:   2 electrons,   2 electrons total;
$n = 2, \ell = 0$:   2 electrons,   4 electrons total.
$n = 2, \ell = 1$:   6 electrons,   10 electrons total;
$n = 3, \ell = 0$:   1 electron,   11 electrons total.
The lightest element with a single electron in the $n = 3$ level is $\boxed{Na(Z = 11).}$

31. We fill the n and ℓ levels, with $2(2\ell + 1)$ electrons per level:
$n = 1, \ell = 0$:   2 electrons,   2 electrons total;
$n = 2, \ell = 0$:   2 electrons,   4 electrons total;
$n = 2, \ell = 1$:   6 electrons,   10 electrons total;
$n = 3, \ell = 0$:   2 electrons,   12 electrons total;
$n = 3, \ell = 1$:   6 electrons,   18 electrons total;
$n = 3, \ell = 2$:   10 electrons,   28 electrons total;
$n = 4, \ell = 0$:   2 electrons,   30 electrons total;
$n = 4, \ell = 1$:   6 electrons,   36 electrons total;
$n = 4, \ell = 2$:   10 electrons,   46 electrons total;
$n = 5, \ell = 0$:   2 electrons,   48 electrons total;
$n = 5, \ell = 1$:   6 electrons,   54 electrons total.
There are a total of 54 electrons, so $\boxed{Z = 54 \ (Xe).}$

33. We specify the energies in a magnetic field by
$E = E_{n,\ell} + m\kappa B$, where $m = \ell, \ell - 1, ..., -(\ell - 1), -\ell$.
The frequency of the spectral line is determined by the energy change:
$$hf = \Delta E = E_{n,\ell i} - E_{n,\ell f} + m_i \kappa B - m_f \kappa B$$
$$= E_{n,\ell i} - E_{n,\ell f} + (m_i - m_f)\kappa B,$$
where $m_i = \ell_i, \ell_i - 1, ..., -(\ell_i - 1), -\ell_i$ and $m_f = \ell_f, \ell_f - 1, ..., -(\ell_f - 1), -\ell_f$.
The number of different spectral lines is given by the number of distinct values of $\Delta m = m_i - m_f$.
(a) For the transition $(n = 2, \ell = 1) \rightarrow (n = 1, \ell = 0)$, we have
$m_i = 1, 0, -1$ and $m_f = 0$,   so $\Delta m = 1, 0, -1$.
There are $\boxed{\text{3 spectral lines.}}$
(b) For the transition $(n = 3, \ell = 2) \rightarrow (n = 2, \ell = 1)$, we have
$m_i = 2, 1, 0, -1, -2$ and $m_f = 1, 0, -1$,   so $\Delta m = 3, 2, 1, 0, -1, -2, -3$.
There are $\boxed{\text{7 spectral lines.}}$
These transitions can be shown on an energy-level diagram.

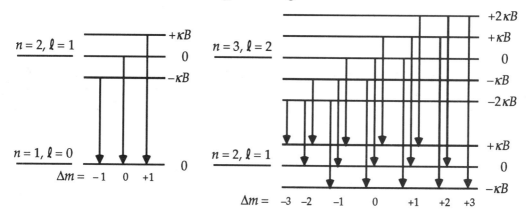

35. If we extend the analysis of Problem 31 through $n = 5$, the maximum number of electrons in each of the n shells is

  $n = 1$: 2 electrons;
  $n = 2$: 8 electrons;
  $n = 3$: 18 electrons;
  $n = 4$: 32 electrons;
  $n = 5$: 50 electrons.

The Z values ($Z < 100$) for the elements with one electron missing from the filled shells are

  $n = 1$ filled $- 1$: $Z = 1$ (H);
  $n = 2$ filled $- 1$: $Z = 9$ (F);
  $n = 3$ filled $- 1$: $Z = 27$ (Co, should be Cl, $Z = 17$);
  $n = 4$ filled $- 1$: $Z = 59$ (Pr, should be Br, $Z = 35$).

The discrepancies are due to the fact that the levels are not filled in the assumed uniform order.

43. The energies of the rotational levels are
  $E_{rot} = \ell(\ell + 1)\hbar^2/2I.$
For the three levels of the excited electronic state, we have
  $E(n = 1, \ell = 0) = 0$, $E(n = 1, \ell = 1) = 2\hbar^2/2I_1 = \hbar^2/I_1$, $E(n = 1, \ell = 2) = 6\hbar^2/2I_1 = 3\hbar^2/I_1.$
For the two levels of the lowest electronic state, we have
  $E(n = 0, \ell = 0) = 0$, $E(n = 0, \ell = 1) = 2\hbar^2/2I_0 = \hbar^2/I_0.$
We find the energies of the photons for the transitions from $hf = hc/\lambda$:
  $(n = 1, \ell = 1) \leftrightarrow (n = 0, \ell = 0)$:
    ΔE_{1100} = $(6.62608 \times 10^{-34}\,\text{J} \cdot \text{s})(2.997925 \times 10^8\,\text{m/s})/(387.4608 \times 10^{-9}\,\text{m})(1.60218 \times 10^{-19}\,\text{J/eV})$
      = 3.199914 eV;
  $(n = 1, \ell = 2) \leftrightarrow (n = 0, \ell = 1)$:
    ΔE_{1201} = $(6.62608 \times 10^{-34}\,\text{J} \cdot \text{s})(2.997925 \times 10^8\,\text{m/s})/(387.3998 \times 10^{-9}\,\text{m})(1.60218 \times 10^{-19}\,\text{J/eV})$
      = 3.200418 eV;
  $(n = 1, \ell = 0) \leftrightarrow (n = 0, \ell = 1)$:
    ΔE_{1001} = $(6.62608 \times 10^{-34}\,\text{J} \cdot \text{s})(2.997925 \times 10^8\,\text{m/s})/(387.5763 \times 10^{-9}\,\text{m})(1.60218 \times 10^{-19}\,\text{J/eV})$
      = 3.198961 eV.
We let Δ be the energy difference between the electronic states. For the three transitions, we have
  $(n = 1, \ell = 1) \leftrightarrow (n = 0, \ell = 0)$:
    $\Delta E_{1100} = \Delta + (\hbar^2/I_1) - 0$,   3.199914 eV $= \Delta + (\hbar^2/I_1)$;
  $(n = 1, \ell = 2) \leftrightarrow (n = 0, \ell = 1)$:
    $\Delta E_{1201} = \Delta + (3\hbar^2/I_1) - (\hbar^2/I_0)$;   3.200418 eV $= \Delta + (3\hbar^2/I_1) - (\hbar^2/I_0)$;
  $(n = 1, \ell = 0) \leftrightarrow (n = 0, \ell = 1)$:
    $\Delta E_{1001} = \Delta + 0 - (\hbar^2/I_0)$;   3.198961 eV $= \Delta - (\hbar^2/I_0)$.
We have three equations for the three unknowns Δ, I_1, and I_0.

If we subtract the second and third equations, we find $I_1 = 1.43 \times 10^{-46}\,\text{kg} \cdot \text{m}^2.$

We then obtain Δ from the first equation and get $\Delta = 3.199428$ eV.

From the third equation, we find $I_0 = 1.48 \times 10^{-46}\,\text{kg} \cdot \text{m}^2.$

The rotational inertia is a measure of the average separation, which can be thought of as the midpoint of the vibrational motion. Because $I_1 < I_0$, the midpoint of the higher energy state is closer to the origin. This implies that the potential created by the electronic cloud is different for the two states. The potential is not that of a perfect harmonic oscillator but has nonlinear terms.

51. The potential energy of the mass is
    $U(r) = mgh = mg\alpha r^2$.
    If we compare this to the potential energy of the harmonic oscillator,
    $\frac{1}{2}kx^2$, we see that the motion in r will be simple harmonic, with a
    force constant $k = 2mg\alpha$. The angular frequency of the motion will be
    $\omega = (k/m)^{1/2} = (2g\alpha)^{1/2}$.
    From the result of Example 42–6, the allowed energies of this system are
    $E_n = n\hbar\omega$,
    so the separation of allowed energies is
    $\Delta E = (\Delta n)\hbar\omega = \hbar(2g\alpha)^{1/2} = (1.055 \times 10^{-34}\,\text{J}\cdot\text{s})[2(9.8\,\text{m/s}^2)(0.25 \times 10^2\,\text{m}^{-1})]^{1/2}$
    $\quad\quad = \boxed{2.3 \times 10^{-33}\,\text{J}\ (1.4 \times 10^{-14}\,\text{eV}).}$

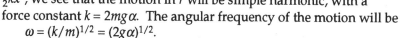

55. We find the dependence of the momentum in the orbit on n from the angular momentum:
    $L = pr = n\hbar$, which gives $p = n\hbar/r = \hbar/na_0$.
    For the ground state and the first excited state, we have
    $p_1 = \hbar/a_0 = (1.055 \times 10^{-34}\,\text{J}\cdot\text{s})/(0.53 \times 10^{-10}\,\text{m}) = 2.0 \times 10^{-24}\,\text{kg}\cdot\text{m/s}$;
    $p_2 = \hbar/2a_0 = (1.055 \times 10^{-34}\,\text{J}\cdot\text{s})/2(0.53 \times 10^{-10}\,\text{m}) = 1.0 \times 10^{-24}\,\text{kg}\cdot\text{m/s}$.
    To know which state is occupied, we must have
    $\Delta p < p_1 - p_2 \approx 1.0 \times 10^{-24}\,\text{kg}\cdot\text{m/s}$.
    This uncertainty in the momentum gives an uncertainty in the position:
    $\Delta x > \hbar/\Delta p = (1.055 \times 10^{-34}\,\text{J}\cdot\text{s})/(1.0 \times 10^{-24}\,\text{kg}\cdot\text{m/s}) = \boxed{1 \times 10^{-10}\,\text{m} \approx 2a_0.}$
    The uncertainty in position is on the order of the separation of radii, so we do not have the classical orbit.

57. (a) We find the force associated with the potential energy from
    $F = -dU/dr = -d[U_0(r/a)^4]/dr = -4U_0(r/a)^3/a$.
    The force provides the centripetal acceleration:
    $4U_0(r/a)^3/a = mv^2/r$, which gives $mv^2 = 4U_0(r/a)^4$.
    The kinetic energy of the particle is
    $K = \frac{1}{2}mv^2 = 2U_0(r/a)^4$, so the total energy is
    $E = K + U = 2U_0(r/a)^4 + U_0(r/a)^4 = \boxed{3U_0(r/a)^4.}$
    (b) The Bohr quantization of the angular momentum is
    $L = mvr = n\hbar$, or $m^2v^2 = n^2\hbar^2/r^2$.
    When we use this in the expression for the kinetic energy, we have
    $K = \frac{1}{2}mv^2 = \frac{1}{2}n^2\hbar^2/mr^2 = 2U_0(r/a)^4$, which gives
    $(r/a)^6 = n^2\hbar^2/4ma^2U_0$, or $(r/a)^4 = (n^2\hbar^2/4ma^2U_0)^{2/3}$.
    The energy is
    $E_n = 3U_0(r/a)^4 = 3U_0(n^2\hbar^2/4ma^2U_0)^{2/3} = \boxed{3(\hbar^2/4ma^2)^{2/3}U_0^{1/3}n^{4/3}.}$

CHAPTER 43 Extended Version

Note: At the atomic scale, it is most convenient to have energies in electron-volts and wavelengths in nanometers. A useful expression for the energy of a photon in terms of its wavelength is

$E = hf = hc/\lambda = (6.63 \times 10^{-34} \text{ J} \cdot \text{s})(3 \times 10^8 \text{ m/s})(10^{-9} \text{ nm/m})/(1.60 \times 10^{-19} \text{ J/eV})\lambda;$

$E = (1.24 \times 10^3 \text{ eV} \cdot \text{nm})/\lambda.$

A factor that appears in the analysis of electron energies is

$\hbar^2/2m_e = (1.055 \times 10^{-34} \text{ J} \cdot \text{s})^2/2(9.11 \times 10^{-31} \text{ kg}) = 6.11 \times 10^{-39} \text{ J}^2 \cdot \text{s}^2/\text{kg}.$

7.   For the energy levels in a box, with $n_1 = n_2 = n_3$, we have

$E = \pi^2(\hbar^2/2m_e)(n_1^2 + n_2^2 + n_3^2)/L^2 = 3\pi^2(\hbar^2/2m_e)n_1^2/L^2.$

For an energy near 1 eV, we have

$(1 \text{ eV})(1.60 \times 10^{-19} \text{ J/eV}) = \pi^2(6.11 \times 10^{-39} \text{ J}^2 \cdot \text{s}^2/\text{kg})n_1^2/(100 \times 10^{-9} \text{ m})^2$, which gives

$\boxed{n_1 = n_2 = n_3 = 94.}$

The actual energy of this state is

$E = 3\pi^2(\hbar^2/2m_e)n_1^2/L^2$

$\quad = 3\pi^2(6.11 \times 10^{-39} \text{ J}^2 \cdot \text{s}^2/\text{kg})(94)^2/(100 \times 10^{-9} \text{ m})^2(1.60 \times 10^{-19} \text{ J/eV}) = \boxed{0.999 \text{ eV.}}$

When n_3 is increased by 1, we find the energy change from

$\Delta E = [\pi^2(\hbar^2/2m_e)/L^2][(n_3 + 1)^2 - (n_3)^2] \simeq [\pi^2(\hbar^2/2m_e)/L^2](2n_3)$

$\quad = [3\pi^2(\hbar^2/2m_e)n_1^2/L^2](2n_3/3n_1^2) = (2/3n_1)E = [2/3(94)](0.999 \text{ eV}) = 7.1 \times 10^{-3} \text{ eV.}$

The energy of the state is

$E' = E + \Delta E = (0.999 \text{ eV}) + (0.0071 \text{ eV}) = \boxed{1.006 \text{ eV.}}$

13.   (a)   The condition for using a nonrelativistic approximation is

$v_F = (\hbar/m_e)(3\pi^2n_e)^{1/3} \leq 0.1c;$

$[(1.055 \times 10^{-34} \text{ J} \cdot \text{s})/(9.11 \times 10^{-31} \text{ kg})](3\pi^2n_e)^{1/3} \leq 0.1(3 \times 10^8 \text{ m/s})$, which gives

$\boxed{n_e \leq 5.9 \times 10^{32} \text{ electrons/m}^3.}$

If we consider each electron to occupy a cube of side d, so d represents the electron spacing, we have

$d^3 = 1/n_e = 1/(5.9 \times 10^{32} \text{ electrons/m}^3)$, or $\quad d \geq 1.2 \times 10^{-11} \text{ m.}$

The size of a typical atom is $\sim 2.5 \times 10^{-10}$ m, so this condition is generally satisfied.

(b)   We use the Bohr model for the innermost electrons, so their radius is

$a = a_0/Z$, where $a_0 = 0.53 \times 10^{-10}$ m.

If we take this for the uncertainty in position, the associated momentum is

$p \simeq \Delta p \simeq \hbar/\Delta x;$

$v = p/m = \hbar Z/ma_0 \leq (0.1)c;$

$(1.055 \times 10^{-34} \text{ J} \cdot \text{s})Z/(9.11 \times 10^{-31} \text{ kg})(0.53 \times 10^{-10} \text{ m}) \leq (0.1)(3 \times 10^8 \text{ m/s})$, which gives $\boxed{Z \leq 14.}$

The innermost electrons are relativistic for all but the smallest atoms.

15.   The energy levels for a one-dimensional potential well are

$E_n = (\pi^2\hbar^2/2m_eL^2)n^2$, $n = 1, 2, 3, \ldots$.

(a)   Because we can put 2 fermions in each energy level, the largest value of n is $\frac{1}{2}N$.

The Fermi energy is the energy for this highest value:

$E_F = (\pi^2\hbar^2/2m_eL^2)(\frac{1}{2}N)^2 = \boxed{(\pi^2\hbar^2/8m_e)(N/L)^2.}$

(b)   For the total energy, we add the energies from each level, with 2 fermions per level:

$E_{\text{total}} = \dfrac{\pi^2\hbar^2}{2m_eL^2} 2\sum_{x=1}^{N/2} x^2.$

We can change the lower limit to 0 without affecting the sum. Because x changes by 1 and $N \gg 1$, we can replace the summation with an integral:

$E_{\text{total}} = \dfrac{\pi^2\hbar^2}{2m_eL^2} 2\int_0^{N/2} x^2 \, dx = \dfrac{\pi^2\hbar^2}{m_eL^2}\dfrac{1}{3}\left(\dfrac{N}{2}\right)^3 = \boxed{\dfrac{\pi^2\hbar^2N^3}{24m_eL^2}.}$

(c)   Because all the levels up to the Fermi level are filled, the smallest amount of energy that can be absorbed must excite an electron at the Fermi level to the next possible level:

$\Delta E = (\pi^2\hbar^2/2m_eL^2)[(\frac{1}{2}N + 1)^2 - (\frac{1}{2}N)^2] \simeq \boxed{(\pi^2\hbar^2/2m_eL^2)N.}$

19. (a) We put the result from Problem 18 in terms of the Fermi energy:

$$p = 2U/3V = 2NE_F/5V$$
$$= 2N(\hbar^2/2m)[3\pi^2(N/V)]^{2/3}/5V = \boxed{(\hbar^2/5m)(3\pi^2)^{2/3}(N/V)^{5/3}.}$$

(b) The electron density in sodium is

$$N/V = (2.65 \times 10^{22} \text{ electrons/cm}^3)(10^6 \text{ cm}^3/\text{m}^3) = 2.65 \times 10^{28} \text{ electrons/m}^3.$$

We use the result from part (a) to find the pressure:

$$p = (2/5)(6.11 \times 10^{-39} \text{ J}^2 \cdot \text{s}^2/\text{kg})(3\pi^2)^{2/3}(2.65 \times 10^{28} \text{ electrons/m}^3)^{5/3} = \boxed{5.5 \times 10^9 \text{ N/m}^2.}$$

21. For a two-dimensional system of free electrons in a plane with sides of length L, the energy values are

$$E = \pi^2\hbar^2(n_1^2 + n_2^2)/2m_eL^2.$$

We estimate the closest distance between electrons d from

$$N_e = (L/d)^2, \text{ which gives } d = N_e/L^2 = n_e^{-1/2}.$$

We set this equal to one-half the de Broglie wavelength that corresponds to the Fermi momentum, $p_F = (2m_eE_F)^{1/2}$:

$$n_e^{-1/2} = \tfrac{1}{2}\lambda_F = \tfrac{1}{2}h/p_F = \hbar\pi/(2m_eE_F)^{1/2}, \text{ which gives}$$
$$\boxed{E_F = (\hbar^2\pi^2/2m_e)n_e.}$$

23. (a) We can express the mass of the neutron star in terms of the number of neutrons or the number of solar masses:

$$M = Nm_n = N_{sm}M_{sun}.$$

We find the radius by equating the magnitudes of the outward degeneracy pressure and the inward gravitational pressure:

$$p_d = p_{grav};$$
$$(\hbar^2/5m_n)(3\pi^2)^{2/3}(N/V)^{5/3} = 0.32GM^2/V^{4/3};$$
$$(\hbar^2/5m_n)(3\pi^2)^{2/3}(M/m_n)^{5/3} = 0.32GM^2V^{1/3};$$
$$V^{1/3} = (\hbar^2/5m_n)(3\pi^2)^{2/3}/0.32GM^{1/3}m_n^{5/3} = (\tfrac{4}{3}\pi)^{1/3}R, \text{ which gives}$$
$$R = (3/4\pi)^{1/3}(\hbar^2/5m_n)(3\pi^2)^{2/3}/0.32GM^{1/3}m_n^{5/3} = [(3\pi/1.6)\hbar^2/(4)^{1/3}Gm_n^3](m_n/N_{sm}M_{sun})^{1/3}$$
$$= [(3\pi/1.6)(1.05 \times 10^{-34} \text{ J} \cdot \text{s})^2/(4)^{1/3}(6.67 \times 10^{-11} \text{ N} \cdot \text{s}^2/\text{kg}^2)(1.67 \times 10^{-27} \text{ kg})^3] \times$$
$$[(1.67 \times 10^{-27} \text{ kg})/N_{sm}(2 \times 10^{30} \text{ kg})]^{1/3}$$
$$= \boxed{12/N_{sm}^{1/3} \text{ km, where } N_{sm} = \text{number of solar masses.}}$$

(b) We find the Fermi energy from

$$E_F = (\hbar^2/2m_n)(3\pi^2n_n)^{2/3} = (\hbar^2/2m_n)(3\pi^2)^{2/3}(N/V)^{2/3}$$
$$= (\hbar^2/2m_n)(3\pi^2)^{2/3}(N_{sm}M_{sun}/m_n)^{2/3}(3/4\pi)^{2/3}/R^2$$
$$= \boxed{(\hbar^2/2m_n)(9\pi/4)^{2/3}(N_{sm}M_{sun}/m_n)^{2/3}/R^2.}$$

When we use the result from part (a), we have

$$E_F = [(1.05 \times 10^{-34} \text{ J} \cdot \text{s})^2/2(1.67 \times 10^{-27} \text{ kg})](9\pi/4)^{2/3}[N_{sm}(2 \times 10^{30} \text{ kg})/(1.67 \times 10^{-27} \text{ kg})]^{2/3} \times$$
$$[N_{sm}^{1/3}/(12 \times 10^3 \text{ m})^2]^2/(1.60 \times 10^{-13} \text{ J/MeV}) = \boxed{60N_{sm}^{4/3} \text{ MeV.}}$$

We find the Fermi momentum from

$$p_F^2 = 2m_nE_F;$$
$$(p_Fc)^2 = 2(m_nc^2)E_F = 2(938 \text{ MeV})(60N_{sm}^{4/3} \text{ MeV}), \text{ which gives} \quad \boxed{p_F = 335N_{sm}^{2/3} \text{ MeV}/c.}$$

(c) If we assume that relativistic expressions are needed when $p_F \approx m_nc$, we have

$$p_Fc \approx m_nc^2;$$
$$335N_{sm}^{2/3} \text{ MeV} = 938 \text{ MeV, which gives } N_{sm} = 4.7.$$

The neutron star mass is $\boxed{4.7M_{sun}.}$

25. For a two-dimensional gas, the energies are given by
$$E = \pi^2(\hbar^2/2m_e)(n_1^2 + n_2^2)/L^2.$$
The possible values of n_1 and n_2 can be plotted as shown.
The Fermi energy is the highest energy of the filled states:
$$(n_1^2 + n_2^2)_{max} = R^2 = 2(m_e E_F L^2/\pi^2\hbar^2).$$
Because each level can hold two electrons, the number of filled
states is twice the area of a quadrant of a circle of radius R in the
two-dimensional space of n_1 and n_2:
$$N = 2(1/4)\pi R^2 = n_e L^2, \text{ which gives } R = (2n_e/\pi)^{1/2}L.$$
When we substitute for R in the expression for the Fermi level, we have
$$(2n_e/\pi)L^2 = 2(m_e E_F L^2/\pi^2\hbar^2), \text{ which gives } \boxed{E_F = (\hbar^2\pi/m_e)n_e.}$$
The electron density determines the distance between electrons:
$$N = n_e L^2 = (L/d)^2, \text{ which gives } d = 1/n_e^{1/2}.$$
For the approximate result, we have
$$d = \lambda_F/2 = h/2p_F;$$
$$1/n_e^{1/2} = \pi\hbar/(2m_e E_F)^{1/2}, \text{ which gives } E_F = \hbar^2\pi^2 n_e/2m_e.$$
The ratio of the approximate result to the exact result is
$$(\text{approximate})/(\text{exact}) = \pi/2 = \boxed{1.57,}$$
which is not as good as for the three-dimensional gas (Problem 24).

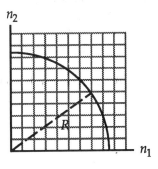

31. (a) Because there is no restriction on the number of bosons in an energy level, the lowest energy state
will have all of the bosons in the $n = 1$ state:
$$E_{lowest} = N\pi^2\hbar^2(1)^2/2mL^2 = \boxed{\pi^2\hbar^2 N/2mL^2.}$$
   (b) The minimum energy that can be absorbed will raise a boson to the $n = 2$ level:
$$\Delta E_{min} = (\pi^2\hbar^2/2mL^2)[(2)^2 - (1)^2] = \boxed{3\pi^2\hbar^2/2mL^2.}$$

33. (a) We find the power of the laser from
$$P = (dN/dt)E = (dN/dt)hc/\lambda$$
$$= (3 \times 10^{17} \text{ s}^{-1})[(1.24 \times 10^3 \text{ eV} \cdot \text{nm})/(663 \text{ nm})](1.60 \times 10^{-19} \text{ J/eV}) = \boxed{0.090 \text{ W.}}$$
   (b) The momentum of a photon is
$$p = h/\lambda.$$
As each photon reflects from the surface, its momentum change is $\Delta p = 2p$. The rate at which the
momentum of the beam changes is $(dN/dt)\Delta p$, which is the magnitude of the force exerted on the
surface. The pressure on the surface is
$$\text{pressure} = (dN/dt)\Delta p/A = 2(dN/dt)h/\lambda\pi r^2$$
$$= 2(3 \times 10^{17} \text{ s}^{-1})(6.63 \times 10^{-34} \text{ J} \cdot \text{s})/(663 \times 10^{-9} \text{ m})\pi(1.75 \times 10^{-3} \text{ m})^2 = \boxed{6.2 \times 10^{-5} \text{ N/m}^2.}$$

41. The energy of the photon is
$$E = hf = \hbar\omega.$$
For the Josephson junction, we have
$$\omega = (2e/\hbar)V, \text{ or } \hbar\omega = 2eV;$$
$$(5 \times 10^{-6} \text{ eV})(1.60 \times 10^{-19} \text{ J/eV}) = 2(1.60 \times 10^{-19} \text{ C})V, \text{ which gives } V = 2.5 \times 10^{-6} \text{ V} = \boxed{2.5 \, \mu\text{V.}}$$

43. We find the density of bosons from
$$N/V = [(0.147 \text{ g/cm}^3)(10^6 \text{ cm}^3/\text{m}^3)/(4 \text{ g/mol})](6.02 \times 10^{23} \text{ bosons/mol}) = 2.23 \times 10^{28} \text{ bosons/m}^3.$$
For T_c, we have
$$T_c = [N/(2.612)(0.886)C]^{2/3} = [N/(2.612)(0.886)2\pi V(2mk/h^2)^{3/2}]^{2/3}$$
$$= [(N/V)/(2.612)(0.886)2\pi]^{2/3}(h^2/2mk)$$
$$= [(2.23 \times 10^{28} \text{ bosons/m}^3)/14.21]^{2/3}[(6.63 \times 10^{-34} \text{ J} \cdot \text{s})^2/2(4)(1.67 \times 10^{-27} \text{ kg})(1.38 \times 10^{-23} \text{ J/K})]$$
$$= \boxed{3.20 \text{ K.}}$$

47. We assume that
$$E_F = \hbar^\alpha n_e^\beta c^\gamma.$$
For the dimensions, we have
$$[E_F] = [\hbar]^\alpha [n_e]^\beta [c]^\gamma;$$
$$[ML^2T^{-2}] = [ML^2T^{-1}]^\alpha [L^{-3}]^\beta [LT^{-1}]^\gamma.$$
When we equate the exponents of the three dimensions, we have
$$1 = \alpha, \quad 2 = 2\alpha - 3\beta + \gamma, \text{ and } -2 = -\alpha - \gamma, \text{ which gives}$$
$$\alpha = 1, \quad \beta = 1/3, \text{ and } \gamma = 1.$$
The relation is
$$\boxed{E_F = \hbar c\, n_e^{1/3}.}$$

51. The maximum energy of the gas of electrons at $T = 0$ K is E_F.
We find the total energy of the electrons from
$$E_{\text{tot}} = \int_0^{E_F} E n(E)\, dE = \frac{V}{2\pi^2}\left(\frac{2m}{\hbar^2}\right)^{3/2}\int_0^{E_F} E^{3/2}\, dE = \frac{V}{2\pi^2}\left(\frac{2m}{\hbar^2}\right)^{3/2}\frac{2}{5}E_F^{5/2}.$$
The total number of electrons is
$$N = \int_0^{E_F} n(E)\, dE = \frac{V}{2\pi^2}\left(\frac{2m}{\hbar^2}\right)^{3/2}\int_0^{E_F} E^{1/2}\, dE = \frac{V}{2\pi^2}\left(\frac{2m}{\hbar^2}\right)^{3/2}\frac{2}{3}E_F^{3/2}.$$
If we divide the two equations and rearrange, we have
$$E_{\text{tot}} = (3/5)NE_F.$$
Note that this means the average energy of an electron is $(3/5)E_F$.

53. We let N be the total number of protons and neutrons. The number of electrons will be $N_e = \frac{1}{2}N$. We can express the mass of the star in terms of N or the number of solar masses:
$$M = Nm_n = N_{\text{sm}}M_{\text{sun}}.$$
(a) We find the radius by equating the magnitudes of the outward degeneracy pressure of the electrons and the inward gravitational pressure:
$$p_d = p_{\text{grav}};$$
$$(\hbar^2/5m_e)(3\pi^2)^{2/3}(\tfrac{1}{2}N/V)^{5/3} = 0.32GM^2/V^{4/3};$$
$$(\hbar^2/5m_e)(3\pi^2)^{2/3}(\tfrac{1}{2}M/m_n)^{5/3} = 0.32GM^2V^{1/3}; \text{ which gives}$$
$$V^{1/3} = (\hbar^2/10m_e)(3\pi^2/2)^{2/3}/0.32GM^{1/3}m_n^{5/3} = (4\pi/3)^{1/3}R; \text{ which gives}$$
$$R = (3/4\pi)^{1/3}(\hbar^2/10m_e)(3\pi^2/2)^{2/3}/0.32GM^{1/3}m_n^{5/3}$$
$$= [(3\pi/3.2)\hbar^2/(4)^{2/3}Gm_em_n^{5/3}](1/N_{\text{sm}}M_{\text{sun}})^{1/3}$$
$$= [(3\pi/3.2)(1.05\times10^{-34}\text{ J}\cdot\text{s})^2/(4)^{2/3}(6.67\times10^{-11}\text{ N}\cdot\text{s}^2/\text{kg}^2)(9.1\times10^{-31}\text{ kg})(1.67\times10^{-27}\text{ kg})^{5/3}]\times$$
$$[1/N_{\text{sm}}(2\times10^{30}\text{ kg})]^{1/3}$$
$$= (7.2\times10^3)/N_{\text{sm}}^{1/3}\text{ km, where } N_{\text{sm}} = \text{number of solar masses.}$$
We find the Fermi energy from
$$E_F = (\hbar^2/2m_e)(3\pi^2 n_e)^{2/3} = (\hbar^2/2m_e)(3\pi^2)^{2/3}(\tfrac{1}{2}N/V)^{2/3}$$
$$= (\hbar^2/2m_e)(3\pi^2/2)^{2/3}(N_{\text{sm}}M_{\text{sun}}/m_n)^{2/3}(3/4\pi)^{2/3}/R^2$$
$$= (\hbar^2/2m_e)(9\pi/8)^{2/3}(N_{\text{sm}}M_{\text{sun}}/m_n)^{2/3}/R^2.$$
When we use the result for the radius, we have
$$E_F = [(1.05\times10^{-34}\text{ J}\cdot\text{s})^2/2(9.1\times10^{-31}\text{ kg})](9\pi/8)^{2/3}[N_{\text{sm}}(2\times10^{30}\text{ kg})/(1.67\times10^{-27}\text{ kg})]^{2/3}\times$$
$$[N_{\text{sm}}^{1/3}/(7.2\times10^6\text{ m})^2]^2/(1.60\times10^{-13}\text{ J/MeV}) = \boxed{0.19N_{\text{sm}}^{4/3}\text{ MeV.}}$$
(b) From part (a), we have
$$\boxed{R = (7.2\times10^3)/N_{\text{sm}}^{1/3}\text{ km, where } N_{\text{sm}} = \text{number of solar masses.}}$$
When we compare this radius for one solar mass to the radius of the sun, we have
$$R/R_{\text{sun}} = (7.2\times10^3\text{ km})/(6.96\times10^5\text{ km}) = \boxed{0.01.}$$

55. Each momentum component is specified by an integer:

$p_i = h/\lambda$, where $\lambda = L/2n_i$,   or $p_i = \hbar\pi n_i/L$.

In a three-dimensional box, the momentum is

$p = (\hbar\pi/L)(n_1^2 + n_2^2 + n_3^2)^{1/2}$.

The energy states are given by

$E = pc = (\hbar c\pi/L)(n_1^2 + n_2^2 + n_3^2)^{1/2}$.

The Fermi energy is the highest energy of the filled states:

$(n_1^2 + n_2^2 + n_3^2)_{max} = R^2 = E_F^2 L^2/\pi^2\hbar^2 c^2$.

We relate the electron density to the radius of the sphere that determines the number of states in the three-dimensional space of n_1, n_2, and n_3:

$N = 2(1/8)\frac{4}{3}\pi R^3 = n_e L^3$, which gives $R = (3n_e/\pi)^{1/3}L$.

When we substitute for R in the expression for the Fermi level, we have

$(3n_e/\pi)^{2/3}L^2 = E_F^2 L^2/\pi^2\hbar^2 c^2$, which gives $\boxed{E_F = \hbar c(3\pi^2 n_e)^{1/3}.}$

We let the average energy of the electron be fE_F, with $f < 1$, so the total energy is $E = fNE_F$.

The degeneracy pressure for this gas is

$p_d = -dE/dV = -(d/dV)[f\hbar c(3\pi^2 N^4/V)^{1/3}] = (f/3)\hbar c(3\pi^2)^{1/3}(N/V)^{4/3}$.

The star will not collapse while

$p_d > p_{grav}$;

$(f/3)\hbar c(3\pi^2)^{1/3}(N/V)^{4/3} > 0.32 GM^2/V^{4/3}$.

We see that this condition is independent of V. If we use $N = 2M/m_{He}$, we have

$(f/3)\hbar c(3\pi^2)^{1/3}(2M/m_{He})^{4/3} > 0.32 GM^2$;

$(f/0.96)(3\pi^2 2^4)^{1/3}(\hbar c/G)/m_{He}^{4/3} > M^{2/3}$, which gives

$\boxed{M < 23.1f^{3/2}(\hbar c/G)^{3/2}/m_{He}^2.}$

Note: A calculation similar to that done in Problems 50 and 51 gives $f = 3/4$ for this system, which gives

$M < 23.1(0.75)^{3/2}[(1.055 \times 10^{-34}\,\text{J}\cdot\text{s})(3.0 \times 10^8\,\text{m/s})/(6.67 \times 10^{-11}\,\text{N}\cdot\text{m}^2/\text{kg}^2)]^{3/2}/[4(1.67 \times 10^{-27}\,\text{kg})]^2$;

$\boxed{M < 3 \times 10^{30}\,\text{kg}}$ (≈ 1.5 solar masses).

CHAPTER 44 Extended Version

Note: It is most convenient to have energies in electron-volts and wavelengths in nanometers. A useful
expression for the energy of a photon in terms of its wavelength is
$E = hf = hc/\lambda = (6.63 \times 10^{-34}\, \text{J} \cdot \text{s})(3 \times 10^8\, \text{m/s})(10^{-9}\, \text{nm/m})/(1.60 \times 10^{-19}\, \text{J/eV})\lambda;$
$E = (1.24 \times 10^3\, \text{eV} \cdot \text{nm})/\lambda.$
A factor that appears in the analysis of electron energies is
$\hbar^2/2m_e = (1.055 \times 10^{-34}\, \text{J} \cdot \text{s})^2/2(9.11 \times 10^{-31}\, \text{kg}) = 6.11 \times 10^{-39}\, \text{J}^2 \cdot \text{s}^2/\text{kg}.$
A factor that appears in the analysis of carrier concentrations is
$m_e k/2\pi\hbar^2 = (9.11 \times 10^{-31}\, \text{kg})(1.38 \times 10^{-23}\, \text{J/K})/2\pi(1.055 \times 10^{-34}\, \text{J} \cdot \text{s})^2 = 1.80 \times 10^{14}\, \text{m}^{-2} \cdot \text{K}^{-1}.$
The value of kT for some common temperatures is
100 K: $kT = (1.38 \times 10^{-23}\, \text{J/K})(100\, \text{K})/(1.6 \times 10^{-19}\, \text{J/eV}) = 0.00863\, \text{eV};$
300 K: $kT = (1.38 \times 10^{-23}\, \text{J/K})(300\, \text{K})/(1.6 \times 10^{-19}\, \text{J/eV}) = 0.0259\, \text{eV}.$

3.   We find the occupation probability from
$$f = 1/\left[1 + e^{(E - E_F)/kT}\right].$$
(a) For $E = 2$ eV, we have
$f(E) = 1/(1 + e^{[(2\, \text{eV}) - (3.23\, \text{eV})]/(0.0259\, \text{eV})}) = 1/(1 + e^{-47}) = \boxed{1.}$
(b) For $E = 3$ eV, we have
$f(E) = 1/(1 + e^{[(3\, \text{eV}) - (3.23\, \text{eV})]/(0.0259\, \text{eV})}) = 1/(1 + e^{-8.9}) = \boxed{0.99986.}$
(c) For $E = 3.2$ eV, we have
$f(E) = 1/(1 + e^{[(3.2\, \text{eV}) - (3.23\, \text{eV})]/(0.0259\, \text{eV})}) = 1/(1 + e^{-1.16}) = \boxed{0.76.}$
(d) For $E = 3.26$ eV, we have
$f(E) = 1/(1 + e^{[(3.26\, \text{eV}) - (3.23\, \text{eV})]/(0.0259\, \text{eV})}) = 1/(1 + e^{+1.16}) = \boxed{0.24.}$
Note that this is [1 − answer to (c)] because E_c is 0.03 eV below E_F and E_d is 0.03 eV above E_F.
(e) For $E = 4$ eV, we have
$f(E) = 1/(1 + e^{[(4\, \text{eV}) - (3.23\, \text{eV})]/(0.0259\, \text{eV})}) = 1/(1 + e^{+30}) = \boxed{1.2 \times 10^{-13}.}$

13. We find the gap width from the intrinsic carrier concentration:
$$n_i^2 = N_c N_v e^{-E_g/kT} = 4\left(\frac{m_e kT}{2\pi\hbar^2}\right)^3 \left(\frac{m_n^* m_p^*}{m_e^2}\right)^{3/2} e^{-E_g/kT};$$
$(2.5 \times 10^{19}\, \text{m}^{-3})^2 = 4\left[(1.80 \times 10\, \text{m}^{-2} \cdot \text{K}^{-1})(300\, \text{K})\right]^3 \left[(0.55)(0.37)\right]^{3/2} e^{-E_g/(0.0259\, \text{eV})}$, which gives
$\boxed{E_g = 0.65\, \text{eV}.}$

15. We find $E_c - E_F$ from the n-carrier concentration:
$$n = N_c e^{-(E_c - E_F)/kT} = 2\left(\frac{m_n^* kT}{2\pi\hbar^2}\right)^{3/2} e^{-(E_c - E_F)/kT};$$
$2.0 \times 10^{23}\, \text{m}^{-3} = 2\left[(0.31)(1.80 \times 10^{14}\, \text{m}^{-2} \cdot \text{K}^{-1})(300\, \text{K})\right]^{3/2} e^{-(E_c - E_F)/(0.0259\, \text{eV})}$, which gives
$\boxed{E_c - E_F = 0.080\, \text{eV}.}$
This is much less than half of $E_g = 1.12$ eV, because the doping has increased the density of n-carriers such
that the Fermi level is not at the midpoint of the gap but closer to the conduction band.

19. The drift speed depends on the current density $j = \sigma E$:

$$v_d = j/ne = \sigma E/ne.$$

When we use the definition of the mobility, we have

$$\mu = v_d/E = \sigma/ne.$$

With both charge carriers, we have

$$\begin{aligned}
\sigma &= \sigma_e + \sigma_p = \mu_n ne + \mu_p pe \\
&= (0.45 \text{ m}^2/\text{V} \cdot \text{s})(2.5 \times 10^{18} \text{ m}^{-3})(1.6 \times 10^{-19} \text{ C}) + (0.35 \text{ m}^2/\text{V} \cdot \text{s})(7.0 \times 10^{13} \text{ m}^{-3})(1.6 \times 10^{-19} \text{ C}) \\
&= \boxed{0.18 \ (\Omega \cdot \text{m})^{-1}.}
\end{aligned}$$

Note that the contribution of the p-carriers in the n-type semiconductor is negligible.

21. (a) From part (b) of Problem 50 in Chapter 43, we have the number of states per unit volume with energy E in a range dE:

$$\frac{1}{2\pi^2}\left(\frac{2m_e}{\hbar^2}\right)^{3/2}\sqrt{E}\ dE = \frac{4\pi}{h^3}\left(2m_e\right)^{3/2}\sqrt{E}\ dE.$$

The Fermi–Dirac distribution gives the probability that a state is occupied. If we multiply the density of states by the occupation probability, we get the energy distribution of the electrons:

$$n(E)\ dE = \frac{4\pi}{h^3}\left(2m_e\right)^{3/2}\frac{\sqrt{E}}{1 + e^{(E - E_F)/kT}}\ dE.$$

For low-energy electrons, we use $E = \frac{1}{2}m_e v^2$ and $dE = m_e v\ dv$ to change to a speed distribution:

$$n(v)\ dv = \frac{4\pi}{h^3}\left(2m_e\right)^{3/2}\frac{\sqrt{m_e v^2/2}}{1 + e^{\left[(m_e v^2/2) - E_F\right]/kT}}m_e v\ dv = \frac{8\pi m_e^3}{h^3}\frac{v^2\ dv}{1 + e^{\left[(m_e v^2/2) - E_F\right]/kT}}.$$

(b) When $kT \ll E_F$, the exponential term in the denominator is much greater than 1, so we have

$$\boxed{n(v)\ dv = \left(\frac{8\pi m_e^3}{h^3}\right)e^{-\left[(m_e v^2/2) - E_F\right]/kT}v^2\ dv.}$$

27.

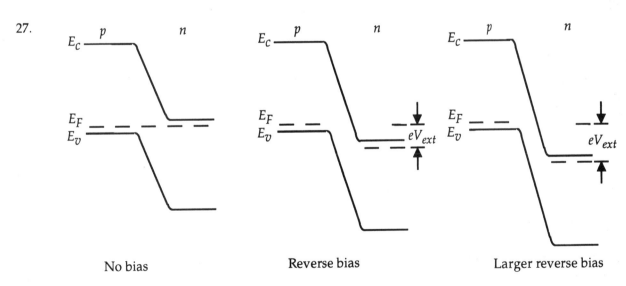

No bias　　　　　　　Reverse bias　　　　　　　Larger reverse bias

(a) From the first two energy diagrams, we see that sufficient reverse biasing will put the conduction band of the n-type next to the valence band of the p-type, so tunneling may occur.

(b) As seen in the third energy diagram, increased reverse biasing decreases the width of the separation, which produces increased tunneling.

29. For the n-side, we find $(E_c - E_F)_n$ from

$$n = N_c e^{-(E_c - E_F)/kT} = 2\left(\frac{m_n^* kT}{2\pi\hbar^2}\right)^{3/2} e^{-(E_c - E_F)_n/kT}.$$

For the p-side, we find $(E_F - E_v)_p$ from

$$p = N_v e^{-(E_F - E_v)/kT} = 2\left(\frac{m_p^* kT}{2\pi\hbar^2}\right)^{3/2} e^{-(E_F - E_v)_p/kT}.$$

We find the potential across the junction from
$$eV_0 = E_g - (E_F - E_v)_p - (E_c - E_F)_n .$$
(a) For $T = 80$ K,

$kT = (1.38 \times 10^{-23} \text{ J/K})(80 \text{ K})/(1.6 \times 10^{-19} \text{ J/eV}) = 0.0069$ eV, so we have

$3.5 \times 10^{22} \text{ m}^{-3} = 2[(0.55)(1.80 \times 10^{14} \text{ m}^{-2}\cdot\text{K}^{-1})(80 \text{ K})]^{3/2} e^{-(E_c - E_F)_n/(0.0069 \text{ eV})}$, which gives

$\boxed{(E_c - E_F)_n = 0.026 \text{ eV};}$

$2.8 \times 10^{23} \text{ m}^{-3} = 2[(0.37)(1.80 \times 10^{14} \text{ m}^{-2}\cdot\text{K}^{-1})(80 \text{ K})]^{3/2} e^{-(E_c - E_F)_n/(0.0069 \text{ eV})}$, which gives

$\boxed{(E_F - E_v)_p = 0.0070 \text{ eV}.}$

For the potential, we have
$$eV_0 = (0.67 \text{ eV}) - (0.026 \text{ eV}) - (0.0070 \text{ eV}) = \boxed{0.64 \text{ eV.}}$$
(b) For $T = 300$ K,

$kT = (1.38 \times 10^{-23} \text{ J/K})(300 \text{ K})/(1.6 \times 10^{-19} \text{ J/eV}) = 0.0259$ eV, so we have

$3.5 \times 10^{22} \text{ m}^{-3} = 2[(0.55)(1.80 \times 10^{14} \text{ m}^{-2}\cdot\text{K}^{-1})(300 \text{ K})]^{3/2} e^{-(E_c - E_F)_n/(0.0259 \text{ eV})}$, which gives

$\boxed{(E_c - E_F)_n = 0.147 \text{ eV};}$

$2.8 \times 10^{23} \text{ m}^{-3} = 2[(0.37)(1.80 \times 10^{14} \text{ m}^{-2}\cdot\text{K}^{-1})(300 \text{ K})]^{3/2} e^{-(E_F - E_v)_p/(0.0259 \text{ eV})}$, which gives

$\boxed{(E_F - E_v)_p = 0.078 \text{ eV}.}$

For the potential, we have
$$eV_0 = (0.67 \text{ eV}) - (0.147 \text{ eV}) - (0.078 \text{ eV}) = \boxed{0.45 \text{ eV.}}$$

33. The energy levels for the well are
$$E_n = (\hbar^2/2m^*)n^2\pi^2/a^2 = (\hbar^2/2m_e)(m_e/m^*)n^2\pi^2/a^2, \quad n = 1, 2, 3, \dots .$$
The energy difference between the ground state and the first excited state is
$$\Delta E = (\hbar^2/2m_e)(m_e/m^*)(2^2 - 1^2)\pi^2/a^2;$$
$(2.28 \times 10^{-2} \text{ eV})(1.6 \times 10^{-19} \text{ J/eV}) = (6.11 \times 10^{-39} \text{ J}^2\cdot\text{s}^2/\text{kg})(m_e/m^*)3\pi^2/(9 \times 10^{-9} \text{ m})^2$, which gives
$$\boxed{m^* = 0.61\, m_e .}$$

35. The current is proportional to the fraction of electrons that cross the barrier, so we have
$$\frac{I_B}{I_A} = \frac{f_B}{f_A} = \frac{e^{-(2/\hbar)a_B\sqrt{2mW}}}{e^{-(2/\hbar)a_A\sqrt{2mW}}} = e^{(2/\hbar)(a_A - a_B)\sqrt{2mW}}.$$

We can evaluate the constants in the exponent:

$(2/\hbar)\sqrt{2mW} = [2/(1.05 \times 10^{-34} \text{ J}\cdot\text{s})]\sqrt{2(9.1 \times 10^{-31} \text{ kg})(2.3 \text{ eV})(1.6 \times 10^{-19} \text{ J/eV})} = 1.56 \times 10^{10} \text{ m}^{-1} = 15.6 \text{ (nm)}^{-1}.$

With a_A and a_B in nanometers, we have

$\dfrac{I_B}{I_A} = 20 = e^{(a_A - a_B)[15.6 \text{ (nm)}^{-1}]}$, which gives $a_A - a_B = 0.19$ nm.

$\boxed{\text{Thus } B \text{ is } 0.19 \text{ nm closer to the tip than point } A.}$

41. With the effective $Z = 1$, we find the radius of the Bohr orbit from
$$r = n^2\hbar^2 4\pi\varepsilon/m^* e^2 = (n^2\hbar^2 4\pi\varepsilon_0/m_e e^2)(\varepsilon/\varepsilon_0)(m_e/m^*) = (\varepsilon/\varepsilon_0)(m_e/m^*)n^2 a_0 .$$
For $n = 1$, we have
$$r = (17.9)(1/0.015)a_0 = \boxed{(1.2 \times 10^3)a_0.}$$

43. We use the result from Problem 41 for the radius:

$$r_1 = (\varepsilon/\varepsilon_0)(m_e/m_n^*)n^2 a_0 = [(15.8)/(0.55)](0.53 \times 10^{-10} \text{ m})(1)^2 = 1.5 \times 10^{-9} \text{ m}.$$

If we assume that the maximum concentration corresponds to touching radii, the volume occupied by an impurity atom is

$$V = \tfrac{4}{3}\pi r_1^3 = \tfrac{4}{3}\pi(1.5 \times 10^{-9} \text{ m})^3 = 1.47 \times 10^{-26} \text{ m}^3, \text{ so the volume density of impurities is}$$

$$\rho_{impurity} = 1/V = 6.8 \times 10^{25} \text{ impurities/m}^3 = \boxed{6.8 \times 10^{19} \text{ impurities/cm}^3.}$$

Typical concentrations are close to this.
The density of host atoms is

$$\rho_{host} = (5.32 \text{ g/cm}^3)(6.02 \times 10^{23} \text{ atoms/mol})/(72.6 \text{ g/mol}) = 4.4 \times 10^{22} \text{ atoms/cm}^3,$$

so the ratio is

$$\rho_{impurity}/\rho_{host} = (6.8 \times 10^{19} \text{ impurities/cm}^3)/(4.4 \times 10^{22} \text{ atoms/cm}^3) = \boxed{1.5 \times 10^{-3} \quad (0.15 \text{ atomic \%}).}$$

47. Because the energy range, $\Delta E = 0.02$ eV, is small, we find the number of available states in this range at the average energy of 12.66 eV from

$$\frac{\Delta N(E)}{V} = \frac{1}{2\pi^2}\left(\frac{2m_e}{\hbar^2}\right)^{3/2}\sqrt{E}\,\Delta E = \frac{4\pi}{h^3}(2m_e)^{3/2}\sqrt{E}\,\Delta E$$

$$= \frac{4\pi}{\left(6.63 \times 10^{-34} \text{ J·s}\right)^3}\left[2\left(9.1 \times 10^{-31} \text{ kg}\right)\right]^{3/2}\sqrt{(12.66\,eV)\left(1.6 \times 10^{-19} \text{ J/eV}\right)}\,(0.02 \text{ eV})\left(1.6 \times 10^{-19} \text{ J/eV}\right)$$

$$= 4.82 \times 10^{26} \text{ m}^{-3}.$$

The Fermi-Dirac distribution gives the probability that a state is occupied. If we multiply the density of states by the occupation probability, we get the electron density.
At $T = 90$ K, we have

$$\Delta n = \frac{1}{1 + e^{(E - E_F)/kT}}\frac{\Delta N(E)}{V}$$

$$= \frac{1}{1 + e^{[(12.66 \text{ eV}) - (10.40 \text{ eV})]/[(0.0259 \text{ eV})(90 \text{ K})/(300 \text{ K})]}}\left(4.82 \times 10^{26} \text{ m}^{-3}\right) \approx \boxed{10^{-100} \text{ electrons/m}^3.}$$

At $T = 300$ K, we have

$$\Delta n = \frac{1}{1 + e^{(E - E_F)/kT}}\frac{\Delta N(E)}{V}$$

$$= \frac{1}{1 + e^{[(12.66 \text{ eV}) - (10.40 \text{ eV})]/(0.0259 \text{ eV})}}\left(4.82 \times 10^{26} \text{ m}^{-3}\right) = \boxed{6.1 \times 10^{-12} \text{ electrons/m}^3.}$$

49. (a) Each photon has an energy

$$E = hf = hc/\lambda = (1.24 \times 10^3 \text{ eV·nm})/(800 \text{ nm}) = 1.5 \text{ eV}.$$

We find the rate at which photons are emitted from

$$(dN/dt) = P/E$$
$$= (1 \times 10^{-3} \text{ W})/(1.5 \text{ eV})(1.6 \times 10^{-19} \text{ J/eV}) \quad \boxed{\approx 4 \times 10^{15} \text{ photons/s.}}$$

(b) To estimate the time it takes for a distance d to pass over the laser beam, we use a radius of 4 cm:

$$d = R\omega t$$
$$0.1 \times 10^{-3} \text{ m} = [(4 \times 10^{-2} \text{ m})(2 \times 10^2 \text{ rev/min})(2\pi \text{ rad/rev})/(60 \text{ s/min})]t, \text{ which gives } t \approx 1 \times 10^{-4} \text{ s}.$$

The number of photons striking the region is

$$N = (dN/dt)t = (4 \times 10^{15} \text{ photons/s})(10^{-4} \text{ s}) \quad \boxed{\approx 4 \times 10^{11} \text{ photons.}}$$

51. (a) The wavelength of an ultraviolet photon has an energy
     $E \simeq (1.24 \times 10^3 \text{ eV} \cdot \text{nm})/(400 \text{ nm}) = 3 \text{ eV}$, which is greater than the energy gap.
     At this wavelength, the radiation can excite electrons from the valence band into the conduction band. The electron-hole pairs created produce a higher density of charge carriers and thus a high conductivity or low resistance. As the wavelength increases, the photon energy decreases. When this energy becomes less than E_g, electron-hole pairs cannot be produced.
     A large drop in carrier concentration occurs, which leads to a rapid increase in the resistance.
   (b) We find the critical wavelength from
     $\lambda = (1.24 \times 10^3 \text{ eV} \cdot \text{nm})/(1.12 \text{ eV}) = \boxed{1.10 \times 10^3 \text{ nm.}}$
   (c) The resistance $\boxed{\text{increases}}$ at the critical wavelength.

53.

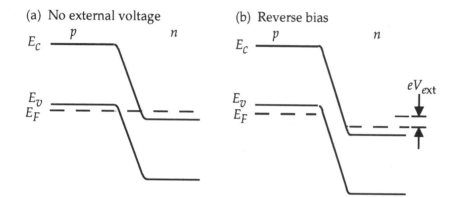

   (b) As the reverse bias increases, there will be more unoccupied states on the n-side across from the large electron densities on the p-side of the boundary. The tunneling current will increase.

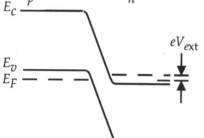
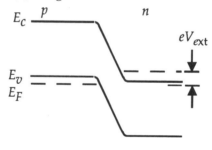

   (c) For a small forward bias, the electrons in the mostly occupied states below E_F on the n-side can tunnel to the mostly unoccupied states above the Fermi level on the p-side. The tunneling current will increase.
   (d) For a large forward bias, the range of occupied states on the n-side that are adjacent to the available states on the p-side will decrease. The tunneling current will decrease.

CHAPTER 45 Extended Version

Note: A useful expression for the energy of a photon in terms of its wavelength is
$$E = hf = hc/\lambda = (6.63 \times 10^{-34}\,\text{J} \cdot \text{s})(3 \times 10^8\,\text{m/s})(10^{-9}\,\text{nm/m})/(1.60 \times 10^{-19}\,\text{J/eV})\lambda;$$
$$E = (1.24 \times 10^3\,\text{eV} \cdot \text{nm})/\lambda.$$
A factor that appears in the analysis of energies is
$$e^2/4\pi\varepsilon_0 = (1.60 \times 10^{-19}\,\text{C})^2(9 \times 10^9\,\text{N} \cdot \text{m}^2/\text{C}^2) = 2.30 \times 10^{-28}\,\text{J} \cdot \text{m} = 1.44\,\text{MeV} \cdot \text{fm}.$$

9. The momentum transfer of the scattered particle is
$$\Delta\mathbf{p} = \mathbf{p}_f - \mathbf{p}_i = [mv(\cos\theta - 1)]\mathbf{i} + (mv\sin\theta)\mathbf{j}.$$
If we "square" this, we have
$$(\Delta p)^2 = (mv)^2(\cos^2\theta - 2\cos\theta + 1 + \sin^2\theta)$$
$$= 2m^2v^2(1 - \cos\theta) = 8mK\sin^2(\theta/2),$$
where K is the kinetic energy of the particle and we
have used a trigonometric identity
$$\sin^2(\theta/2) = \tfrac{1}{2}(1 - \cos\theta).$$
We use this in the expression for the cross section:

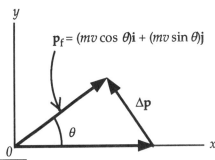

$$\sigma = (Z_1Z_2e^2/4\pi\varepsilon_0)^2/16K^2\sin^4(\theta/2)$$
$$= (Z_1Z_2e^2/4\pi\varepsilon_0)^2/16[(\Delta p)^2/8m]^2 = \boxed{(Z_1Z_2e^2/4\pi\varepsilon_0)^2[4m^2/(\Delta p)^4].}$$

13. If we ignore the small difference between the hydrogen and neutron masses, the difference in binding
energies, and thus the change in electrostatic energy, is determined by the difference in isotopic masses.
We use the result from Problem 12:
$$\Delta E = (\Delta M)c^2(931.5\,\text{MeV}/uc^2) = (3/5)(e^2/4\pi\varepsilon_0 r_0 A^{1/3})(Z_2^2 - Z_1^2).$$
For ^{7}Li and ^{7}Be, we have
$$(7.016930\,\text{u} - 7.016005\,\text{u})c^2(931.5\,\text{MeV}/uc^2) = (3/5)[(1.44\,\text{MeV} \cdot \text{fm})/r_0(7)^{1/3}][(4)^2 - (3)^2],$$
which gives $\boxed{r_0 = 3.67\,\text{fm.}}$
For ^{9}B and ^{9}Be, we have
$$(9.01333\,\text{u} - 9.012183\,\text{u})c^2(931.5\,\text{MeV}/uc^2) = (3/5)[(1.44\,\text{MeV} \cdot \text{fm})/r_0(9)^{1/3}][(5)^2 - (4)^2],$$
which gives $\boxed{r_0 = 3.50\,\text{fm.}}$
For ^{11}B and ^{11}C, we have
$$(11.011433\,\text{u} - 11.009305\,\text{u})c^2(931.5\,\text{MeV}/uc^2) = (3/5)[(1.44\,\text{MeV} \cdot \text{fm})/r_0(11)^{1/3}][(6)^2 - (5)^2],$$
which gives $\boxed{r_0 = 2.16\,\text{fm.}}$
For ^{21}Ne and ^{21}Na, we have
$$(20.99765\,\text{u} - 20.99385\,\text{u})c^2(931.5\,\text{MeV}/uc^2) = (3/5)[(1.44\,\text{MeV} \cdot \text{fm})/r_0(21)^{1/3}][(11)^2 - (10)^2],$$
which gives $\boxed{r_0 = 1.86\,\text{fm.}}$
For ^{23}Na and ^{23}Mg, we have
$$(22.9941\,\text{u} - 22.9898\,\text{u})c^2(931.5\,\text{MeV}/uc^2) = (3/5)[(1.44\,\text{MeV} \cdot \text{fm})/r_0(23)^{1/3}][(12)^2 - (11)^2],$$
which gives $\boxed{r_0 = 1.75\,\text{fm.}}$
We see that r_0 approaches 1.2 fm as Z gets larger; however, there are no mirror nuclei at large Z, because
the number of neutrons increases faster than the number of protons.

19. We find the energy required to assemble $Z = 6$ protons in carbon from
$$E_C = 3Z^2(e^2/4\pi\varepsilon_0)/5r_0 A^{1/3}$$
$$= 3(6)^2[(2.30 \times 10^{-28}\,\text{J} \cdot \text{m})]/(1.6 \times 10^{-13}\,\text{J/MeV})5(1.2 \times 10^{-15}\,\text{m})12^{1/3} = 11\,\text{MeV}.$$
The mass of a carbon nucleus is
$$Am_nc^2 = (12)(1.0\,\text{u})c^2(931\,\text{MeV}/uc^2) = 1.1 \times 10^4\,\text{MeV}.$$
For the fraction, we have
$$E_C/Am_nc^2 = (11\,\text{MeV})/(1.1 \times 10^4\,\text{MeV}) = \boxed{1.0 \times 10^{-3}.}$$

23. For the reaction $\alpha + {}^{14}\text{N} \rightarrow p + {}^{17}\text{O}$, we find the Q value from
$Q = [M({}^4\text{He}) + M({}^{14}\text{N}) - M({}^1\text{H}) - M({}^{17}\text{O})]c^2$
$\quad = [(4.002603 \text{ u}) + (14.003074 \text{ u}) - (1.007825 \text{ u}) - (16.999131 \text{ u})]c^2(931.5 \text{ MeV}/uc^2) = -1.19 \text{ MeV}.$
We find the final kinetic energies of the particles from
$Q = K_f - K_i;$
$-1.19 \text{ MeV} = K_f - (5.00 \text{ MeV}),$ which gives $\boxed{K_f = 3.80 \text{ MeV}.}$

25. For all particles, $K \ll Mc^2$, so we can use a nonrelativistic treatment: $K = p^2/2M$. We simplify the masses to multiples of the neutron mass. For momentum conservation of the assumed one-dimensional reaction, we have
$p_{2\text{H}} = p_{4\text{He}} + p_n,$ or $p_{4\text{He}} = p_{2\text{H}} - p_n;$ $\sqrt{2(4m_n)K_{4\text{He}}} = \sqrt{2(2m_n)K_{2\text{H}}} - \sqrt{2(m_n)K_n}.$
When we eliminate common factors and square, we get
$K_{4\text{He}} = \tfrac{1}{2}K_{2\text{H}} - \sqrt{\tfrac{1}{2}K_{2\text{H}}K_n} + \tfrac{1}{4}K_n.$
For the conservation of energy we have
$Q + K_{2\text{H}} = K_{4\text{He}} + K_n = \tfrac{1}{2}K_{2\text{H}} - \sqrt{\tfrac{1}{2}K_{2\text{H}}K_n} + \tfrac{1}{4}K_n + K_n;$
$17.6 \text{ MeV} + 2.0 \text{ MeV} = \tfrac{1}{2}(2.0 \text{ MeV}) - \sqrt{\tfrac{1}{2}(2.0 \text{ MeV})K_n} + \tfrac{5}{4}K_n;$ or $K_n - \left(0.80\sqrt{\text{MeV}}\right)\sqrt{K_n} - 14.9 \text{ MeV} = 0.$
This is a quadratic equation for $K_n^{1/2}$, which has the positive solution
$K_n^{1/2} = 4.28 \text{ (MeV)}^{1/2},$ or $\boxed{K_n = 18.3 \text{ MeV}.}$
Note that this is the maximum kinetic energy possible for the neutron, because the final products are in the direction of the incident beam.

27. For the reaction $\alpha + {}^{11}\text{B} \rightarrow n + {}^{14}\text{N}$, we find the Q value from
$Q = [M({}^4\text{He}) + M({}^{11}\text{B}) - m_n - M({}^{14}\text{N})]c^2$
$\quad = [(4.002603 \text{ u}) + (11.009305 \text{ u}) - (1.008665 \text{ u}) - (14.003074 \text{ u})]c^2(931.5 \text{ MeV}/uc^2) = -0.157 \text{ MeV}.$
We find the kinetic energy of the neutron from
$Q = K_f - K_i = K_n + K_N - K_\alpha - K_B;$
$0.157 \text{ MeV} = K_n + (0.8 \text{ MeV}) - (5.3 \text{ MeV}) - 0,$ which gives $\boxed{K_n = 4.7 \text{ MeV}.}$

29. For the reaction ${}_4^9\text{Be} + {}_2^4\text{He} \rightarrow {}_6^{12}\text{C} + n,$ we find the Q value from
$Q = [M({}^9\text{Be}) + M({}^4\text{He}) - M({}^{12}\text{C}) - m_n]c^2$
$\quad = [(9.012183 \text{ u}) + (4.002603 \text{ u}) - (12.000000 \text{ u}) - (1.00866 \text{ u})]c^2(931.5 \text{ MeV}/uc^2) = +5.71 \text{ MeV}.$
If we ignore the recoil of the C atom, we find the maximum kinetic energy of the neutron from
$Q = K_f - K_i = K_n + K_C - K_{\text{He}} - K_{\text{Be}};$
$5.71 \text{ MeV} = K_n + 0 - (4.78 \text{ MeV}) - 0,$ which gives $\boxed{K_n = 10.49 \text{ MeV}.}$
The average kinetic energy is $\approx 5 \text{ MeV}.$

33. For the positron-emission process
$_{Z+1}^A X \rightarrow {}_Z^A X' + e^+ + \nu,$
we need to add $Z + 1$ electrons to the nuclear mass of X to be able to use the atomic mass. On the right-hand side we use Z electrons to be able to use the atomic mass of X'. Thus we have 1 electron mass and the β-particle mass, which means that we must include 2 electron masses on the right-hand side. The Q value will be
$Q = M({}^A X) - [M({}^A X') + 2m_e],$
which must be positive because the kinetic energy increases in the decay. Thus
$M_P > M_D + 2m_e.$

37. We find the number of atoms of each isotope:
    $N_U = [(1 \times 10^6 \text{ g})/(238 \text{ g/mol})](6.02 \times 10^{23} \text{ atoms/mol}) = 2.5 \times 10^{27}$.
    $N_{Th} = [(3 \times 10^6 \text{ g})/(232 \text{ g/mol})](6.02 \times 10^{23} \text{ atoms/mol}) = 7.8 \times 10^{27}$.
    Because one year is very short, compared to the half life of either isotope, we can assume that the activity is constant over the year, so $\Delta N/\Delta t = N\lambda = 0.693 N/t_{1/2}$. Each decay produces one radon atom. Because the half life of either radon isotope is much greater than one year, we can ignore the decay of the radon and find the total number of radon atoms from the number of parents that have decayed:
    $dN_{Rn}/dt = 0.693[(N/t_{1/2})_U + (N/t_{1/2})_{Th}]$
    $= 0.693\{[(2.5 \times 10^{27})/(4.51 \times 10^9 \text{ yr})] + [(7.8 \times 10^{27})/(1.39 \times 10^{10} \text{ yr})]\} = 7.7 \times 10^{17} \text{ atoms/yr}$.
    Radon is an inert monatomic gas. We find the volume production using the ideal gas law:
    $p(dV/dt) = (dN_{Rn}/dt)kT$;
    $(1.01 \times 10^5 \text{ N/m}^2)(dV/dt) = (7.7 \times 10^{17} \text{ atoms/yr})(1.38 \times 10^{-23} \text{ J/K})(293 \text{ K})$, which gives

    $\boxed{dV/dt = 3.1 \times 10^{-8} \text{ m}^3/\text{yr} = 0.031 \text{ cm}^3/\text{yr}.}$

41. Because animal cells are constantly being replaced, the concentration of ^{14}C in the living animal will be the concentration in the atmosphere until the animal dies. After that time no carbon atoms will be replaced, so the concentration will decrease with the half-life of ^{14}C.
    We find the elapsed time from
    $N(t) = 0.15N_0 = N_0 e^{-t/\tau}$, or
    $\ln 0.15 = -t/\tau = -0.693t/T_{1/2}$, which gives
    $t = T_{1/2}(-\ln 0.15)/0.693 = (5730 \text{ yr})(1.90)/(0.693) = 1.57 \times 10^4 \text{ yr ago}$.
    We find the year the mammoth lived from
    date $= (1.57 \times 10^4 \text{ yr}) - (0.20 \times 10^4 \text{ yr}) \approx$ $\boxed{13{,}700 \text{ B. C.}}$

43. For the <u>electron-emission</u> process, we have
    $^{80}_{35}\text{Br} \to {}^{80}_{36}\text{Kr} + e^- + \bar{\nu}$.
    From Problem 32, the Q value of this decay is
    $Q = M(^{80}\text{Br}) - M(^{80}\text{Kr})$
    $= [(79.918528 \text{ u}) - (79.916376 \text{ u})](931.5 \text{ MeV/u}) = 2.00 \text{ MeV}$.
    Because the Q value is positive, the decay is possible.
    For the <u>positron-emission</u> process, we have
    $^{80}_{35}\text{Br} \to {}^{80}_{34}\text{Se} + e^+ + \nu$.
    From Problem 33, the Q value of this decay is
    $Q = M(^{80}\text{Br}) - [M(^{80}\text{Se}) + 2m_e]$
    $= [(79.918528 \text{ u}) - [(79.916521 \text{ u})](931.5 \text{ MeV/u}) + 2(0.511 \text{ MeV})] = 0.85 \text{ MeV}$.
    Because the Q value is positive, the positron-emission process is possible.
    For the <u>electron capture</u>, we have
    $^{80}_{35}\text{Br} + e^- \to {}^{80}_{34}\text{Se} + \nu$.
    From Problem 34, the Q value of this decay is
    $Q = M(^{80}\text{Br}) - M(^{80}\text{Se})$
    $= [(79.918528 \text{ u}) - (79.916521 \text{ u})](931.5 \text{ MeV/u}) = 1.87 \text{ MeV}$.
    Because the Q value is positive, the decay is possible.

45. Because each decay of an a nucleus produces a b nucleus, the change in the number of a nuclei,

$dN_a = -(N_a/\tau_1)\,dt$,

with a positive sign, gives the number of b nuclei produced in dt. In the same time interval dt, the number of b nuclei will decrease because of its own decay. The net change in the number of b nuclei is

$dN_b = -(N_b/\tau_2)\,dt + dN_a = -(N_b/\tau_2)\,dt + (N_a/\tau_1)\,dt$.

Because each decay of a b nucleus will produce a c nucleus, etc., there will be two similar terms for each of the successive decays, with the number of a particular nucleus decreasing because of its own decay and increasing because of the decay of its parent. If we put the expressions in terms of rates, we have

$dN_a/dt = -N_a/\tau_1$;
$dN_b/dt = -(N_b/\tau_2) + (N_a/\tau_1)$;
$dN_c/dt = -(N_c/\tau_3) + (N_b/\tau_2)$;
$dN_d/dt = -(N_d/\tau_4) + (N_c/\tau_3)$; ... .

51. (a) We find the number of fissions required to produce the electrical energy from

$N_f(200\text{ MeV}) = P_{nucl}t = (P_{elect}/0.25)t = [(440\text{ MW})/0.25]t = (1760\text{ MW})t$.

We assume that all the ^{235}U undergoes fission during the year; the number of fissions is the number of ^{235}U atoms, 6% of the number of U atoms. If m is the mass of the uranium in grams, we have

$N_f = 0.06(m/M)N_A$.

Because most of the uranium is ^{238}U, we use $M = 238$ g/mol when we combine the equations:

$0.06(m/M)N_A(200\text{ MeV}) = (1760\text{ MW})t$;

$0.06[m/(238\text{ g/mol})](6.02 \times 10^{23}\text{ atoms/mol})(200\text{ MeV})(1.6 \times 10^{-13}\text{ J/MeV}) =$
$(1760 \times 10^6\text{ W})(1\text{ yr})(3.16 \times 10^7\text{ s/yr})$, which gives

$m = 1.15 \times 10^7\text{ g} = \boxed{11.5\text{ metric tons.}}$

(b) We find the energy released to the environment from

$E_{lost} = E_{nucl} - E_{elect} = (E_{elect}/0.25) - E_{elect} = 3E_{elect}$
$= 3(440 \times 10^6\text{ W})(1\text{ yr})(3.16 \times 10^7\text{ s/yr}) = \boxed{4.2 \times 10^{16}\text{ J.}}$

(c) We find the fission rate from

$N_f(200\text{ MeV}) = (P_{elect}/0.25)t$;

$n_f = N_f/t = (440 \times 10^6\text{ W})/0.25(200\text{ MeV})(1.6 \times 10^{-13}\text{ J/MeV}) = \boxed{5.5 \times 10^{19}\text{ fissions/s.}}$

53. (a) If we compare the sums of the left and right sides of the reactions, we have

$4p \rightarrow {}^4\text{He} + 2e^+ + 2\nu + 3\gamma$.

The total energy produced in one cycle is $\boxed{24.8\text{ MeV released.}}$

(b) We find the number of carbon nuclei from the mass of carbon present:

$N_C = [(0.001)(3 \times 10^{30}\text{ kg})(10^3\text{ g/kg})/(12\text{ g/mol})](6.02 \times 10^{23}\text{ nuclei/mol}) = 1.5 \times 10^{53}\text{ nuclei}$.

Because each cycle involves one ^{12}C atom, we find the energy produced in 1 year from

$E = (1.5 \times 10^{53}\text{ nuclei})(24.8\text{ MeV/cycle})(1\text{ yr})/(5 \times 10^6\text{ yr/cycle}) = \boxed{7.5 \times 10^{47}\text{ MeV} = 1.2 \times 10^{34}\text{ J.}}$

59. (a) When the nuclei get close enough to be inside the electron cloud, there will be a repulsive Coulomb force between the nuclei.

(b) At the closest approach, the helium nuclei will be at rest with the centers 4 fm apart and they will have potential energy.
We find the required kinetic energy of approach from the conservation of energy:

$K_1 + U_1 = K_2 + U_2$;
$2K + 0 = 0 + (Z^2e^2/4\pi\varepsilon_0)/r_{min}$;
$2K = 2^2(1.44\text{ MeV/fm})/(4\text{ fm})$, which gives $\boxed{K = 0.7\text{ MeV for each helium atom.}}$

63. The volume of the nucleus is

$V = 4\pi R^3/3 = 4\pi r_0^3 A/3$.

The expression for the total energy of Z protons is

$E_p = V(\hbar^2\pi^3/10m_p)(3Z/V\pi)^{5/3} = (\hbar^2\pi^3/10m_p)(3/\pi)^{5/3}Z^{5/3}/V^{2/3}$
$= (\hbar^2\pi^3/10m_p)(3/\pi)^{5/3}Z^{5/3}/(4\pi r_0^3 A/3)^{2/3} = \boxed{(\hbar^2/10m_p r_0^2)(3^7\pi^2Z^5/16A^2)^{1/3}.}$

Because the number of neutrons is $N = A - Z$, we have

$\boxed{E_n = (\hbar^2/10m_n r_0^2)[3^7\pi^2(A - Z)^5/16A^2]^{1/3}.}$

CHAPTER 46 Extended Version

Note: A useful expression for the energy of a photon in terms of its wavelength is
$E = hf = hc/\lambda = (6.63 \times 10^{-34} \text{ J} \cdot \text{s})(3 \times 10^8 \text{ m/s})(10^{-9} \text{ nm/m})/(1.60 \times 10^{-19} \text{ J/eV})\lambda;$
$E = (1.24 \times 10^3 \text{ eV} \cdot \text{nm})/\lambda.$
A factor that appears in the analysis of energies is
$e^2/4\pi\varepsilon_0 = (1.60 \times 10^{-19} \text{ C})^2(9 \times 10^9 \text{ N} \cdot \text{m}^2/\text{C}^2) = 2.30 \times 10^{-28} \text{ J} \cdot \text{m} = 1.44 \text{ MeV} \cdot \text{fm}.$

3.   For every particle except the target proton, we have $E = pc$. We let p_i be the initial momentum of the electron, p_f be the final momentum of the electron, and q be the final momentum of the proton, with mass M. For energy conservation we have
$p_i c + Mc^2 = p_f c + qc.$
For momentum conservation we have
$p_i + 0 = -p_f + q.$
When we eliminate p_f, we get
$$\boxed{q = p_i + \tfrac{1}{2}Mc.}$$
If the energy of the incident electron is doubled, the momentum is doubled, so we have
$q'/q = (2p_i + \tfrac{1}{2}Mc)/(p_i + \tfrac{1}{2}Mc).$
The momentum transferred increases by a factor of $\boxed{(4p_i + Mc)/(2p_i + Mc).}$

5.   For every particle except the target proton, we have $E = pc$. We let p_i be the initial momentum of the electron, p_f be the final momentum of the electron, and q be the final momentum of the proton, with mass M. For energy conservation we have
$p_i c + Mc^2 = p_f c + qc.$
For momentum conservation we have
x-direction: $p_i + 0 = q \cos \theta;$
y direction: $0 + 0 = -p_f + q \sin \theta.$
We combine these to get
$p_i^2 + p_f^2 = q^2$, which is evident from the diagram,
and then use the energy result to eliminate p_f to get
$$\boxed{\begin{aligned} q &= [p_i^2 + (p_i + Mc)^2]/2(p_i + Mc), \\ \cos \theta &= 2p_i(p_i + Mc)/[p_i^2 + (p_i + Mc)^2]. \end{aligned}}$$
If $p_i \gg Mc$, we have
$$\boxed{\cos \theta \simeq 1,\ \theta \simeq 0,\ \text{and}\ q \simeq p_i.}$$
If we expand $(p_i + Mc)^2 = p_i^2 + 2Mcp_i + (Mc)^2$, and neglect the last term, which is not as strong a restriction on p_i, we have
$\cos \theta = 2p_i(p_i + Mc)/[p_i^2 + p_i^2 + 2Mcp_i] \simeq 1$, $\theta \simeq 0$, and $q \simeq p_i$, which is the same result.

9.   We have $E = pc$ for the electrons. In the center-of-mass frame, the electrons have opposite momentum p. At the minimum energy, the two particles produce two electrons and the proton–antiproton pair at rest, which we can consider to be an object with mass $2m_e + 2m_p \simeq 2m_p$. In the laboratory frame, the incident electron momentum is p_{lab} and the target electron momentum is 0. The two particles produce an object with mass $M = 2m_p$ and momentum P. For energy conservation we have
$p_{lab}c + m_e c^2 = [P^2 c^2 + (Mc^2)^2]^{1/2}.$
For momentum conservation we have
$p_{lab} + 0 = P.$
When we use this to eliminate P in the energy equation and square both sides, we have
$(p_{lab}c)^2 + (m_e c^2)^2 + 2p_{lab}cm_e c^2 = (p_{lab}c)^2 + (Mc^2)^2$, which gives
$p_{lab}c = [(Mc^2)^2 - (m_e c^2)^2]/2m_e c^2$
$\simeq \{[2(0.938 \text{ GeV})]^2 - 0\}/2(0.511 \times 10^{-3} \text{ GeV}) = 3.44 \times 10^3 \text{ GeV} = \boxed{3.44 \text{ TeV.}}$

19. The quark content of each of the neutrons is (udd). The π^0 can interact through either the u quark with $(u\bar{u})$ or the d quark with $(d\bar{d})$.

    The quark content can be $\boxed{(u\bar{u}) \text{ or } (d\bar{d}) \text{ or a combination of these two.}}$

25. We have $E = pc$, except for the stationary proton. In the laboratory frame, the incident proton momentum is p and the target proton momentum is 0. If the two particles produce an object with mass M and momentum P, for energy conservation we have
$$pc + m_p c^2 = [P^2 c^2 + (Mc^2)^2]^{1/2}.$$
For momentum conservation we have
$$p + 0 = P.$$
When we use this to eliminate P in the energy equation and square both sides, we have
$$(pc)^2 + (m_p c^2)^2 + 2pcm_p c^2 = (pc)^2 + (Mc^2)^2, \text{ which gives } (Mc^2)^2 \simeq 2pcm_p c^2.$$
In the center-of-mass frame, each proton has momentum q. If the two particles produce an object at rest with mass M, for energy conservation we have
$$2qc = Mc^2.$$
When we use the above result for Mc^2, we have
$$
\begin{aligned}
qc &= \tfrac{1}{2}(2pcm_p c^2)^{1/2} \\
&= \tfrac{1}{2}[2(300 \text{ GeV})(0.94 \text{ GeV})]^{1/2} = 11.9 \text{ GeV}, \text{ or } \boxed{q = 11.9 \text{ GeV}/c.}
\end{aligned}
$$

31. We have $E = pc$ for the photon. In the center-of-mass frame, the minimum energy of the photon for the reaction $\gamma + p \rightarrow p + \bar{p} + p$ creates three particles at rest, which we can treat as one particle with mass $M = 3m_p$. In the laboratory frame, this particle will have momentum P.
For energy conservation we have
$$p_\gamma c + m_p c^2 = [P^2 c^2 + (3m_p c^2)^2]^{1/2}.$$
For momentum conservation we have
$$p_\gamma + 0 = P.$$
When we eliminate P in the energy equation and square both sides, we have
$$
\begin{aligned}
(p_\gamma c)^2 + (m_p c^2)^2 + 2p_\gamma cm_p c^2 &= (p_\gamma c)^2 + (3m_p c^2)^2, \text{ which gives} \\
p_\gamma c &= [(3m_p c^2)^2 - (m_p c^2)^2]/2m_p c^2 = 4m_p c^2 \\
&= 4(0.938 \text{ GeV}) = \boxed{3.75 \text{ GeV.}}
\end{aligned}
$$

33. The energy of the laser photon is
$$
\begin{aligned}
E_0 = p_0 c &= hc/\lambda \\
&= (1.24 \times 10^3 \text{ eV} \cdot \text{nm})/(590 \text{ nm}) = 2.10 \text{ eV} = 2.10 \times 10^{-9} \text{ GeV}.
\end{aligned}
$$
For maximum energy transfer, we assume that the X-ray recoils backward, the direction of the incident electron. If the initial and final momenta of the electron are p_i and p_f, for conservation of momentum we have
$$p_i - p_0 = p_f + p_\gamma.$$
For energy conservation we have
$$[p_i^2 c^2 + (m_e c^2)^2]^{1/2} + p_0 c = [p_f^2 c^2 + (m_e c^2)^2]^{1/2} + p_\gamma c.$$
When we use the approximation suggested in the hint, we have
$$p_i c + [(m_e c^2)^2/2p_i c] + p_0 c = p_f c + [(m_e c^2)^2/2p_f c] + p_\gamma c.$$
If we use the result from momentum conservation to eliminate p_γ, we get
$$(m_e c^2)^2/2p_f c = [(m_e c^2)^2/2p_i c] + 2p_0 c, \text{ or}$$
$$
\begin{aligned}
p_f c &= p_i c(m_e c^2)^2/[4p_0 cp_i c + (m_e c^2)^2] \\
&= (8 \text{ GeV})(0.511 \times 10^{-3} \text{ GeV})^2/[4(2.10 \times 10^{-9} \text{ GeV})(8 \text{ GeV}) + (0.511 \times 10^{-3} \text{ GeV})^2] = 6.4 \text{ GeV}.
\end{aligned}
$$
We find the energy of the X-ray photon from
$$
\begin{aligned}
p_\gamma c &= p_i c - p_0 c - p_f c \\
&= (8 \text{ GeV}) - (2.10 \times 10^{-9} \text{ GeV}) - (6.4 \text{ GeV}) = \boxed{1.6 \text{ GeV.}}
\end{aligned}
$$
The photon recoils backward, so the angle is $\boxed{180°.}$

37. For a galaxy that is a distance x from Earth, we find the change in the distance in a time dt from
$dx = u\, dt$, which becomes
$dt = dx/u = dx/Hx$.
To find the time, we integrate this from D to $1.1D$, with H constant:
$$\int_0^T dt = \int_D^{1.1D} \frac{dx}{Hx} = \frac{1}{H}\int_D^{1.1D} \frac{dx}{x};$$
$$T - 0 = \frac{1}{H}\ln\left(\frac{1.1D}{D}\right), \text{ which gives } \quad T = \frac{1}{H}\ln(1.1).$$
When we use the value of the Hubble constant, we get
$T = \ln(1.1)/H = \ln(1.1)/(2.5 \times 10^{-18}\text{ s}^{-1}) = 3.8 \times 10^{16}\text{ s} = \boxed{1.2 \times 10^9 \text{ yr.}}$

43. (a) If we let a be the number of moles of hydrogen and b be the number of moles of helium that were formed, the number of protons is
$N_p = (a + 2b)N_A$, and
the number of neutrons is
$N_n = (2b)N_A$.
From the initial distribution, we have
$N_n/N_p = 1/4 = 2b/(a + 2b)$, which gives $a/b = 6$.
(b) If we take $m_p = m_n$, we have
$m_{\text{hydrogen}}/m_{\text{helium}} = a/4b = 6/4 = 1.5$.

47. The magnetic force is radial, so the speed in the circle is constant. At relativistic speeds, the momentum is $\mathbf{p} = m\gamma\mathbf{u}$, and $d\mathbf{p}/dt = m\gamma\, d\mathbf{u}/dt$, where $\gamma = 1/[1 - (u^2/c^2)]^{1/2}$. The rate of change of the velocity is the centripetal acceleration u^2/R. From the magnetic force, we have
$euB = m\gamma u^2/R$, or
$R = m\gamma u/eB = p/eB$
$= (3.2 \times 10^5 \text{ GeV}/c)(1.6 \times 10^{-10} \text{ J/GeV})/(1.6 \times 10^{-19} \text{ C})(9.5\text{ T})(3 \times 10^8 \text{ m/s})$
$= 1.1 \times 10^5 \text{ m} = \boxed{1.1 \times 10^2 \text{ km.}}$

49. We have $E = pc$ for the electron and positron. In the center-of-mass frame, the total momentum is zero. The reaction $e^- + e^+ \rightarrow N(\overline{\pi} + \pi)$ will produce the maximum number of mesons $2N$ if all the mesons are at rest. For energy conservation we have
$p_e c + p_p c = 2Nm_\pi c^2$.
$(40\text{ GeV}) + (40\text{ GeV}) = 2N(0.140\text{ GeV})$, which gives $\boxed{2N = 570 \text{ mesons}}$ (285 pairs).

51. The lifetime of the state is inversely proportional to the probability:
$\tau \propto (4\pi R^3/3)/(4\pi L^3/3) = (R/L)^3$.
The radius of the ground-state orbit in a hydrogenlike atom is
$R = a_0/Z$.
For a nucleus with Z protons, the volume occupied by the protons is proportional to the number of protons, so for the "size" of the nucleus we have
$L \propto L_0 Z^{1/3}$.
If we form the ratio of lifetimes, we have
$\tau/\tau_c = (R/a_0)^3(L_0/L)^3 = (1/Z)^3(1/Z^{1/3})^3 = 1/Z^4$.

53. For the reaction $e^- \to e^- + \gamma$, the isolated electron is at rest. For the photon, we have $E_\gamma = p_\gamma c$.
For energy conservation we have
$$m_e c^2 = [(p_e c)^2 + (m_e c^2)^2]^{1/2} + E_\gamma.$$
For momentum conservation we have
$$0 = p_e - p_\gamma.$$
When we eliminate p_e in the energy equation and rearrange, we have
$$m_e c^2 - E_\gamma = [E_\gamma^2 + (m_e c^2)^2]^{1/2}.$$
When we square both sides, we have
$$(m_e c^2)^2 + E_\gamma^2 - 2E_\gamma m_e c^2 = E_\gamma^2 + (m_e c^2)^2, \text{ which gives } E_\gamma = 0.$$
Thus no photon is emitted, so the assumed reaction cannot occur.

55. If we have N_{0n} neutrons at $t = 0$, we find the number of neutrons that remain after 1000 s from
$$N_n = N_{0n}\, e^{-t/\tau} = N_{0n}\, e^{-(1000\ \mathrm{s})/(890\ \mathrm{s})} = 0.325 N_{0n}.$$
Because the decay of each neutron produces a proton, the number of protons will increase:
$$N_p = N_{0p} + (N_{0n} - N_n) = N_{0p} + N_{0n}(1 - e^{-t/\tau}).$$
The ratio becomes
$$N_n/N_p = (N_{0n}\, e^{-t/\tau})/[N_{0p} + N_{0n}(1 - e^{-t/\tau})] = (N_{0n}/N_{0p})\, e^{-t/\tau}/[1 + (N_{0n}/N_{0p})(1 - e^{-t/\tau})]$$
$$= (0.2)(0.325)/[1 + (0.2)(1 - 0.325)] = 0.057 \quad \boxed{\approx 1{:}17.5.}$$

57. (a) We can use the results for hydrogen if we replace the electron mass with the reduced mass of the proton-antiproton system, which is
$$m^* = m_p m_p/(m_p + m_p) = \tfrac{1}{2}m_p.$$
The binding energy is the negative of the ground-state energy:
$$\begin{aligned}
E_b &= (m^*/2\hbar^2)(e^2/4\pi\varepsilon_0)^2 = (m_p/4\hbar^2)(e^2/4\pi\varepsilon_0)^2 \\
&= (m_p/2m_e)(m_e/2\hbar^2)(e^2/4\pi\varepsilon_0)^2 = (m_p/2m_e)E_1 \\
&= [(1.67 \times 10^{-27}\ \mathrm{kg})/2(9.1 \times 10^{-31}\ \mathrm{kg})](13.6\ \mathrm{eV}) = \quad \boxed{1.25 \times 10^4\ \mathrm{eV.}}
\end{aligned}$$
(b) We find the radius of the ground state from
$$\begin{aligned}
r &= (\hbar^2/m^*)(4\pi\varepsilon_0/e^2) = 2(\hbar^2/m_p)(4\pi\varepsilon_0/e^2) \\
&= (2m_e/m_p)(\hbar^2/m_e)(4\pi\varepsilon_0/e^2) = (2m_e/m_p)a_0 \\
&= [2(9.1 \times 10^{-31}\ \mathrm{kg})/(1.67 \times 10^{-27}\ \mathrm{kg})](0.53 \times 10^{-10}\ \mathrm{m}) = 5.8 \times 10^{-14}\ \mathrm{m} = \quad \boxed{58\ \mathrm{fm.}}
\end{aligned}$$

Discussion of Odd-numbered Questions

Chapter 1

1. We can make a vector in any number of dimensions by making sure that it has a component in each of the dimensions. Thus, if we have a four-dimensional space we can form a set of Cartesian axes, x, y, z, and w, say, and a vector **V** will have an x-component V_x, a y-component V_y, a z-component V_z, and a w-component V_w. The displacement would then be defined as a four-component vector that is the difference between two four-component position vectors.

3. *My* height is pretty close to 6 feet—let's take 6.00 ft as a value to work with. That is equivalent to (6.00 ft)×(12 in/ft) = (72.0 in)×(2.54 cm/in) = 183 cm. To find the height in meters, we divide the number of centimeters by 100: 1.83 m. In every case we have used three significant figures. Incidentally, your height will vary by an amount of the order of 1 cm over the course of a day—you are taller in the morning, after a night in bed. So unless you want to be precise about the time of day, you can reliably state your height to only three significant figures.

5. An essential feature of any standard is that it stay the same: the human pulse fills this need very poorly. A pendulum does much better in this regard—the main reason it might change has to do with changes in its length caused by temperature changes, and this is relatively easier to control with temperature control and control over the material and construction of the pendulum. A second feature desirable in a standard is that it be easily available to all, everywhere. Again, a human pulse serves poorly. Individual pulse rates and steadiness vary greatly, and it would be inconvenient to choose one person. While not ideal, a pendulum is again fairly satisfactory. Our ability to construct identical pendulums is good, and variations in the value of g over the surface of the earth are fairly small. The pendulum clock did in effect form a time standard for many years.

7. First, we are really asking if three vectors *in a plane* can form a null vector. That is because we can start with the plane formed by two of them. If the third vector is not in that plane, then the resultant cannot be zero. So if the vectors do not all lie in the same plane then we cannot form the null vector. Now suppose that they do. Our question can now be rephrased: Can three vectors of equal length in a plane be arranged with tails touching heads such that they form a closed figure? Imagine three rods of equal length, with the ends of two of them hinged to the third one, one at each end of the third rod. The two outer rods have one end free to move, and there is only one position in which their free ends coincide to make a closed triangle. This happens when the three rods form an equilateral triangle.

9. The vector sum of **i** and **j** is the vector **V** = **i** + **j**. The magnitude of **V** is
$V = \sqrt{\mathbf{V}^2} = \sqrt{(\mathbf{i}^2 + \mathbf{j}^2)} = \sqrt{(1+1)} = \sqrt{2}$. The result is not unity! For the answer to the
second part of this question, read the discussion of question 1-7.

11. Let's compare the estimate with a more accurate calculation of a year, the time it takes
for Earth to orbit the sun. The number of full days in the year is 365.25—the extra quarter
day is accounted for in our calendars by leap year, a calendar year with an additional day
that occurs once every four calendar years. Thus the number of seconds is
365.25 d = (365.25 d)×(24 hr/d)×(60 min/hr)×(60 s/min) = 3.156×10^7 s. With π =
3.1415..., we see the approximation is accurate to (3.156 − 3.142)/(3.142) = 0.005 =
0.5%.

13. The mass of the worm is proportional to its volume, and if it increases its linear
dimensions by a factor of ten, its mass, and with it its need for oxygen, will increase by a
factor of 10^3, or a thousand. In contrast, its surface area will increase by a factor of 10^2,
and this is how much the oxygen-absorbing capacity of the worm will increase. Between
the factors of 1000 and 100 there is a deficit of a factor of 10. The giant salamander of
China, which can weigh more than 100 lbs and which spends long periods underwater by
absorbing oxygen through its skin, is an extremely wrinkled animal. In fact, almost all
large animals depend on a huge lung or gill area.

Chapter 2

1. You should be worried about something that might happen to bring the car in front of
you to a stop. Your stopping time depends on two factors: Your fixed reaction time, and
the time required for your brakes to bring your car to a stop. At a higher initial speed, you
travel farther during the time it takes you to react and apply the brakes, and you travel
farther in the time it takes your brakes to bring you to a stop. Both factors, then, argue for
increasing spacing with increasing speed.

3. We have seen that for a fixed acceleration g_x, the relation between fall distance d and
fall time t is, assuming the falling object starts from rest, $d = (1/2)g_x t^2$. Thus for fixed d, $t =$
$(2d/g_x)^{1/2}$. The variation from planet to planet, that is, with g_x, is then $t \propto (g_x)^{-1/2}$. The
larger g_x, the smaller the fall time. The speed of the object at the end of the fall is $v = g_x t =$
$(2dg_x)^{1/2}$. The speed increases with g_x according to $v \propto (g_x)^{1/2}$.

5. Certainly if the (negative) acceleration has a constant magnitude, the velocity cannot
remain positive. Indeed, if the initial velocity has magnitude v_0, and the acceleration has
the constant magnitude a, then the velocity varies with time according to $v = v_0 - at$, and v
= 0 at a time $t = v_0/a$; for times greater than this the velocity is negative. However, the
acceleration could steadily decrease in magnitude, while remaining negative, such that the
velocity could remain positive. A physical example occurs when a rocket is sent away
from Earth with enough initial speed to leave the Solar System—we say that its initial
speed exceeds the "escape speed." If we say that "up" is the positive direction, then the
acceleration is negative while the velocity is positive. As the object moves away, the force

238

of gravity on it, and hence its acceleration, decreases in magnitude while remaining negative. For a fast enough start, the object never comes to rest or turns around. Incidentally, the escape speed from Earth is about 11.2 km/s.

7. As long as the motion is in the positive direction, so that the velocity always is positive, there will be no difference between the average speed and the magnitude of the average velocity. This corresponds to Table 2-1. Once negative velocities can occur, even along a straight line, then the average velocity can have any magnitude, including zero, while the average speed is always greater than zero if there is any movement at all. For more complicated motions, the two quantities are not closely related. For example, when a runner goes exactly once around a track, the average velocity is zero—the net displacement is zero—while the average speed is not.

9. The ancient Greek mathematicians never learned the admittedly subtle notion of a *limit*—that the summation of a larger number of smaller and smaller terms, in this case the terms corresponding to the smaller and smaller subdivisions in time and distance traveled, can add to a finite result, in this case the finite time for the runner to catch a tortoise a finite distance ahead of him. Zeno's paradox certainly doesn't correspond to our experience!

11. Neglecting the effects of air resistance, all three beanbags have exactly the same constant acceleration, namely the acceleration of gravity g, *all the time* they are in the air. There isn't much to compare here.

13. Let's assume that the amount of time spent going up equals the amount of time spent coming back down. That means that we can measure the total time of many jumps, an easy thing to do with precision, divide by the total number of jumps to find the time of one jump, and then divide by two to find the time Δt to fall, say. Our meter stick allows us to measure the height h of a jump; with this information, we can find the average acceleration a by applying the formula $h = a\Delta t^2/2$, or $a = 2h/\Delta t^2$.

Chapter 3

1. If the athlete behaved like a projectile once he or she has left the ground, then we know from our thinking about projectiles that there are two factors: the initial speed and the initial angle. The athlete approaches the take-off spot with a short sprint, just long enough so that his or her speed is at its maximum value. At this point the athlete has to translate the motion so that it has a vertical component, and some of the initial horizontal velocity is translated into vertical motion, with a net loss in speed. There is a trade-off here, and the human body is such that too much initial speed is lost if the athlete tries to take off at a 45° angle, even though this angle maximizes the range for a *given* initial speed. Thus a long jump champion leaves the take-off spot at an angle smaller than 45°, trading the reduced range associated with a smaller angle for a higher initial speed. A second factor that certainly pays a role, even if a somewhat smaller one, is wind resistance. A following

wind can add inches to the jump, and if the following wind is too strong a jump cannot qualify as a record.

As our discussion shows, the most important contribution to the length of the jump is the ability to convert the highest possible initial speed to a vertical motion that gets as close as possible to a takeoff at 45°. It is known that good sprinters generally make good long-jumpers, but not all sprinters hold long-jump records.

3. If the motion is entirely linear, any nonzero acceleration leads to a changing speed. In fact any component of the acceleration parallel to the velocity will change the velocity's magnitude. Only if the acceleration is entirely perpendicular to the motion does the speed remain unchanged, although the direction of the velocity must change. We studied a version of this case in some detail: uniform circular motion.

5. The upper trajectory is more subject to effects due to wind, both because it is higher and because the flight time is longer for that trajectory. The lower trajectory has a shorter flight time—see Eq. (3-41)—and this may allow the pass to reach the receiver before defenders can react properly. This trajectory is superior when the receiver is between the passer and a defender. However, the higher trajectory may be better in that the ball may be beyond the reach of intermediate defenders. The upper trajectory is superior when the receiver is beyond the defender, so that this is the trajectory normally chosen for long passes.

7. The released ball has the horizontal component of the car's velocity at the moment the ball is released. The ball retains that horizontal component. When the car moves with a constant velocity, it "tracks" the ball, so that the ball comes back down into the car. If the car slows down, it will fall behind the horizontal position of the ball, and the ball will land ahead of the car.

9. There are an infinite number of velocities that give the same maximum height. The relation $h = \dfrac{v_0^2 \sin^2(\theta_0)}{2g}$ can be satisfied for a whole range of values of v_0 provided the angle is adjusted properly in the range 0° to 90°. The same argument applies to the time of flight.

11. Let's say you toss the ball at the moment that you are moving due west. You throw it straight up in your own frame, so the ball has a component of horizontal motion to the west equal to your own at the moment you toss it. The ball continues to move due west in a frame of reference fixed to Earth, and this motion takes it obliquely away from the center of the merry-go-round. However, you are moving in a circle, and so change your direction continuously. To you, the ball will appear to move in part away from the center of the merry-go-round and in part to fall behind you. This makes for a rather complicated curving motion that you describe as accelerating. To an observer fixed to Earth it is evident that this "acceleration" is just a manifestation of the fact that you are in an accelerating frame.

13. In question 12, the projectile is fired from the North Pole and so has no initial motion associated with Earth's rotation. Here there is an initial motion that is provided by the motion of a point in Washington, D.C.

Chapter 4.

1. After the string has been cut the only force acting on the mass is gravity, and the mass moves as a projectile under its influence. The *only* memory of the fact that the mass was swinging as a pendulum before the string was cut is that the velocity of the mass when the string is cut is the initial velocity of the projectile motion. If the string is cut when the mass is moving horizontally—the bottom of the swing—then the projectile motion is one in which the initial velocity is horizontal, indistinguishable from the motion of a mass pushed off the edge of a flat tabletop with a non-zero initial velocity.

3. The net force on you is the vector sum of the force of gravity and the contact force from the elevator floor, and it is the net force that describes your acceleration. If the elevator falls freely, your acceleration is the acceleration of gravity, which implies that the net force on you is just gravity; consequently there is no contact force from the floor. If the elevator accelerates upward, you do too, so that the net force must be upward, and this can only be due to an increased contact force from the elevator floor.

5. A constant speed does not mean a constant velocity! The direction of the velocity of the satellite is changing constantly as it circles Earth, so that by Newton's *second* law there must be a net force on the satellite, and Newton's first law is not appropriate. The force responsible for the changing velocity (acceleration) is the gravitational force due to Earth.

7. A force-free environment is a very elusive idea. As the satellite travels from Earth to the Moon, it is influenced by, among other things, the gravitational force of the Sun, of Earth, and of the Moon. Even at the special point between Earth and the Moon where the vector sum of their gravitational forces on the satellite is zero, the force due to the Sun is still present. This is still true for the spaceship traveling from Earth to Mars. It is true that the forces due to Earth and to Mars will be quite small for much of that journey, but the Sun's force will remain substantial, and affects the spaceship for its entire flight.

9. In our everyday experience, "touch" is involved with friction, normal forces from the floor, and the rope pulling on an object, whether it is wrapped around a pulley or not. We would classify neither the force of gravity nor forces due to magnets as contact forces.

11. We can say that there is no net force on the parachutist if he or she is moving with constant velocity. Thus gravity must be balanced by an equal and opposite force, and for a parachutist that force is the drag force between the air and the parachute. Indeed, a parachute is designed to make the drag force as large as possible!

13. Newton's remark was more wishful thinking than one based on real knowledge. We know today that the stars are not fixed, and that the mythical inertial frame is an ideal. We

can approach this ideal more or less well according to circumstances. We can take the systems mentioned as representing a set of representative circumstances. The physics lab is probably pretty well fixed to Earth, so that we might have to worry only about earthquakes and Earth's rotation. The major effect of Earth's rotation is to modify the acceleration of gravity by a small, fixed amount. The space shuttle in orbit is in free-fall, so that gravity does not appear to be acting. In the sense that contact forces are the most serious consequences of a noninertial frame, the Shuttle is a pretty good platform. A ship at sea, on the other hand, with its pitching and rolling is a very poor approximation to an inertial system. But a building on Mars, which is geologically more stable than Earth, would be a good approximation.

15. Taken as a whole, an automobile accelerates because an external force acts on it. This force is one that is exerted on it at the point where the tires touch the road: friction. An automobile is not a point object, and the friction between tire and road is the end result of a number of forces that act inside: the force of hot gases on pistons; the force due to those pistons that turns a crankshaft; the transmission, which in effect allows the crankshaft to act on wheels and make them turn. And when a tire attempts to rotate, the nonzero coefficient of friction between rubber and road introduces the friction force that moves the automobile. Incidentally, there are many other forces of residual friction and drag that tend to slow the car down, so that friction between tire and road is necessary just to keep the automobile moving at a constant speed, at least on a flat or rising road.

17. The forces between ball and bat are, of course, contact forces. The bat and ball are both moving when they meet, but the bat is more massive than the ball, and moreover it has other forces acting on it due to the batter's hands. Thus the forces that act between ball and bat tend to have a greater effect on the ball than on the bat; that is, they accelerate the ball to a greater degree. Neither bat nor ball is rigid, and during the collision the ball compresses and the bat both compresses and bends to some degree. The bat is itself accelerated, but the forces due to the batter's hands tend to resist this acceleration, and the better he or she can do this the more the ball is accelerated. Home run hitters tend to have muscular arms and bodies. Because the bat is not a point object, its response to the forces on it due to the ball are complicated, leading to effects such as the "stinger" that every batter can describe when the ball hits a certain region on the bat.

19. The contact force at the bowl's surface is perpendicular to the motion of the marble, so that it does not itself cause the speed to change. In the absence of friction only gravity would do that. Friction is of course present, and it acts in such a way that the marble slows due to it. In later chapters we shall see that energy is a very efficient way to think about this system, and about why, in the absence of friction, the ball would rise up on the side of the bowl, to the height from which it started.

21. The scale reads the contact force between your feet and the floor. If the elevator were not accelerating, the scale would read the magnitude of the force of gravity on you. But as we described in the answer to question 3, the contact force between floor and feet would

increase if the elevator accelerated upward, and this would give an increased reading on the scale.

23. Not very. However, if the professor had said that the normal force on the table due to the mass had magnitude (5 kg)(9.8 m/s2) he or she would be right. That the normal force on the mass due to the table has this magnitude is a consequence of Newton's first law, which states that the net force on the mass is zero. Thus the normal force on the mass must balance the force of gravity acting on it.

25. This question is meant to confuse! But you won't be if you keep in mind the fact that the force on one object makes that object accelerate, and that the forces that are equal and opposite in the question act on different objects. If these were the only forces acting, the cart would accelerate to the horse as the horse accelerated to the cart. Of course there are many more forces acting, not the least important of which is friction between the horse's hooves and the ground.

27. Either someone has turned on a large magnet or the bus has suddenly started to move forward. In order for you to move along with the bus, a force must then act on you, and this is the force of friction between your feet and the bus floor. If these forces are not strong enough, or if there are effects having to do with your extended size and the point of application of the force, then you will not accelerate forward as much as the bus; in other words, you will be thrown backwards *with respect to the bus*.

29. Let's assume that the forces involved are steady. One side being stronger than the other means that the net force on one group is greater than the net force on the other. (Note that these forces involve not only direct forces on the rope but friction between feet and ground.) The motion of both groups is the same because they are connected by the taught rope, and this motion is matched by the motion of the handkerchief. This motion will be steady acceleration towards the winning side.

31. Sounds like a great idea until you think about it a little. If you pull upward on your bootstraps, your bootstraps exert an equal and opposite force downward on your hands. The internal forces—and these are indeed internal forces—cancel, and cannot effect your motion as a whole. If you want to make your bootstraps rise, it is a fine idea to pull on them; if you want to make yourself rise, the only way to make the method work is to annul Newton's third law. Only the Baron was able to do this; of course the Baron was also a notorious liar, so you might draw your own conclusions.

Chapter 5

1. The tightrope walker has an upward force acting on him or her that must balance the force of gravity. This force is a contact force from the wire that must come from the tension in the wire. If the wire were perfectly horizontal, the tension would have only a horizontal component. In order for the tension to have a vertical component, the wire must be aligned with a nonzero component along the vertical direction.

3. If the string is itself massless, then the tension must equal the weight of the object suspended from it. Only if the string has a mass significant compared to the mass of the object will the tension vary along the string. The tension in the string at a particular point must match the weight of the object plus the weight of all the string below that point. In that case the tension except at the very top of the string is greater than the weight of the object.

5. When a rope has mass, then the tension above a certain point must be large enough to move not only the mass attached to the end, but the mass of the rope below that point as well. The rope must be strong enough to support this additional tension. Friction in a pulley produces much the same effect. The tension must increase to the point where not only the weight of an object being accelerated against gravity can be overcome, but the force due to friction as well—remember, the friction acts against the motion. A properly lubricated axle can help a great deal when a pulley is in use.

7. A cyclist making a turn must make use of a centripetal force, one that is perpendicular to his direction of motion. The only external force able to do this on a flat road is the force of friction between tire and road, and in leaning into the curve the cyclist is attempting to increase this force by trying to put the bike into a position where it would "slide out" if there were no friction. If the curve is banked, then the cyclist can make use of the normal force that keeps the bicycle from moving into the road surface. By tilting the bicycle so that it is perpendicular to the road, the cyclist *avoids* introducing a sideways friction force which in this case would be undesirable.

9. For a massless rope, the tension in the rope that balances the weight of a suspended object is the same everywhere. That is because the rope itself, and hence any part of it, has no weight that itself might need balancing. In this case, the spring scale would reveal the same tension everywhere. On the other hand, if the rope is not massless, then the tension in the rope at a particular point must match the weight of the object plus the weight of all the rope below that point. Thus the tension would be larger and larger as we followed the rope up away from the object, and this is what the inserted spring would reveal.

11. A centerboard has a single purpose: to keep the boat from moving sideways through the water when the wind blows at right angles to the boat's direction. It does this by experiencing a drag force from the water that is largest when the boat moves sideways through the water. A keel plays this role plus one more: it is a very massive object and an enormous force is required to lift it. Thus the keel also tends to keep the boat bottom down. The keel is important when the wind forces are so large that the boat may be overturned, and this is typically associated with larger boats carrying sails of comparatively large area. On the other hand, a keel is so massive that it has a significant effect on the acceleration of the boat. Thus a centerboard is adequate when overturning is not a problem and is advantageous when rapid accelerations are desired.

13. Drag forces on boats are a type of friction that depends on the surface area of the boat in contact with the water. The more people in a scull, the lower she sits in the water. This increases the area of surface in contact with the water, and as a consequence the drag increases as well.

15. Not only can there be a normal force on the car, but we strongly recommend that you only take rides in roller coasters where there is such a force at the top of a loop. If there were not, gravity would be the only force acting, and if this were true even slightly away from the top of the loop the car would not execute the circular motion that corresponds to contact with the track! Note that the force required for circular motion at the top of the loop is a downward force, and this can come both from gravity and the normal force.

17. Any circular motion requires a centripetal force, and the boat following a great circle is no exception. For the boat, circular motion along the great circle requires a force directed to Earth's center. The net effect of gravity and the normal (upward) force on the boat, plus the net vertical component of wind forces, must combine to give a net accelearation v^2/R, where v is the speed of the boat and R is Earth's radius. Since this is a very small number, the excess of gravity over the other forces must be very tiny.

19. A centripetal force is needed to keep you moving in a circle along with the car. Your seat belt or a contact force from the car door and seat provide this force. When it is not present, you would move in a straight line; that is, you would be thrown from the car.

21. Newton's third law states that the forces on you and your partner are equal and opposite. (We are here ignoring the vertical forces due to gravity and the normal force form the ice; these cancel in any case for each partner, and all the motion discussed here is horizontal.) If you have equal masses, then the result, according to Newton's second law, is that you will experience an equal and opposite acceleration. If partner B is half as massive as partner A, then partner B will experience an acceleration of twice the magnitude as that of partner A.

23. The net force on Tarzan is such that he moves in a circle. The tension in the vine is the only force that can act to move Tarzan in this way. The tension at the bottom of the swing must also compensate for the entire force of gravity on Tarzan, and moreover, the speed is greatest at the bottom of the swing, because the movement has a tangential component due to gravity that increase Tarzan's speed to the bottom. Thus the tension must be greatest at the bottom, and this is the point of greatest danger, meaning the point where the tension is most likely to exceed the vine's breaking strength.

25. We can think about this most easily by first visualizing the situation from the outside. The die slides smoothly downward in a straight line, because without friction there is no force that will start the die moving around the central axis. Let's say that in looking straight down we see the die sliding along the x-axis. As it does so an observer turning with the bowl who starts by looking outward from the vertical axis along the x-axis will have rotated away from the x-axis, say towards the y-axis, if the rotation seen from above

is counterclockwise. That observer will see the die curving away from his or her left hand as it descends the bowl's wall. Such curving motion suggests a force acting, a phenomenon associated with the fact that the rotating observer is in a noninertial frame.

Chapter 6

1. The real motion of a baseball depends to a large extent on the fact that it rotates and that it is not smooth. But we can proceed here by ignoring these complications and thinking about the drag force as a small effect compared to the gravity-induced projectile motion. In the motion of a *thrown* baseball (as opposed to that of a pop-up), the change in vertical velocity is small, as is the distance the ball falls. Then we need to worry only about the horizontal motion. Here we can say two things: (a) If we don't know the drag force, we can at least assume that it is constant, because the speed does not change much. We could then find the drag force by dividing the change in horizontal kinetic energy by the distance traveled. (b) If we do know the drag force, we can find the work done by that force by integrating it along a straight path and thereby predict the change in the horizontal kinetic energy. We can do this in small steps using computers, and this approach allows us to take into account the small changes in the drag force associated with the small changes in the horizontal velocity, as well as small changes in the vertical velocity.

3. We do expend energy when we hold a bag of groceries stationary, but the reason for this is associated with the biochemical processes that allow us to hold our muscles flexed. At the molecular level a tensed muscle is not "locked." Rather, the muscle fibers continuously are released and must be re-flexed. Work is done each time there is a tiny movement of individual muscle fibers.

5. The net work done on the piano is the same in each case. We are neglecting friction, and the work done by gravity is independent of the path taken by the piano to get from one place to another. Note that when the crew carries the piano, they must also do work on themselves to raise themselves from ground level to the third floor.

7. Let's assume for simplicity that the ship is moving uniformly. Then the net force on both the man and the smokestack is zero, and no *net* work is done on anything. However, the force exerted by the man on the smokestack does indeed do work. This work requires no exertion on his part. The force of friction that the deck exerts on his shoes is in the direction of the ship. Because there is no net force on the man, the smokestack exerts a force equal and opposite to that of deck friction on the man. By Newton's third law, the force the man exerts on the smokestack is equal and opposite to the force the smokestack exerts on him; that is, the force the man exerts on the smokestack is equal to the deck friction on the man. The work done by the man on the smokestack is the force he exerts on the smokestack times the distance moved, equal to the work done on the man by deck friction. The energy for this comes from the engine, and the man only acts as a conduit for the work done. You might think about the case where the ship starts to move and everything accelerates.

9. Once the parachutist reaches terminal speed the work done on him or her is zero, because there is no change in his or her kinetic energy. Work is done only during the period over which the parachutist reaches terminal speed. In the measure that the distance over which terminal speed is reached depends only weakly on the starting height, the net work done is also independent of the starting height. Note that the net work is the sum of the work done by gravity and the drag force, which is why the work done by gravity is different from the net work.

11. The only influences on the bag are the force of gravity and the contact force from your hands. These cancel if you hold the bag stationary and there is no net work done on the bag. The forces similarly cancel if the bag is moving upward with constant velocity. No net work is done on the bag in that case as well. If you and the bag are *accelerating* upward, then there is indeed net work done on the bag. You must exert an increased force on the bag, one larger in magnitude than its weight. Remember that a floor scale reads an increased amount when you stand on it in an upward-accelerating elevator; in effect your arms must support the bag against what appears to you to be an increased weight.

13. There are two stages to consider here. The first stage is the one where the speed of the participants increases. At this stage there must be a tangential acceleration; the net force has a component along the displacement, and work is done on the participants. The appropriate tangential force is friction between floor and walls and the participants. The work-energy theorem is realized because the kinetic energy of the participants increases. In the second stage, the participants have reached their final speed and the floor has fallen away. The contact force between the walls and the participants keeps the participants moving along their circular path. This force does no work; the force is perpendicular to the displacement. Without net work there should be no change in kinetic energy, and indeed the speed is constant even if the velocity is not.

15. You might measure the maximum height reached by the first acrobat; that tells you speed and hence the kinetic energy he or she was given when he or she was flipped. The work done on the first acrobat is, by the work-energy theorem, that kinetic energy.

17. The forces acting on the sled are gravity (its weight), the tension from the rope, the contact force from the ground, and friction from the ground. If the motion is horizontal, then only the forces with horizontal components do work. Friction is purely horizontal, while tension has a horizontal component given by its magnitude times the cosine of the angle θ. As long as the sled moves with a constant velocity, the *net* work is zero.

19. The work-energy theorem offers a simple way to determine the average drag force. By the work-energy theorem, the work done by the drag force is equal to the loss of kinetic energy, which is the entire amount of kinetic energy the diver has when she enters the water. This work can be expressed as the net displacement in the water—the depth reached by the diver—times the average force. Measure the depth reached and the speed

with which the diver enters the water, and you can calculate the average drag force. Did you understand the signs of the quantities that enter here?

21. (a) Drag forces are like friction, always opposed to the direction of motion and hence nonconservative. (b) Again a drag force, again nonconservative. (c) The force here is the pressure of the expanding gases behind the bullet. These are somewhat more complicated and not entirely conservative, as we shall see when we study hot gases in chapters 17-20. (d) The trampoline is like a spring. The forces it is responsible for are conservative.

Chapter 7

1. Drag forces are like friction in that they always oppose the motion and therefore depend on velocity (its direction, that is). They are therefore not conservative. When drag forces act, energy is lost to heating and turbulence in the medium responsible for the drag.

3. The use of the conservation of energy in the analysis of motion depends on finding changes in the various terms of the energy. Changes in a particular energy term are independent of any constant in that term.

5. The conservation of energy is a principle that covers the entire physical world, including biological systems. In the case of sugar ingestion followed by exercise, there are many places where the chemical energy locked up in the sugar and the oxygen of the air that you breathe might go. This includes the energy of your motion, the heating of your muscles and blood and then of the surrounding air as you cool, and energy in the chemical products that are produced, including those contained in your expelled breath, perspiration, and waste products collected by your kidneys and other internal organs.

7. It is possible to test experimentally whether the potential energy *mgh* is converted to kinetic energy in projectile motion, and that the potential energy depends on height alone. The nonconservative effects of the atmosphere can be removed by conducting such experiments in a vacuum. More extensive tests are possible in observing motion of satellites far enough from Earth that the variation of the gravitational force, and the correspondingly modified potential energy, comes in.

9. When friction is present, energy is lost as the motion continues. We can think of a small amount of energy removed from the total with each cycle. But the energy of the motion determines to what height along the slide the moving object goes. With every cycle, then, the maximum height reached decreases, and the back-and-forth motion becomes more and more reduced towards the minimum of the potential, that is, the minimum in the "bowl." Finally, the object will comes to rest at the bottom. Certainly this is in accord with our experience.

11. Zeros of the potential energy have absolutely nothing to do with zeros of the force. Zeros of the potential energy depend on arbitrary additive constants in that energy, not on any property of the force. These zeros have no physical significance. Zeros of the force are

located at places where the *derivative* of the potential energy is zero; that is, at locations where the potential energy is not changing. Force zeros have a very real physical significance.

13. Two types of energy sources have been used in mechanical clocks. In one, weights are raised, attaining a potential energy associated with gravity. The weights are allowed to fall slowly, so that forces such as the tension in the rope from which the weights hang can make the clock go. In the second type, springs are flexed ("wound") in order to make the potential energy of a compressed spring available to be converted to the kinetic energy of the running clock.

15. As long as there is no source of energy loss, the forces are entirely conservative, and with each bounce all the initial potential energy is converted to kinetic energy and then back to potential energy. As the marble reaches the top of its trajectory after a given bounce it has regained all the initial potential energy it had originally. Since this potential energy is proportional to its initial height, the marble will go back to its initial height with each bounce.

Chapter 8

1. The comet moves under the influence of the gravitational force of the Sun. We can think of the Sun-comet system as a two-body system that undergoes a collision; Newton's third law guarantees that net momentum is conserved throughout. The net momentum is the momentum of the comet plus that of the Sun. If the momentum of the comet is in one direction coming in and another going out, the corresponding change in momentum must be compensated by an equal and opposite momentum change for the Sun. The Sun is so much more massive than any comet that the motion of the Sun would probably not be perceptible to our instruments, just as the motion of Earth is not perceptible in a collision with a falling apple.

3. The simplest object to visualize is a horseshoe, for which symmetry makes it clear that the center of mass is somewhere within the open portion. If we take a symmetrical hollow sphere the center of mass is at the center, in the hollow portion. While you might say that that point is still within the sphere, imagine now drilling a tunnel to the hollow from the outside and gradually increasing the size of the tunnel. At some point the tunnel will be large enough so that you would no longer say that the center of mass is actually within the hollowed sphere.

5. The center of mass of a object curved like a horseshoe lies with in the open part of the object. That is true for a curved human body as well. If the jumper can manipulate his or her body so that it is "draped" over the bar, then the center of mass can be below the bar even if a part of the body is over the bar.

7. Momentum is conserved only in the system of vase and Earth; the downward motion of the vase is accompanied by a tiny upward motion of Earth. However, the net horizontal

momentum of the vase is indeed conserved if the floor is smooth, so that pieces do not "catch" on the floor. If the vase falls straight down, with no initial horizontal motion, then the net horizontal momentum of the pieces of the vase must add to zero.

9. It is a good assumption to take the collision of the racket and ball to be elastic, roughly independent of the tension in the strings, for the typical range of stringing tension. More power but less control is associated with looser stringing. The ball remains in contact with the strings for a longer period and the player can push on the ball for a longer period—note that the racket and ball do not form an isolated system, because the player's muscles act on the racket. In contrast, more control but less power is associated with tighter stringing. This time the contact period is shorter and the player has less time in which variation in the directionality of the shot can come in. Incidentally, if the racket is *too* loosely strung, then the elasticity of the collision is lost, and so is power.

11. There is no absolute meaning to "short" in the context of collisions; the physical principles are the same for all periods over which the force acts. For us it is more a matter of when the time is so short that our ordinary senses cannot see a finite time interval. Scientific instruments can of course greatly improve our ability to see what is happening during the time over which "impulsive" forces act.

13. The acceleration of the more massive car is less than that of the less massive car in a collision. It is the acceleration that breaks bones or otherwise injures, so that the more massive car is the safer. Of course other factors play a role, such as the ease with which the car is crumpled. Energy that goes into crumpling a fender is better for you than energy that goes into crumpling a part of you!

15. Suppose the shell is fired to the right. Then, to conserve momentum in the closed system of car and shell, the car will recoil to the left. When the cannon ball hits the right-hand wall of the car (and assuming is does not pass through) the car will stop: since the initial total momentum is zero, the final total momentum must also be zero. In the process the car will have moved a little to the left, showing an outside observer that something has happened inside.

17. If the parachute is moving quickly, then in a fixed time interval it will collide with more air molecules than it would if it were moving slowly. Thus the momentum transfer in that interval will be greater at higher speeds, and this corresponds to a greater drag force on the chute at higher speeds.

Chapter 9

1. When you rotate the fingers of your right hand in the sense of the turntable rotation, the thumb will point down, to the ground, and this is the direction of the angular velocity. With a left-hand rule, the angular velocity would point upward. The choice of a right-hand rule is purely conventional, but once the convention is chosen, one must stick with it for

torque. The important things have to do with how torques change the angular velocity, and these will be the same whether one sticks with a right-hand or left-hand rule.

3. Let's consider an extreme example of the situation described here. On the one hand, take a sphere in which all of the mass M is concentrated within a small radius R_1 about the center. Then the rotational inertia is $2/5\ MR_1^2$. On the other hand, take another sphere in which the same mass M is concentrated within a thin shell a larger distance R_2 from the center. The second sphere has rotational inertia $2/3\ MR_2^2$. The second rotational inertia is much larger than the first, chiefly because of the factor R_2^2/R_1^2 is large. Gram for gram, mass distributed farther out contributes more to the rotational inertia.

5. The ball picks up a rotation as it rolls down on the side with friction. The conservation of energy states that the translational plus rotational kinetic energy the ball has at the bottom equals the initial potential energy, proportional to the initial height. If on the far side there is no friction, then the ball will not slow down its rotation as it rises. At the top of its rise it is still rotating, and the kinetic energy to be turned into potential energy consists only of the translational kinetic energy at the bottom. This is less than the initial potential energy; hence, the ball rises to a lower height than the height from which it started.

7. If the mass distribution in the sphere is spherically symmetric, as the question describes, then we have seen that the speed attained by the sphere as it rolls down a ramp depends on its rotational inertia. A measurement of this speed will therefore do the job.

9. As long as the object is flat, the method described on page 303 will allow you to find the center of mass of your map. Suspend your map from any point; the torque on the map due to gravity, which acts on the center of mass, will be zero only when the center of mass is directly below the suspension point. When you drop a plumb bob from the suspension point as well, you can trace its line on the map and be confident that the center of mass will be along this line. Drop the map and the bob from a second point and draw a second line. The center of mass will be at the intersection of the two lines.

11. Once the geometry of the flywheel is fixed, the only way to increase the energy is to increase the angular speed of the wheel. The angular speed, however, cannot be increased indefinitely, because the centripetal forces that keep all the parts rotating together depend on the construction of the flywheel and the strength of the materials that compose it. This aspect is strictly limited, and at some point as the angular speed increases, the wheel will fly apart.

13. A great deal of torque is required to rotate a large rudder against the force of water currents, and even with an appropriate gearing it is best to enable a large torque on the wheel. Since the arm that turns the wheel is capable only of exerting a fixed force, the torque can be made larger by increasing the radius of the wheel and hence the moment arm for the torque.

15. The stretched spring supplies the centripetal force that keeps the mass moving with a centripetal acceleration. This acceleration is proportional to $\omega^2 r \propto r/T^2$, where r is the radius of rotation, ω is the angular speed, and T is the revolution time, or period. If the revolution time is decreased by some means, then the acceleration is increased and the force must increase as well. This means that the spring is even further stretched and hence that the radius of the circular motion increases. Note that an increased radius also tends to mean an increased acceleration, so that this becomes a fairly complicated problem. If there is a final equilibrium radius then it can only be found by knowing the external force decreasing the period and by using the dynamical equation of motion.

17. While it is swinging, the rock is moving under the influence of the string tension and gravity. If the string breaks, gravity alone acts on it, and it moves as a projectile, with an initial velocity given by the value of its velocity at the moment the string breaks.

19. The answer to the question actually depends on how the torque varies with time, so let us assume that the torque is a constant torque τ. Let us also label the two wheels as 1 and 2, with the larger rotational inertia belonging to wheel 1: $I_1 = 2I_2$. The angular velocity after time t, starting from rest, is $\omega_i = t\tau/I_i$; that is, the wheel with the larger rotational inertia is rotating less quickly. Because the rotational kinetic energy of wheel i is $K_i = I_i\omega_i^2/2 = t^2\tau^2/(2I_i)$, we have $K_1/K_2 = [t^2\tau^2/(2I_1)]/[t^2\tau^2/(2I_2)] = [I_2/I_1]$. The wheel with the larger rotational inertia attains only half the energy of the wheel with the smaller rotational inertia.

Chapter 10

1. The dimensions of angular momentum are those of mass × speed × distance; that is, [M × (L/T) × L] = [M (L/T)^2 × T]. Because the factor M (L/T)^2 is the dimension of energy [(mass) × (speed)^2], the dimensions of angular momentum are indeed those of energy × time, one joule·second is indeed an SI unit for angular momentum.

3. Because there is no strong source for a torque on the ball once it starts its flight, the spin maintains its orientation and so does the ball. This suggests the major reasons for putting spin on the ball. First, the effects of air resistance are more predictable if the ball maintains a fixed orientation. In addition, the receiver can more reliably judge the flight of the ball with its fixed orientation, and it is easier to catch.

5. As the astronauts spin their arms, their bodies spin in an opposite direction in order to maintain their total angular momentum at a constant value (typically zero in this case). When the arms stop spinning, so do their bodies. The fact that after having rotated their bodies the astronauts have changed their orientations in no way contradicts the principle of angular momentum conservation.

7. When you use the bent leg technique for lifting, there is little torque on any part of you. In contrast, a bent back must counter the torque you experience from the weight of the lifted object, and it is the muscles of the back and abdomen that must supply the

countering torque. The result may be pulled back muscles, or worse, damage to the spinal column.

9. Unless the diver is prepared to use countering revolutions with his or her body (analogous to the technique used by the dropped cat and described in the text), an initial rotation is helpful. The diver will strive to acquire this by using the contact force from the diving board to provide a torque on his or her center of mass. Once there is an initial angular momentum, the diver can use the flexibility of his or her body to change the rate of rotation.

11. The hard boiled egg is a solid, and when it is set in rotation internal forces set all the parts in rotation as a rigid body. The large angular momentum is maintained for a long time and the spin resembles that of a top. If the egg has not been cooked, the interior is a liquid, and the shear forces that tend to put the entire egg in rotation when the shell is spun are not entirely effective. The shell may initially spin while the inside does not. The angular momentum is far smaller than it is if the egg is hard-cooked, and the spin rapidly stops. This is a very amusing experiment to try.

13. Yes. Imagine a uniform bar lying horizontally on a table with the bar's orientation away from you. Then imagine two blows that strike the bar simultaneously, one from the left on the part of the bar closest to you and one from the right on the part of the bar farthest from you. If the blows have equal strength, the net impulse is zero; the center of mass of the bar will not move. But the bar will certainly rotate as the result of a nonzero angular impulse.

15. We did indeed see that the conservation of the classical angular momentum when there is no net torque is a consequence of Newton's laws, as is the conservation of linear momentum. When quantum physics is involved, the conservation of angular momentum takes on a separate significance.

17. The angular momentum associated with the propeller spin is, by application of a right-hand rule, directly away from the pilot. (We reference vector quantities from an origin at the center of the airplane, at the location of the pilot.) In turning to the right, the spin angular momentum is also rotating in that direction; that is, there is a change of angular momentum to the right. To effectuate this change, there must be a torque to the right, and hence a force $\mathbf{F}$ such that $\mathbf{r} \times \mathbf{F}$ is to the right. Since $\mathbf{r}$ points forward, the force on the nose, where the propeller is located, must be up. Thus the pilot must arrange his ailerons and rudder to introduce a force that will push the nose up, and this will aid the turn.

19. If we think of the woman and platform as a single isolated system, there are no external forces, so asking for the source of the torque is a bit of a red herring. The system of woman plus platform is one in which the angular momentum is constant. As the woman moves to the center, the rotational inertia of the system decreases, so the platform plus woman must speed up their joint rotation. On the other hand we may think of the platform alone and the woman alone. The only force acting between these two separate systems is

the friction at the shoes. If the woman stays at a given radius, then friction from the platform acts centripetally to keep the woman moving in uniform circular motion; by Newton's third law, friction from the woman acts radially outward on the platform and this supplies no torque. However, if the woman moves to the inside of the platform, then her velocity would tend to decrease if the motion continued with the same period, meaning that friction from the platform acts tangentially to slow her down; again by Newton's third law, friction must then act tangentially on the platform, and in a direction that speeds it up. The result is the same as in the description of the woman and platform as a single system.

21. We refer to the figure, and think of the origin for the calculation of torques as placed at the center of the flywheel. (a) Without the flywheel, a wave approaching from the side of the boat would lift the side of the boat, so that there is a force that points upward acting at the side of the boat. With the flywheel we want to find the corresponding torque and see how it changes an existing ω. The torque is $\mathbf{r} \times \mathbf{F}$, which for $\mathbf{r}$ pointing from the flywheel center to the side of the boat is oriented to the front of the boat. A vector $\Delta\omega$ added to the existing ω reorients ω towards the bow direction, and the boat rotates about a vertical axis to the right with respect to the bow rather than tilting along the long axis of the boat. (b) If a force acts upward at the bow, then $\mathbf{r}$ runs from the flywheel to the bow, and the torque acts in a direction opposite to the existing ω; this has no effect on the motion of the boat.

23. The friction that acts on the ball has two effects: it introduces a torque on the ball and it acts on its center of mass to accelerate the ball. The force is horizontal. We can see whether it is forward or backward by looking at the direction of spin and deducing the effect of the torque. If the spin is forward, then friction must act in the backward direction, and this is sensible. Such a force will accelerate the center of mass in such a way that the horizontal component of the ball's velocity decreases. Let's now assume that the vertical forces acting when the ball hits the floor are elastic, so that the vertical component of the ball reverses but suffers no change in magnitude. The angle the ball's path makes with the horizontal has a tangent that is the ratio of the vertical to the horizontal component of the velocity. If the horizontal component has decreased in magnitude but the vertical component has not, then the angle will be increased on the bounce.

Chapter 11.

1. The forces that act on the mountaineer are the force of friction between the boots and the steep slope, the force of gravity, and the normal force. The normal force is perpendicular to the direction of the slope, while friction acts along the slope. For a stationary climber, the vertical component of friction together with the vertical component of the normal force balances the force of gravity, and the horizontal component of the force of friction must cancel the horizontal component of the normal force. Because the climber is an extended object, we must take into consideration torques. The climber could fall backwards, or he could fall forward in such a way that his feet slide out from under him. If the climber leans towards the slope so that his center of gravity is in front of the vertical, towards the slope, then there is a torque about the point of contact which would

tend to make the climber fall toward the slope. He can, of course, put out his hands to keep from falling, but if he still leans into the mountain, the normal force and hence the force of friction is reduced. It is not enough for his hands to keep him from leaning into the mountain; the hands actually have to keep him from sliding down, and this may be very difficult if the rock surface is smooth. Thus standing away from the mountain is the safer procedure.

3. If a rope with a mass at the end is not vertically aligned, the force of gravity on the mass produces a torque about the point of suspension. This rotates the system in the direction of the vertical. If there is no friction, the rope oscillates about the vertical position.. With friction and/or drag, the system will come to rest in a (vertical) equilibrium position.

5. The equilibrium is stable, since a displacement from the rest position leads to torques that tend to rotate the rocking chair and its occupant towards the rest position.

7. One of the methods consists of suspending the plywood sheet from any two points and, using a plumb bob, drawing vertical lines through the points. The intersection point gives the center of mass, as was discussed in question 9-9. The second method involves balancing the plywood sheet on one finger. At the point of balance the net torque about the point of contact is zero, and this implies that this special point is the center of mass.

9. This is a case of stable equilibrium. A small displacement of the pendulum from the equilibrium angle will lead to a swinging motion about that equilibrium angle, and not to a runaway motion.

11. No external forces act on the station as an extended body. Once the station starts rotating, the conservation of angular momentum ensures that the rotation continues. One angular velocity is as good as any other, provided there are no disruptive forces entering once the angular velocity becomes too large. Thus the equilibrium is neutral.

13. It is not possible to do this. The vertical component of the tension must balance the weight of the beam if there is no friction. Since the tension acts at the far end of the beam and gravity acts at the midpoint of the beam, there is a net torque about the point of suspension. Equilibrium conditions cannot be satisfied, and the beam rotates by slipping along the frictionless wall.

15. For a straight beam, the net torque about the pivot is zero no matter what the angle of inclination is. A slow change in angle will lead to another equilibrium position. Thus the equilibrium is neutral.

Chapter12.

1. In this chapter we mentioned friction and spring forces. Other nonfundamental forces include: drag forces in fluids or gases, the tension in ropes or solid rods, the normal force

that opposes gravity for objects at rest. These forces are all associated on the microscopic level with the electromagnetic forces that hold matter together.

3. Tidal forces arise from the difference between the attraction to the Moon (and Sun) on one side of Earth and that on the other. Near the North Pole this difference is very tiny; it would actually vanish at the pole if the Sun and Moon moved in the plane of the equator and there would be no tidal effects at the pole.

5. If the satellite is launched near the equator, in the direction of Earth's rotation, then it starts off with a velocity relative to the center of Earth that is the sum of the velocity of the launch pad due to Earth's rotation and the velocity it acquires as a result of the acceleration produced by the burning fuel. This can be a substantial help. The science-fiction writer Arthur C. Clarke has pointed out that the ideal launching pad would be from the top of a very large mountain on the equator in Sri Lanka (Ceylon).

7. The parabolic path is a result of a constant gravitational field perpendicular to a plane. Over distances that are short compared with Earth's radius, this is an accurate description of gravity. However, it ceases to be accurate for paths that are long enough for Earth's curvature to matter.

9. Angular momentum is conserved for *any* central force, and the text shows that this is equivalent to the equal areas in equal times law. The r-dependence is irrelevant for this purpose.

11. Any satellite in a circular orbit will move in a large circle centered on Earth. All points on Earth's surface move in circular orbits about the N-S axis of Earth. The orbit of a satellite and the "orbit" of a point on the surface can only overlap at more than two points if the satellite orbit lies above the equator. A geosynchrous satellite must be placed over the equator.

13. In the text you learned that a spherically symmetric mass distribution exerts the same force on an external object as the same mass concentrated at the center point of the distribution. Let's consider the simplest case of inhomogeneity, namely a small pointlike lump somewhere inside Earth. The mass distribution may be viewed as a superposition of a spherically symmetric mass distribution centered at Earth's center plus a point mass somewhere inside Earth. Since forces add vectorially, the force may be viewed as one due to two masses, one large, centered at Earth's center, and one small, at the lump location. The small mass will perturb the orbit in ways that are different for different orientations of the orbit, and from the study of the orbit the location and magnitude of the small mass may be inferred. More general inhomogeneities may be constructed as a superposition of lumps, and the orbits will then be ones in the field of a massive point mass at Earth's center accompanied by a number of smaller masses (some of which may have to be chosen negative!) in the vicinity of the large one. It should be noted that a spherical mass distribution—a lump that is extended in a spherically symmetric way—is equivalent to a point of the same total mass, but unless we know the density of the spherical distribution,

we will not know its size. Thus the above construction still leaves us ignorant of some aspects of Earth's interior. That is why seismology is essential for the detailed investigation of Earth's mass distribution.

15. If we assume uniform density for Earth, then surface measurements of g may yield information about the shape of the surface. The answer to question 13 also applies to this question.

17. The Sun also causes (smaller) tides, so that if the Moon were not present the slowing down would take longer, but it would still occur.

19. For a circular orbit the force is always perpendicular to the displacement, so that no work is done at any instant. For (closed) elliptical orbits, the net work done is also zero, since otherwise the object in orbit would lose or gain energy, giving rise to a change in the orbit.

21. Suppose your spaceship is at rest in a region of negative potential energy. (This could happen if the spaceship left Earth with a velocity somewhat less than the terminal velocity, and reached the turn-around point). Can you eject some material and put your spaceship in a region of larger potential energy? From momentum conservation we can see that by ejecting some material in the direction of Earth, the spaceship could acquire a velocity in a direction away from Earth. This can only happen if there is some energy stored in the spaceship. It may be chemical or nuclear, but that energy can be converted into kinetic energy. The spaceship will then move to a region of higher potential energy, losing some of its newly acquired kinetic energy in the process. No law of nature is violated by this scenario.

Chapter 13

1. The text gives spring motion and the motion of the pendulum as examples of simple harmonic motion. Other examples include properly constructed organ pipes (so that only one frequency is excited), the vibrations of molecules, and electrical oscillators. In the latter case, the oscillations involve the movement of electric charges in wires.

3. The maximum kinetic energy, that is, the total energy, is proportional to the square of the amplitude. Therefore a doubling of the amplitude implies a quadrupling of the maximum kinetic energy.

5. At the top of the mountain g is less than its sea-level value. Since the pendulum period T is proportional to $\sqrt{(L/g)}$ the period will grow as the value of g decreases. However, because the air is thinner, the drag force on the pendulum decreases at higher amplitude. There are thus two competing effects, and it is hard to predict which one will win out.

7. The halving of the mass has no effect, so only the doubling of the length is relevant, increasing the period by a factor of $\sqrt{2}$.

9. Without going into details assocaited with the shape of a sitting versus standing person, we note that a standing person has a center of gravity higher than a sitting one, and effectively the pendulum that is formed by the swinging person is shortened when the person stands. This reduces the period.

11. The period of a pendulum is proportional to $\sqrt{L}$, and if the length is decreased by a factor of 3/2, the period is decreased by a factor of $\sqrt{(3/2)}$. The energy, in terms of the maximum angle of swing θ_0 is proportional to $L\theta_0^2$, or A^2/L, where A is the amplitude of the (small) swing. Thus E/ω is proportional to ET; that is, to $A^2/\sqrt{L}$. This is invariant, so that A must vary as $L^{1/4}$. If L is decreased by a factor of 3/2, then the amplitude is decreased by a factor of $(3/2)^{1/4} \approx 1.1$.

13. A force diagram shows that if the ropes do not stretch, the suspension is equivalent to a single pendulum, with motion transverse to the line that joins the two points of suspension. This motion is simple harmonic. In fact, any small amplitude motion about a point of stable equilibrium is simple harmonic, as long as the bottom of the potential energy curve can be approximated by a parabola.

15. It is the harmonic driving force that provides power. This power oscillates with the same frequency as the force, so that energy flows in and out of the system from the external oscillator. The oscillating force may increase the amplitude of the mass on the spring if the phase is right, and that is the phenomenon of resonance. When resonance occurs, the transfer of energy from the external oscillator becomes very efficient!

17. The spring exerts a harmonic force on the mass, and by Newton's third law, the mass exerts an equal and opposite harmonic force on the spring, and therefore on the point of attachment to the sled. Thus the sled experiences the force of gravity, the normal reaction of the plane and also a harmonic force. The harmonic force on the sled ensures that the sled will not accelerate smoothly down the hill.

Chapter 14.

1. Sound waves are longitudinal; there it is the compression and rarefaction of air that propagates. Ocean waves are more transverse than longitudinal; what propagates is the displacement of (incompressible) water from the flat surface that is created by gravity.

3. The train generates waves in the steel rails. The rails transfer the sound waves efficiently along a one-dimensional system rather than dispersing them through space, as occurs for sound in air. The fall-off with distance associated with the conservation of energy is much less in the one-dimensional rails than in air, so that the approach of a train is detected earlier through the vibrations of the rails than through the vibrations in the air. Another effect leads to the same conclusion: the speed of sound in materials is much greater than the speed of sound in air.

5. The speed of sound is larger if the density of the gas is less. Since helium is less dense than air, the speed of sound is larger in helium than in air. On the other hand, the wavelength of the voice is determined by the dimensions of the throat and is therefore independent of the medium in which the sound is generated. For fixed wavelengths, the frequency of the sound is proportional to the velocity, and it is therefore higher in helium than in air.

7. A node is a point at which the string does not move. Nothing would change if it were clamped.

9. Nodes are, by definition, fixed points in standing waves, so the answer is yes.

11. When a child talks into one end of the can, the sound waves set the closed end of the can into oscillation. These oscillations translate into waves along the string. When the string is taut, the waves move the bottom of the can at the receiving end, and this motion causes the compression and rarefaction of the air in the can which is a sound wave.

Chapter 15

1. Yes. For example, the trigonometric identity
$$\sin(kx - \omega t) = \sin kx \cos \omega t - \cos kx \sin \omega t$$
decomposes a sinusoidal traveling wave as a superposition of two standing waves.

3. Pulses represent solutions of a linear wave equation, and the sums or differences of the pulses are also solutions. This means that the pulses pass through each other and maintain their integrity. When the pulses are identical in shape, then in the first case they will reinforce each other on approach, and when they are centered on one another a single peak with double the amplitude can be seen. When the pulses are of opposite sign, then at the time that their centers coincide they cancel each other out completely. Subsequently the pulses reappear with their original shape and sign on opposite sides of each other. When the pulses are at 90° to each other, they pass by each other with a local distortion that forms a tilted pulse when their centers are at the same spot on the string.

5. The sharpness of the pulse is determined by the thickness of the string and by how much the inside of the curve that forms the edge can be compressed.

7. The pebble causes a depression in the surface of the water. Water flows into the depression, leaving a trough that continues to propagate. This is the wave front of the water wave. The wave that enters the depression builds up to fill it and goes beyond that, so that there is now a peak in the middle. This water flows outwards, giving rise to a second wave, and another trough, which in its turn gets filled, and so on. Thus one gets a series of waves rather than a single front.

9. The energy density of a wave is proportional to the square of the wave function $y(x, t)$. If the wave function consists of two waves, so that $y(x, t) = y_1(x, t) + y_2(x, t)$, then the

energy density is proportional to the square of the sum of the two wave functions. Wherever the wave functions cancel, the energy density is zero. If the wave functions cancel in one place, they will build up constructively in another place. There the energy density will be larger than the energy density of the component waves. The total energy will always be conserved.

Chapter 16

1. Hot air is lighter (less dense) than cold air. A given volume of hot air will weigh less than the same volume of cold air. Thus the buoyancy, which is the difference between the weight of the hot air and the cold air that it displaces, may be enough to lift the balloon.

3. The method depends on the difference in density between body fat and the rest of human tissue. The weight in air represents the weight of tissue plus the weight of fat. The weight in water will be smaller by the weight of the water displaced. The ratio of that weight to the weight in air will vary linearly with the proportion of fat between a number for an idealized fat-free person, and a number for an idealized pure-fat person. A table is presumably available for these measurements.

5. If there is no viscosity, then there is no drag, and the acceleration is g, no matter what the mass. This is, of course, a highly unrealistic situation.

7. Mercury is the densest fluid at normal temperatures. With it one can construct a barometer of reasonable height (approximately 760 mm). A liquid with a density ten times smaller would make for a 20 ft barometer! Water makes a 32 ft barometer.

9. The solid body of water which includes the water in the sink and the water in the tube is equivalent, as far as air pressure is concerned, to a cup of water, open on top, with a small hole in the bottom. The excess pressure at the top causes the flow. If the tube is turned up so that the water level in it is higher than the water level in the sink, then there will be flow from the tube into the sink.

11. Bernoulli's equations come into play in sailing in the same way that they do in the lift of a slow (nonsupersonic) airplane. Curved sails are indeed a familiar feature of sailing vessels.

13. The Dead Sea is very salty, and the density of its water is much higher than ordinary sea water. Thus the weight of the water displaced by the part of the body under water, is larger than in the sea, and the body is pushed upwards and floats higher. This works only as long as the (largely) water content in the body is plain water. A swimmer pickled in the Dead Sea for long enough will not float as well.

15. The fluid in a centrifuge is accelerated towards the center, and the force responsible for that is the normal force. The fluid acts as if it were on a section of floor in the presence of gravity, with the floor corresponding to the outer circle of the centrifuge. There we

know that the denser fluid is at the bottom, and the less dense fluid on top. Thus the oil will be closer to the center of the centrifuge.

17. A long snorkel requires you to bring air to a depth where its pressure is large; that is, where it is highly compressed. This requires work, and even more work if the depth is greater. Beyond a certain depth your lungs are not capable of performing the necessary work.

19. There is buoyancy equal to the weight of the water displaced by the kilogram weight. However, the displaced water has been moved up in the jar, and its weight just cancels the buoyancy. Thus the reading on the scale, which measures the normal force necessary to balance the weights of everything standing on it, is unchanged.

Chapter 17

1. To an adequate approximation, we can assume that the air is at 1 atm of pressure and at a temperature of some 300K. To estimate our lung volume, we can span our hands across our chest—about 25 cm—and estimate a lung volume of roughly $(25 \text{ cm})^3 = (0.25 \text{ m})^3 = 0.015 \text{ m}^3$. At this point we can estimate the number of molecules using the ideal gas law, $pV = NkT$, or $N = pV/(kT)$. Using SI units throughout, $N = (10^5 \text{ Pa})(0.015 \text{ m}^3)/((1.4 \times 10^{-23} \text{ J/K})(300 \text{ K})) = 4 \times 10^{23}$ molecules. A lot! This makes it clear why it is much easier to float when you take a deep breath.

3. The pressure increases as the depth increases. There are more molecules in a given volume for a higher pressure (assuming constant temperature), so the matter in the tank is used up more quickly.

5. When the car is driven, friction with the road increases the temperature of the tires, and of the air within them, considerably. The tire volume is fixed, so accordingly the pressure increases as the temperature of the air within the tire increases. You may then think the pressure within is adequate when a cold measurement would show that it is not.

7. See the answer to question 5. When winter arrives, the pressure of the air in the tires is exceptionally low. If you check the air in too-cold tires and boost the pressure at that point to the nominal value, the pressure may be dangerously high when the tires are heated by a trip.

9. Not really. The importance of the metric system for length has to do with the construction of standardized—interchangeable—parts, but there is little about temperature that depends on it in the same way that the size of a screw shank depends on metric length. Of course, it is useful to you to be able to understand what it means when the weather forecast you hear on your European trip says that the maximum temperature tomorrow will be 25.

11. Yes. Indeed, the early ideas of ideal gases and their behavior predates the acceptance of the existence of atoms and molecules by a good deal. The recognition that molecular behavior lies behind temperature clarifies the meaning of the quantities involved and allows us to go farther than the simple definition of temperature.

13. In the case of molecules within a container that we could, say, hold within our hands, each individual collision is so feeble on our scale that it would take special instruments to detect it. In addition, the vast numbers and chaotic nature of the collisions with the walls mask the individual collisions. To see a container shake, the momentum transfer of individual collisions would have to be much larger, and the numbers of collisions much less.

15. Absolutely nothing is wrong with this scale from the scientific point of view. It may be bad publicity, however, when someone arrives in their winter gear because they heard that the maximum temperature that day was going to be 35 degrees!

Chapter 18

1. The air in the chamber of the pump is undergoing adiabatic compression. The stroke is sufficiently rapid that there is little heat flow leaving through the walls. Thus work is being done on the gas, and one of its effects is to increase the thermal energy, and hence the termperature. In a diesel engine, this is carried to an extreme, with adiabatic compression of the gas in the cylinder raising the temperature enough to ignite fuel.

3. The bucket of water does indeed have an increased energy as a result of its having been carried uphill, but that energy is in the form of potential energy, not thermal energy. One way to convert that energy to thermal energy would be to drop the bucket back to the original height. The collision with the ground is inelastic, and one result would be an increased temperature for the water.

5. An adiabatic process is a controlled, or reversible one. Each infinitesimal step in the process can be "undone" by reversing the external conditions. This is certainly not true of irreversible processes such as free expansion, even if free expansion involves no heat flow.

7. It is best to keep the temperature of an operating engine stable in order to avoid thermal expansion effects or, worse, destruction of the lubricating oil or the engine itself. A cooling fluid with a large heat capacity is best, because then a relatively large heat flow can be absorbed by the cooling fluid with a relatively small temperature change.

9. Let's first think about the formula. It is one that describes the coming to equilibrium of two masses of water at different temperatures. From the discussion in Chapter 17, we know that the equilibrium temperature will be intermediate to the two initial temperatures. The cooler temperature rises and the higher one drops, and that explains the sign in the formula. In order to effect the change, there is a transfer of thermal energy between one mass and the other, and because no work is done in the process, all that thermal energy is

in the form of heat flow. The heat flow ΔQ out of the hotter mass is absorbed by the cooler mass. But both masses are water, so the temperature change of each is related by the formula $\Delta Q = cm\Delta T$, where c is the specific heat of water. The equality of the heat flow magnitudes means that $-cm_1\Delta T_1 = cm_2\Delta T_2$, and this gives the formula of the question. The statement about mercury and water involves the same process and reasoning as in the first part of the question; however, considerably more mercury at a higher temperature is required in a mixture with water at a lower temperature in order to reproduce the observed intermediate temperature. This is simply a statement that the heat capacity of mercury is much smaller than that of water. The formula $-c_{Hg}m_1\Delta T_1 = c_{water}m_2\Delta T_2$ applies, and a larger mass of mercury compensates the lower heat capacity.

11. In vaporization, the molecules start close together and end up far apart. In melting, the molecules remain close together even if there is some rearrangement. It takes more energy to overcome the intermolecular forces and carry the molecules to large separations than it does to bring an ordered lattice to disorder, and this corresponds to the much larger size of the heat of vaporization.

13. It is the evaporation of your perspiration that cools you off. Thermal energy from your body supplies the latent heat of vaporization to send water molecules into the atmosphere as water vapor.

15. As long as the transformation is reversible, so that the thermodynamic variables being used for the plot have a definite value at each stage of the transformation, you can plot the transformation on a plot of one variable versus another. Of course if too many variables are changing the plot is incomplete. For example, we assumed throughout that the number of molecules involved in the thermal transformation of an ideal gas remains fixed. That means any two variables will do, the third being determined by the equation of state. But if chemical reactions are involved, the number can change, and that means more than one plot is necessary to specify the entire process.

17. There is much more water in the suburbs than in the city, both in the form of open water and water brought up from beneath the ground by trees. Tree leaves in particular present a large surface from which evaporation can take place, and thermal energy from the surroundings is used to supply the latent heat of vaporization. Trees are for this reason and others one of the most valuable elements of the urban landscape.

19. The difference between a car being driven and one standing still is that in the first case far more air flows across the radiator surface and is available to carry thermal energy away from the engine. As far as winter versus summer, there are two factors to take into account. On the one hand, winter air is dryer than summer air and therefore has a substantially lower heat capacity. This factor alone makes it harder to cool the engine in winter. On the other hand, the typical temperature change in the vicinity of the radiator can be much larger in the winter than in the summer, and this tends to make it easier to cool the engine in winter. We can at least state that a dry winter climate makes overheating more likely than a damp winter climate.

Chapter 19

1. If the cylinder is insulated, the transformation is adiabatic. From a thermodynamic point of view, temperature goes up if the volume decreases. The reason for this is that work is done on the gas, but there is no heat flow. Thus the thermal energy increases and the temperature goes up. From the molecular point of view, even a slow inward movement of the piston adds some momentum to the molecules recoiling from a collision with it, and hence adds to the molecules' kinetic energy.

3. It is indeed true that from the molecular point of view thermal energy consists of rotational as well as translational kinetic energy. In thermal equilibrium, these various components of the energy are shared "equally," or more precisely as the equipartition theorem demands. The giving of a rotation to a given molecule in a given collision is of no special interest; it is only averages that have visible consequences. There will be as many collisions in which molecules slow down because their rotation speeds increase as collisions in which rotational kinetic energy is turned into translational kineitc energy.

5. The walls have a temperature. This temperature is shared with the gas because a wall is not a fixed uniform solid but is itself made of molecules, and collisions with the wall will reflect whether there is some motion in the wall associated with temperature. Thus there is an equilibrium temperature between the walls of a container and the molecules within it. It is true that once the numbers of molecules of a gas become very low, the departures from average quantities become larger and temperature is less well defined.

7. See the answer to question 3. The equipartition theorem tells us that rotational energy is as closely tied to temperature as translational kinetic energy.

9. See the answer to question 5. The stove's thermal energy is transferred to molecules by collisions between moleucles and the stove walls, as well as by molecular absorption of electromagnetic radiation emitted by the stove—as we saw in Chapter 17, the emitted radiation is characteristic of the stove temperature.

11. The question of mean free path has to do with the probability of meeting another molecule. This does depend on the average spacing, but it also depends on molecular size. Imagine that rather than being the size they are, billiard balls were reduced to the size of gnats. The average spacing between balls would remain the same, but you would have a hard time hitting a ball with another under those circumstances and the mean free path would be very long.

13. There is a play of temperature and gravity in the question of a planet's holding an atmosphere. The higher the temperature, the higher the average speed, and the higher the likelihood that a given moleucle will have enough kinetic energy to escape. On the other hand, the escape speed is higher for a larger planet, so that for a given temperature the fraction of molecules that can escape is smaller. All planets lose atmospheric molecules to

some degree, but less massive planets lose them much more quickly. Two additional effects enter: First, the average temperature drops when the faster molecules leave, because the average speed has dropped. Most planets have mechanisms for reheating, including internal energy and solar energy. Second, larger planets are typically more active, and evaporation from liquids or events such as volcanic eruptions will supply new atmospheric molecules. The Moon is small, so that a larger fraction of atmospheric molecules have enough energy to escape, and it is quiet, so that it has no way of replenishing its atmosphere.

15. The pane of glass is bombarded on both sides, and the effects cancel so that there is no net force on the pane.

17. There are always some molecules at the surface of a hot fluid that have enough energy to overcome the forces that would normally keep them bound to the liquid. We might refer to these as the "hot" molecules. Some of them get away—this is evaporation—and the liquid, which then has fewer highly energetic molecules, is cooler. Most of the escaped molecules collide with air molecules above the surface; they either return to the liquid or, if not, leave some energetic air molecules in the vicinity of the surface which is relatively likely to return to the liquid. When we blow across the surface of the liquid, we help the "hot" molecules get away from the surface. We are supplying another mechanism for thermal energy transport away from the cup of coffee, namely convection.

19. The heated air around the candle is less dense than the cool air farther out, and as we know from Chapter 16, bouyant forces therefore act on it to make it rise. This makes room for more oxygen to arrive at the candle wick and burning can continue. Without the force of gravity to produce buoyancy, the mechanism fails and the candle would indeed snuff itself out.

Chapter 20

1. Thermal energy flows spontaneously only from a reservoir at higher temperature to a reservoir at lower temperature. Thus the ocean temperature should be higher than the air temperature. Because water has such a large heat capacity, the air can become a lot colder than the ocean in the winter. Thermal energy flowing from the water to the air moderates the climate, so that temperature changes are less extreme in locations near the ocean.

3. The second law of thermodynamics is certainly assoiated with the "randomization" in physical processes. However, the second law is one of those laws of physics in whose name many unrelated phenomena are explained. The words used by Herman Daly convey negative images, and this was undoubtedly part of Daly's purpose. But it is highly doubtful that there is a real connection between the second law and human history, which, to misuse the analogy ourselves, is hardly an equilibrium process!

5. The efficiency of the ideal Carnot cycle does *not* depend on any details of the substances used, and this includes the particular equation of state obeyed by a working

gas. The efficiency for a van der Waals gas as the working fluid is the same as the efficiency for an ideal gas.

7. No. In doing his cleanup job, he has involved a good deal of energy, some of it thermal energy. This energy has come from and gone into the environment at large in various ways, and the question of whether the entropy has increased must include all the pieces. In addition, the systems involved, especially the professor himself, are not yet in thermal equilibrium with their surroundings.

9. Just a few minutes ago I watched some cream poured into my coffee, mixing with it. The entropy involved is the entropy of mixing discussed in the text. In addition, I can see the exhaust from a bus spreading through the air. As it spreads through an increased volume its entropy increases, and here too there is mixing.

11. A light bulb absorbing electromagnetic radiation from the air and turning that radiation into electric current is consistent with the first law, as is the spontaneous rejuvenation of a living cell that uses energy in its environment to reverse, one by one, the chemical processes that make it age.

13. The close-up view shows individual molecular events, as opposed to the large-scale view that shows vast numbers of such events. Only the view that shows the large numbers has a "direction in time," and that is because in this case the statistical likelihood of collections of events plays a role. The arrow of time in macroscopic events is compatible with the reversibility of individual events because it reflects the huge difference in the number of ways in which certain events and their inverses can happen compared to their inverses. An egg can break into pieces in a huge number of ways. However, a broken egg, with its shell fragment scattered and white and yolk all over the floor, can reconstitute itself only in a very small number of ways—the air vibrations that carried away the "plop" sound, for example, would have to come back at just the right time and place.

15. Intensive thermodynamic variables are ones that depend on the amount of material, or equivalently of the number of molecules, N; extensive ones are independent of N. Thus we want to know if entropy is proportional to N or not. A glance at Eq. (20-27) shows that entropy is indeed proportional to amount. More intuitively, entropy is a measure of the "amount" of disorder, and two identical boxes of a mole of gas have twice as much disorder as one of them.

Chapter 21

1. Cement's small coefficient of thermal conductivity makes the material unsuitable for boiler construction. Most directly, a boiler should have a large coefficient of thermal conductivity because the thermal energy in the burning fuel must be rapidly transferred to water in the pipes of the boiler. Second, the material must be stable under large temperature extremes, and concrete is not: it will crack when it is heated on one side. This is closely related to the thermal conductivity, because a small coefficient of thermal

conductivity means the temperature is not uniform throughout and the thermal expansion or contraction is not very uniform.

3. Wet clothing takes thermal energy from your body when it evaporates through the heat of vaporization. Thus you can become easily chilled on a cold day if you have wet clothing. With a diver's wet suit, there is first no problem of evaporation. The suit is designed to trap a layer of water near your skin. Once this layer has been warmed towards your body temperature, it serves to keep the colder water outside the suit away from you.

5. A layer of still air is an excellent thermal insulator. First, the fact that it is a gas automatically keeps the coefficient of thermal conductivity down (see Table 21-4); fewer molecules are available to transport energy through sequential collisions. The fact that it is still removes the possibility of transport of energy through convection. Of course, there has to be a mechanism to keep the air still, such as trapping it in pockets.

7. As the technique of making diamonds suggests, diamond is a form of solid carbon in which the carbon atoms are more tightly bound to each other than in graphite. Thus to decompose diamond, it is necessary to add energy to each atom to allow it to dissociate from its neighbor. One way to do this would be to add thermal energy; that is, to cook the diamond. Don't try this with a valuable gemstone!

9. Fluids certainly exist—think of thick molasses—with increasing degrees of resistance to shear. Glass is a fluid with a large viscosity, and this is associated with resistance to shear. Thus the macroscopic resistance to shear is not a clear distinction between fluids and solids.

11. Layers of leaves trap air in pockets where it cannot move. Your body heat will warm the air near you, and since air is such a poor thermal conductor, the layer of air will not cool off rapidly. If there were no leaves, then the air molecules near you would have a much better chance of diffusing away, or, with the slightest breeze, of being blown away, to be replaced by air at a lower temperature. See the discussion of question 5.

13. We can think of a cold atmosphere freezing the layer of water with which it is in contact, for example the top layer of a lake. If ice were more rather than less dense than liquid water the ice layer would sink to the bottom of the lake, opening a fresh layer of water to the atmosphere, more freezing, and more sinking ice. In effect, ice would build up from the bottom of the lake, which would end up frozen throughout, a disaster for the life within. The situation in the ocean might appear more complex, because the fact that it is deep means there is a significant pressure increase to the bottom, and when real solid ice is subject to higher pressure it melts. But this would not likely be the case if solid ice were more dense than liquid water, because compression tends to increase the density and thus would preserve the ice in its solid form. In fact, the water at the bottom of the ocean might freeze simply as a result of the large pressure! The result would be the one we described

for the lake. Happily, floating ice acts as an insulator for the liquid water beneath, preserving the liquid water for the life within.

15. As in the remarks on question 5 and question 11, a skier makes use of the excellent thermal insulation properties of trapped air in wearing layers of clothing. It is also possible to remove the clothing one layer at a time if the skier feels too warm.

Chapter 22.

1. The forces on a test charge due to the two identical positive charges will have the same magnitude if the distances from the test charge to the two charges are the same. Thus the points of zero net force must lie on the perpendicular bisector of the line joining the two identical charges. However, the forces have directions, and these directions are opposite only at the midpoint of the line joining the two identical charges. Only at that point will the test charge experience a zero net force.

3. The spark is caused by a transfer of the charge carried by the walker to the door knob. This can happen only if there is a build-up of charge on the walker. In the winter the air is drier; that is, there is less water vapor in the air. Since water molecules are efficient at picking up and carrying off charges on an object in their presence, it is more difficult for you to build up a significant charge in humid air. In the winter there is a better chance for a charge build-up, and therefore of a dramatic discharge.

5. The basic principle of the quantitative operation of the electroscope is outlined in Problem 22-5. The measurement of the angle made by a gold leaf can be translated into a measurement of the charge. In order to measure accurately the charge you carry you might want to stand on an insulating mat as you touch the metal top of the electroscope. That both controls the situation and ensures that your charge doesn't leak off elsewhere just as you are trying to measure it.

7. Let's call the quarks with charge $2e/3$ *u*–quarks, and those with charge $-e/3$ *d*–quarks (this is the standard nomenclature). Then the following compose all the possible combinations of three quarks, together with their charges:

uuu	$2e$
uud	e
udd	0
ddd	$-e$

The second combination has charge corresponding to that of the proton, while the third combination has charge corresponding to that of the neutron. The other two combinations do not occur as stable particles.

9. The net force on an object at the center of the circle is indeed zero. However, this point is not a stable equilibrium point. If the positive charge moves away from the center along the axis of the circular charge distribution, all the forces act to repel it, so that it will accelerate away from the center. The center is thus a point of unstable equilibrium,

analogous to a ball resting on the top of a hill. The slightest displacement will cause it to move away from its starting point.

11. Let ball 1 have an initial charge of -4.8 (in units of 10^{-19} C); balls 2,3,4 are initially uncharged. If we assume that the balls are identical, then touching 1 and 2 will give each one of them a charge of -2.4 units, while balls 3 and 4 remain uncharged. If now ball 1 (or 2) is made to touch both 3 and 4 simultaneously, then the three balls each get one third of the available charge, that is -0.8 units.

13. If the electrical charge of a fundamental particle such as the electron depended on its velocity, then we would have a chance to measure the tiny parameter κ only because of a departure from neutrality. Gravity is so weak that for all practical purposes two electrically neutral blocks of material do not exert forces on one another. Under the hypothesis, such objects would only be neutral because any surplus of charge due to a different motion of the electrons and the positive ions would be neutralized by ambient charges. However, we could put the two blocks in a vacuum. If at that point there is no force between the objects, then presumably the values of κ for the electrons and ions are such that at the given temperature the net charge of each object is zero. That is because a given temperature for the electrons corresponds, on average, to a certain value of v_e^2 for the electrons and, by equipartition, another value of v_I^2 for the ions determined by the relation $m_e v_e^2 = m_I v_I^2$. But now all we have to do is to keep the objects in the vaccum and raise the temperature. Each object will now acquire the same net charge, different from zero, and the repulsion should be detectable. We know that any such parameter κ must be very tiny, if it is not zero, because the existence of a temperature-dependent inverse square force has not been observed to the accuracy of our instruments.

15. No, because the data only show that for electrons and for protons the charge of each does not change. It is entirely possible that in electron-proton collisions at high temperatures (high energies) at some stage of the development of the quasar, some new net charge is produced from a neutral environment. This would correspond to charge non-conservation. As long as the amount of new charge is small, and as long as the processes producing these new charges do not affect the processes which cause the radiation that we observe, charge conservation would not be observed by the study of the colors of the quasar light.

17. This is exactly analogous to the motion of a mass inside the earth. We found in Chapter 12 that a mass inside the earth undergoes harmonic motion about the center. A point charge of one sign embedded in a spherically symmetric charge distribution of the opposite sign will also oscillate about the center of the sphere. The frequency can be found when the force inside the sphere can be calculated. This will be done in Chapter 24.

Chapter 23

1. The moving truck picks up electric charge as it moves. Rubber tires are good insulators; the charge will not automatically flow to the ground. There is danger that when enough

charge builds up, a breakdown can occur with the formation of a spark, and such a spark is extremely dangerous when gasoline is present. For just this reason, tires are made today with materials that conduct well, and a dragging chain is no longer necessary.

3. The introduction of a gravitational field $\mathbf{g} = \text{limit}_{m \to 0} \mathbf{F}/m$ is indeed useful for the same reasons that the introduction of an electric field is useful. The field resembles that of the electric field in that in the absence of matter (charge) the field lines are continuous, and their density represents the strength of the field. It differs in that mass comes in only one sign: gravity is a uniquely *attractive* force, so that field lines have only one end on matter. The other end must be at infinity, since there is no mass of opposite sign for the line to attach to. In other words, there is no analogue of an overall neutral charge distribution, in which lines start in part of the distribution and end elsewhere.

5. Not really, since we may think of a negative charge $-Q$ distributed uniformly over a spherical surface at infinity to accompany our single positive point charge Q. The charge density everywhere is zero, so that this depiction has no practical consequence other than the satisfying notion that the universe involving single charges is still electrically neutral.

7. In order to visualize a sphere with an induced dipole moment, think of the induction of a positive charge $+Q$ at the sphere's north pole and a negative charge $-Q$ at the south pole in response to an external field oriented along the north-south axis. Suppose that we now suddenly change the external field so that it is now perpendicular to the north-south axis. The charges will move in response so that now the dipole moment is oriented along the direction of the new external field. But since these charges are not attached to the conductor—they move freely on the conducting surface—their motion does not induce a rotation of the sphere. With a long rod, the situation is different. Even if the charges are free to move within the conductor, the shape of the conductor itself restricts the movement of the charges. Thus there will be equal and opposite forces on the two ends, tending to rotate the rod. At the same time, the original inducing field is now gone, and the charges rush back to each other under the influence of the coulomb forces between them. Whether there is an actual motion of the rod depends on how rapidly the charges move back together compared to how rapidly the new field acts.

9. The total charge of the water molecule is zero. The charge distribution shown in the figure suggests that the electric field is that of two dipoles touching at one end. The superposition of two electric dipole fields is again an electric dipole field (they both fall as $1/r^3$), except under very special circumstances in which there is a cancellation, so that only the $1/r^4$ terms are left. This is not the case here.

11. The velocity field has features common to the electric field. Sources (like faucets) correspond to positive charges, and sinks (like drains) correspond to negative charges. The velocity field is represented by a vector at every point in space, just like the electric field. The major difference is that in a liquid there is something that actually moves along the lines (look back at Chapter 16), whereas the electric field lines do not represent motion except in the sense that a test charge would accelerate along the tangent of a field line.

The electric field is thus more like an "acceleration field," something which is of little interest in the study of fluids.

13. From the figure we can see that the forces will align the small dipole in such a way that the attraction is maximized, or such that the potential energy is minimized. In other words, the small dipole will align its electric dipole moment to be antiparallel to the large fixed dipole's electric dipole moment.

15. Both the force of gravity and the electrical force are independent of the height. If the gravitational force of attraction is stronger than the repulsive force, the pellet will fall down, albeit with an acceleration smaller than that due to gravity. If the repulsive force is stronger, then the particle will accelerate away from the plate, and go upwards.

Chapter 24

1. The value of temperature at each point forms a scalar field. Since there is no directionality associated with temperature, flux, which measures an abstract sort of flow across a surface, cannot be associated with a temperature field. However, given a temperature field, one can calculate at each point a vector, termed the *temperature gradient*, representing the change in temperature. This vector field has the components $(\partial T/\partial x, \partial T/\partial y, \partial T/\partial z)$, where you will recall that the symbol ∂ refers to partical differentiation. Because this field has directionality, it is possible to define a flux for it.

3. Suppose we consider a charge free region, and there is a break (discontinuity) in a field line. It is then possible to construct a gaussian surface that envelops the tip of the break in the field line. There will be a net flux across that surface, but on the other hand, there is no charge in the region. Thus a break in an electric field line in a charge-free region violates Gauss' law. The only way to satisfy Gauss' law is to insist that when a field line ends, it ends on a charge.

5. The reconciliation follows by considering a pill-box gaussian surface on the first plate, with its flat ends, of area A, extending just outside of the plate itself. The charge density is not changed, so that the Q/ε_0 part of Gauss' law is unchanged. The flux through the end surfaces, however, is changed. On the side of the plate away from the second, negatively charged, plate, the flux through the end of the pill-box is $(\sigma/2\varepsilon_0)A$ from the positive plate, and $(-\sigma/2\varepsilon_0)A$ from the negative plate. These cancel. The flux through the end surface of the pill-box between the plates is $(\sigma/2\varepsilon_0)A$ from the positive plate and a like amount from the negative plate. These add to a total of $(\sigma/\varepsilon_0)A$. There is no conflict.

7. Gauss' law for fluid flow involves the fluid flux, given by $\Phi = \int \mathbf{v} \cdot d\mathbf{A}$. This flux describes the rate at which the fluid crosses the surface. For a closed surface there will be a net outflow of fluid only if there is a source of fluid somewhere within the enclosed volume. Thus Gauss' law will read $\Phi = S$, where S is the rate at which fluid is "created" inside the surface by a source, in m^3/s. If there are sources (faucets) in the region, then S is

positive; if there are sinks (drains) in the region, then S is negative. Evaporation acts as a sink; that is, a negative contribution to the flux. Looking at the net flux, it is impossible to separate evaporation from any other type of sink. In the case of electricity, the analog of evaporation would be the disappearance of electric charge. There are deep principles that argue against that, and therefore one would not expect S to change with time unless charges actually cross the boundary of the enclosed surface.

9. Symmetry does allow us to state that the electric field is parallel to the vector normal to the surface of the torus. Thus Gauss' law gives us a value for the integral $\int E\ dA$. Because of the curvature of the surface, E is not the same on the inner part of the torus as on the outer part, and therefore the integral cannot be converted to the form $E\int dA = EA$.

11. With Gauss' law we can show that there is no charge in the region of uniform electric field. Take a Gaussian surface in the shape of a can with the two ends perpendicular to the constant field direction. The net flux through the surface is zero, and so the net charge inside the region is zero. The surface can be anywhere within the large region, so that there is no net charge anywhere. If you are worried that this does not rule out equal positive and negative charges inside the region, just make the can smaller. No matter how small the can's volume, there is no net charge.

13. Let's use Gauss' law together with a Gaussian surface in the form of a tiny pill-box whose flat ends are perpendicular to the z–axis. Since the z–component of E vanishes and the other components are independent of z, the net flux through this Gaussian surface is independent of z, and the net charge can only depend on x and y. Thus the charge density must also be independent of z.

Chapter 25

1. The volt is defined so that 1 N/C = 1 V/m. Cross-multiplying yields 1 V–C = 1 N–m = 1 J.

3. If charge is placed inside the hollow space in a spherical metal shell, there will be an electric field. The field lines will join the charges to the inner surface of the shell where induced charges appear so as to yield a net charge inside any surface entirely within the spherical shell. To make a constant field in a small region, insert a small uniformly charged plane within the space; near that plane the field is constant.

5. The question is analogous to asking for the source of the energy which moves a test-mass in a gravitational field that arises from a distribution of masses. The test charge—like the test mass—has potential energy by virtue of being in the field of the existing charge distribution (mass distribution) and some of this potential energy is converted to kinetic energy in giving the test charge (test-mass) some motion. The source of the potential energy is the work that had to be done to assemble the charge distribution (mass distribution) by bringing in the constituent charges (constituent masses) from infinity.

7. The surface of a conductor will always be an equipotential in equilibrium. If a charge is placed on an electric surface, there is a short time during which it is localized. After that short time interval it distributes itself over the surface so that there is no force on any part of the charge, and equilibrium is achieved. The mention of a "short time" indicates that the statement that a conductor forms an equipotential is specifically true for *static* fields. The question then becomes more a matter of finding the time scale that distinguishes static from nonstatic fields.

9. The point of the demonstration is to place charge on the person. Once that happens, the individual hairs share the charge and repel each other much like the leaves of an electroscope. If the person is not on an insulated mat, then the charge from the generator will flow through the person as current, with potentially painful or even fatal results.

11. Knowledge of the potential at a point does not allow us to determine the electric field. The simplest way to see this is to observe that potential energy contains an arbitrary constant. In contrast, if we know the potential at two adjacent points, then we know a potential difference, and this has physical meaning. In fact, the difference in the potential can allow us to find the electric field in the direction of the vector that connects the two adjacent points, as can be seen from Eq. (25-9); the points b and a are taken near each other.

13. The electric fields are very large at sharp corners of any charged object, and with large fields breakdown becomes much more likely. Smooth spherical surfaces minimize this possibility by minimizing the presence of points.

15. Nothing of physical significance would change, since potential energy, and thus electric potential, are not specified to within an additive constant.

17. The potential at a fixed point tells us nothing about the electric field because it is only *differences* in potentials that give us information about the electric field. (See the discussion of question 11.) If, however, we know the value of the potential in the vicinity of the points where it is zero, we can do better.

Chapter 26

1. The fact that there are two ways of writing the units of permittivity changes nothing. The flexibility may allow for ease in the cancellation of units in different equations, but this is purely a matter of convenience.

3. The reminder provides the proof. Consider a closed path which consists of a segment between the plates, leading from one to the other, and a segment which closes the path outside of the plates. If there is a voltage drop in the interior region, there must be a voltage rise outside that region. This would be impossible if the electric field vanished outside.

5. The reason the capacitance per unit length goes to zero is that in the limit under consideration, the potential difference becomes infinitely large. It takes infinite work to concentrate an infinite amount of charge along an infinitely long line.

7. The fact that unlike charges attract and like charges repel forces the dielectric constant to be larger than or equal to unity.

9. When V is held fixed, Q is proportional to C. When the plates are pushed together, C increases, and so must Q. Another way to see this is to note that if the potential difference between the plates is fixed as the distance between the plates decreases, then the electric field must grow. This can only happen if the surface charge density, and thus the total charge, grows. What about energy? For fixed V, the energy is proportional to C (or Q). Thus pushing the plates together increases the energy, and positive work must be done to push the plates together. Alternatively, note that the energy is proportional to E^2 times the interior volume. The volume decreases as the distance beteen the plates decreases, but the electric field grows in the same way. Thus E^2 grows quadratically with the separation, and the total energy is proportional to the separation.

11. The two plates of a large charged capacitor carry charges Q and $-Q$; these charges may be large. When a wire connects the plates the charge will flow through the wire, generally in a very short time. This could be dangerous if the person making the connection is careless and allows some of the charge to flow through him or her. More generally, a large amount of energy has been stored in the capacitor and is dissipated when the charges flow out. It is always a potential danger when a large amount of energy is dissipated.

13. Two oppositely charged nonconductors will give rise to an electric field between them, with the field lines going from the positive to the negative charges. This configuration will store electrical energy and act in that sense just like a capacitor. The difficulty is that it is hard to put a large charge on insulators, and it is also hard to discharge them. That is why conductors are much more useful.

15. The question is whether the vacuum can be polarized; that is, whether positive and negative charges can be separated from it. In terms of what we know, there are no charges in the vacuum, and therefore an electric field applied to it will not induce a charge separation. When you study quantum mechanics you will learn that the full answer to this question is quite different from the answer given here.

Chapter 27

1. The current is a measurement of the rate at which charge passes. Since charge is conserved, the rate at which electrons pass various points along a beam is the same no matter how the individual electrons in it may have been accelerated; otherwise you would say that some charge has been lost or gained along the way. In the part of the beam that

has been sped up, the electrons have become more widely spaced. In this way the rate of passage remains the same.

3. The resistance is inversely proportional to the area, and for a fixed current the power dissipated, I^2R is therefore inversely proportional to the area. The thinner wire will get hotter.

5. It follows from Eq. (27-9) that for constant current and area the drift velocity only depends inversely on the density of free electrons. This is a characteristic of the material making up the conductor.

7. In the free electron model, the temperature dependence of the resistivy [Eq. (27-25)] has to do with the time between collisions between electrons and the obstacles—fixed ions—that give rise to the drag force on the electrons. At very low temperatures the time between random collisions would increase; however, the accelerating field would still be present and there would still be a current. Thus the free electron model as it stands, with the accelerations due to the field as a small perturbation on the random motion, would have to be replaced with a "pinball" model, in which the only motion is due to the field and the electrons must "navigate," through multiple collisions, the forest of fixed ions. There would still be conduction even at T = 0 in this picture, whereas the free electron model would predict none.

9. When a switch is thrown and charge flows, it is because there is an electric field in the wire. Free charge in the wire—mainly electrons—will move, but the wire itself remains neutral, because as many charges as leave a segment of wire from one end enter it from the other end.

11. There is no electric field inside a conductor when there is equilibrium. When charges are flowing due to an continuously applied potential we do not have equilibrium, and charges can flow inside the conductor.

13. When the power $P = VI$ becomes too large, there is too much power dissipated for the heat to be conducted away, and melting of the resistor material occurs. Since generally the potential V is held fixed, the current becomes too large if the resistance is too small.

15. The resistance is proportional to the length of the resistor. This means that if two wires are tied together, the resistance is the sum of the resistances of the individual wires. This is consistent with the rule $R_{eq} = R_1 + R_2$ for two resistors in series.

17. The term "high wattage" refers to the large amount of power dissipated in the bulb. More power is dissipated when the resistance is larger, and this is done in a light bulb by making the filament long and thin. A curled up filament allows for a longer filament in a small space.

Chapter 28

1. Water is an excellent conductor, and if the appliance falls into the tub there is a danger that a large current will flow through the body of the person in the tub, causing burns and sometimes heart failure.

3. The current that flows into the battery is the same as the current flowing out of the battery. Whether there is a potential drop Ir just before the current reaches the battery or whether the drop occurs just after the current leaves the battery is irrelevant, since either way there will be the same contribution to the loop rule, and that is all that counts.

5. (a) We may view a wire of area A as consisting of n parallel strands of area A/n. Since $1/(\text{resistance}) = R^{-1}$ is proportional to the area, we see that with $A = n\ (A/n)$, $R^{-1} =$ the sum of R_{strand}^{-1}. (b) We may view a wire of length L as consisting of n sequential segments of length L/n. Since R is proportional to length, with $L = n\ (L/n)$, $R =$ the sum of R_{segment}.

7. This is impossible, since by choosing the direction in which the emf is positive, one could create a situation in which one would be creating energy. Such "perpetual motion" machines violate energy conservation.

9. An unfair question. The circuit in a flash does not give the falling exponential characteristic of a pure RC circuit. Rather it uses solid state devices to tailor the release of energy from a capacitor, typically of size 1000 μF, so that a current that is basically flat for a period of about 0.01 s results. If we work backwards and ask what value of R in a pure RC circuit would give a time constant $\tau = RC$ of 0.01 s with a capacitor or 1000 μF, we would find $R = \tau/C = (0.01 \text{ s})/(10^{-3} \text{ F}) = 10\ \Omega$, a very reasonable value.

11. Let's construct a potential energy diagram with a fluid analogy in mind. The batteries "raise" the liquid, increasing its potential energy. A resistance corresponds to a drop of potential energy given by IR. When the current goes through resistors in series, it is as if it cascaded down several downward slopes. If current goes through two resistors in parallel, it splits up so "at the bottom" the two currents are reunited at the same potential. Consider now this diagram turned upside down. The batteries (with reversed polarities) now lower the potential, and IR must raise them. Since R is unchanged by the reversal, I must change sign.

13. Technically speaking this is certainly true. However, it is useless information. The value of the current itself depends on the rest of the circuit as well as on the value of the internal resistance, and so does the value of the "shifted" emf. This is not a very useful way to think about a circuit. In contrast, the original emf is a constant which at least for an ideal battery does not vary with current.

15. The teenagers provide a path for the current parallel to the wire. If the wire has no resistor along it, the resistance is low compared to that of the teenager, and most of the current flows through the wire. With the resistor, more of the current flows through the

teenager, with more serious consequences. They are both dumb, but the one with the resistor is dumber.

Chapter 29

1. The wire is neutral because there are equal numbers of negative and positive charge carriers in each segment. Under the influence of the internal electric field that gives rise to a flowing current, only one of these charges will ordinarily move; if they both move, it will be in opposite directions. In the first case $q(\mathbf{v} \times \mathbf{B})$ is not zero; in the second case, there are two terms, but both $q_+(\mathbf{v_+} \times \mathbf{B})$ and $q_-(\mathbf{v_-} \times \mathbf{B})$ are in the same direction, and there is no cancellation.

3. Both effects could cause such a deflection. Assuming we cannot look closely at the surroundings of the CRT, we could change the original accelerating voltage and see what happens. For magnetic effects there should be a change, since the velocity of the electron changes. For an electric deflection, due to a transverse electric field, the deflection angle will change as well, but in a different fashion. If the field causing the deflection is not constant in space, distinguishing the effects in this way will not be simple. Still another approach might be to place a soft ferromagnetic material in the vicinity and see if the effect is greatly exagerrated; that would suggest a magnetic origin.

5. Had we used a left-hand rule, we would just write $\mathbf{F} = -q(\mathbf{v} \times \mathbf{B})$ and our calculations of physical effects would be in agreement with experiment.

7. Consider two magnets with the axes from S to N parallel. There is a repulsive force between the magnets; if we rotate one magnet so that the axes are antiparallel, there is an attraction between the magnets. It is hard to see how induced charges would give rise to such an effect.

9. The magnetic dipole moment of a current loop is proportional to the product NA, where N is the number of turns and A is the area of the loop. The largest area for a given perimeter is that of a circle, so we conclude we should form circles. Now suppose the wire length is L, and that we use the wire to form a cylindrical coil of N turns. Each loop has circumference L/N, and hence a radius $L/(2\pi N)$ and area $\pi[L/(2\pi N)]^2 = L^2/(4\pi N^2)$. The magnetic moment is therefore proportional to $NA = L^2/(4\pi N)$. This decreases with N for a fixed value of L. Thus for a given fixed current and a fixed length of wire, a single loop has the largest magnetic dipole moment.

11. The loops are equivalent to bar magnets, and when these are parallel, there is a repulsion. If the directions of the currents are opposite, the loops attract.

13. The origin of the discrepancy lies in the fact that one end of a bar magnet points to the Earth's geographic North Pole. It was natural for a sailor to call that end of the bar magnet a "north pole." In doing so he was automatically labeling Earth's geographic North Pole as a magnetic south pole.

15. We know geographic N from observations of where the Sun rises. A bar magnet floating on cork in a pail of water will align along Earth's magnetic field lines, which allows us to label the end pointing to geographic north as the magnetic north pole. We now have a compass; that is, a device that points north.

17. Dimensionally $[F] = [q][v][B]$. Since $I = dq/dt$, $[q] = [I][t]$. Thus $[F] = [I][vt][B] = [I][L][B]$, or $[B] = [F]/([I][L])$. With B measured in the Systeme Internationale, this relation reads 1 T = 1N/A·m. Your classmate is correct.

Chapter 30

1. The compass needle aligns along the magnetic field lines. Thus in this case the compass needle will maintain an orientation that follows the circular path of the compass itself.

3. The magnetic field in the yoke of a U-shaped magnet "closes" the U-shape. The wire passing through the yoke lies roughly perpendicular to this magnetic field. When a current passes through the wire, the magnetic field exerts a force on it proportional to the vector product $\mathbf{l} \times \mathbf{B}$, where $\mathbf{l}$ is parallel to the wire. If $\mathbf{l} \times \mathbf{B}$ points upward, the wire will make an upward jump that will be quite sizeable if the force is large and the mass is small. If $\mathbf{l} \times \mathbf{B}$ points downward, the wire will move downward—assuming there is space for that to happen.

5. No, because the wires are electrically neutral.

7. It is much easier to make precision measurements of forces than it is to count charges.

9. The change can be summarized by the replacement of the permittivity of free space by a permittivity associated with the material, $\varepsilon_0 \to \varepsilon$. In particular, the electric field will contain a factor $1/\varepsilon$ rather than $1/\varepsilon_0$. This change in turn carries through in the discussion of the displacement current to the point where the displacement current derived in the text is multiplied by $\varepsilon/\varepsilon_0$.

11. The dimensions of electric field are [N/C]; the dimensions of magnetic field are [N/C] times the dimensions of inverse velocity, so that $[E/B] = [v]$, the dimensions of a velocity.

Chapter 31

1. By Gauss' law for magnetism, the magnetic flux through a closed surface is always zero. Thus there will be no induced electric field. Another way of seeing this is to view the sphere as consisting of two adjacent hemispheres separated by the equator. Suppose for example that **B** is oriented perpendicular to the plane of the equator and increases with time. The normal to the surface of the lower atmosphere points in a direction opposite to that of **B**, hence by Faraday's law there is an induced electric field on the equatorial boundary of the lower hemisphere due to a *decreasing* **B·dA**. On the upper hemisphere

the normal to the surface is aligned with the direction of **B**, there is an *increasing* **B·dA**, and the induced electric field on the equatorial boundary of the upper hemisphere will be the same as that of the lower atmosphere, but of opposite sign. The two contributions cancel.

3. By twisting the wires together that might be part of a closed circuit, the effects of Faraday's law are drastically reduced. You can contrast this case with a case in which the wires are widely separated: this opens a large area and hence the possibility for a large magnetic flux. Pulling the wires together reduces this possibility; twisting them reduces it even further, because then there are a series of very small areas, and successive areas are oppositely oriented, reducing the possibility of a flux even further.

5. (a) No: the flux entering from the right increases, and by Lenz' law the current must be such as to counteract that. (b) Yes: the induced current gives rise to a field that opposes the increase in the flux due to the approaching loop. (c) No: when the switch is closed, there is a rising flux in the direction to the right. The induced current must counter that. (d) No: the loops is oriented so that there is no magnetic flux through it from the straight wire, hence no current is induced.

7. As long as the loop is in a region in which the magnetic field is uniform, and it does not change its orientation, then the magnetic flux is constant and there is no induced emf. If the loop enters or leaves the region of constant field, then there will be a change in flux, and there will be an emf, unless the plane of the loop is parallel to the direction of the magnetic field vector.

9. There will be no change in any physically measurable quantity. The reason is the invariance of the laws of electromagnetism under Galilean transformations: the laws of motion are the same in all inertial frames.

11. Consider, for definiteness, a flow of plasma directed from the top to the bottom of this page, up in the plane of the paper, and a magnetic field directed into the plane of the paper. The force on moving positive charges will be to the left. As a result, these charges will be deposited on the left-hand boundary, with a consequent an electric field to the right. This electric field will increase as charges accumulate until it is strong enough to prevent more charges from accumulating. This happens when $\mathbf{E} + \mathbf{v} \times \mathbf{B} = 0$. Thus $v = E/B$, independent of the density of the carriers and the magnitude of their charge. The same analysis leads to the same speed if the carriers are negative.

13. When the circuit is closed, there is a change in the current in the wire, and therefore a change of magnetic flux in the iron core. This induces an emf in the aluminum ring, and the current generated by this emf must be in a direction opposite to that in the coil. Since two antiparallel currents repel, the aluminum ring jumps. The repulsion can be thought of as a direct manifestation of Lenz' law.

15. A harmonically varying voltage across the solenoid gives rise to a harmonically varying current through the solenoid, and therefore a harmonically varying magnetic flux. This induces an emf in the copper ring, which in its turn gives rise to a current flowing in the copper. The current in the copper ring will always be opposite in direction to the current in the coil by Lenz' law. Since opposite currents repel, the copper ring has a tendency to be accelerated in a direction opposite to the direction of gravity. The copper ring must be of such a radius and thickness that the two forces cancel.

Chapter 32

1. The forces between an electron and a nucleus leave their center of mass stationary. But the nucleus is thousands or tens of thousands of times more massive than an electron, so the system's center of mass is essentially at the nucleus. The situation is much the same as Earth's orbit of the sun: the sun is so much more massive than Earth that for all practical purposes Earth orbits a stationary sun.

3. As long as the surface over which we integrate the magnetic flux lies within a material of constant permeability μ, the relation $\mathbf{B} = \mu\mathbf{H}$ allows us to remove μ from within the integral and express Gauss' law in terms of an integral over $\mathbf{H}$; only if μ varies over the surface integrated in the expression of Gauss' law do we need to keep μ within the integral.

5. Diamagnetism is a universal phenomenon, based on Faraday's law, so we expect it to be present in iron as well as in any other atom. However, it can be very difficult to detect if paramagnetism or ferromagnetism is present. High temperatures tend to decrease paramagnetic and ferromagnetic effects, since those effects depend on order. Thus one way to see diamagnetism is to search for it at high temperatures.

7. If we construct the latch of a magnetically hard material, the constant banging the material takes when the door is closed will not demagnetize it. Similarly, you wouldn't want your little brother to undo the effect of the latch by playing with external magnets nearby, and a magnetically hard material protects the latch from this possibility.

9. Paramagnetism occurs when atoms or molecules with permanent magnetic moments align with an external magnetic field. The alignment becomes less effective as the temperature increases, because the constant jostling associated with finite temperatures increases as the temperature increases, and this jostling tends to destroy the alignment. Diamagnetism is unaffected by the microscopic motion of thermal energy, and thus dominates over paramagnetism as the temperature increases.

11. You can't. The rods will attract and repel as magnets even if one is initially unmagnetized, because the one that is initially magnetized will tend to magnetize the other. Moreover, Newton's third law holds in any case, and that law does not permit you to assign the source of the force to one or the other rod. If you had a bulb and a wire, you

could make a loop and push each rod through the loop. The rod that is magnetized would induce a current that would light the bulb.

13. If you had charge carriers of both signs, you could have the negative carriers circulate in one direction while the positive carriers circulate in the other in such a way that net angular momentum is zero. However, the current is all in the direction of the positive charge carrier movement, so there is a net current circulating and a net magnetic moment. If you have only one sign of charge to work with, it is not possible to have a circulating current (magnetic moment) without a net circulating momentum (angular momentum).

15. The source of the original magnetic field that gave lodestones their permanent magnetic properties is Earth's magnetic field. This makes a correlation of the magnetization of lodestones of different age (in the sense of the time at which the rock was formed from, say, lava) an important geological tool. This tool traces the history of Earth's magnetic field, which is found to change rapidly at certain moments in Earth's history. These changes suggest significant abrupt changes at Earth's core, the region where Earth's magnetic field is generated.

Chapter 33

1. The principle that governs the existence of mutual inductance is a general one. However, it is possible to arrange the circuits so that quantity M, the mutual inductance itself, is zero. Because the orientation of the two circuits is usually not time dependent, this means arranging the circuits so that the magnetic flux from one does not link the other. For example, two small planar loops are only weakly linked if they oriented so that their planes are perpendicular.

3. (b) will have the largest mutual inductance; the magnetic field generated by one coil passes almost directly through the other. (a) will have the next-largest mutual inductance; if we think of the coil as generating a field like that of a permanent magnet, the magnetic field from one comes back around to close on itself, and in doing so passes through the other. However, the field is more "spread," so the linking is weaker. (c) will have the weakest linking; for example, the field comes up through the lower coil but no flux from it passes through the upper coil (see the response above to question 1).

5. Yes. It is direct to calculate the magnetic flux passing through the loop due to the magnetic field generated around the straight wire. This flux will be proportional to the current in the straight wire, and hence gives the mutual inductance. By the "mutual" property of the mutual inductance, that also tells us the emf generated in the straight wire. Where is the loop associated with the straight wire? That wire must close somewhere, and that is what forms the loop of which the straight wire is a part.

7. You would have to be able to concoct a switch in addition to the items listed. You could then arrange the items in an RL circuit of the type shown in Fig 33–10. The time dependence of the rise in the potential across the resistor or the inductor would allow us

to solve for the value of the inductance. If L were too small compared to R, you would have to exercise more ingenuity in the switch construction and in the use of your timer, because the exponential rise (or fall) time R/L would be too small to observe with your unaided senses.

9. It flows in the fields as well as in actual movement of charge. We have already seen that there is energy in electric and magnetic fields. The energy in the inductor is purely magnetic while the energy in the capacitor is purely electric. The exchange takes place as the current flows; the movement of charges will generate both electric and magnetic fields. And of course a moving set of charges carries kinetic energy.

11. Yes, with an energy density proportional to the field squared at any given location. In the limit that the solenoid is ideal, this field becomes zero and the magnetic energy density outside disappears. For a real solenoid, even one that is close to ideal, the energy density outside may be small, but the integral of that energy over all space is not negligible.

13. The inductive and capacitive energies of the RLC circuit are analogous, respectively, to the kinetic and potential energies of the mechanical spring. This can be seen directly from Table 33-3, which shows that $LI^2/2$ is analogous to $m(dx/dt)^2/2$ and $Q^2/2C$ corresponds to $kx^2/2$. Energy flows back and forth between kinetic and potential terms in the damped mechanical spring as it does between inductor and capacitor in the RLC circuit. In each case there is in addition a mechanism for energy loss.

15. If we think of regions with higher (lower) net field as regions of higher (lower) pressure, then the region between two parallel wires carrying current in the same direction is a region of lower pressure, because the contributions from each wire to the field in that region between them tend to cancel whereas the contributions in the regions to the outside of both wires tends to add. Hence the attraction is expressed as the influence of higher external and lower internal pressure. If the wires carry current in opposite directions, the fields tend to add in the region between the wires and cancel in the region outside the two wires, so a net internal pressure tends to push the wires apart. This analogy can be carried farther by thinking about the forces between current sheets, in which case you can show that the magnetic forces are perpendicular to the surfaces of the sheets and proportional to the area, just like a pressure. This idea is heavily used in magnetohydrodynamics, which concerns electrically conducting fluids and is applicable to many plasma physics problems.

Chapter 34

1. The core material plays a major role in linking the two coils, so it should efficiently amplify the magnetic field as well as confine it to the linking region. This means that the energies within these materials are substantial, and there are forces present that require that the core have a sufficient mechanical strength. Cores used in demonstrations for introductory physics often involve thick wires tied into a spiral for mechanical strength.

3. Superconducting properties break down when high fields are present—for example, see the Chapter 32 discussion of the magnetic properties of superconductors. So what ultimately prevents infinities in driven response when superconductors are involved is that they cannot remain superconductors!

5. The current alternates, sometimes negative and sometimes positive. In the harmonic form, every time it has a certain positive value there is another time when the current has the same value, only negative. The average of the current itself is simply zero, and this does not tell us much about important features we want to study.

7. The lamp is a resistor, and the "brightness" with which it burns is a measure of the energy loss within it, hence of the energy loss in a driven RC circuit. Recall also that a capacitor acts like an open switch for direct current, but passes alternating current more and more easily as the frequency increases. With these statements we can see that (a) is false except when the generator frequency is zero (direct current). We would also reckon that (b), but not (c), is true. We can verify this by recalling Eq. (34–46), which shows that for a driven RC circuit with an emf amplitude V_0, the power loss is given by

$$\frac{1}{2} \frac{V_0^2 R}{(1/\omega C)^2 + R^2}.$$

As ω increases, this loss increases.

9. The maximum current is most often determined by capacity of the appliance to dissipate thermal energy, and that in turn is a function of the internal resistance. The result of too much current through a resistor is so much thermal energy generated that the device melts! Other factors that may enter have to do with the stability of semiconductor materials and of microcircuits to withstand the changes generated by elevated temperatures.

11. Resistors in series are equivalent to a resistance that is the sum of the individual resistances, so that the formula in question is indeed correct in this case. And when individual capacitors are placed in series, their inverses add to an equivalent inverse capacitor, so that once again the impedances add. A similar result holds for inductance. We can see all this from Eq. (34–56). The reason the individual impedances do not add when more than one element is involved has to with the fact that the different terms in effect enter with different phases. The phase does not matter when only one type of element is involved, but the problem is made more difficult when the phase is present. In fact, when the impedance is written using complex numbers, there is a very useful method to handle phase complications: *impedances connected in series are indeed equivalent to a single impedance whose (complex) value is the sum of the (complex) individual impedances.* In this language Eq. (34–56) is an expression for the magnitude of the impedance.

13. Just because a capacitor's impedance for low frequency is large does not mean it is infinite, and just because an inductor's impedance for high frequency is large does not

mean it is infinite either. There is plenty of room for current to pass for frequencies that are neither zero nor infinitely large.

15. There are two features of appliance plugs worth noting. Each are associated with safety. First, some appliances have three-prong plugs. The third prong is a grounding line. It connects to ground in a wall receptacle which connects that prong to a metal pipe or a grounding cable somewhere. Its purpose is to make sure that a loose wire or extraneous piece of metal in the appliance will not make the outer case "hot," but rather connects the hot wires to ground and thereby causes a fuse to blow. The second feature is something that nearly all new appliances with two-prong plugs have, and that is that one of the prongs takes a flat spade form that can only enter the wall receptacle in one sense. The spade-accepting receptacle is connected to ground, while the second receptacle is "hot," with a potential that varies from positive to negative. Thus the spade-shaped prong plays much the same role as the third prong in the first case we discussed. Unfortunately, older buildings may not be equipped to take these plugs, and in these cases many people will file the spade shape down. User beware!

17. It is always possible to use transformers to change the voltage amplitude when it becomes desirable. The possible amplitude of a generator has to do with factors such as the speed of rotating turbine blades and the strength of the magnetic field within which the rotors turn. Most often this amplitude is considerably less than the high voltages that make for efficient transport of electrical energy, and transformers are used at the transition area between generating plant and power lines.

Chapter 35

1. No force is associated with uniformly moving electric charges, so that no work is done, and no energy is expended. When charges accelerate, however, then forces must act, work will be done, and energy will be expended.

3. If the electric field in an electromagnetic wave has a definite orientation, so does the magnetic field, because the wave is transverse with $\mathbf{E} \cdot \mathbf{B} = 0$. Thus the magnetic field characterizes the polarization just as well as the electric field. The electric field is used as the measure of the polarization because its effect on matter is more direct.

5. To answer this question, we might want to think about the effects of electric charge. The very existence of magnetic monopoles will not change the nature of electromagnetic waves any more than the very existence of electric charge does. They would merely supply new ways to generate those waves. How about the presence of magnetic charge in free space where electromagnetic waves propagate? The fields of an electromagnetic wave cause free electric charge in its path to accelerate; these charges will accordingly reradiate. One effect is to change the effective propagation speed, others have to do with the direction of propagation. We would expect the presence of magnetic monopoles to have similar effects.

7. Yes. What is important in the ejection of mass is that the mass carries momentum. The principle of the conservation of momentum ensures the propulsion of the rocket. We have seen in this chapter that light also carries momentum, so that it could fulfill the same role as the ejected mass. Of course, whether the quantitative result is the same is another question entirely.

9. No. Sound is a longitudinal wave, while polarization is associated with transverse waves, and in particular with the choice of a special transverse direction out of the entire transverse plane that is possible. Thus sound does not exhibit polarization.

11. There is really nothing to check, because there is little difference between electromagnetic waves and Faraday induction. Faraday's law states that a time-dependent magnetic field gives rise to an electric field, and as we saw in this chapter, this interdependence is what gives rise to the electromagnetic waves. Of course the Maxwell generalization of Ampere's law is also crucial.

13. Let's suppose that electromagnetic wave, propagating in the $+z$–direction, is linearly polarized, with the electric field initially increasing in the $+x$–direction; that means that the magnetic field will be increasing in the $+y$–direction. Let's also put the positive (negative) charge of the dipole at positive (negative) x to start. The forces on the dipole charges accelerate the positive charge to the right and the negative charge to the left; that is, they tend to spread the charges in the dipole. Once these charges move, the magnetic field exerts forces on them. The force on the positive charge is to the positive z-direction, while the force on the negative charge is also to the positive z-direction. There is no net torque, and *a linearly polarized electromagnetic wave carries no angular momentum*. However, when we consider a circularly polarized wave, in which a superposition of two linearly polarized waves makes a wave consisting of electric and magnetic fields remaining constant in magnitude while rotating in the xy–plane, an analysis like the one we described above shows that the wave carries angular momentum with a z–component.

15. Yes, we will indeed pick up less of a signal from a sending antenna if there is a receiving antenna between us and the sending antenna, at least if the receiving antenna expends energy in any way, such as driving speakers. The conservation of energy ensures us that this must be true.

Chapter 36

1. Light propagates within your house or other spaces mainly through multiple reflections from all the surfaces that are present. The wave-like bending around corners plays virtually no role at household scales.

3. A fish can see a fisherman before the fisherman can see the fish. This has nothing to do with the optical properties of their respective eyes or brains, but rather with the fact that water has a higher index of refraction than air, so that shallow rays from a fish towards a fisherman undergo total internal reflection rather than reaching the fisherman's eyes. On

the other hand, all the rays from the fisherman can arrive at the eyes of the fish. We hope no fish are insulted by our insinuation that they cannot think.

5. First, the depth of the fish appears to be less than its true depth, because the rays from the fish are bent towards the horizontal as the leave the water. The fish also appears shorter, because the angle between rays from the head and the tail decrease. An easy way to see this is to think of the limiting case where the rays from the fish are almost at the critical angle for total internal reflection; in that case both rays are very close to the horizontal and the fish is seen highly shortened and only a little below the surface.

7. The changing color of the Sun has to do with the amount of atmosphere its rays pass through, which increases as the Sun falls to the horizon. The atmosphere scatters the shorter wavelengths more strongly than the longer wavelengths, leaving the longer wavelengths a more predominant part of the Sun's light as it approaches the horizon. The squashing has to do with increased refraction of the rays that travel through more of the atmosphere; that is, that come from the bottom part of the Sun. These rays are refracted more strongly towards Earth's surface, so that the eye interprets them as coming from a position higher up, i.e., closer to the geometric center of the Sun's disk. The form the disk takes is accordingly shortened at the bottom, or flattened.

9. Think of a series of horizontal layers of air, ever hotter as you approach the ground. Light from, say, the frond of the palm tree penetrates one of those layers and each time it does so it bends more away from the perpendicular to the horizontal boundary layer. This is characteristic of the lower (hotter) layer having a smaller index of refraction than the upper (cooler) layer. Thus we conclude that light travels faster in hotter air. This conclusion is in accord with our understanding of, say, hot air balloons (see Chapter 16), in which we found that hot air is less dense than cool air. Since the index of refraction has its origin in rescattering from atoms, we would certainly expect that a system with a lower density of rescattering atoms has a lower index of refraction than one with more rescattering atoms per unit volume. You might want to think about how, once the ray is moving horizontally, it starts to move upward again. The answer to *that* has to do with total internal reflection.

11. The blue sky is the result of the scattering of sunlight back to the eye by atmospheric molecules. In the vacuum of outer space, there is no scattering; light from the sun that is not directed towards an astronaut cannot scatter from some other part of the sky to be redirected to the astronaut, and the sky is dark.

13. This is best answered by drawing a series of lines—rays in reverse—from your eye and, starting with a vertical line, moving down in angle. (The light actually comes from the lifeguard, of course, but we can certainly think about lines from your eye that follow the rays from the lifeguard.) As the line moves down you will encounter the lifeguard's head. It looks higher than it really is, because the ray from the lifeguard is bent to the vertical as it enters the water. When the line from your eye is at the critical angle for total internal reflection, the line grazes the surface outside; the corresponding ray comes from the

lifeguard's waist. As the line drops further, you receive light from below the lifeguard's waist that has reflected from the water surface and goes to your eye having reflected once. You see a section of the lower half of the lifeguard as if in a mirror, and that part will be *upside down*. Finally, the angle of the line from your eye drops enough so that it runs directly to the lifeguard's waist, and from that point on you see the entire bottom half of the lifeguard in its correct proportion and orientation.

15. If the two surfaces of the glass are parallel, then it is not hard to show that any ray that enters the glass at some angle of incidence leaves it at exactly the same angle. However, the ray is displaced from the entering line by an amount that increases as the index of refraction increases. (Problem 30 treats this quantitatively.) If there is dispersion, then the amount of displacement does depend on the color. However, the angle with which the rays of different colors leave the glass is the same; that is, they make a parallel bundle.

Chapter 37

1. This sign is meant for drivers who see a vehicle with flashing lights arrive from behind and read the message "AMBULANCE" in their rear-view mirrors. Try it!

3. *Any* portion of the lens will give the same image, because all that counts is the curvature of the surfaces, not how big those surfaces are. We have found it convenient to draw principle rays that require the use of a "complete" lens, but remember that *any* ray will end up where the principle rays do in an ideal lens.

5. The dentist will typically want an enlarged view of your tooth in circumstances where the mirror is close to the tooth. Figure 37–10 shows that what the dentist will use for this purpose is a concave mirror. As the discussion on p 997 indicates, a convex mirror always gives an image reduced in size from the object.

7. All these lenses obey Eq. (37–2), which states that when an object is placed at the focal point ($s = f$), the image is at infinity ($i = \infty$). Eq. (37–6) then gives the magnification to be infinite. Ray-tracing will give the same result: The image is infinitely far away and infinitely large. The crucial feature that makes this true is that rays from an edge of the ball that lies above the focal point will trace out a path that is not quite parallel to the optical axis. At infinity, where the rays from the off-axis part of the object arrive at an image, they are infinitely far above the optical axis. Of course, if there were an image of finite size infinitely far away, its angular size would be zero and you could not see it!

9. The amount of refraction depends on the indices of refraction of both media involved. Although we derived the lens-maker's formula assuming one of these media had an index of refraction equal to one, we could rederive the formula assuming the exterior medium to be water, $n = 1.33$. We would then find a different result for the focal length, the distance at which rays from infinity are brought to a focus.

11. The camera couldn't care less about whether the object being photographed is virtual or real. It uses rays coming into the lens, and whether those rays actually come from a real object or only are aligned as if they are coming from a virtual image do not come into play in the formation of the final image on the film plane.

13. Not very. A ray coming in along the axis continues along the axis, and generally the principal rays all overlap. The one exception is the case that the originating point is the focus; here we can use the principal ray that comes from the focus and goes out parallel to the axis.

15. Let's call the direction out from the eye the z–direction. We would start by tailoring the glass so that there is a central point O in the xy–plane and, unless an astigmatism is involved, we do what ever is done as a function only of the distance out from the central point, that is, radially. Is there any tailoring to be done in the z–direction? Yes; if not, any ray would be passing though a flat plane of glass of fixed index of refraction and would be displaced but not changed in direction. If, say, it is necesssary for eyesight correction to make rays from infinity converge closer to the lens, then we would want the index of refraction of the glass to decrease as we move through the glass toward the eye. This would bend the rays towards the point O, which is the effect desired.

Chapter 38

1. As discussed in the book, a single source of incoherent light with two small slits yields two sources of coherent light, but at the cost of a great reduction in intensity. A laser does not suffer from this disadvantage.

3. The bright spots occur at places where the amplitude of the electromagnetic field is largest. In principle we could see the bright spots if we positioned ourselves so that the retina is placed exactly where the constructive interference takes place. This is not very practical, since the placement would have to be done to a very high precision; movement by just a half wavelength changes everything. A screen is necessary.

5. Antireflecting coatings must have a thickness that is a quarter wavelength of the light transmitted, or an integer fraction of that. For attainable thicknesses it turns out that only one wavelength in the visible range of light will have a wavelength for which there is the required destructive interference.

7. The energy density at a given point is proportional to $|\mathbf{E}(r, t)|^2$, and if the field vanishes there, so does the the energy density. Energy conservation merely demands that the total energy—that is, $|\mathbf{E}(r, t)|^2$ integrated over all space—be conserved.

9. The sum of two fields that takes the form $E_1\,\mathbf{i} + E_2\,\mathbf{j}$ cannot form an interference pattern. The square of the magnitude of this quantity is $E_1^2 + E_2^2$, with no cross terms; cross terms are essential for the existence of interference terms.

11. There is some reflection of light from the top of the curved piece of glass, and in principle there is interference between the light reflected at the top surface and that reflected at the curved surface, for example. However the path difference in the lens (in contrast to that in the narrow air gap) is many thousands of wavelengths. Since incoming light is never perfectly monochromatic, and since one would not necessarily go to the trouble of machining the top surface with the same precision as the curved surface, such interference patterns are normally washed out.

13. The positions of maxima of interference patterns can generally be measured to very high precision. Therefore a tiny change in wavelength can be translated into a measurable change in the positions of the maxima.

15. As indicated in the text, total destructive interference between light waves requires not only the proper phase relationship but also (nearly) equal amplitudes. The condition for this is worked out in Problem 38-48.

17. Actually lenses used in optical instruments are coated on both sides; the front coating is used to eliminate reflection at the front surface, and the back coating is used to eliminate the secondary reflection from the back side of the lens. The coating on the back surface will not eliminate reflection from the front surface for the reasons that appear in the answer to Question 38-11; similarly the coating on the front will not work to eliminate reflection from the back side.

Chapter 39

1. For a fixed wavelength, the angle of diffraction decreases as the obstacle size increases. thus a thinner support pole will give rise to an increased diffraction pattern. In the limit of a support pole that is very small compared to the wavelength, the waves will pass around the pole as if it were not there.

3. The bright spot that we refer to as the Poisson spot is due to the constructive interference between the coherent light coming from the rim of the obstacle. In contrast to a flat disk, a coherent light source that is close to a bowling ball will have rays of grazing incidence on the side of the ball that is invisible from the screen. The waves would have to diffract a good way around the ball, and the result is that the intensity is highly reduced. There is little hope of seeing the Poisson spot unless the source is very far away, effectively at infinity. By the same token, light that originates on an equatorial circle of a sphere cannot be focused onto a near screen unless a lens that focuses parallel rays onto the screen is used. Without a lens, the screen too must be far away.

5. This has been discussed in question 38-9.

7. The relationship between the angle subtended by the central bright spot that limits the resolution and the wavelength of the light is $\theta_{min} = \lambda/D$, where D is the size of the aperture. If an instrument has fixed aperture, the angular size of the bright spot is therefore

larger if the the wavelength is larger. Red light has a longer wavelength than blue light, so that the resolution is worse if red light is used.

9. Figure 39-9 illustrates the differences. The locations of the principal maxima are exactly the same in the two cases. The intensities of these maxima are much larger for large N than for the double slit, while their widths are much smaller. It is these differences that make the N-slit pattern so much more useful for the measurement of wavelength.

11. There is a component of diffraction in what happens to ocean waves that pass around you, but most of it is just hydrodynamic flow. To see pure diffraction effects due to an obstacle, you need waves that involve a disturbance of a stationary medium. If you were to stand in a still pond into which somebody dropped a stone, your body would diffract the resulting surface waves.

13. A rough estimate is best made with the formulas that describe the diffraction due to a single slit. For visible light, the wavelength λ is around 4×10^{-7} m. For a diffraction angle θ such that $\sin\theta$ is of the order of one, the intensity of the diffracted light is roughly I_0 $(\lambda/d)^2$, where I_0 is the intensity of the undiffracted light and d is the slit width. For any reasonable slit width this is a very tiny factor.

Chapter 40

1. A perfectly rigid object would be one for which, if the object were to be pushed at one end, the other end would move instantaneously. In this way a signal could be made to propagate faster than with the speed of light, in contradiction to special relativity. There are no perfectly rigid objects.

3. The statement $M = 937$ MeV/c^2 translates into $Mc^2 = 937$ MeV. This is quite acceptable: both sides have the dimensions of energy.

5. If there were an ether, and if a single measurement showed no sign of movement through the ether, one could conclude that at the time of measurement Earth was stationary relative to the ether, but nothing more.

7. The straightforward answer is that it would have the shape of an ellipse—think of a circle drawn on graph paper on which the spacing on the x-axis is smaller by a factor of $\sqrt{1-v^2/c^2}$ than that on the y-axis. Actually the situation is a bit more complicated. If the measurement were to be carried out by means of a super-fast camera taking a snapshot, then the golf ball would still look spherical. This is because the radiation from different parts of the golf-ball has different distances to travel, and these effects alter the naive expectation. A simple discussion can be found in *Physics in the Twentieth Century*, Victor F. Weisskopf, MIT Press (1972).

9. This observation would indeed contradict the laws of relativity if it represented a signal that were transmitted with $v > c$. But the observation can be explained in other ways.

Imagine somebody with a flashlight standing on planet X and turning on her heels, keeping the flashlight pointing straight ahead of her. The light from the flashlight travels with the speed c. It reaches a huge interstellar cloud many light years away and reflects back from that cloud to a distant observer. The observer sees a spot sweeping across the cloud very rapidly indeed, and may conclude by dividing the distance the spot moves by the time it takes to move that he has observed faster-than-light motion. But there is no actual causal motion here. In particular, no message can be sent from one end of the interstellar cloud to the other end in this way, and relativity is not violated.

11. Radiation travels with the maximum speed c, independent of frequency. A particle moving faster than that would force us to give up the special theory of relativity. There is an out: the theory of relativity is based on a maximum speed. If radiation consisted of particles that had a tiny mass, then one could imagine particles with an even tinier mass and the same energy traveling a little bit faster.

13. The theory of relativity relates electric and magnetic fields, without relying on the mechanism by which currents are carried. Whether the charges drift or whether they are hand-delivered, the relationship is still the same.

How things look from a different frame is interesting: In the laboratory frame, the wire is neutral and negative charges drift from left to right. In this frame the positive and negative charge densities are the same. From the point of view of an observer moving along the wire, however, negative charges move with one velocity and positive ones with another. This changes their charge densities, since the volume of a cylinder aligned along the direction of motion is decreased by a factor $\sqrt{1 - v^2/c^2}$. Thus to the moving observer the wire is no longer neutral. In addition to the magnetic field generated by the moving charges (currents) there will be an electric field generated by the non-zero charge density. The combination of the two forces will change the forces on a test particle some distance from the wire in such a way that the resulting motion will not distinguish between the two frames.

15. A single line cannot specify a spectrum. We need to see a pattern of wavelengths—a good portion of the entire spectrum—to be able to identify the type of atom that is radiating, and then compare the observed spectrum with the spectrum of the same atom observed in the laboratory. This allows us to determine the shift.

17. If the elevator is not accelerating, then the light beam moves in a straight line. An observer cannot distinguish between the situation of a uniformly moving elevator with a truly horizontal ray, or a stationary elevator and a slightly tilted light beam. When there is acceleration, then the beam "falls" and the motion is along a parabolic curve. Light does fall in this way in an elevator moving with constant velocity if the elevator is in a gravitational field.

Chapter 41 (regular)

1. When atoms in distant objects are excited, and the electrons make transitions to different states, photons with definite wavelengths are emitted. The pattern of these wavelengths form a spectrum characteristic of the particular atom and can therefore be used to identify the species of atom. In addition, relative intensities can be used to specify the proportions of different speproportions of different species of atoms present and radiating.

3. The behavior of a photon that makes it wavelike is behavior that involves interference and diffraction phenomena. When a photon has a very short wavelength, then only the presence of objects of that size or smaller can provide a diffraction pattern. Objects that are this small become rarer as the frequency goes up, until the photon exhibits no wavelike aspects at high frequencies.

5. There are really two levels to this question. At the first level, we could imagine that the exclusion principle were simply abolished. In that case all the electrons in atoms would drop to the ground state, at least for temperatures that correspond to energies much less than the energy difference between the first excited atomic state and the ground state, several thousand degrees. The distinction between metals and non-metals, and indeed all chemistry and material properties would cease to operate, and no one would be around to ask silly questions such as this one. We could imagine a second level, in which somehow metals were formed, with free electrons in the material, and the exclusion principle no longer applies to these electrons. This is as if we had free electrons in a box, with a set of interrupting pins that represent the crystalline structure of the ions. Quantum mechanics would still apply, so that closely-spaced levels with a gap might be present, but because the exclusion principle no longer holds, the electrons will all sit in or near the bottom level, with a Maxwell-Boltzmann distribution indicating that at finite temperatures some of the electrons would be a little above the bottom level. The description of currents would then correspond nicely to the Drude model described in Chapter 27.

7. An electron sees a charge $Z = 2$, reduced somewhat by screening due to the second electron. We thus have an effective hydrogen-like atom with Z lying somewhere between 1 and 2. Since the radius is proportional to $1/Z$, we expect it to be smaller than that of hydrogen. Basically the attractive force due to the nucleus has been increased, and that brings the electron in closer.

9. By analogy with light we expect to require waves of wavelength λ less than or of the order of d. This corresponds to h/p of order d, or p of order h/d. Nonrelativistically the energy would take the form $p^2/2m_e \approx h^2/2m_e d^2$. As d decreases still further, p increases to the point where relativistic forms for the energy are required. In the limit of $d \rightarrow 0$, the extreme relativistic form for the energy enters, $E = pc = hc/d$. In either case, the required energy increases as d becomes smaller.

11. The formula $u(f, T) = (8\pi f^2/c^3)kT$ represents the energy per unit volume for frequency in the vicinity of f. The total radiant energy per unit volume may be obtained by integrating

this over all values of *f*. Since $\int f^2 \, df$ is infinite when the upper limit is infinite, we know that something is wrong.

13. Particles don't "know" that they are moving. In fact, in the rest frame of the particle the lifetime is a fixed number. It is only an observer moving with uniform speed relative to the particle who measures the time to be dilated.

15. The answer is yes, but the probability of this happening is very tiny. The number is exp(−*S*), where *S* is the product of a typical energy and a typical time, divided by *h*. Since *h* is of order 10^{-34} J–s, we can estimate that for typical macroscopic quantities such as 1 J and 1 s, the probability is $\exp(-10^{34})$. That is awfully small!

17. We apologize that this question is not really answerable from a reading of this chapter alone. It is, however, an important question. In fact, although it might seem from $\Delta t \approx \hbar/\Delta E \approx \hbar m/p\Delta p \approx m\Delta x/p$ that the uncertainties in *x* and *t* are linked, they are not, and it is possible to make simultaneous measurements of *x* and *t*. The reason is that in quantum mechanics the variables *t* and *x* represent rather different quantities. *t* is just a label for the passage of time, whereas *x* is a dynamical variable that refers to the location of a particular physical system. The relation $\Delta t \approx \hbar/\Delta E$ differs from $\Delta x \approx \hbar/\Delta p$. The latter is built into the formulation of quantum mechanics. The former is just a transcription of the classical relation between pulse duration and band width ($\Delta t \, \Delta f \approx 1$) into quantum language that relates *E* and *f* by *E* = *hf*.

19. The fact that the speed of light is a definite quantity (in fact it is nowadays a defined quantity) does not contradict anything in the quantum theory. The velocity of light enters into photon kinematics through *E* = *pc*, and any uncertainties in measurements do not affect the ratio *E*/*p*.

21. Such an atom—it does, in fact, exist—is very weakly bound, because in addition to the attraction of the nucleus the second electron is repulsed by the first one. However, there is a wave function that allows the atom to exist. To put it in classical terms, the net effect is to have the two electrons on opposite sides of the proton; the electrons are closer to the proton than to each other, so that each electron has a net attraction to the proton. For a third electron, the exclusion principle would require an orbit to be outside that of the first two electrons. In that case, the third electron sees a net negative charge, and it is repulsed, so that the atom is not stable.

Chapter 41 (extended)

1. The behavior of a photon that makes it wavelike is behavior that involves interference and diffraction phenomena. When a photon has a very short wavelength, then only the presence of objects of that size or smaller can provide a diffraction pattern. Objects that are this small become rarer as the frequency goes up, until the photon exhibits no wavelike aspects at high frequencies.

3. Let's call N the number of photons the lit cigarette emits per second, and let's suppose these photons are emitted equally in all directions. Then the number of photons that hit a pupil of estimated diameter 3 mm (radius 3/2 mm) at 500 m is

$$N\,(\pi(3/2 \times 10^{-3} \text{ m})^2)/(4\pi(500 \text{ m})^2) \approx 10^{-11}\,N.$$

To estimate N we need to estimate the temperature, which we take to be of order 10^3 K, and use the blackbody formula to estimate the rate of energy radiation in the range (say) of 500 - 700 nm wavelength. The energy carried by an individual photon of frequency f is hf. Thus the rate of photon emission can be obtained by dividing the rate of energy emission by hf, with f corresponding to the midpoint of the above range of wavelengths.

5. We can take a naive point of view and say that the Doppler shift describes, for example, a change of wavelength of a moving source of waves. Since by the de Broglie formula the momentum is inversely proportional to the wavelength, this would be a laborious way to describe the fact that when an electron is ejected from a stationary source with a certain energy in the rest frame of the source, you will measure it to have a different momentum if you are moving with respect to the source.

7. If half have decayed in the first time interval T, then half of the remainder, or $(1/2)^2$, will decay in the second time interval T, half of what is left, or $(1/2)^3$, in the next time interval T, and so forth. The answer to the question posed is that it requires an additional time T for another factor of two in the number of nuclei to decay. It is also interesting to think about how long it takes for *all* the nuclei to disappear. After n time intervals T the number remaining is $(1/2)^n$, so that, strictly speaking, it will take forever for an infinite number of initial nuclei to disappear. If there are a finite number of nuclei N to begin with, then the number of time intervals T will be such that $(1/2)^n \approx 1/N$.

9. Particles don't "know" that they are moving. In fact, in the rest frame of the particle the lifetime is a fixed number. It is only an observer moving with uniform speed relative to the particle who measures the time to be dilated.

11. The answer is yes, but there is not much probability of this happening. The number is $\exp(-S)$, where S is the product of a typical energy and a typical time, divided by h. Since h is of order 10^{-34} J–s, we can estimate that for typical macroscopic quantities such as 1 J and 1 s, the probability is $\exp(-10^{34})$. That is awfully small!

13. We apologize that this question is not really answerable from a reading of this chapter alone. It is, however, an important question. In fact, although it might seem from $\Delta t \approx \hbar/\Delta E \approx \hbar m/p\Delta p \approx m\Delta x/p$ that the uncertainties in x and t are linked, they are not, and it is possible to make simultaneous measurements of x and t. The reason is that in quantum mechanics the variables t and x represent rather different quantities. t is just a label for the passage of time, whereas x is a dynamical variable that refers to the location of a particular physical system. The relation $\Delta t \approx \hbar/\Delta E$ differs from $\Delta x \approx \hbar/\Delta p$. The latter is built into the formulation of quantum mechanics. The former is just a transcription of the classical relation between pulse duration and band width ($\Delta t\,\Delta f \approx 1$) into quantum language that relates E and f by $E = hf$.

15. The fact that the speed of light is a definite quantity (in fact, it is nowadays a defined quantity) does not contradict anything in the quantum theory. The velocity of light enters into photon kinematics through $E = pc$, and any uncertainties in measurements do not affect the ratio E/p.

Chapter 42

1. If $n = 0$ then, according to the quantization conditions and Fig. (42–2), the wave corresponds to a totally flat curve, a wave with infinite wavelength and zero momentum. By the uncertainty principle $p = 0$ can only occur if x is undefined over an infinite interval. Since in a finite box $\Delta x = b$, we cannot have $p = 0$.

3. The largest energy jump possible—if we insist that the atom remains bound and if we do not take into account selection rules—corresponds to a jump from the ground state to the state for which $n \to \infty$. This corresponds to the largest frequency, and therefore to the shortest wavelength. By considering jumps between two arbitrarily large n values, we involve energy shifts, and hence emission frequences, that are arbitrarily small. This means arbitrarily large wavelengths are possible.

5. There are really two levels to this question. At the first level, we could imagine that the exclusion principle were simply abolished. In that case all the electrons in atoms would drop to the ground state, at least for temperatures that correspond to energies much less than the energy difference between the first excited atomic state and the ground state, several thousand degrees. The distinction between metals and non-metals, and indeed all chemistry and material properties would cease to operate, and no one would be around to ask silly questions such as this one. We could imagine a second level, in which somehow metals were formed, with free electrons in the material, and the exclusion principle no longer applies to these electrons. This is as if we had free electrons in a box, with a set of interrupting pins that represent the crystalline structure of the ions. Quantum mechanics would still apply, so that closely-spaced levels with a gap might be present, but because the exclusion principle no longer holds, the electrons will all sit in or near the bottom level, with a Maxwell-Boltzmann distribution indicating that at finite temperatures some of the electrons would be a little above the bottom level. The description of currents would then correspond nicely to the Drude model described in Chapter 27.

7. In the same state, all true hydrogen atoms, ones with a single proton as the nucleus, are quite literally indistinguishable from each other—there is no way to dab paint on them or otherwise label them. However, the three hydrogen-like atoms 1H, 2H, 3H are distinguished from one another by the fact that their nuclei have different masses, which are, to good approximation M, $2M$ and $3M$, where M is the proton (or neutron) mass. This means that the reduced mass, which is the quantity that actually appears in all expressions for bound-state energies, is slightly different for the different species, and therefore the energies are slightly different.

9. In physics the term "fine structure" is actually used in a more precise way than is implied in this rather vague question. We could remark that there are small differences in the energy levels of molecules due to the vibrational and rotational terms in the energy. In atoms, the fact that electrons move and that both electrons and nuclei have magnetic moments also introduce small magnetic corrections to the simple Coulomb terms that we consider in this chapter.

11. For He$^+$ there is only one electron, and that electron "sees" a charge with $Z = 2$ in the nucleus. For neutral helium there are two electrons, and one of them acts to screen some of the nuclear charge. The effective charge lies somewhere between $Z = 1$ and $Z = 2$. Since the energies are proportional to Z^2, there is roughly a factor of two difference between the transition energies.

13. When the potential energy is minimum, the force vanishes.

15. The spins will be antiparallel. In that state, the two electrons can be in the same $n = 1$ spatial orbit. If the spins were parallel, the exclusion principle implies that they would have to be in different orbits, and one of them would have to be $\ell = 1$, corresponding to a higher energy state.

Chapter 43

1. If the Pauli exclusion principle were not applicable, all electrons in atom would be in the lowest ($n = 1$, $\ell = 0$) orbital state, at least at temperatures low compared to the O(eV) energies of the atom. They would not be easily detached, and chemical reactions, including the possibility of "free" electrons in crystalline metals, would not exist at temperatures below many thousands of degrees.

3. The fact that the decay of an excited atom with the emission of a photon has a rate that is enhanced by a factor of N^2 in the presence of N other photons of the same momentum and polarization is directly due to the congregating effect of bosons.

5. E_F is the energy of the electron (or pair of electrons) that has the highest energy among all the electrons that are stacked, two per energy level, in a box at $T = 0$. If the temperature is raised from $T = 0$, the electrons will acquire some additional kinetic energy. If T is small, then there is no way for the electrons that have energies low compared with E_F to absorb the thermal energy. The reason is that there are no energy levels that are empty in their energy neighborhood. Only electrons with energies near the Fermi energy are affected. Once the temperature is increased to the Fermi temperature, even the lowest-energy electrons acquire enough energy to lift them above the Fermi energy.

7. The energy diagram for the He-Ne system shows the laser transitions, as well as at least one other one that is not a laser transition connecting the n = 3, $\ell = 1$ state to the n = 2, $\ell = 0$ state. In addition, there are higher states in helium that are excited by the electrical

discharge, and these can populate higher states in neon; the decay of these states will give rise to incoherent light.

9. Inert gases have their electrons strongly bound, meaning they have no electrons to give up that behave as if they were free in the material. A crystal made from these atoms would not be likely to act as a metal.

11. You could study the spectra of atoms that have two electrons in the outermost shell. There will be transitions involving states in which the electrons have their spins parallel, and transitions involving states in which the spins are antiparallel. The Pauli exclusion principle suggests that these states must be rather different. Evidence for two "ladders" corresponding to these possibilities will show that the exclusion principle is involved. In any case, if there were no exclusion principle in distant galaxies, then the atoms there would have all of their electrons in the lowest state, and there would be no resemblance between their spectra and spectra on Earth. As a matter of fact, the spectra are identical (except for Doppler shifts), and there is no doubt that the exclusion principle is valid elsewhere.

13. The electrons are very unlikely to be in a state with n large and ℓ small when they first approach the protons. Thus in their de-excitation they must get rid of a lot of angular momentum, and this generally has to be done one unit of $\hbar$ at a time. Thus the sequence of transitions is much more likely.

15. X-ray photons have energies in the range of 100 eV and up. The outer electrons in atoms have binding energies of the order of 10s of eV. Thus X-ray transitions must involve electrons from the inner shells of the atom. An inner shell electron must be excited to an unoccupied level from which it can make a transition to a metastable level. It is from this level that the stimulated emission will take place. This points to several difficulties in constructing X-ray lasers.
(1) The pumping mechanism must involve large energy input. It is not obvious that it is possible to arrange the energy input so that only the wanted inner shell electrons are excited to the wanted level. It may be that an enormous amount of energy will have to be fed in to get a reasonable number of the wanted transitions. For gamma-ray lasers, where the energy is even higher, of the order of 10 keV, it has been estimated that power densities only present in nuclear explosions would be necessary.
(2) The cavity in which the X-rays are to be reflected back and forth is also a problem. Ordinary mirrors will not merely reflect X-rays.
(3) The excited electrons are supposed to fall back to the ground state in order to generate the appropriate photons. However, once a vacancy in an inner shell exists, electrons from other states have a reasonable probability of making a transition into that "hole," either with the emission of a photon which does not match the laser photon, or, without emission of any photon at all, by the process known as internal conversion.
(4) The decay time from a highly excited state, the type of state that will give short-wavelength radiation when it decays, becomes shorter as the energy spacing increases.

Thus it is hard to have metastability for X-ray emission. The problem of making X-ray lasers is a very active and, so far, frustrating field of research.

17. Non-metals are quite different from metals. If we consider d = $n_e^{-1/3}$ as the distance between free electrons in metals, we get numbers of the order of 0.1 nm. In water, in contrast, with $A = 18$ and density $\rho = 1$ gm/cm^3, the typical distance between the nuclei is of the order of 30 nm. The electrons will stick out a bit, so that the separation between electrons belonging to different adjacent nuclei may be of the order of 10 nm. Thus there is much more "space" between electrons, and one would expect the net long distance forces between atoms (electron-electron and ion-ion repulsions, electron-ion attractions), known as Van der Waals forces, to be much more important for relatively small compressions. When insulators are compressed to the point where the electrons start pushing against each other, the exclusion effects will become important. In the normal regime, we expect insulators to be much more compressible. Having said this, we must point out that water is highly incompressible! In reality this is a rather complicated subject.

Chapter 44

1. In undoped material the number density of *n*-carriers is the same as the number of *p*-carriers.

3. The derivation that shows the location of the Fermi energy lies in the middle of the gap relies on the equality of factors contining the density of *n*-carriers in the conduction band, N_c, to that of *p*-carriers in the valence band, N_v. These quantities do indeed depend on the effective masses of the *n*- and *p*-carriers, respectively, as Eq. (44-12) shows. These effective masses generally differ, but since the derivation involves an equality between exponents of two terms, the inequality in the pre-factors of the exponentials—the density factors $n(E)$—only changes the relation logarithmically. More concretely, the equality of the exponents is disturbed by a term of the form $(3/2)kT \ln(m_n*/m_p*)$. Although the ratio of the effective masses can be quite large, the logarithm never is. Thus for most intrinsic semiconductors the mid-point location of the Fermi energy is a good approximation.

5. Atoms form the surface of a conductor or semiconductor. When we say there is a hole on the surface, we mean that an atom on the surface has a missing valence electron. Generally such a vacancy will be spontaneously filled by stray electrons, but the process can be hastened by directing a beam of low energy electrons to the surface.

7. When we say that a hole travels from the *p*-side to the *n*-side in a *pn*-junction, we really mean that an electron that was on the *n*-side moves to a vacant site on the *p*-side, leaving a vacancy on the *n*-side. The vacant sites on the *p*-side are just below the top of the valence band, while the excess electrons on the *n*-side are just above the bottom of the conduction band. In the process of reaching equilibrium, the electrons make a transition to a lower energy level, and this makes it appear that the holes have climbed up in energy.

9. We argued in the case of the *n*-type semiconductor that the electrons are very loosely bound to the donor ion, both because the effective mass is smaller and because the dielectric constant of the medium that the electron moves in is very large. In band theory language, the energy gap represents the minimum energy required to ionize the atom of the semiconductor. An electron that belongs to the donor atom needs much less energy to become free, and thus contribute to the conduction. Thus the energy level in which the donor electron is situated is just a little below the conduction band. Let us now consider an acceptor atom that replaces an atom in a *p*-type semiconductor. The acceptor will have one fewer valence electron, and it thus provides room for a stray electron to fall into. The acceptor atom thus provides an empty level for an electron from the valence band to jump up to, if there is enough excitation energy to get it there. The difference in energy between the top of the valence band—that is, the highest filled level in the semiconductor material—and the unoccupied valence level in the acceptor atom is usually quite small, so that in band theory language we say that there is a low-lying acceptor level just above the top of the valence band. The electrons that move to these acceptor levels leave vacancies, i.e. holes, just below the top of the valence band.

11. The energy always comes from the voltage source. The variation in the current that flows into the base amplifies the current into the collector by acting as a "gatekeeper" for the current that is generated by the biasing voltage, but does not itself provide energy.

Chapter 45

1. When alpha particles are emitted by a source, their deflection by an electric field determines the ratio Q/K, where K is the kinetic energy and Q is the charge. In this way one first determines that the alphas are monoenergetic. When these alpha particles are further deflected in a magnetic field, the radius of curvature determines P/Q, where P is the momentum of the particle. For ordinary energies, $K = P^2/2M$. Hence we can determine K and the ratio Q/M from these two experiments (usually combined into a single one, as discussed in Chapter 29). One could then use Rutherford scattering from a nucleus whose Z-value is known to determine the ratio Q/K^2. This then gives us Q and therefore M.

3. Since the total A is conserved (nucleons may redistribute themselves, but the number of nucleons is fixed), there is no energy difference proportional to A between a parent droplet and two daughter droplets.

5. To produce positive ions we only need to strip one or more electrons from the neutral atom. The binding of each electron is typically in the 10 eV range, so that this does not require much energy. To make a negative ion we have two choices: one is to add an electron, and hope that it stays bound to the atom. This can happen—for example H⁻, a hydrogen atom with an extra electron, does exist—but the additional electron is so loosely bound that the ion will lose it under the influence of the slightest perturbation. The other choice is to strip a proton from the nucleus. This can also be done, but it takes typically 10 MeV to do that, so that a nuclear collision reaction is necessary. This is hard to achieve. Furthermore, since the nuclear charge is now reduced by unity, one of the electrons

becomes superfluous, so to speak, and the resulting atom is once again very loosely bound.

7. To study the inner structure of a nucleus, we need high energy projectiles. In a high energy collision, the alpha particle has a good change of being disintegrated. In any case, a close collision involves the nuclear forces, which are not known to the same precision as the electromagnetic forces. An electron only interacts with the nucleus through electromagnetic forces, and there are far fewer unknowns.

9. The nuclides have the structure $^{15}N = 7p + 8n$, $^{16}O = 8p + 8n$ and $^{17}F = 9p + 8n$. If we want to take one proton out we need to consider Coulomb effects, the symmetry term, and the pairing term in Eq. (45-17). The changes in A are small over the range from 15 – 17. For Coulomb effects, with an average $A = 16$, the repulsion effect gives a contribution $0.3Z^2$. Thus in ^{17}F the last proton experiences a binding of $0.3(9^2 – 8^2)$, which is of the order of 5 MeV, while stripping a proton from ^{16}O similarly costs 4.5 MeV. However, the symmetry term for ^{16}O gives the proton an additional binding of order of magnitude 1.5 MeV, and the pairing term gives a contribution of the order of 4 MeV. Thus ^{16}O has the largest proton separation energy.

11. Radiation adversely affects blood cells. A radioactive material in the bone marrow affects the place where these cells are created, and there the effect is particularly strong.

13. The reason is that a detailed examination of the semi-empirical mass formula (and, more importantly, experiment) shows that there is a minimum energy that neutrons must have to cause ^{238}U to fission. This energy is approximately 1 MeV.

15. Generally only one Z value gives an energy minimum for a given A, so that for adjacent Z values there will be some sort of radioactive decay that leads to a stable daughter nucleus. In fission a number of divisions of (A, Z) into (A_1, Z_1) and (A_2, Z_2) (with $A = A_1 + A_2$ and $Z = Z_1 + Z_2$) are possible: all that is necessary is that $M(A, Z) > M(A_1, Z_1) + M(A_2, Z_2)$. In any case, it is very unlikely that both combinations (A_1, Z_1) and (A_2, Z_2) will be stable. Most daughter nuclei will be unstable.

Chapter 46

1. It is evident that any meridian from the North Pole to the equator hits the equator at right angles. Thus if two such lines make an angle of α with each other, the sum of the interior angles in the triangle is $2\pi + \alpha$. The fact that this exceeds 2π is evidence that Earth's surface has positive curvature.

3. There are reactions such as $n \rightarrow p + e^- +$ anti-neutrino, which convert a neutron into a proton and two other particles, with the total charge conserved. In the reactions such as this the number of protons + the number of neutrons is conserved rather than the number of protons alone: this is an expression of baryon conservation. The proton is the lightest baryon, and the possibility that it can decay at an extraordinarily slow rate means that any

other baryon can also decay at a comparable slow rate by decaying into a proton + byproducts. There is no conflict.

5. The initial state has zero charge, zero baryon number and zero lepton number. Any final state must have the same quantum numbers, since to a very good approximation these are good conservation laws. The final states most easily accommodated will consist of photons (which have no baryon number, no lepton number and are neutral), some other particles with the same characteristics as the photon, and particle- antiparticle pairs.

7. No particles that carry lepton number and are hadrons (that is, strongly interacting) have been discovered. Mesons such as the pion, the K meson and so on are hadrons; protons, neutrons and others that are baryons are also hadrons.

9. The advantages of using high energy cosmic rays is that one can attain energies not accessible in the laboratory. The disadvantages are that such particles do not really form beams; that is, they cannot be controlled, and the rate at which these particles arrive decreases very rapidly with their energy. The advantages outweigh the disadvantages to the extent that active experiments are based on cosmic rays.

11. In principle the answer is yes. However the universe is very large, and when you look through a telescope far enough away, you see the universe as it was tens of billions of years ago. The back of your head was not there at that time.

13. The center of mass energy of a particle of mass m and laboratory energy E striking a particle of mass M at rest in the laboratory can be worked out with the help of some algebra. The result is that the center of mass energy is $E^* = \sqrt{(mc^2)^2 + (Mc^2)^2 + 2Mc^2E}$, and this is larger when m is larger. Note that at very high energies ($E \gg M, m$), it does not matter what the value of m is.